MW01089903

HANDBOOK OF ELECTRIC POWER CALCULATIONS

Other Electrical Power Engineering Books of Interest

Cadick • ELECTRICAL POWER SYSTEMS SAFETY HANDBOOK

Denno • ELECTRIC PROTECTIVE DEVICES

Dugan, et al. • ELECTRICAL POWER SYSTEMS QUALITY

Fink and Beaty • STANDARD HANDBOOK FOR ELECTRICAL ENGINEERS

Hanselman • BRUSHLESS PERMANENT-MAGNET MOTOR DESIGN

Kundur • POWER SYSTEM STABILITY AND CONTROL

Kurtz and Shoemaker • THE LINEMAN'S AND CABLEMAN'S HANDBOOK

Kusko • EMERGENCY/STANDBY POWER SYSTEMS

Linden • HANDBOOK OF BATTERIES

Lundquist • ON-LINE ELECTRICAL TROUBLESHOOTING

Miller and Malinowski • POWER SYSTEM OPERATION

Pete • ELECTRIC POWER SYSTEMS MANUAL

Prabhakara, et al. • INDUSTRIAL AND COMMERCIAL POWER SYSTEMS HANDBOOK

Smeaton • SWITCHGEAR AND CONTROL HANDBOOK

Subramanyam • ELECTRIC DRIVES

Taylor • POWER SYSTEM VOLTAGE STABILITY

Wang and Macdonald • MODERN POWER SYSTEM PLANNING

HANDBOOK OF ELECTRIC POWER CALCULATIONS

H. Wayne Beaty

Third Edition

McGRAW-HILL
New York San Francisco Washington, D.C. Auckland Bogotá
Caracas Lisbon London Madrid Mexico City Milan
Montreal New Delhi San Juan Singapore
Sydney Tokyo Toronto

McGraw-Hill

A Division of The ***McGraw-Hill*** *Companies*

1 2 3 4 5 6 7 8 9 0 DOC/DOC 9 0 9 8 7 6 5 4 3 2 1 0

P/N 0-07-136299-1
Part of
ISBN 0-07-136298-3

The sponsoring editor for this book was Stephen Chapman and the production supervisor was Sherri Souffrance. It was set in Times Roman by Progressive Publishing Alternatives.

Printed and bound by R. R. Donnelley & Sons Company.

This book is printed on recycled, acid-free paper containing a minimum of 50% recycled, de-inked fiber.

CONTENTS

CONTRIBUTORS

Amick, Charles L. *Lighting Consultant.* (SECTION 20: LIGHTING DESIGN)

Chowdhury, Badrul H. *Associate Professor, Electrical and Computer Engineering, University of Missouri-Rolla.* (SECTION 11: LOAD-FLOW ANALYSIS IN POWER SYSTEMS)

Galli, Anthony W. *Project Engineer, Newport News Shipbuilding.* (SECTION 1: BASIC NETWORK ANALYSIS)

Hollander, Lawrence J. *Dean of Engineering Emeritus, Union College.* (SECTION 3: DC MOTORS AND GENERATORS; SECTION 6: SINGLE-PHASE MOTORS; SECTION 10: ELECTRIC-POWER NETWORKS; SECTION 13: SHORT-CIRCUIT COMPUTATIONS)

Ilic, Marija *Senior Research Scientist, Electrical Engineering and Computer Science, Massachusetts Institute of Technology.* (SECTION 12: POWER SYSTEMS CONTROL)

Khan, Shahriar *Electrical Design Engineer, Schlumberger Ltd.* (SECTION 2: INSTRUMENTATION)

Liu, Yilu (Ellen) *Associate Professor, Electrical Engineering Department, Virginia Tech University.* (SECTION 4: TRANSFORMERS)

Mazzoni, Omar S. *President, Systems Research International, Inc.* (SECTION 7: SYNCHRONOUS MACHINES)

Migliaro, Marco W. *Chief Electrical and I&C Engineer, Nuclear Division, Florida Power & Light.* (SECTION 7: SYNCHRONOUS MACHINES)

Oraee, Hashaam *Professor, Electrical & Computer Engineering, Worcester Polytechnic Institute.* (SECTION 5: THREE-PHASE INDUCTION MOTORS)

Rivas, Richard A. *Associate Professor, Universidad Simón Bolívar.* (SECTION 9: OVERHEAD TRANSMISSION LINES AND UNDERGROUND CABLES)

Robertson, Elizabeth *President, Lyncole XIT Grounding.* (SECTION 14: SYSTEM GROUNDING)

Sauer, Peter W. *Professor, Electrical Engineering, University of Illinois at Urbana-Champaign.* (SECTION 16: POWER SYSTEM STABILITY)

Schneider, Alexander W., Jr. *Senior Engineer, Mid-America Interconnected Network.* (SECTION 16: POWER SYSTEM STABILITY)

Shaalan, Hesham *Assistant Professor, Georgia Southern University.* (SECTION 8: GENERATION OF ELECTRIC POWER; SECTION 17: COGENERATION)

Sheble, Gerald B. *Professor, Iowa State University.* (SECTION 19: ELECTRIC ENERGY ECONOMIC METHODS)

Stocking, David R. *Lyncole XIT Grounding.* (SECTION 14: SYSTEM GROUNDING)

PREFACE

The *Handbook of Electric Power Calculations* provides detailed step-by-step calculation procedures commonly encountered in electrical engineering. The *Handbook* contains a wide array of topics and each topic is written by an authority on the subject. The treatment throughout the *Handbook* is *practical* with very little emphasis on theory.

Each of the 20 Sections follows this format:

- Clear statement of the problem.
- Step-by-step calculation procedure.
- Inclusion of suitable graphs and illustrations to clarify the procedure.
- Use of SI and USCS equivalents.

This relatively simple, yet comprehensive format adds greatly to the use of the *Handbook* by the engineer or technician. Arithmetic and algebra are employed in the solution of the majority of the problems. Each section contains a list of references or a bibliography that is pertinent to the subject matter.

This edition also includes a CD that has calculation procedures available for inclusion of other parameters, which will allow you to calculate problems with your specific numbers inserted.

Grateful acknowledgment is given to each of the authors for their contribution to this 3rd edition of the *Handbook*.

H. Wayne Beaty

SECTION 1

BASIC NETWORK ANALYSIS

A. Wayne Galli, Ph.D.
Project Engineer
Newport News Shipbuilding

SERIES-PARALLEL DC NETWORK ANALYSIS

A direct-current circuit (network) contains 19 resistors arranged as shown in Fig. 1.1. Compute the current through and the voltage drop across each resistor in this circuit.

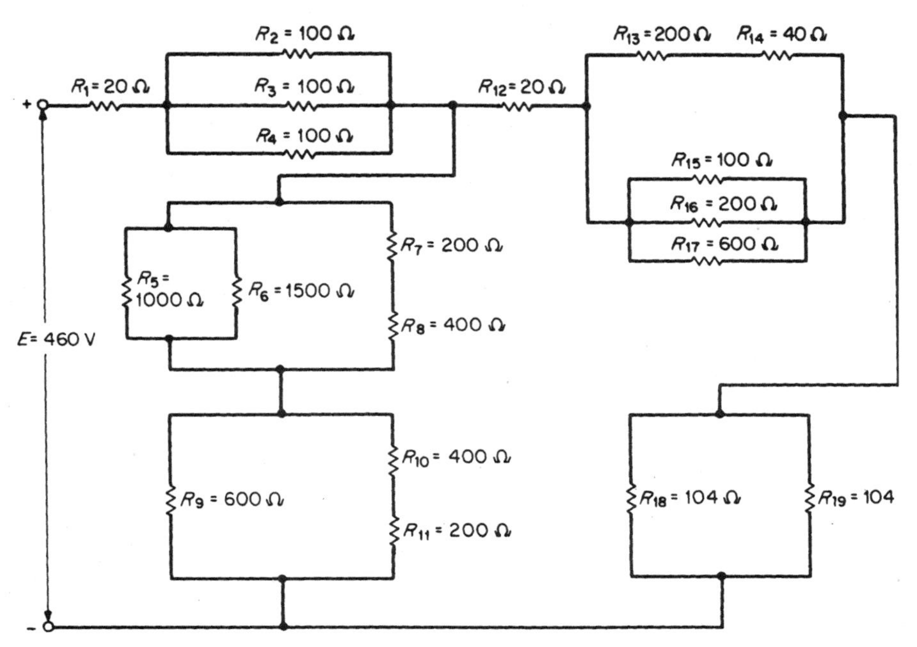

FIGURE 1.1 A series-parallel dc circuit to be analyzed.

Calculation Procedure

1. Label the Circuit

Label all the sections. Mark the direction of current through each resistor (Fig. 1.2). The equivalent resistance of the series-parallel combination of resistors can be found by successive applications of the rules for combining series resistors and parallel resistors.

2. Combine All Series Resistors

In a series circuit, the total or equivalent resistance R_{EQS} seen by the source is equal to the sum of the values of the individual resistors: $R_{EQS} = R_1 + R_2 + R_3 + \cdots + R_N$.

Calculate the series equivalent of the elements connected in series in sections *DE, CG,* and *GF:* R_{EQS} (section *DE*) $= R_{13} + R_{14} = 200 + 40 = 240\ \Omega$, R_{EQS} (section *CG*) $= R_7 + R_8 = 200 + 400 = 600\ \Omega$, and R_{EQS} (section *GF*) $= R_{10} + R_{11} = 400 + 200 = 600\ \Omega$. Replace the series elements included in sections *DE, CG,* and *GF* by their equivalent values (Fig. 1.3).

3. Combine All Parallel Resistors

In the case of a parallel circuit of two unequal resistors in parallel, the total or equivalent resistance R_{EQP} can be found from the following product-over-sum equation: $R_{EQP} = R_1 \| R_2 = R_1R_2/(R_1 + R_2)$, where $\|$ stands for *in parallel with.* The equivalent parallel resistance is always less than the smaller of the two resistors.

In section *CG*, $R_5 \| R_6 = (1000 \times 1500)/(1000 + 1500) = 600\ \Omega$. Section *CG* now consists of two 600-Ω resistors in parallel. In a case of a circuit of *N* equal resistors in

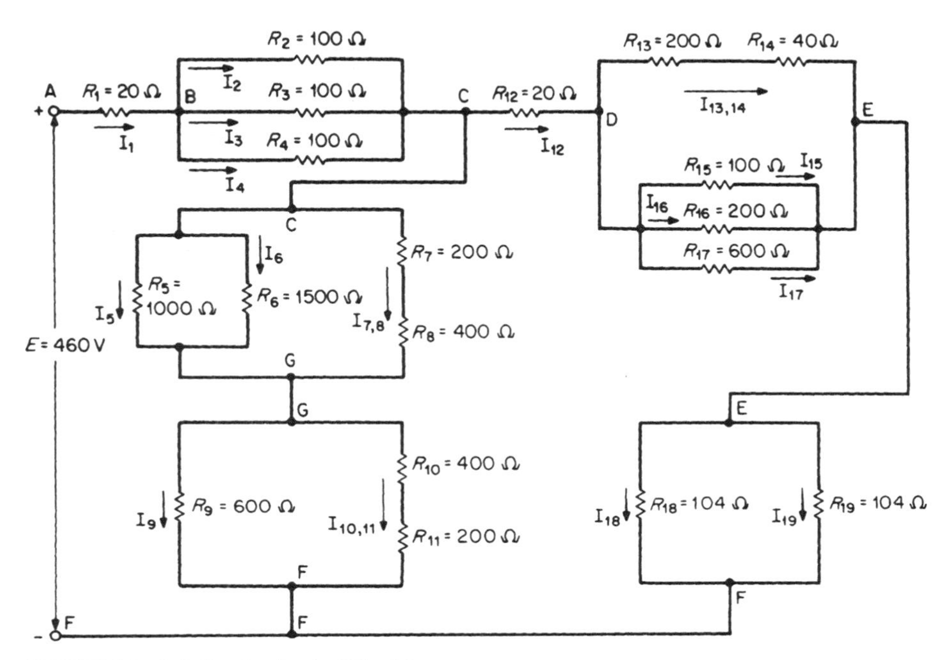

FIGURE 1.2 Labeling the circuit of Fig. 1.1.

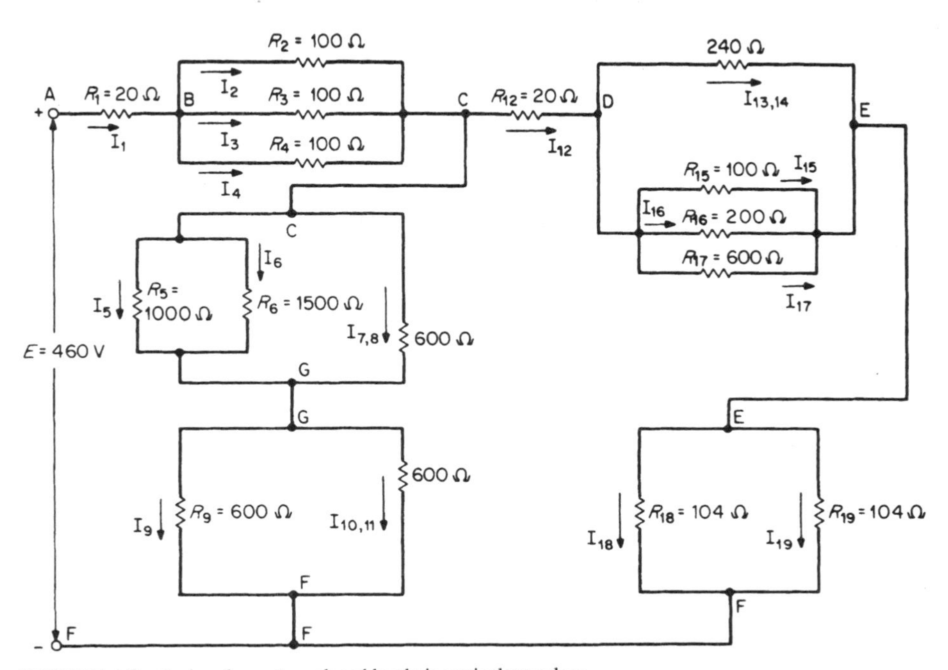

FIGURE 1.3 Series elements replaced by their equivalent values.

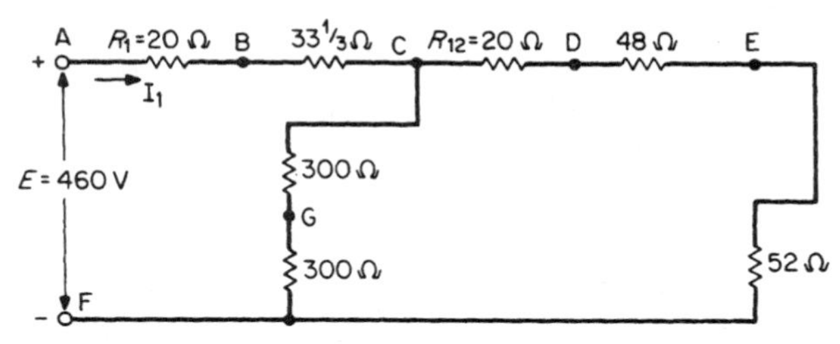

FIGURE 1.4 Parallel elements replaced by their equivalent values.

parallel, the total, or equivalent, resistance R_{EQP} can be determined from the following equation: $R_{EQP} = R/N$, where R is the resistance of each of the parallel resistors and N is the number of resistors connected in parallel. For section CG, $R_{CG} = 600/2 = 300\,\Omega$; for section BC, $R_{BC} = 100/3 = 33\frac{1}{3}\,\Omega$; for section EF, $R_{EF} = 104/2 = 52\,\Omega$; for section GF, $R_{GF} = 600/2 = 300\,\Omega$.

In a circuit of three or more unequal resistors in parallel, the total, or equivalent resistance R_{EQP} is equal to the inverse of the sum of the reciprocals of the individual resistance values: $R_{EQP} = 1/(1/R_1 + 1/R_2 + 1/R_3 + \cdots + 1/R_N)$. The equivalent parallel resistance is always less than the smallest-value resistor in the parallel combination.

Calculate the equivalent resistance of the elements connected in parallel in section DE: $R_{15} \| R_{16} \| R_{17} = 1/(1/100 + 1/200 + 1/600) = 60\,\Omega$. Calculate R_{DE}: $R_{DE} = 240 \| 60 = (240)(60)/(240 + 60) = 48\,\Omega$. Replace all parallel elements by their equivalent values (Fig. 1.4).

4. Combine the Remaining Resistances to Obtain the Total Equivalent Resistance

Combine the equivalent series resistances of Fig. 1.4 to obtain the simple series-parallel circuit of Fig. 1.5: $R_{AB} + R_{BC} = R_{AC} = R_{EQS} = 20 + 33\frac{1}{3} = 53\frac{1}{3}\,\Omega$, $R_{CG} + R_{GF} = R_{CF} = R_{EQS} = 300 + 300 = 600\,\Omega$, $R_{CD} + R_{DE} + R_{EF} = R_{CF} = R_{EQS} = 20 + 48 + 52 = 120\,\Omega$. Calculate the total equivalent resistance R_{EQT}: $R_{EQT} = 53\frac{1}{3} + (600 \| 120) = 153\frac{1}{3}\,\Omega$. The final reduced circuit is illustrated in Fig. 1.6.

5. Compute the Total Line Current in Fig. 1.6 Using Ohm's Law

$I_1 = E/R_{EQT}$, where I_1 = total line current, E = line voltage (power-supply voltage), and R_{EQT} = line resistance or total equivalent resistance seen by power supply. Substituting values yields: $I_1 = E/R_{EQT} = 460/153\frac{1}{3} = 3$ A.

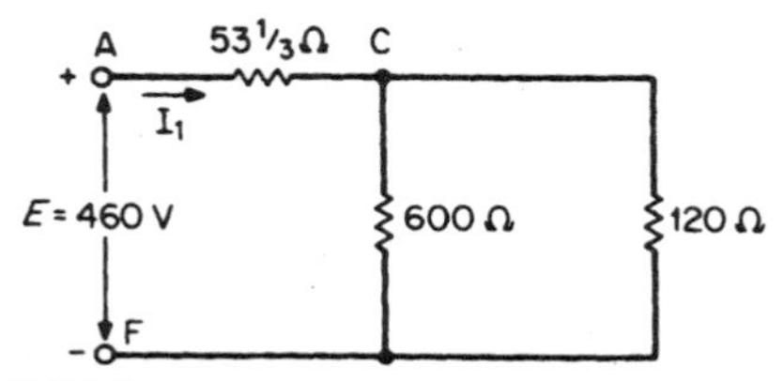

FIGURE 1.5 Circuit of Fig. 1.4 reduced to a simple series-parallel configuration.

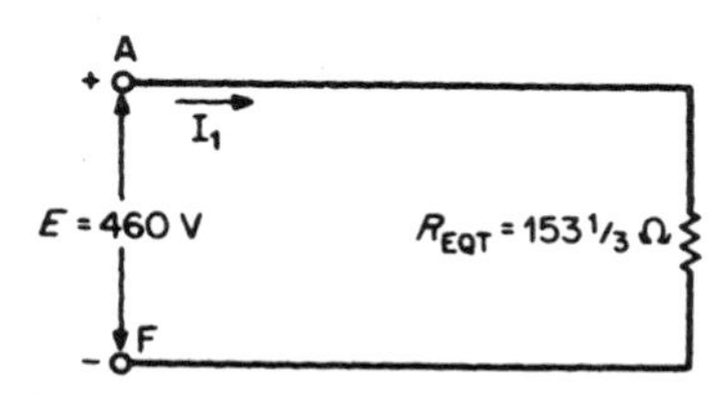

FIGURE 1.6 Final reduced circuit of Fig. 1.1.

6. Compute the Current Through, and the Voltage Drop Across, Each Resistor in the Circuit

Refer to Figs. 1.2 and 1.4. Analysis of R_1 yields: $I_1 = 3$ A (calculated in Step 5); $V_1 = V_{AB} + I_1R_1 = (3)(20) = 60$ V; and for R_2, R_3, and R_4 we have: $V_{BC} = V_2 = V_3 = V_4 = I_1R_{BC} = (3)(33^1/_3) = 100$ V. Current $I_2 = I_3 = I_4 = 100/100 = 1$ A. Hence, V_{CF} can be calculated: $V_{CF} = E - (V_{AB} + V_{BC}) = 460 - (60 + 100) = 300$ V. The current from C to G to F is $300/600 = 0.5$ A.

Kirchoff's current law (KCL) states: The algebraic sum of the currents entering any node or junction of a circuit is equal to the algebraic sum of the currents leaving that node or junction: ΣI entering $= \Sigma I$ leaving. Applying KCL at node C, we find $I_{12} = 3 - 0.5 = 2.5$ A. Therefore, $V_{12} = V_{CD} = I_{12}R_{12} = (2.5)(20) = 50$ V.

The voltage-divider principle states that the voltage V_N across any resistor R_N in a series circuit is equal to the product of the total applied voltage V_T and R_N divided by the sum of the series resistors, R_{EQS}: $V_N = V_T(R_N/R_{EQS})$. This equation shows that V_N is directly proportional to R_N and $V_{CG} = V_{GF} = 300 \times (300/600) = 150$ V. Hence, $I_7 = I_8 = 150/600 = 0.25$ A, $V_7 = I_7R_7 = (0.25)(200) = 50$ V, $V_8 = I_8R_8 = (0.25)(400) = 100$ V, $I_{10} = I_{11} = 150/600 = 0.25$ A, $V_{10} = I_{10}R_{10} = (0.25)(400) = 100$ V, $V_{11} = I_{11}R_{11} = (0.25)(200) = 50$ V.

The current-divider principle states that in a circuit containing N parallel branches, the current I_N in a particular branch R_N is equal to the product of the applied current I_T and the equivalent resistance R_{EQP} of the parallel circuit divided by R_N: $I_N = I_T(R_{EQP}/R_N)$. When there are two resistors R_A and R_B in parallel, the current I_A in R_A is $I_A = I_T[R_B/(/R_A + R_B)]$; the current I_B in R_B is $I_B = I_T[R_A/(R_A + R_B)]$. When R_A is equal to R_B, $I_A = I_B = I_T/2$. Refer to Figs. 1.2 , 1.3, and 1.4 for the remaining calculations: $(R_5 \| R_6) = R_7 + R_8 = 600\ \Omega$.

From the preceding equations, the value of the current entering the parallel combination of R_5 and R_6 is $I_5 + I_6 = 0.5/2 = 0.25$ A. $I_5 = 0.25 \times (1500/2500) = 0.15$ A, and $I_6 = 0.25 \times (1000/2500) = 0.10$ A. Ohm's law can be used to check the value of V_5 and V_6, which should equal V_{CG} and which was previously calculated to equal 150 V: $V_5 = I_5R_5 = (0.15)(1000) = 150$ V and $V_6 = I_6R_6 = (0.10)(1500) = 150$ V.

The current entering node G equals 0.5 A. Because $R_9 = R_{10} + R_{11}$, $I_9 = I_{10} = I_{11} = 0.5/2 = 0.25$ A. From Ohm's law: $V_9 = I_9R_9 = (0.25)(600) = 150$ V, $V_{10} = I_{10}R_{10} = (0.25)(400) = 100$ V, $V_{11} = I_{11}R_{11} = (0.25)(200) = 50$ V. These values check since $V_{GF} = V_9 = 150\text{ V} = V_{10} + V_{11} = 100 + 50 = 150$ V.

The remaining calculations show that: $V_{DE} = I_{12}R_{DE} = (2.5)(48) = 120$ V, $I_{13} = I_{14} = 120/240 = 0.5$ A, $V_{13} = I_{13}R_{13} = (0.5)(200) = 100$ V, and $V_{14} = I_{14}R_{14} = (0.5)(40) = 20$ V. Since $V_{15} = V_{16} = V_{17} = V_{DE} = 120$ V, $I_{15} = 120/100 = 1.2$ A, $I_{16} = 120/200 = 0.6$ A, and $I_{17} = 120/600 = 0.2$ A.

These current values check, since $I_{15} + I_{16} + I_{17} + I_{13,14} = 1.2 + 0.6 + 0.2 + 0.5 = 2.5$ A, which enters node D and which leaves node E. Because $R_{18} = R_{19}$, $I_{18} = I_{19} = 2.5/2 = 1.25$ A and $V_{EF} = V_{18} = V_{19} = (2.5)(52) = 130$ V.

Kirchoff's voltage law (KVL) states that the algebraic sum of the potential rises and drops around a closed loop or path is zero. This law can also be expressed as: $\Sigma V_{\text{rises}} = \Sigma V_{\text{drops}}$. As a final check $E = V_{AB} + V_{BC} + V_{CD} + V_{DE} + V_{EF}$ or $460\text{ V} = 60\text{ V} + 100\text{ V} + 50\text{ V} + 120\text{ V} + 130\text{ V} = 460$ V.

Related Calculations. Any reducible dc circuit, that is, any circuit with a single power source that can be reduced to one equivalent resistance, no matter how complex, can be solved in a manner similar to the preceding procedure.

BRANCH-CURRENT ANALYSIS OF A DC NETWORK

Calculate the current through each of the resistors in the dc circuit of Fig. 1.7 using the branch-current method of solution.

Calculation Procedure

1. Label the Circuit

Label all the nodes (Fig. 1.8). There are four nodes in this circuit, indicated by the letters *A*, *B*, *C*, and *D*. A node is a junction where two or more current paths come together. A branch is a portion of a circuit consisting of one or more elements in series. Figure 1.8 contains three branches, each of which is a current path in the network. Branch *ABC* consists of the power supply E_1 and R_1 in series, branch *ADC* consists of the power supply E_2 and R_2 in series, and branch *CA* consists of R_3 only. Assign a distinct current of arbitrary direction to each branch of the network (I_1, I_2, I_3). Indicate the polarities of each resistor as determined by the assumed direction of current and the passive sign convention. The polarity of the power-supply terminals is fixed and is therefore not dependent on the assumed direction of current.

2. Apply KVL and KCL to the Network

Apply KVL around each closed loop. A closed loop is any continuous connection of branches that allows us to trace a path which leaves a point in one direction and returns to that same starting point from another direction without leaving the network.

Applying KVL to the minimum number of nodes that will include all the branch currents, one obtains: loop 1 (*ABCA*): $8 - 2I_1 - 4I_3 = 0$; loop 2 (*ADCA*): $24 - I_2 - 4I_3 = 0$. KCL at node C:$I_1 + I_2 = I_3$.

3. Solve the Equations

The above three simultaneous equations can be solved by the elimination method or by using third-order determinants. The solution yields these results: $I_1 = -4$ A, $I_2 = 8$ A, and $I_3 = 4$ A. The negative sign for I_1 indicates that the actual current flows in the direction opposite to that assumed.

Related Calculations. The above calculation procedure is an application of Kirchoff's laws to an irreducible circuit. Such a circuit cannot be solved by the method used in the previous calculation procedure because it contains two power supplies. Once the branch currents are determined, all other quantities such as voltage and power can be calculated.

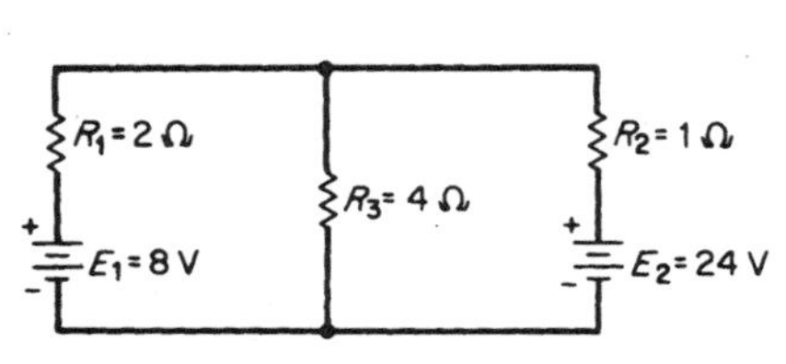

FIGURE 1.7 Circuit to be analyzed by branch currents.

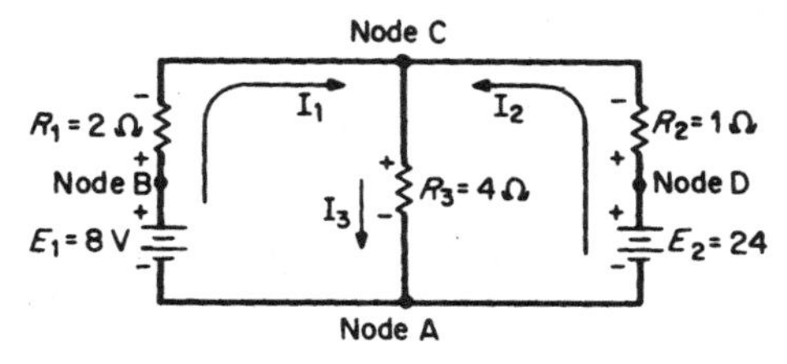

FIGURE 1.8 Labeling the circuit of Fig. 1.7.

MESH ANALYSIS OF A DC NETWORK

Calculate the current through each of the resistors in the dc circuit of Fig. 1.9 using mesh analysis.

Calculation Procedure

1. Assign Mesh or Loop Currents

The term *mesh* is used because of the similarity in appearance between the closed loops of the network and a wire mesh fence. One can view the circuit as a "window frame" and the meshes as the "windows." A mesh is a closed pathway with no other closed pathway within it. A loop is also a closed pathway, but a loop may have other closed pathways within it. Therefore, all meshes are loops, but all loops are not meshes. For example, the loop made by the closed path *BCDAB* (Fig. 1.9) is not a mesh because it contains two closed paths: *BCAB* and *CDAC*.

Loop currents I_1 and I_2 are drawn in the clockwise direction in each window (Fig. 1.10). The loop current or mesh current is a fictitious current that enables us to obtain the actual branch currents more easily. The number of loop currents required is always equal to the number of windows of the network. This assures that the resulting equations are all independent. Loop currents may be drawn in any direction, but assigning a clockwise direction to all of them simplifies the process of writing equations.

2. Indicate the Polarities within Each Loop

Identify polarities to agree with the assumed direction of the loop currents and the passive sign convention. The polarities across R_3 are the opposite for each loop current. The polarities of E_1 and E_2 are unaffected by the direction of the loop currents passing through them.

3. Write KVL around Each Mesh

Write KVL around each mesh in any direction. It is convenient to follow the same direction as the loop current: mesh I: $+8 - 2I_1 - 4(I_1 - I_2) = 0$; mesh II: $-24 - 4(I_2 - I_1) - I_2 = 0$.

4. Solve the Equations

Solving the two simultaneous equations gives the following results: $I_1 = -4$ A and $I_2 = -8$ A. The minus signs indicate that the two loop currents flow in a direction opposite to that assumed; that is, they both flow counterclockwise. Loop current I_1 is therefore 4 A in the direction of *CBAC*. Loop current I_2 is 8 A in the direction *ADCA*. The true

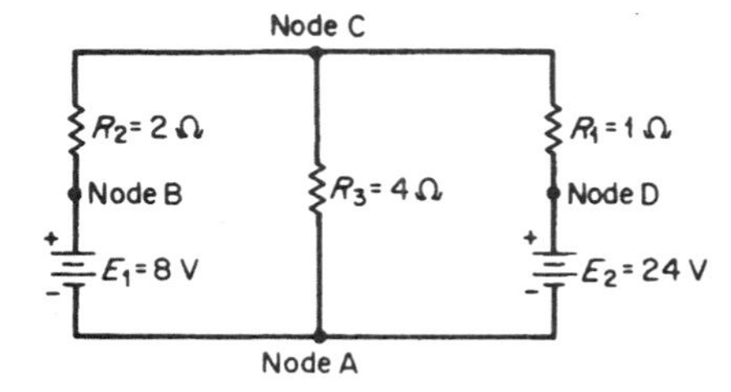

FIGURE 1.9 Circuit to be analyzed using mesh analysis.

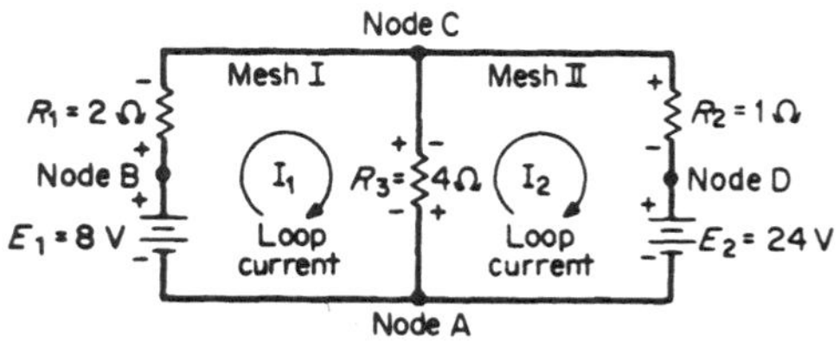

FIGURE 1.10 Labeling the circuit of Fig. 1.9.

direction of loop current I_2 through resistor R_3 is from C to A. The true direction of loop current I_1 through resistor R_3 is from A to C. Therefore, the current through R_3 equals $(I_2 - I_1)$ or $8 - 4 = 4$ A in the direction of CA.

Related Calculations. This procedure solved the same network as in Fig. 1.8. The mesh-analysis solution eliminates the need to substitute KCL into the equations derived by the application of KVL. The initial writing of the equations accomplishes the same result. Mesh analysis is therefore more frequently applied than branch-current analysis. However, it should be noted that mesh analysis can only be applied to planar circuits.

NODAL ANALYSIS OF A DC NETWORK

Calculate the current through each of the resistors in the dc circuit of Fig. 1.11 using nodal analysis.

Calculation Procedure

1. Label the Circuit

Label all nodes (Fig. 1.12). One of the nodes (node A) is chosen as the reference node. It can be thought of as a circuit ground, which is at zero voltage or ground potential. Nodes B and D are already known to be at the potential of the source voltages. The voltage at node C (V_C) is unknown.

Assume that $V_C > V_B$ and $V_C > V_D$. Draw all three currents I_1, I_2, and I_3 away from node C, that is, toward the reference node.

2. Write KCL at Node C

$I_1 + I_2 + I_3 = 0.$

3. Express Currents in Terms of Circuit Voltages Using Ohm's Law

Refer to Fig. 1.12: $I_1 = V_1/R_1 = (V_C - 8)/2$, $I_2 = V_2/R_2 = (V_C - 24)/1$, and $I_3 = V_3/R_3 = V_C/4$.

4. Substitute in KCL Equation of Step 2

Substituting the current equations obtained in Step 3 into KCL of Step 2, we find $I_1 + I_2 + I_3 = 0$ or $(V_C - 8)/2 + (V_C - 24)/1 + V_C/4 = 0$. Because the only unknown is V_C, this simple equation can be solved to obtain $V_C = 16$ V.

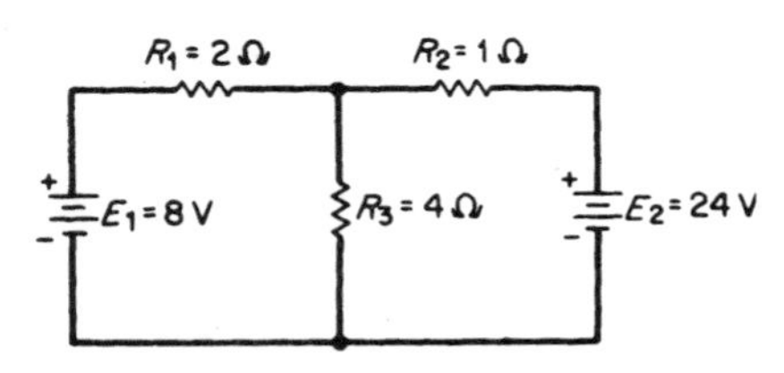

FIGURE 1.11 Circuit to be analyzed by nodal analysis.

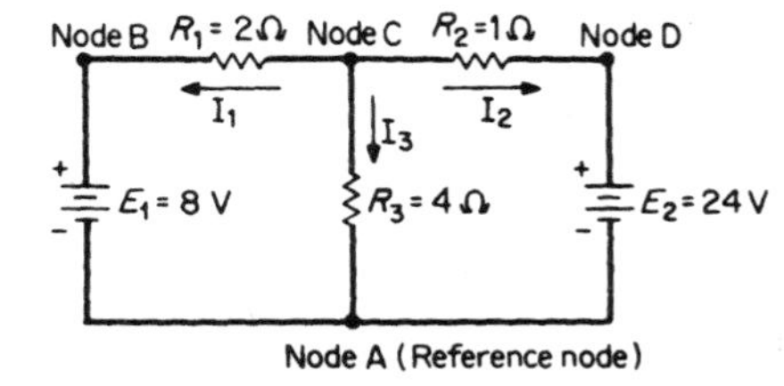

FIGURE 1.12 Labeling the circuit of Fig. 1.11.

5. Solve for All Currents

$I_1 = (V_C - 8)/2 = (16 - 8)/2 = 4$ A (true direction) and $I_2 = (V_C - 24)/1 = (16 - 24)/1 = -8$ A. The negative sign indicates that I_2 flows toward node C instead of in the assumed direction (away from node C). $I_3 = V_C/4 = 16/4 = 4$ A (true direction).

Related Calculations. Nodal analysis is a very useful technique for solving networks. This procedure solved the same circuits as in Figs. 1.7 and 1.9.

DIRECT-CURRENT NETWORK SOLUTION USING SUPERPOSITION THEOREM

Calculate the value of the current through resistor R_3 in the dc network of Fig. 1.13*a* using the superposition theorem. The superposition theorem states: In any linear network containing more than one source of electromotive force (emf) or current, the current through any branch is the algebraic sum of the currents produced by each source acting independently.

Calculation Procedure

1. Consider the Effect of E_A *Alone (Fig. 1.13*b*)*

Because E_B has no internal resistance, the E_B source is replaced by a short circuit. (A current source, if present, is replaced by an open circuit.) Therefore, $R_{TA} = 100 + (100 \| 100) = 150\ \Omega$ and $I_{TA} = E_A/R_{TA} = 30/150 = 200$ mA. From the current-divider rule, $I_{3A} = 200\text{ mA}/2 = 100$ mA.

2. Consider the Effect of E_B *Alone (Fig. 1.13*c*)*

Because E_A has no internal resistance, the E_A source is replaced by a short circuit. Therefore, $R_{TB} = 100 + (100 \| 100) = 150\ \Omega$ and $I_{TB} = E_B/R_{TB} = 15/150 = 100$ mA. From the current-divider rule, $I_{3B} = 100\text{ mA}/2 = 50$ mA.

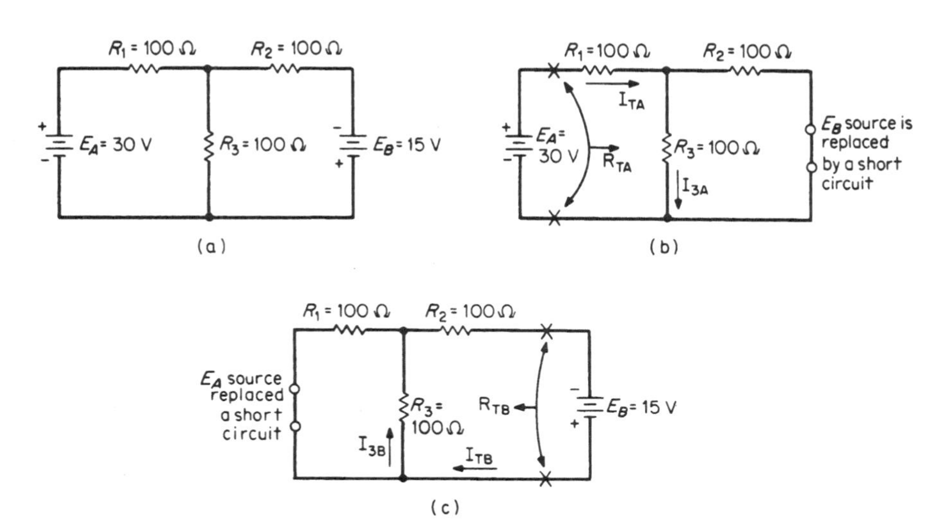

FIGURE 1.13 Application of the superposition theorem: (*a*) current in R_3 to be determined; (*b*) effect of E_A alone; and (*c*) effect of E_B alone.

3. Calculate the Value of I_3

The algebraic sum of the component currents I_{3A} and I_{3B} is used to obtain the true magnitude and direction of I_3: $I_3 = I_{3A} - I_{3B} = 100 - 50 = 50$ mA (in the direction of I_{3A}).

Related Calculations. The superposition theorem simplifies the analysis of a *linear network* only having more than one source of emf. This theorem may also be applied in a network containing both dc and ac sources of emf. This is considered later in the section.

DIRECT-CURRENT NETWORK SOLUTION USING THEVENIN'S THEOREM

Calculate the value of the current I_L through the resistor R_L in the dc network of Fig. 1.14*a* using Thevenin's theorem.

Thevenin's theorem states: Any two-terminal linear network containing resistances and sources of emf and current may be replaced by a single source of emf in series with a single resistance. The emf of the single source of emf, called E_{Th}, is the open-circuit emf at the network terminal. The single-series resistance, called R_{Th}, is the resistance between the network terminals when all of the independent sources are replaced by their internal resistances.

Calculation Procedure

*1. Calculate the Thevenin Voltage (Fig. 1.14*b*)*

When the Thevenin equivalent circuit is determined for a network, the process is known as "thevenizing" the circuit.

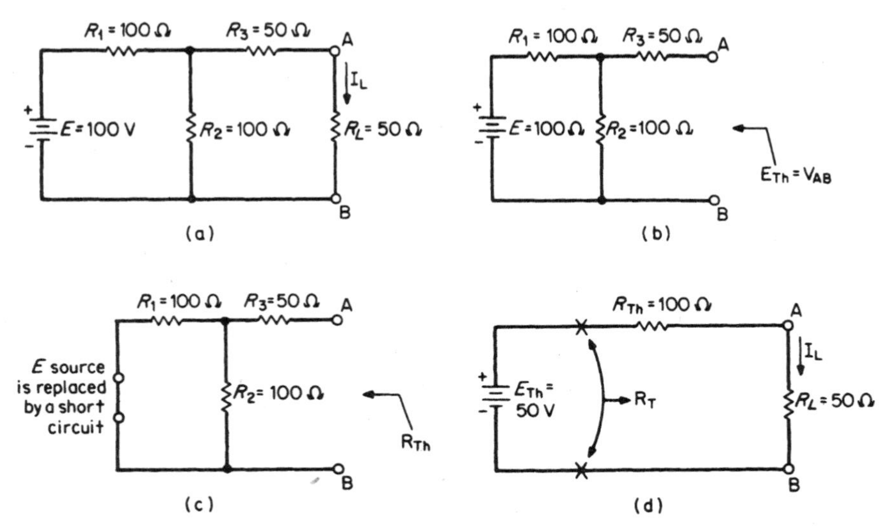

FIGURE 1.14 Application of Thevenin's theorem: (*a*) current I_L to be determined; (*b*) calculating E_{Th}; (*c*) calculating R_{Th}; and (*d*) resultant Thevenin equivalent circuit.

The load resistor is removed as shown in Fig. 1.14*b*. The open-circuit terminal voltage of the network is calculated; this value is E_{Th}. Because no current can flow through R_3, the voltage E_{Th} (V_{AB}) is the same as the voltage across resistor R_2. Use the voltage-divider rule to find E_{Th}: $E_{Th} = (100\text{ V}) \times [100/(100 + 100)] = 50\text{ V}$.

2. *Calculate the Thevenin Resistance (Fig. 1.14c)*

The network is redrawn with the source of emf replaced by a short circuit. (If a current source is present, it is replaced by an open circuit.) The resistance of the redrawn network as seen by looking back into the network from the load terminals is calculated. This value is R_{Th}, where $R_{Th} = 50\ \Omega + (100\ \Omega)\,\|\,(100\ \Omega) = 100\ \Omega$.

3. *Draw the Thevenin Equivalent Circuit (Fig. 1.14d)*

The Thevenin equivalent circuit consists of the series combination of E_{Th} and R_{Th}. The load resistor R_L is connected across the output terminals of this equivalent circuit. $R_T = R_{Th} + R_L = 100 + 50 = 150\ \Omega$, and $I_L = E_{Th}/R_T = 50/150 = 1/3$ A.

Related Calculations. With respect to the terminals only, the Thevenin circuit is equivalent to the original linear network. Changes in R_L do not require any calculations for a new Thevenin circuit. The simple series Thevenin circuit of Fig. 1.14*d* can be used to solve for load currents each time R_L is changed. The Thevenin theorem is also applicable to networks with dependent sources. Additionally, node-voltage analysis and mesh-current analysis may be applied to determine V_{Th}. In rare cases, with only dependent sources present, one may have to assume a fictitious 1 A or 1V "injection" source at the terminals.

DIRECT-CURRENT NETWORK SOLUTION USING NORTON'S THEOREM

Calculate the value of the current I_L through the resistor R_L in the dc network of Fig. 1.15*a* using Norton's theorem.

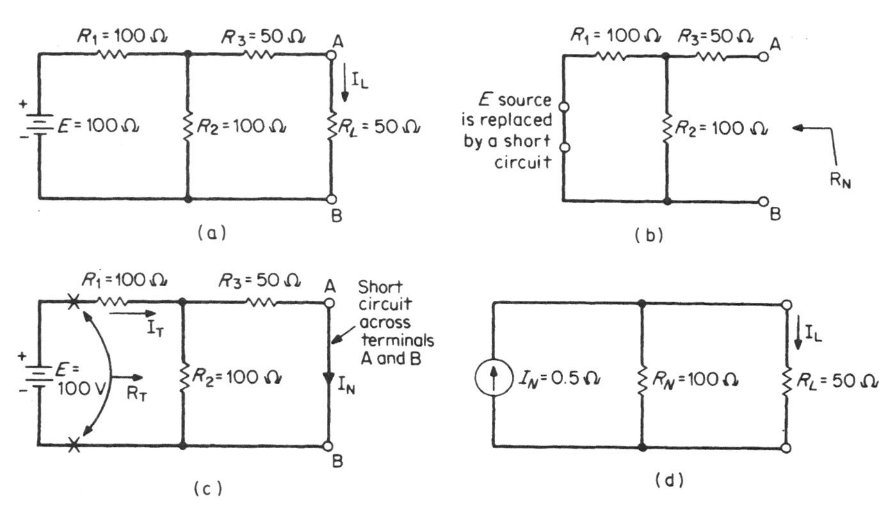

FIGURE 1.15 Application of Norton's theorem: (*a*) current I_L to be determined; (*b*) calculating R_N; (*c*) calculating I_N; and (*d*) resultant Norton equivalent circuit.

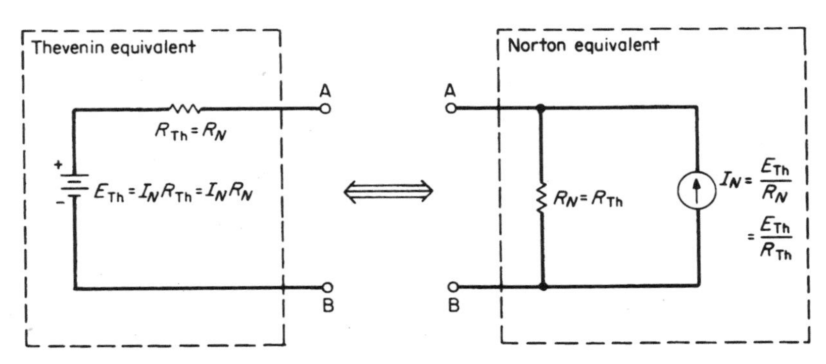

FIGURE 1.16 Source conversion equations.

Norton's theorem states: Any two-terminal linear dc network can be replaced by an equivalent circuit consisting of a constant-current source I_N in parallel with a resistor R_N.

Calculation Procedure

1. Calculate the Norton Parallel Resistance, R$_N$ *(Fig. 1.15*b*)*

The load resistor is removed (Fig. 1.15*b*). All sources are set to zero (current sources are replaced by open circuits, and voltage sources are replaced by short circuits). R_N is calculated as the resistance of the redrawn network as seen by looking back into the network from the load terminals A and B: $R_N = 50\ \Omega + (100\ \Omega \| 100\ \Omega) = 100\ \Omega$. A comparison of Figs. 1.14*c* and 1.15*b* shows that $R_N = R_{Th}$.

2. Calculate the Norton Constant-Current Source, I$_N$ *(Fig. 15*c*)*

I_N is the short-circuit current between terminals A and B. $R_T = 100\ \Omega + (100\ \Omega \| 50\ \Omega) = 133\,^1\!/_3\Omega$ and $I_T = E/R_T = (100/133\,^1\!/_3) = {}^3\!/_4$ A. From the current-divider rule: $I_N = ({}^3\!/_4\ \text{A})\,(100)/(100 + 50) = 0.5$ A.

*3. Draw the Norton Equivalent Circuit (Fig. 1.15*d*)*

The Norton equivalent circuit consists of the parallel combination of I_N and R_N. The load resistor R_L is connected across the output terminals of this equivalent circuit. From the current-divider rule: $I_L = (0.5\ \text{A})[100/(100 + 50)] = {}^1\!/_3$ A.

Related Calculations. This problem solved the same circuit as in Fig. 1.14*a*. It is often convenient or necessary to have a voltage source (Thevenin equivalent) rather than a current source (Norton equivalent) or a current source rather than a voltage source. Figure 1.16 shows the source conversion equations which indicate that a Thevenin equivalent circuit can be replaced by a Norton equivalent circuit, and vice versa, provided that the following equations are used: $R_N = R_{Th}$; $E_{Th} = I_N R_{Th} = I_N R_N$, and $I_N = E_{Th}/R_N = E_{Th}/R_{Th}$. The conversion between Thevenin and Norton equivalents is generally known as a source transformation.

BALANCED DC BRIDGE NETWORK

Calculate the value of R_x in the balanced dc bridge network of Fig. 1.17.

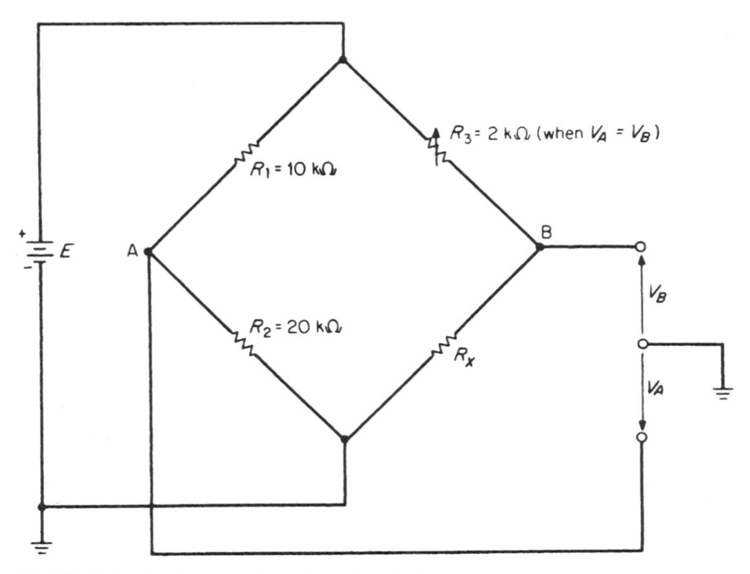

FIGURE 1.17 Analysis of a balanced dc bridge.

Calculation Procedure

1. *Solve for* $\mathbf{R_x}$

The bridge network is balanced when R_3 is adjusted so that $V_A = V_B$. Then: $R_1/R_2 = R_3/R_x$. Solving for R_x, we find $R_x = R_2R_3/R_1 = (20)(2)/10 = 4\ \text{k}\Omega$.

Related Calculations. This network topology is classically known as a Wheatstone Bridge and is used to precisely measure resistances of medium values in the range of 1 ohm to 1 mega-ohm. There is a potential drop across terminals A and B when the bridge is not balanced, causing current to flow through any element connected to those terminals. Mesh analysis, nodal analysis, Thevenin's theorem, or Norton's theorem may be used to solve the unbalanced network for voltages and currents. Using the same topology of the circuit (Fig. 1.17), but replacing the dc source with an ac source and the four resistors with properly biased diodes, one obtains a simple rectifier circuit for converting ac input to a unidirectional output.

UNBALANCED DC BRIDGE NETWORK

Calculate the value of R_{EQT} in the unbalanced dc bridge network of Fig. 1.18.

Calculation Procedure

1. *Convert the Upper Delta to an Equivalent Wye Circuit*

Delta-to-wye and wye-to-delta conversion formulas apply to Fig. 1.19. The formulas for delta-to-wye conversion are: $R_1 = R_AR_C/(R_A + R_B + R_C)$, $R_2 = R_BR_C/(R_A + R_B +$

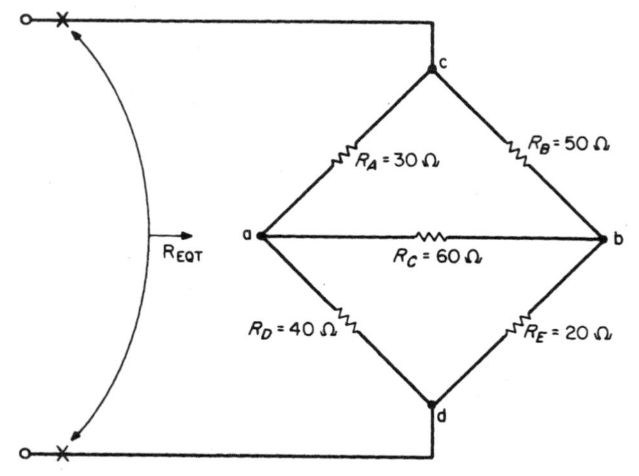

FIGURE 1.18 Analysis of an unbalanced bridge.

R_C), and $R_3 = R_B R_A/(R_A + R_B + R_C)$. The formulas for wye-to-delta conversion are: $R_A = (R_1R_2 + R_1R_3 + R_2R_3)/R_2$, $R_B = (R_1R_2 + R_1R_3 + R_2R_3/R_1)$, and $R_C = (R_1R_2 + R_1R_3 + R_2R_3)/R_3$.

The upper delta of Fig. 1.18 is converted to its equivalent wye by the conversion formulas (see Fig. 1.20): $R_1 = [(30)(60)]/(30 + 50 + 60) = 12.9\ \Omega$, $R_2 = [(50)(60)]/(30 + 50 + 60) = 21.4\ \Omega$, and $R_3 = [(50)(30)]/(30 + 50 + 60) = 10.7\ \Omega$. From the simplified series-parallel circuit of Fig. 1.20*b*, it can be seen that: $R_{EQT} = 10.7 + [(12.9 + 40) \| (21.4 + 20)] = 33.9\ \Omega$.

Related Calculations. Delta-to-wye and wye-to-delta conversion is used to reduce the series-parallel equivalent circuits, thus eliminating the need to apply mesh or nodal analysis. The wye and delta configurations often appear as shown in Fig. 1.21. They are then referred to as a tee (T) or a pi (π) network. The equations used to convert from a tee to a pi network the exactly the same as those used for the wye and delta transformation.

ANALYSIS OF A SINUSOIDAL WAVE

Given: the voltage $e(t) = 170 \sin 377t$. Calculate the average or dc (E_{dc}), peak (E_m), rms (E), angular frequency (ω), frequency (f), period (T), and peak-to-peak (E_{pp}) values.

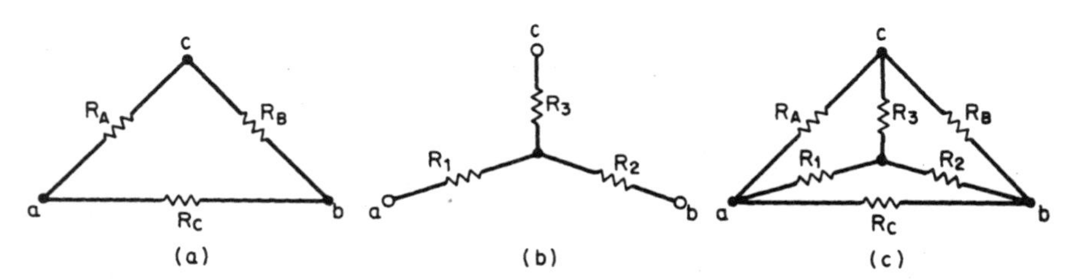

FIGURE 1.19 (*a*) Delta circuit; (*b*) wye circuit; and (*c*) delta-to-wye conversions.

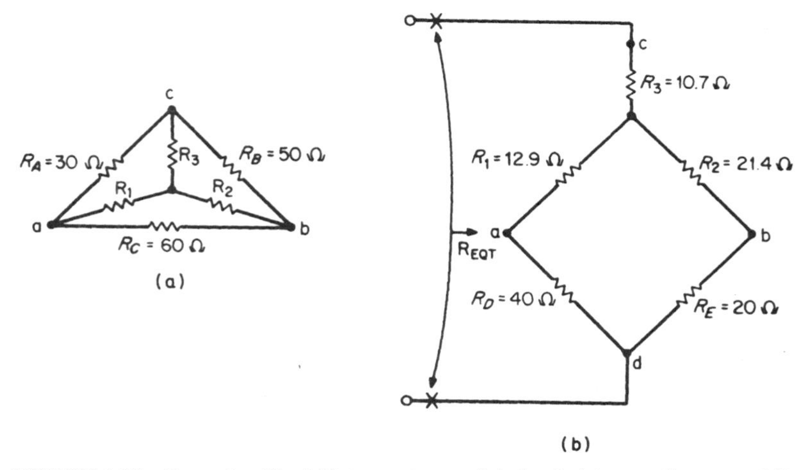

FIGURE 1.20 Converting Fig. 1.18 to a series-parallel circuit: (*a*) converting upper delta to a wye circuit and (*b*) resultant series-parallel circuit.

Calculation Procedure

1. Calculate Average Value

$E_{dc} = 0$ because the average value or dc component of a symmetrical wave is zero.

2. Calculate Peak Value

$E_m = 170$ V, which is the maximum value of the sinusoidal wave.

3. Calculate rms Value

$E = 0.707E_m$ where E represents the rms, or effective, value of the sinusoidal wave. Therefore $E = (0.707)(170) = 120$ V. Note that the 0.707 value is for a pure sine (or cosine) waveform. This comes from the relation $E = \sqrt{\frac{1}{T}\int_t^{t+T} e^2(t)dt}$.

4. Calculate Angular Frequency

The angular frequency ω equals 377 rad/s.

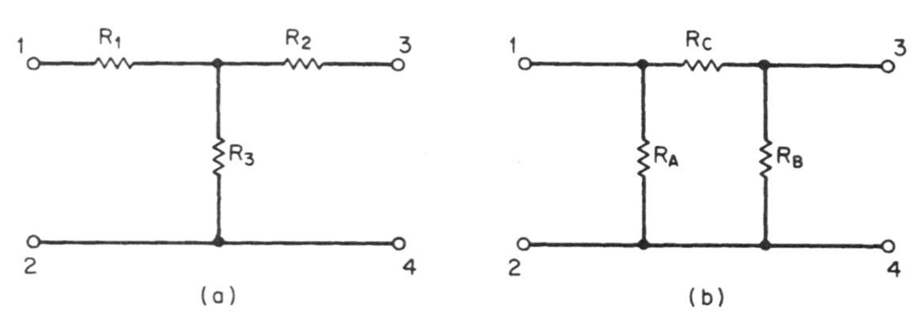

FIGURE 1.21 Comparison of wye to tee and delta to *pi* circuits: (*a*) wye or tee configuration and (*b*) delta to *pi* configuration.

5. Calculate Frequency

$f = \omega/2\pi = 377/(2 \times 3.1416) = 60$ Hz.

6. Calculate Period

$T = 1/f = 1/60$ s.

7. Calculate Peak-to-Peak Value

$E_{pp} = 2E_m = 2(170) = 340$ V.

Related Calculations. This problem analyzed the sine wave, which is standard in the United States, that is, a voltage wave that has an rms value of 120 V and a frequency of 60 Hz.

ANALYSIS OF A SQUARE WAVE

Find the average and rms values of the square wave of Fig. 1.22.

Calculation Procedure

1. Calculate the Average Value

The average value, or dc component, of the symmetrical square wave is zero; therefore: $V_{dc} = V_{avg} = 0$.

2. Calculate the rms Value

The rms value is found by squaring the wave over a period of 2 s. This gives a value equal to 100 V², which is a constant value over the entire period. Thus, the mean over the period is V². The square root of 100 V² equals 10 V. Therefore, the rms value is $V = 10$ V.

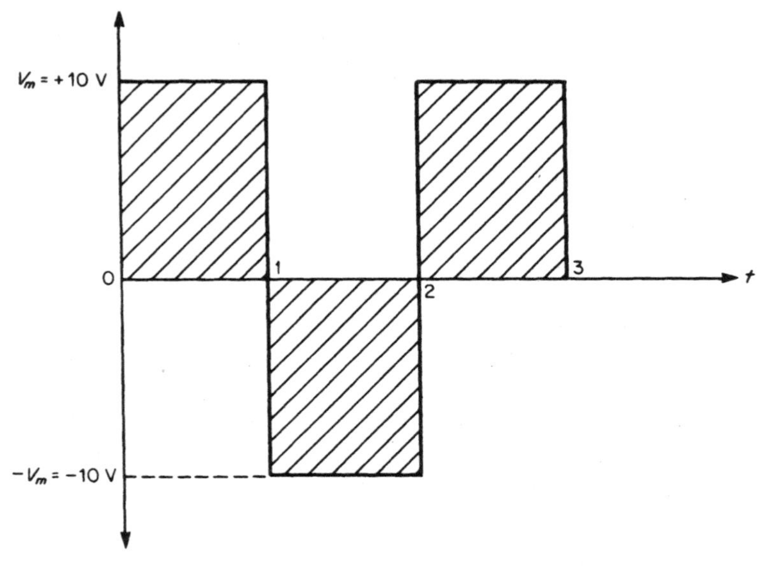

FIGURE 1.22 Square wave to be analyzed.

Related Calculations. The equation for the wave of Fig. 1.22 is: $v(t) = (4V_m/\pi)(\sin \omega t + \frac{1}{3} \sin 3\omega t + \frac{1}{3} \sin 5\omega t + \cdots + 1/n \sin n\omega t)$. This equation, referred to as a *Fourier series*, shows that a symmetrical square wave beginning at $t = 0$ has no dc component, no even harmonics, and an infinite number of odd harmonics.

ANALYSIS OF AN OFFSET WAVE

Find the average and rms values of the offset wave of Fig. 1.23.

Calculation Procedure

1. Calculate the Average Value

$V_{avg} = V_{dc} =$ net area/T, where net area = algebraic sum of areas for one period and T = period of wave. Hence, $V_{avg} = V_{dc} = [(12 \times 1) - (8 \times 1)]/2 = 2$ V.

2. Calculate the rms Value

$$V(\text{rms value}) = \sqrt{\frac{\text{area}[v(t)^2]}{T}} = \sqrt{\frac{(12 \times 1)^2 + (8 \times 1)^2}{2}}$$

$$= \sqrt{104} = 10.2 \text{ V.}$$

Related Calculations. Figure 1.23 is the same wave as Fig. 1.22 except that it has been offset by the addition of a dc component equal to 2 V. The rms, or effective, value of a periodic waveform is equal to the direct current, which dissipates the same energy in a given resistor. Since the offset wave has a dc component equal to 2 V, its rms value of 10.2 V is higher than the symmetrical square wave of Fig. 1.22.

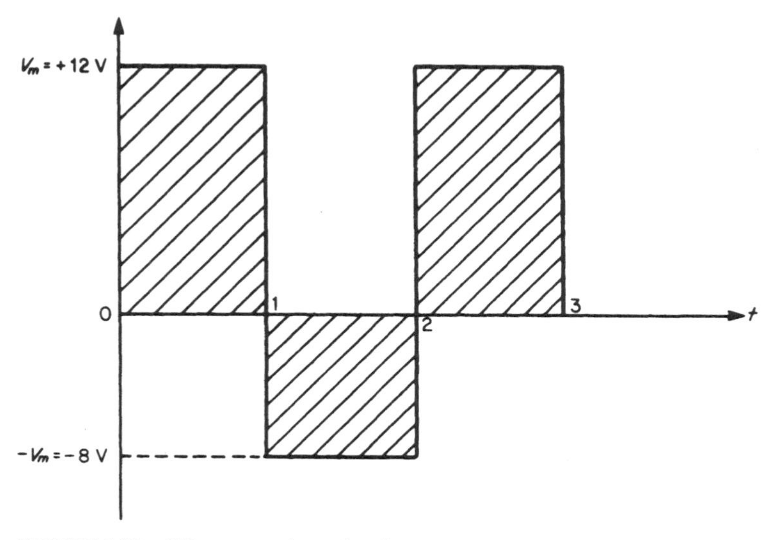

FIGURE 1.23 Offset wave to be analyzed.

CIRCUIT RESPONSE TO A NONSINUSOIDAL INPUT CONSISTING OF A DC VOLTAGE IN A SERIES WITH AN AC VOLTAGE

The input to the circuit of Fig. 1.24 is $e = 20 + 10 \sin 377t$. (*a*) Find and express i, v_R, and v_C in the time domain. (*b*) Find I, V_R, and V_C. (*c*) Find the power delivered to the circuit. Assume enough time has elapsed that v_C has reached its final (steady-state) value in all three parts of this problem.

Calculation Procedure

1. Determine the Solution for Part a

This problem can be solved by the application of the superposition theorem, since two separate voltages, one dc and one ac, are present in the circuit. Effect of 20 V dc on circuit: when v_C has reached its final (steady-state) value $i = 0$, $v_R = iR = 0$ V, and $v_C = 20$ V. Effect of ac voltage ($10 \sin 377t$) on circuit: $X_C = 1/\omega C = 1/(377)(660 \times 10^{-6}) = 4\ \Omega$. Hence, $\mathbf{Z} = 3 - j4 = 5\underline{/-53^\circ}\ \Omega$. Then, $\mathbf{I} = \mathbf{E}/\mathbf{Z} = (0.707)(10)\underline{/0^\circ}/5\underline{/-53^\circ} = 1.414\underline{/+53^\circ}$ A.

Therefore, the maximum value is $I_m = 1.414/0.707 = 2$ A and the current in the time domain is $i = 0 + 2 \sin(377t + 53^\circ)$. $\mathbf{V_R} = \mathbf{I}R = (1.414\underline{/+53^\circ})(3\underline{/0^\circ}) = 4.242\underline{/+53^\circ}$ V. The maximum value for $\mathbf{V_R}$ is $4.242/0.707 = 6$ V, and the voltage v_R in the time domain is $v_R = 0 + 6 \sin(377t + 53^\circ)$.

$\mathbf{V_C} = \mathbf{I}X_C = (1.414\underline{/+53^\circ})\ (4\underline{/-90^\circ}) = 5.656\ \underline{/-37^\circ}$. The maximum value for $\mathbf{V_C} = 5.656/0.707 = 8$ V, and the voltage v_C in the time domain is $v_C = 20 + 8 \sin(377t - 37^\circ)$.

2. Determine the Solution for Part b

The effective value of a nonsinusoidal input consisting of dc and ac components can be found from the following equation:

$$V = \sqrt{V_{\text{dc}}^2 + \frac{(V_{m1}^2 + V_{m2}^2 + \cdots + V_{mn}^2)}{2}}$$

where V_{dc} = voltage of dc component and V_{m1}, etc. = maximum value of ac components. Therefore $|\mathbf{I}| = \sqrt{0^2 + 2^2/2} = 1.414$ A, $|\mathbf{V_R}| = \sqrt{0^2 + 6^2/2} = 4.24$ A, and $|\mathbf{V_C}| = \sqrt{20^2 + 8^2/2} = 20.8$ V.

3. Determine the Solution for Part c

$P = I^2R = (1.414)^2(3) = 6$ W.

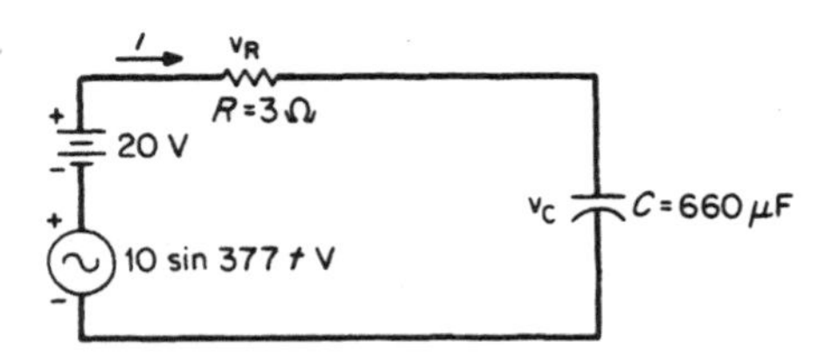

FIGURE 1.24 Analysis of circuit response to a nonsinusoidal input.

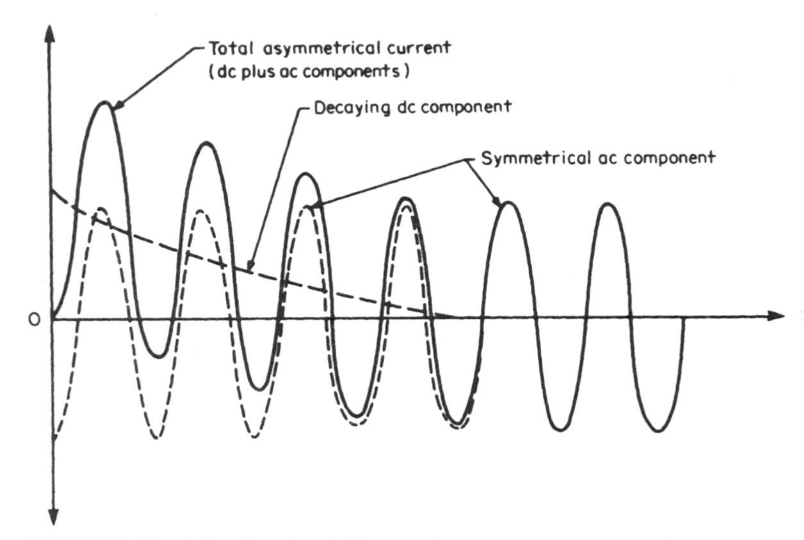

FIGURE 1.25 A decaying sinusoidal wave.

Related Calculations. This concept of a dc component superimposed on a sinusoidal ac component is illustrated in Fig. 1.25. This figure shows the decay of a dc component because of a short circuit and also shows how the asymmetrical short-circuit current gradually becomes symmetrical when the dc component decays to zero.

STEADY-STATE AC ANALYSIS OF A SERIES RLC CIRCUIT

Calculate the current in the circuit of Fig. 1.26*a*.

Calculation Procedure

1. Calculate Z

Angular frequency $\omega = 2\pi f = (2)(3.1416)(60) = 377$ rad/s. But $X_L = \omega L$; therefore, $X_L = (377)(0.5) = 188.5\ \Omega$. Also, $X_C = 1/\omega C = 1/[(377)(26.5) \times 10^{-6}] = 100\ \Omega$. Then $\mathbf{Z} = R + j(X_L - X_C) = R + jX_{\text{EQ}}$, where $X_{\text{EQ}} = X_L - X_C =$ net equivalent reactance.

In polar form, the impedance for the series RLC circuit is expressed as $\mathbf{Z} = \sqrt{R^2 + X_{\text{EQ}}^2}\,\underline{/\tan^{-1}(X_{\text{EQ}}/R)} = |\mathbf{Z}|\underline{/\theta}$. $\mathbf{Z} = 100 + j(188.5 - 100) = 100 + j88.5 = \sqrt{(100)^2 + (88.5)^2}\,\underline{/\tan^{-1}(88.5/100)} = 133.5\underline{/41.5^\circ}\ \Omega$. The impedance triangle (Fig. 1.26*b*) illustrates the results of the preceding solution.

Apply KVL to the circuit: $\mathbf{E} = V_R + jV_L - jV_C = V_R + jV_X$ where $V_X = V_L - V_C =$ net reactive voltage.

2. Draw the Phasor Diagram

The phasor diagram of Fig. 1.26*c* shows the voltage relations with respect to the current as a reference.

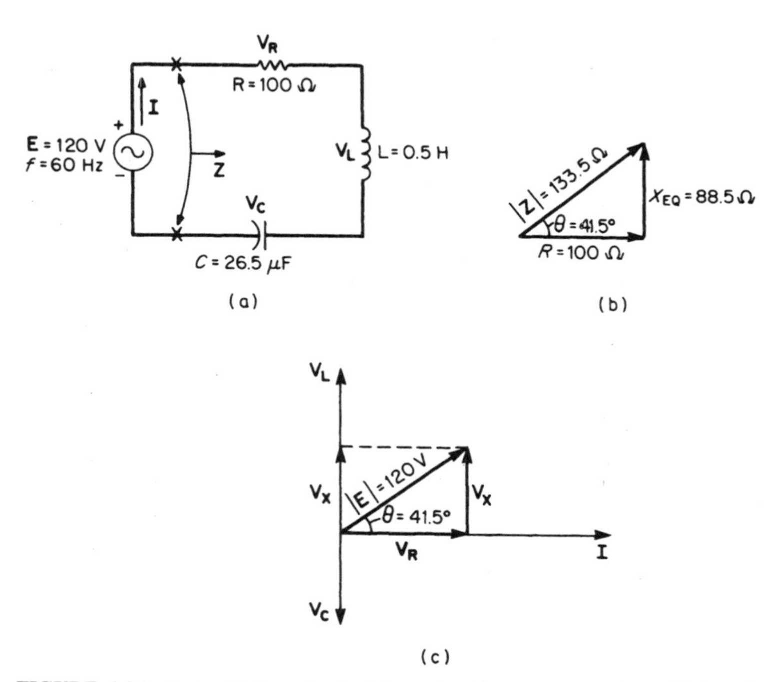

FIGURE 1.26 Series *RLC* ac circuit: (*a*) circuit with component values; (*b*) impedance triangle; and (*c*) phasor diagram.

3. Calculate I

From Ohm's law for ac circuits, $|\mathbf{I}| = 120/133.5 = 0.899$ A. Because **I** is a reference it can be expressed in polar form as $\mathbf{I} = 0.899\underline{/0^\circ}$ A. The angle between the voltage and current in Fig. 1.26*c* is the same as the angle in the impedance triangle of Fig. 1.26*b*. Therefore $\mathbf{E} = 120\underline{/41.5^\circ}$ V.

Related Calculations. In a series *RLC* circuit the net reactive voltage may be zero (when $\mathbf{V}_L = \mathbf{V}_C$), inductive (when $\mathbf{V}_L > \mathbf{V}_C$), or capacitive (when $\mathbf{V}_L < \mathbf{V}_C$). The current in such a circuit may be in phase with, lag, or lead the applied emf. When $\mathbf{V}_L = \mathbf{V}_C$, the condition is referred to as *series resonance*. Voltages $\mathbf{V}_L$ and $\mathbf{V}_C$ may be higher than the applied voltage **E**, because the only limiting opposition to current is resistance *R*. A circuit in series resonance has maximum current, minimum impedance, and a power factor of 100 percent.

STEADY-STATE AC ANALYSIS OF A PARALLEL RLC CIRCUIT

Calculate the impedance of the parallel *RLC* circuit of Fig. 1.27*a*.

Calculation Procedure

1. Calculate the Currents in **R, L,** *and* **C**

In a parallel circuit, it is convenient to use the voltage as a reference; therefore $\mathbf{E} = 200\underline{/0^\circ}$ V. Because the *R*, *L*, and *C* parameters of this circuit are the same as in

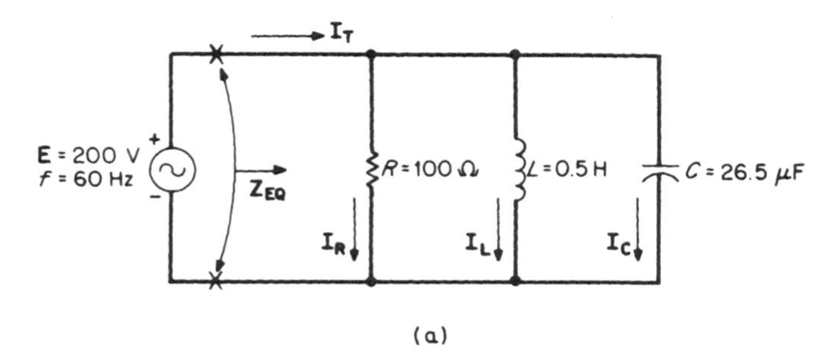

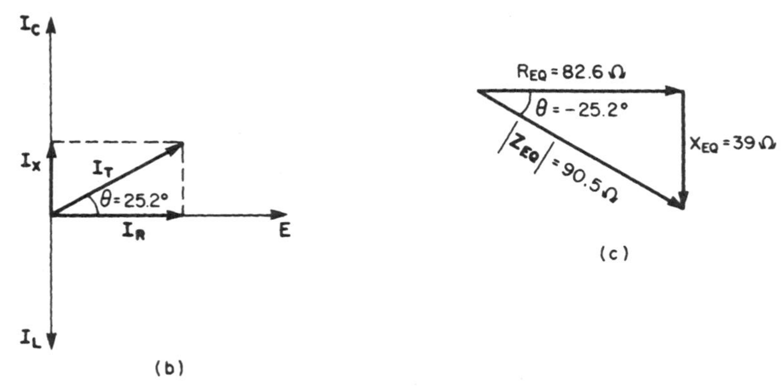

FIGURE 1.27 Parallel *RLC* circuit: (*a*) circuit with component values; (*b*) phasor diagram; and (*c*) impedance triangle.

Fig. 1.26*a* and the frequency (60 Hz) is the same, $X_L = 188.5\ \Omega$ and $X_C = 100\ \Omega$. From Ohm's law: $\mathbf{I_R} = \mathbf{E}/R = 200\angle 0°/100\angle 0° = 2\angle 0°$ A. $\mathbf{I_L} = \mathbf{E}/X_L = 200\angle 0°/188.5\angle 90° = 1.06\angle -90° = -j1.06$ A, and $\mathbf{I_C} = \mathbf{E}/X_C = 200\angle 0°/100\angle -90° = 2\angle 90° = +j2$ A. But $\mathbf{I_T} = I_R - jI_L + jI_C$; therefore, $\mathbf{I_T} = 2 - j1.06 + j2 = 2 + j0.94 = 2.21\angle 25.2°$ A.

2. Calculate Z_{EQ}

Impedance is $\mathbf{Z_{EQ}} = \mathbf{E}/\mathbf{I_T} = 200\angle 0°/2.21\angle 25.2° = 90.5\angle -25.2°\ \Omega$. $\mathbf{Z_{EQ}}$, changed to rectangular form, is $\mathbf{Z_{EQ}} = 82.6\ \Omega - j39\ \Omega = R_{EQ} - jX_{EQ}$. Figure 1.27*b* illustrates the voltage-current phasor diagram. The equivalent impedance diagram is given in Fig. 1.27*c*. Note that $\mathbf{Z_{EQ}}$ can also be calculated by

$$\mathbf{Z_{EQ}} = \frac{1}{\dfrac{1}{R} + \dfrac{1}{jX_L} - \dfrac{1}{jX_C}}$$

since $Z_L = jX_L$ and $Z_C = -jX_C$.

Related Calculations. The impedance diagram of Fig. 1.27*c* has a negative angle. This indicates that the circuit is an *RC* equivalent circuit. Figure 1.27*b* verifies this observation because the total circuit current $\mathbf{I_T}$ leads the applied voltage. In a parallel *RLC* circuit the net reactive current may be zero (when $\mathbf{I_L} = \mathbf{I_C}$), inductive (when $\mathbf{I_L} > \mathbf{I_C}$), or capacitive (when $\mathbf{I_L} < \mathbf{I_C}$). The current in such a circuit may be in phase with, lag, or lead the applied emf. When $\mathbf{I_L} = \mathbf{I_C}$, this condition is referred to as *parallel resonance*. Currents $\mathbf{I_L}$ and $\mathbf{I_C}$ may be much higher than the total line current, $\mathbf{I_T}$. A circuit in parallel resonance has a minimum current, maximum impedance, and a power factor of 100 percent. Note in Fig. 1.27*b* that $\mathbf{I_T} = I_R + jI_X$, where $I_X = I_C - I_L$.

ANALYSIS OF A SERIES-PARALLEL AC NETWORK

A series-parallel ac network is shown in Fig. 1.28. Calculate $\mathbf{Z_{EQ}}$, $\mathbf{I_1}$, $\mathbf{I_2}$, and $\mathbf{I_3}$.

Calculation Procedure

1. Combine All Series Impedances

The solution to this problem is similar to that for the first problem in the section, except that vector algebra must be used for the reactances. $\mathbf{Z_1} = 300 + j600 - j200 = 300 + j400 = 500\underline{/53.1°}\ \Omega$, $\mathbf{Z_2} = 500 + j1200 = 1300\underline{/67.4°}\ \Omega$, and $\mathbf{Z_3} = 800 - j600 = 1000\underline{/-36.9°}\ \Omega$.

2. Combine All Parallel Impedances

Using the product-over-the-sum rule, we find $\mathbf{Z_{BC}} = \mathbf{Z_2Z_3}/(\mathbf{Z_1} + \mathbf{Z_3}) = (1300\underline{/67.4°})\,1000\underline{/-36.9°})/[(500 + j1200) + (800 - j600)] = 908\underline{/5.7°} = 901 + j90.2\ \Omega$.

3. Combine All Series Impedances to Obtain the Total Impedance, Z_{EQ}

$\mathbf{Z_{EQ}} = \mathbf{Z_1} + \mathbf{Z_{BC}} = (300 + j400) + (901 + j90.2) = 1201 + j490 = 1290\underline{/22.4°}\ \Omega$.

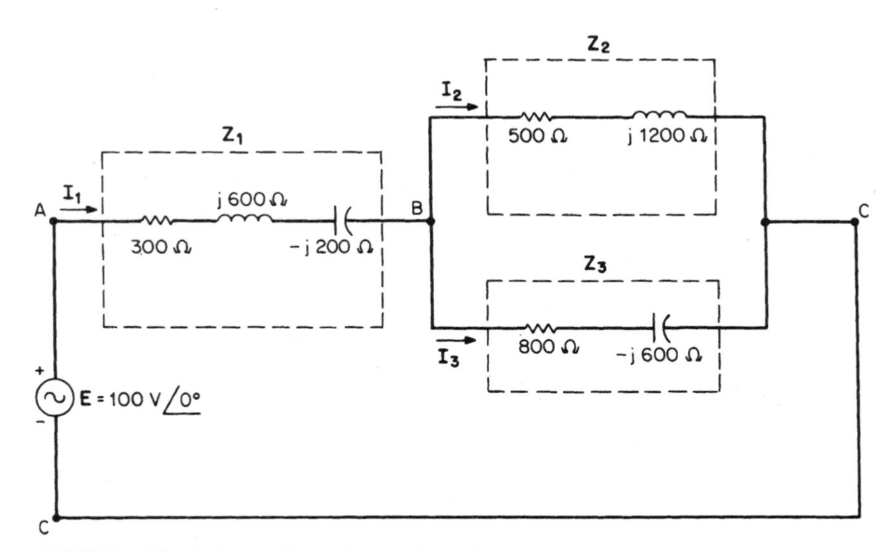

FIGURE 1.28 Series-parallel ac circuit to be analyzed.

4. Calculate the Currents

$\mathbf{I}_1 = \mathbf{E}/\mathbf{Z}_{EQ} = 100\angle 0°/1290\angle 22.4° = 0.0775\angle -22.4°$ A. From the current-divider rule: $\mathbf{I}_2 = \mathbf{I}_1\mathbf{Z}_3/(\mathbf{Z}_2 + \mathbf{Z}_3) = (0.0775\angle -22.4°)(1000\angle -36.9°)/[(500 + j1200) + (800 - j600)] = 0.0541\angle -84.1°$ A. $\mathbf{I}_3 = \mathbf{I}_1\mathbf{Z}_2/(\mathbf{Z}_2 + \mathbf{Z}_3) = (0.0775\angle -22.4°)(1300\angle 67.4°)/[(500 + j1200) + (800 - j600)] = 0.0709\angle 20.2°$ A.

Related Calculations. Any reducible ac circuit (i.e., any circuit that can be reduced to one equivalent impedance $\mathbf{Z}_{EQ}$ with a single power source), no matter how complex, can be solved in a similar manner to that just described. The dc network theorems used in previous problems can be applied to ac networks except that vector algebra must be used for the ac quantities.

ANALYSIS OF POWER IN AN AC CIRCUIT

Find the total watts, total VARS, and total volt-amperes in the ac circuit of Fig. 1.29*a*. Recall that watts, VARS, and volt-amperes are all dimensionally the same, that is, the product of voltage and current. However, we use the designators of watts (W) to represent real power (instantaneous or average), volt-amperes-reactive (VARS) to represent reactive power, and volt-amperes (VA) to represent complex (or apparent) power.

Calculation Procedure

1. Study the Power Triangle

Figure 1.30 shows power triangles for ac circuits. Power triangles are drawn following the standard of drawing inductive reactive power in the $+j$ direction and capacitive reactive power in the $-j$ direction. Two equations are obtained by applying the Pythagorean theorem to these power triangles: $S^2 = P^2 + Q_L^2$ and $S^2 = P^2 + Q_C^2$. These equations can be applied to series, parallel, or series-parallel circuits.

The net reactive power supplied by the source to an *RLC* circuit is the difference between the positive inductive reactive power and the negative capacitive reactive power: $Q_X = Q_L - Q_C$, where Q_X is the net reactive power, in VARS.

2. Solve for the Total Real Power

Arithmetic addition can be used to find the total real power. $P_T = P_1 + P_2 = 200 + 500 = 700$ W.

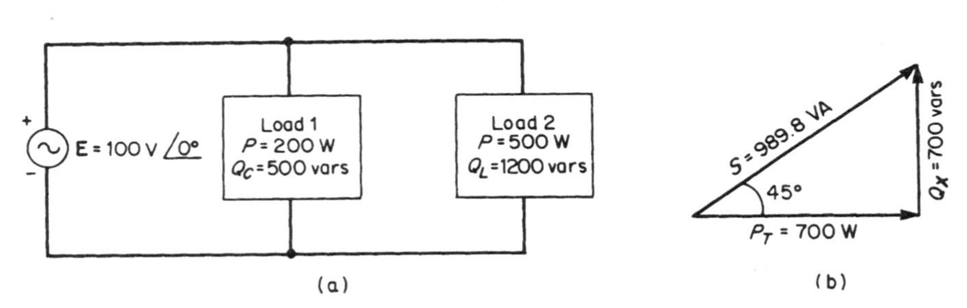

FIGURE 1.29 Calculating ac power: (*a*) circuit and (*b*) power triangle.

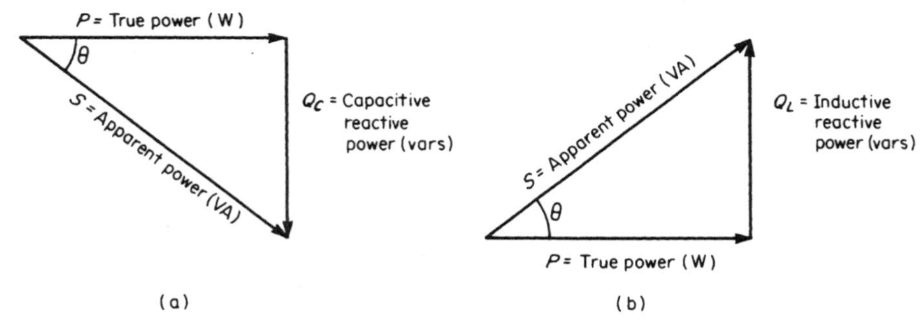

FIGURE 1.30 Power triangles for (*a*) *RC* and (*b*) *RL* equivalent circuits.

3. Solve for the Total Reactive Power

$Q_X = Q_L - Q_C = 1200 - 500 = 700$ VARS. Because the total reactive power is positive, the circuit is inductive (see Fig. 1.29*b*).

4. Solve for the Total Volt-Amperes

$S = \sqrt{P_T^2 + Q_X^2} = \sqrt{(700)^2 + (700)^2} = 989.8$ VA.

Related Calculations. The principles used in this problem will also be applied to solve the following two problems.

ANALYSIS OF POWER FACTOR AND REACTIVE FACTOR

Calculate the power factor (pf) and the reactive factor (rf) for the circuit shown in Fig. 1.31.

Calculation Procedure

1. Review Power-Factor Analysis

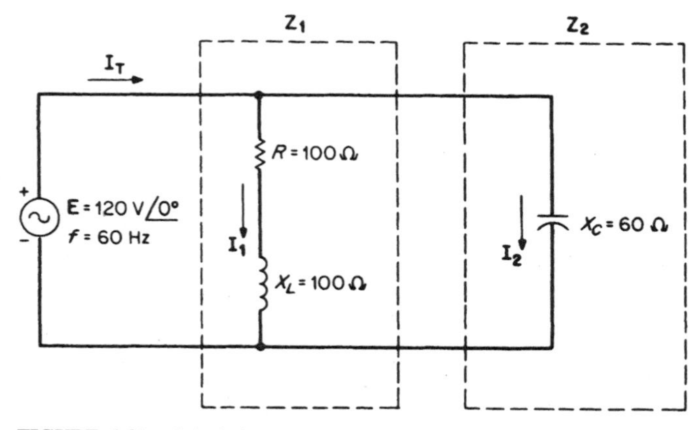

FIGURE 1.31 Calculating power and reactive factors of circuit.

The power factor of an ac circuit is the numerical ratio between the true power P and the apparent power S. It can be seen by referring to the power triangles of Fig. 1.30 that this ratio is equal to the cosine of the power-factor angle θ. The power-factor angle is the same as the phase angle between the voltage across the circuit (or load) and the current through the circuit (or load). pf $= \cos\theta = P/S$.

2. Review Reactive-Factor Analysis

The numerical ratio between the reactive power and the apparent power of a circuit (or load) is called the reactive factor. This ratio is equal to the sine of the power-factor angle (see Fig. 1.30). rf $= \sin\theta = Q/S$.

3. Calculate the Power and Reactive Factors

$\mathbf{Z}_1 = R + jX_L = 100 + j100 = 141.4\angle 45°$. $\mathbf{I}_1 = \mathbf{E}/\mathbf{Z}_1 = 120\angle 0°/141.4\angle 45° = 0.849\angle -45°$ A. $\mathbf{I}_1 = (0.6 - j0.6)$ A. $\mathbf{I}_2 = \mathbf{E}/X_C = 120\angle 0°/60\angle -90° = 2\angle -90° = (0 + j2)$ A. $\mathbf{I}_T = \mathbf{I}_1 + \mathbf{I}_2 = (0.6 - j0.6) + (0 + j2) = (0.6 - j1.4)$ A $= 1.523\angle 66.8°$ A. $S = |\mathbf{E}||\mathbf{L}_T| = (120)(1.523) = 182.8$ VA. Power factor $= \cos\theta = \cos 66.8° = 0.394$ or 39.4 percent; rf $= \sin\theta = \sin 66.8° = 0.92$ or 92 percent.

Related Calculations. Inductive loads have a lagging power factor; capacitive loads have a leading power factor. The value of the power factor is expressed either as a decimal or as a percentage. This value is always less than 1.0 or less than 100 percent. The majority of industrial loads, such as motors and air conditioners, are inductive (lagging power factor). Thus, power engineers often refer to capacitors or capacitive loads as sources of reactive power.

POWER-FACTOR CORRECTION

Calculate the value of the capacitor needed to obtain a circuit power factor of 100 percent (Fig. 1.32).

Calculation Procedure

1. Calculate the Motor Current

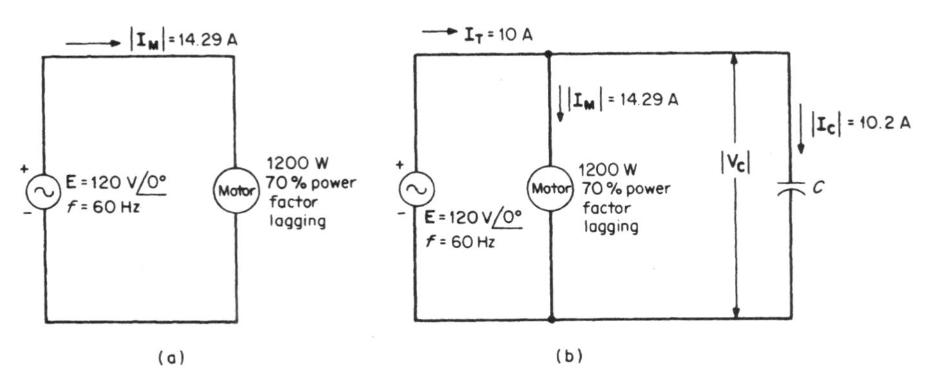

FIGURE 1.32 Power-factor correction: (*a*) given circuit and (*b*) adding a capacitor (C) in parallel to improve power factor.

$S = P/\cos\,\theta = 1200/0.7 = 1714$ VA. Hence, the motor current $|\mathbf{I}|$ is: $|\mathbf{I}| = S/|\mathbf{E}| =$ (1714 VA) (120 V) = 14.29 A. The active component of this current is the component in phase with the voltage. This component, which results in true power consumption, is: $|\mathbf{I}|\cos\,\theta = (14.29\text{ A})(0.7) = 10$ A. Because the motor has a 70 percent power factor, the circuit must supply 14.29 A to realize a useful current of 10 A.

2. Calculate the Value of **C**

In order to obtain a circuit power factor of 100 percent, the inductive apparent power of the motor and the capacitive apparent power of the capacitor must be equal. $Q_L = |\mathbf{E}||\mathbf{I}|\sqrt{1 - \cos^2\theta}$ where $\sqrt{1 - \cos^2\theta}$ = reactive factor. Hence, $Q_L = (120)(14.29)\sqrt{1 - (0.7)^2} = 1714\sqrt{0.51} = 1224$ VARS (inductive). Q_C must equal 1224 VARS for 100 percent power factor. $X_C = V_C^2/Q_C = (120)^2/1224 = 11.76\ \Omega$ (capacitive). Therefore, $C = 1/\omega X_C = 1/(377)(11.76) = 225.5\ \mu$F.

Related Calculations. The amount of current required by a load determines the sizes of the wire used in the windings of the generator or transformer and in the conductors connecting the motor to the generator or transformer. Because copper losses depend upon the square of the load current, a power company finds it more economical to supply 10 A at a power factor of 100 percent than to supply 14.29 A at a power factor of 70 percent.

A mathematical analysis of the currents in Fig. 1.32*b* follows: $|\mathbf{I}_C| = Q_C/|\mathbf{V}_C| = (1220$ VARS)/(120 V) = 10.2 A = $(0 + j10.2)$ A. θ (for motor) $= \cos^{-1} 0.7 = 45.6°$; therefore, $\mathbf{I_M} = 14.29\underline{/45.6°} = (10 - j10.2)$ A. Then $\mathbf{I_T} = \mathbf{I_M} + \mathbf{I_C} = (10 - j10.2) + (0 + j10.2) = 10\underline{/0°}$ A (100 percent power factor). Typically, power factor correction capacitors are rated in kVARS (kilo-VARS) and may be installed in switched banks to provide a range of pf correction.

MAXIMUM POWER TRANSFER IN AN AC CIRCUIT

Calculate the load impedance in Fig. 1.33 for maximum power to the load.

Calculation Procedure

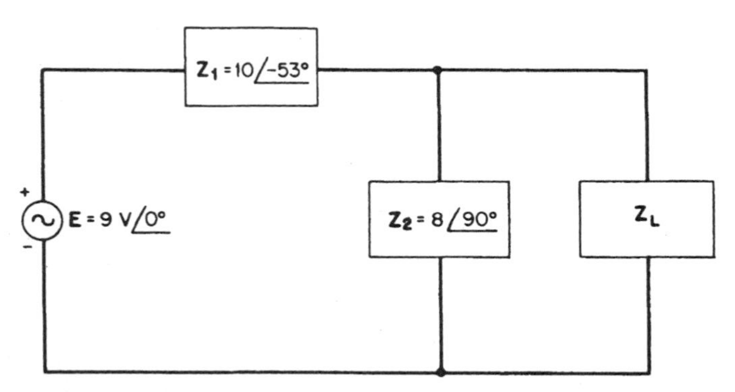

FIGURE 1.33 Finding value of Z_L for maximum power transfer.

1. Statement of the Maximum Power Theorem

The maximum power theorem, when applied to ac circuits, states that maximum power will be delivered to a load when the load impedance is the complex conjugate of the Thevenin impedance across its terminals.

2. Apply Thevenin's Theorem to the Circuit

$\mathbf{Z}_{Th} = \mathbf{Z}_1\mathbf{Z}_2/(\mathbf{Z}_1 + \mathbf{Z}_2) = (10\angle -53°)(8\angle 90°)/[6 - j8) + j8] = 13.3\angle -37°\ \Omega$, or $\mathbf{Z}_{Th} = 10.6 + j8\ \Omega$ where $R = 10.6\ \Omega$ and $X_L = 8\ \Omega$. Then, $\mathbf{Z}_L$ must be $13.3\angle -37° = 10.6 - j8\ \Omega$, where $R_L = 10.6\ \Omega$ and $X_C = -8\ \Omega$.

In order to find the maximum power delivered to the load, $\mathbf{E}_{Th}$ must be found using the voltage-divider rule: $\mathbf{E}_{Th} = \mathbf{E}\mathbf{Z}_2/(\mathbf{Z}_1 + \mathbf{Z}_2) = (9\angle 0°)(8\angle 90°)/[(6 - j8) + j8] = 12\angle 90°$ V. $P_{max} = |\mathbf{E}^2_{Th}|/4R_L$; therefore $P_{max} = (12)^2/(4)(10.6) = 3.4$ W.

Related Calculations. The maximum power transfer theorem, when applied to dc circuits, states that a load will receive maximum power from a dc network when its total resistance is equal to the Thevenin resistance of the network as seen by the load.

ANALYSIS OF A BALANCED WYE-WYE SYSTEM

Calculate the currents in all lines of the balanced three-phase, four-wire, wye-connected system of Fig. 1.34. The system has the following parameters: $\mathbf{V}_{AN} = 120\angle 0°$ V, $\mathbf{V}_{BN} = 120\angle -120°$ V, $\mathbf{V}_{CN} = 120\angle 120°$ V, and $\mathbf{Z}_A = \mathbf{Z}_B = \mathbf{Z}_C = 12\angle 0°\ \Omega$.

Calculation Procedure

1. Calculate Currents

$\mathbf{I}_A = \mathbf{V}_{AN}/\mathbf{Z}_A = 120\angle 0°/12\angle 0° = 10\angle 0°$ A. $\mathbf{I}_B = \mathbf{V}_{BN}/\mathbf{Z}_B = 120\angle -120°/12\angle 0° = 10\angle -120°$ A. $\mathbf{I}_C = \mathbf{V}_{CN}/\mathbf{Z}_C = 120\angle 120°/12\angle 0° = 10\angle 120°$ A. $\mathbf{I}_N = \mathbf{I}_A + \mathbf{I}_B + \mathbf{I}_C$; hence, $\mathbf{I}_N = 10\angle 0° + 10\angle -120° + 10\angle 120° = 0$ A.

Related Calculations. The neutral current in a balanced wye system is always zero. Each load current lags or leads the voltage by the particular power factor of the load. This system, in which one terminal of each phase is connected to a common star point, is often called a star-connected system.

ANALYSIS OF A BALANCED DELTA-DELTA SYSTEM

Calculate the load currents and the line currents of the balanced delta-delta system of Fig. 1.35. The system has the following load parameters: $\mathbf{V}_{AC} = 200\angle 0°$ V, $\mathbf{V}_{BA} = 200\angle 120°$ V, $\mathbf{V}_{CB} = 200\angle -120°$ V, and $\mathbf{Z}_{AC} = \mathbf{Z}_{BA} = \mathbf{Z}_{CB} = 4\angle 0°\ \Omega$.

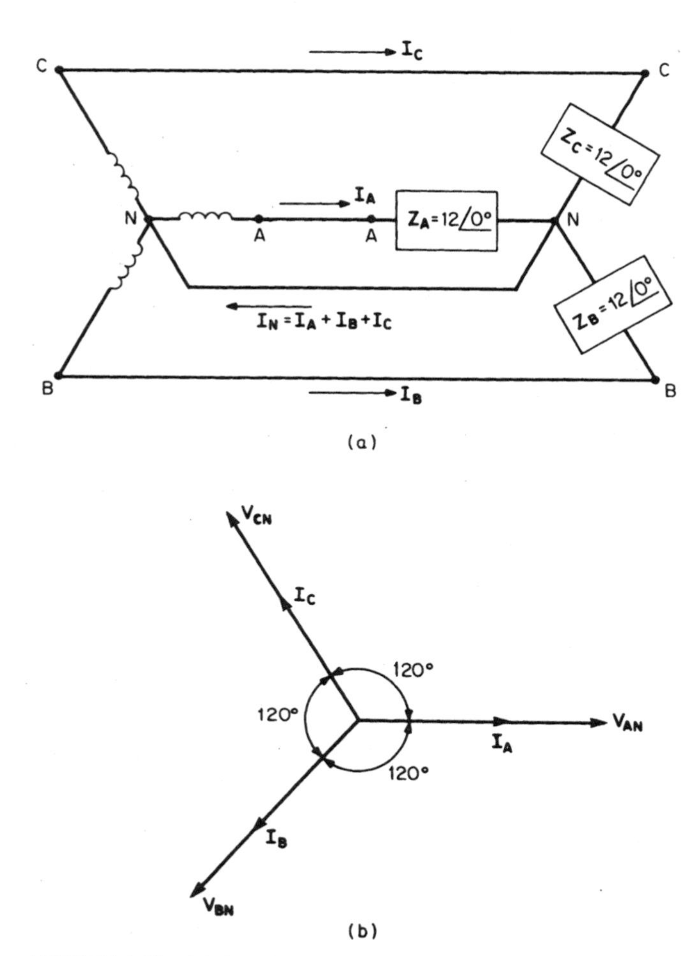

FIGURE 1.34 A balanced three-phase, four-wire, wye-connected system: (*a*) circuit and (*b*) load-phasor diagram.

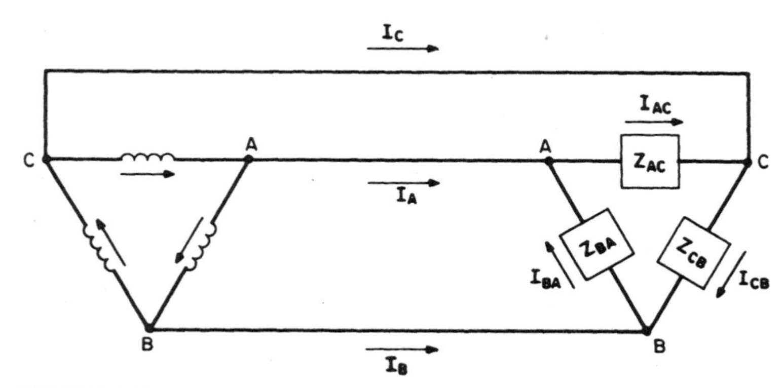

FIGURE 1.35 A balanced delta-delta system.

Calculation Procedure

1. Solve for the Load Currents

$\mathbf{I}_{AC} = \mathbf{V}_{AC}/\mathbf{Z}_{AC} = 200\angle 0°/4\angle 0° = 50\angle 0°$ A, $\mathbf{I}_{BA} = \mathbf{V}_{BA}/\mathbf{Z}_{BA} = 200\angle 120°/4\angle 0° = 50\angle 120°$ A, and $\mathbf{I}_{CB} = \mathbf{V}_{CB}/\mathbf{Z}_{CB} = 200\angle 120°/4\angle 0° = 50\angle -120°$ A.

2. Solve for the Line Currents

Convert the load currents to rectangular notation: $\mathbf{I}_{AC} = 50\angle 0° = 50 + j0$, $\mathbf{I}_{BA} = 50\angle 120° = -25 + j43.3$, and $\mathbf{I}_{CB} = 50\angle -120° = -25 - j43.3$. Apply KCL at load nodes: $\mathbf{I}_A = \mathbf{I}_{AC} - \mathbf{I}_{BA} = (50 + j0) - (-25 + j43.3) = 86.6\angle -30°$ A, $\mathbf{I}_B = \mathbf{I}_{BA} - \mathbf{I}_{CB} = (-25 + j43.3) - (-25 - j43.3) = 86.6\angle -90°$ A, $\mathbf{I}_C = \mathbf{I}_{CB} - \mathbf{I}_{AC} = (-25 - j43.3) - (50 + j0) = 86.6\angle -150°$ A.

Related Calculations. In comparing a wye-connected system with a delta-connected system, one can make the following observations:

1. When a load is wye-connected, each arm of the load is connected from a line to the neutral. The impedance $\mathbf{Z}$ is shown with a single subscript, such as $\mathbf{Z}_A$.
2. When a load is delta-connected, each arm of the load is connected from a line to line. The impedance $\mathbf{Z}$ is shown with a double subscript such as $\mathbf{Z}_{AC}$.
3. In a wye-connected system, the phase current of the source, the line current, and the phase current of the load are all equal.
4. In a delta-connected system, each line must carry components of current for two arms of the load. One current component moves toward the source, and the other

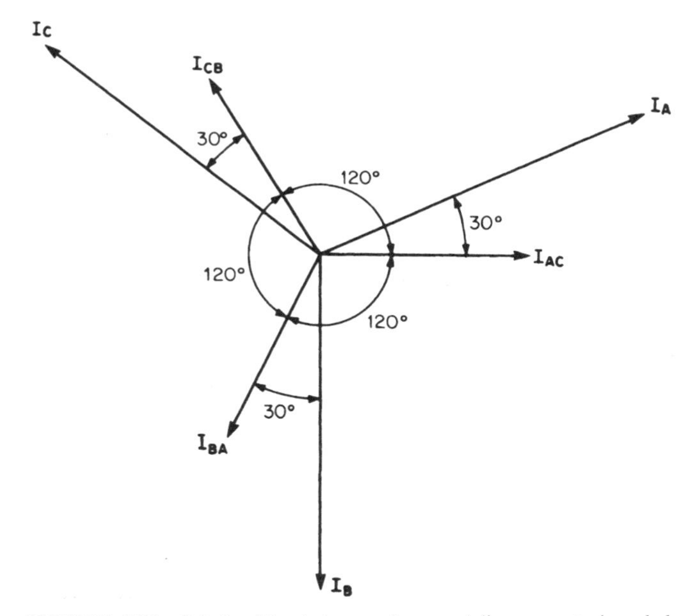

FIGURE 1.36 Relationships between phase and line currents in a balanced delta-connected system.

current component moves away from the source. The line current to a delta-connected load is the phasor difference between the two load currents at the entering node.

5. The line current in a balanced delta load has a magnitude of $\sqrt{3}$ times the phase current in each arm of the load. The line current is 30° out of phase with the phase current (Fig. 1.36).
6. The line-line voltage in a balanced, wye-connected, three phase source has a magnitude of $\sqrt{3}$ times the line-neutral voltage. The line-line voltage is 30° out of phase with the line-neutral voltage.

RESPONSE OF AN INTEGRATOR TO A RECTANGULAR PULSE

A single 10-V pulse with a width of 100 μs is applied to the *RC* integrator of Fig. 1.37. Calculate the voltage to which the capacitor charges. How long will it take the capacitor to discharge (neglect the resistance of the pulse source)?

Calculation Procedure

1. Calculate the Voltage to Which the Capacitor Charges

The rate at which a capacitor charges or discharges is determined by the time constant of the circuit. The time constant of a series *RC* circuit is the time interval that equals the product of *R* and *C*. The symbol for time constant is τ (Greek letter tau): $\tau = RC$, where *R* is in ohms, *C* is in farads, and τ is in seconds.

The time constant of this circuit is: $\tau = RC = (100\ \text{k}\Omega)(0.001\ \mu\text{F}) = 100\ \mu\text{s}$. Because the pulse width equals 200 μs (2 time constants), the capacitor will charge to 86 percent of its full charge, or to a voltage of 8.6 V. The expression for *RC* charging is: $v_C(t) = V_F(1 - e^{-t/RC})$, where V_F is the final value. In this case the final value, $V_F = 10$ V, would be reached if the pulse had a width of 5 or more time constants. See the *RC* time constant charging table (Table 1.1).

2. Calculate the Discharge Time

The capacitor discharges back through the source at the end of 200 μs. The total discharge time for practical purposes is 5 time constants or (5)(100 μs) = 500 μs. The expression for *RC* discharging is: $v_C(t) = V_i(e^{-t/RC})$, where V_i is the initial value. In this case, the initial value before discharging is 8.6 V. Table 1.2 shows the *RC* time constant discharge characteristics.

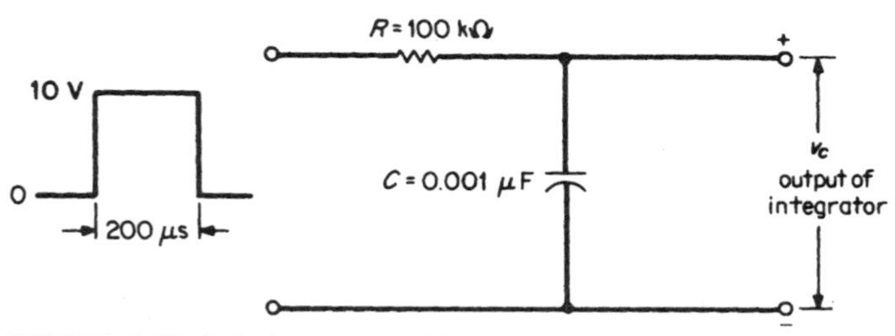

FIGURE 1.37 Pulse input to an *RC* integrator.

TABLE 1.1 *RC* Time Constant Charging Characteristics

τ	% Full charge
1	63
2	86
3	95
4	98
5	99*

*For practical purposes, 5 time constants are considered to result in 100 percent charging.

TABLE 1.2 *RC* Time Constant Discharging Characteristics

τ	% Full charge
1	37
2	14
3	5
4	2
5	1*

*For practical purposes, 5 time constants are considered to result in zero charge or 100 percent discharge.

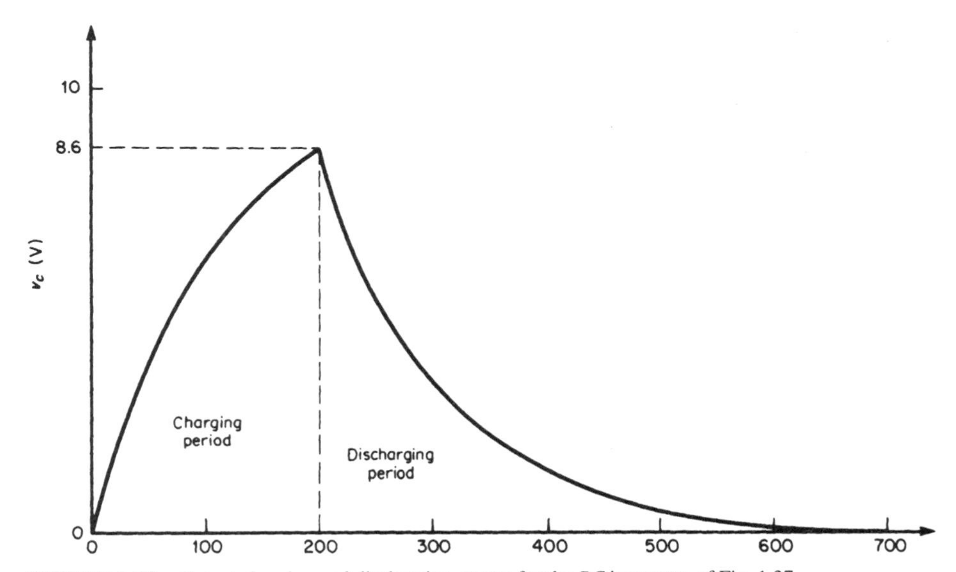

FIGURE 1.38 Output charging and discharging curves for the *RC* integrator of Fig. 1.37.

Related Calculations. Figure 1.38 illustrates the output charging and discharging curves.

BIBLIOGRAPHY

Hayt, William J., and Jack Kemmerly. 1993. *Engineering Circuit Analysis,* 5th ed. New York: McGraw-Hill.

Nilsson, James W., and Susan A. Riedel. 1999. *Electric Circuits,* 6th ed. Englewood Cliffs, N.J.: Prentice Hall.

Stanley, William D. 1999. *Network Analysis with Applications,* 3rd ed. Englewood Cliffs, N.J.: Prentice Hall.

SECTION 2

INSTRUMENTATION

Shahriar Khan, Ph.D.
Department of Electrical Engineering
Texas A&M University

VOLTAGE MEASUREMENT

The line voltage of a three-phase 4160-V power line supplying an industrial plant is to be measured. Choose the appropriate voltmeter and potential transformer for making the measurement.

Calculation Procedure

1. Select Voltmeter

Self-contained ac voltmeters with scales ranging from 150 to 750 V are available. Where higher voltages are to be measured, a potential transformer is required to produce a voltage suitable for indication on a meter with 150-V full-scale indication. A 150-V meter is therefore selected.

2. Select Potential Transformer

By dividing the line-to-line voltage by the voltmeter full-scale voltage, one obtains an approximate value of transformer ratio: 4160 V/150 V = 27.7. Select the next higher standard value, 40:1. To check the selection, calculate the secondary voltage with the potential transformer: 4160 V/40 = 104 V.

3. *Connect Transformer and Voltmeter to the Line*

The potential transformer and voltmeter are connected to the three-phase line, as shown in Fig. 2.1.

Precautions

Any leakage in the insulation between windings in potential transformer can cause high voltage to appear on low voltage side.

Related Calculations. Instruments used for measuring electrical quantities in utility or industrial service, such as voltmeters, ammeters, and wattmeters, are referred to as switchboard instruments. Instruments used for measurement of nonelectrical quantities, such as pressure, temperature, and flow rate, involve more complex techniques that include a sensor, transmission line, and receiver or indicator. In these systems, the receiver may also perform a recording function.

Very often, the instrumentation system is part of a process-control system. In such cases, the instrument that receives and indicates also serves as a controller. These industrial measurement systems may involve pneumatic, electrical analog (voltage or current), or electrical digital signal-transmission techniques.

Question

Describe a system for monitoring the voltage of a power system digitally (Fig. 2.2). Assume that up to the 11th harmonic is of interest, and that the analog-to-digital converter has a range of 0 to 5 V.

Calculation Procedure

A potential transformer (PT) will step down the voltage of the power system to a value compatible to digital and analog electronic systems. In this case, a standard PT is used to step down the voltage to 10 V peak-to-peak.

The value of the PT output has to be made compatible with the range of the A/D converter. We assume the PT produces a maximum of 5 V peak (or 10 V peak-to-peak). An op-amp summer circuit is then connected, such that the voltage remains compatible with the analog-to-digital converter at all times.

The summer should step down the input by half and add a reference of 2.5. This will keep the input of the A/D converter within 0 to 5 V at all times.

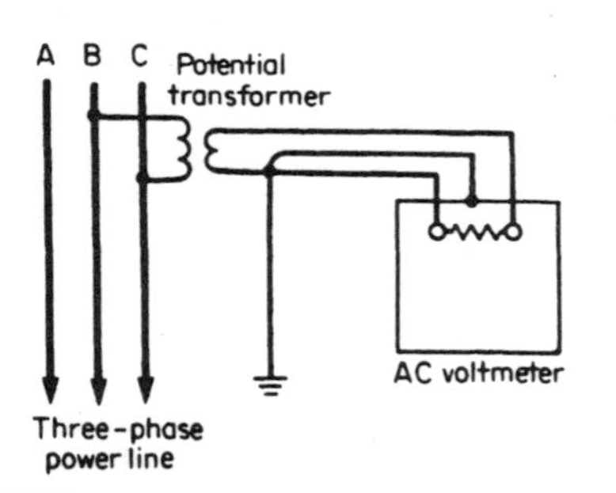

FIGURE 2.1 Voltmeter connections to a three-phase power line.

The following values in the op-amp summer circuit satisfy the preceding requirements:

$$R_f = 10\text{ K}$$

$$R_i = 20\text{ K}$$

$$R_{\text{ref}} = 10\text{ K}$$

An antialiasing filter is used such that the cutoff frequency is equal to half the

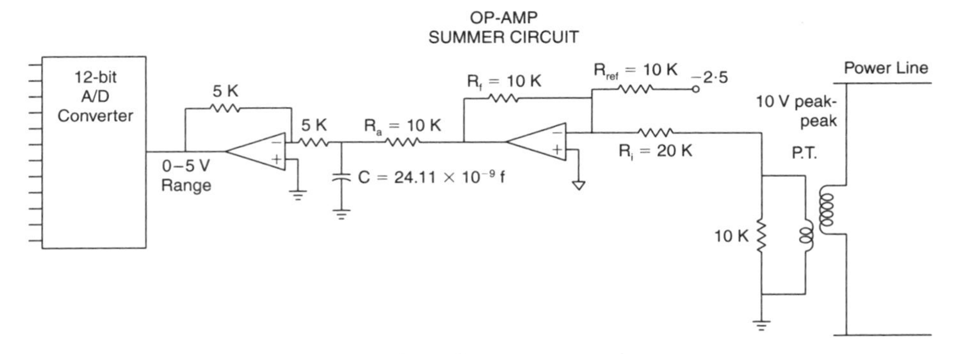

FIGURE 2.2 Digitally monitoring the voltage of a power system.

sampling frequency of the A/D converter. Since the 11th harmonic is of interest to the digital monitoring system, the sampling frequency should equal at least twice the 11th harmonic.

$$2 \times 11 \times 60 = 1320 \text{ samples per second}$$

The RC filter shown has a cutoff frequency of $1/2 \times \text{pi} \times R_aC$, which should equal

$$11 \times 60 = 660 \text{ Hz.}$$

If R_a is chosen as 10 K,

$$C = 2.411 \times \text{E-8 farad.}$$

If R is too small, it may overload the op-amp. If R is too large, the output may be noisy. The value of R is also determined by the availability of capacitors.

Lastly, the negative voltage is reconverted to positive by using an inverter circuit.

The A/D converter will now give a 12-bit digital value for the voltage 1320 times per second.

CURRENT MEASUREMENT

Current is to be measured in a single-phase line that supplies a 240-V, 20-kW load with a 0.8 power factor (pf). Select an appropriate ammeter and current transformer.

Describe the limitations of the current transformer under short-circuit conditions. Describe the applicability of fiber-optic transducers.

Calculation Procedure

1. Select Ammeter

Direct-reading ammeters are available with full-scale readings ranging from 2 to 20 A. Measurement of larger currents requires the use of a current transformer. Standard practice is to use a 5-A full-scale ammeter with the appropriate current transformer. Ammeters so used are calibrated in accordance with the selected transformer.

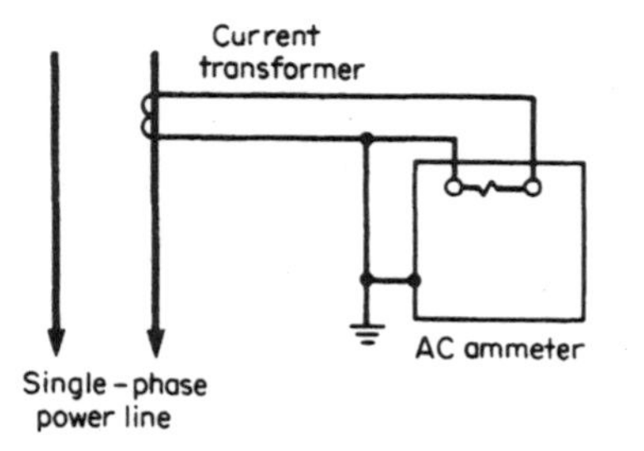

FIGURE 2.3 Ammeter connections to a single-phase power line.

2. Calculate Current

$I = P(V \times \text{pf}) = 20{,}000/(240 \times 0.8) = 104$ A.

3. Select Current Transformer

Because the current is greater than 20 A, a current transformer is required. A transformer is chosen that can accommodate a somewhat higher current; a 150:5 current transformer is therefore selected. The ammeter is a 5-A meter with its scale calibrated from 0 to 150 A.

4. Connect Ammeter and Transformer to Line

The ammeter is connected to the line, through the current transformer, as in Fig. 2.3.

Precautions

The secondary of the transformer is to be closed at all times. Otherwise, the operator side would step-up the voltage and a dangerous voltage may be generated.

Limitations

Under short-circuit conditions, the large currents may saturate the core and distort the output waveform. The core may also be unable to accurately respond to the rapid transients associated with short circuits.

Optical Transducers

Recently, there has been some progress in using optical transducers to replace current transformers. This is based on the principle of measuring the magnetic field in the vicinity of the current-carrying conductor. Advantages include that the fiber-optic transmission of such signals is immune to noise and that such transducers are very compact devices. The sensors also isolate the measurement devices from the high-tension side.

POWER MEASUREMENT USING A SINGLE-PHASE WATTMETER

The power consumption of a load, estimated to be 100 kVA, is to be measured. If the load is supplied by a 2400-V single-phase line, select a suitable wattmeter to make the measurement.

Calculation Procedure

1. Select Wattmeter

Single-phase, as well as three-phase, wattmeters often require current and/or potential transformers as accessories. Wattmeters are generally designed for 120- or 480-V operation with a maximum current rating of 5 A. For this application, a 120-V, 5-A wattmeter is selected.

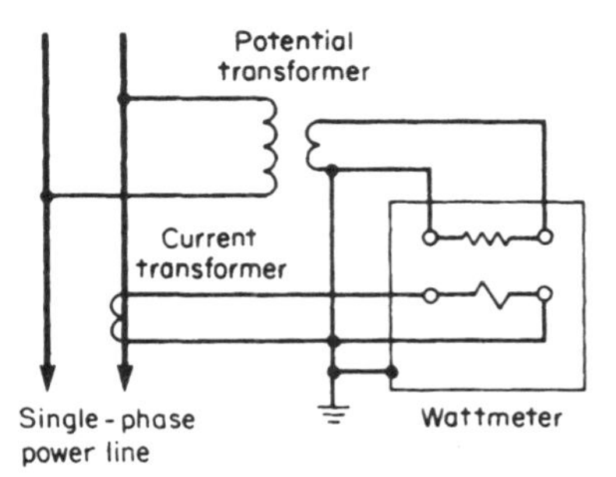

FIGURE 2.4 Single-phase wattmeter circuit.

2. Select Current Transformer

The line current is 100,000/2400 = 41 A. A current transformer is therefore required; a 50:5-A rating is suitable.

3. Select Potential Transformer

The line voltage of 2400 V is required to be stepped down to 120 V. A 20:1 potential transformer is chosen.

4. Connect Wattmeter and Transformers to Line

The wattmeter and both transformers are connected to the 2400-V line as indicated in Fig. 2.4.

POWER MEASUREMENT USING A THREE-PHASE WATTMETER

The power consumption of a load, estimated to be 1500 kVA, is to be measured. The load is supplied by a three-phase, three-wire line, 12,000-V line-to-line. Specify a suitable wattmeter for the measurement.

Calculation Procedure

1. Select Wattmeter

Three-phase wattmeters intended for use with three-wire, three-phase lines are available, as well as others intended for use with four-wire, three-phase lines. For this application, a 120-V, 5-A, three-wire wattmeter is a good choice.

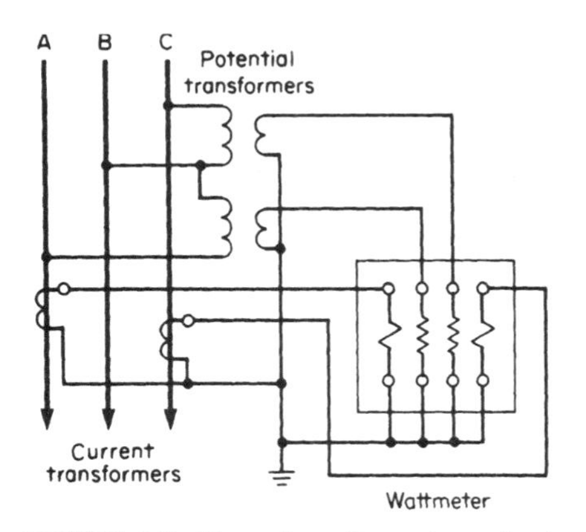

FIGURE 2.5 Three-phase, three-wire wattmeter circuit.

2. Select Current Transformer

The line current is 1,500,000/(1.73 × 12,000) = 72 A. A current transformer with a 100:5-A ratio is chosen.

3. Select Potential Transformer

The line-to-line voltage of 12,000 V requires that a 12,000/120, or 100:1, potential transformer be used.

4. Connect Wattmeter and Transformers to Line

The wattmeter and both transformers are connected to the three-wire line, as shown in Fig. 2.5.

POWER MEASUREMENT ON A FOUR-WIRE LINE

A 500-kVA load is supplied by a three-phase, four-wire, 4160-V line. Select a suitable wattmeter for measuring power consumption.

Calculation Procedure

1. Select Wattmeter

A four-wire type having a 120-V, 5-A rating is chosen.

2. Select Current Transformer

The line current is 500,000/(1.73 × 4160) = 72 A. Therefore, a 75:5-A ratio is chosen for each of the current transformers.

3. Select Potential Transformer

The line-to-neutral voltage is 4160/1.73 = 2400 V. Therefore, a 20:1 ratio is selected for the two potential transformers.

4. Connect the Wattmeter and Transformers to the Lines

The wattmeter and transformers are connected to the 4160-V, three-phase, four-wire line, as in Fig. 2.6.

REACTIVE-POWER MEASUREMENT

A varmeter is used to measure the reactive power in an industrial plant that is supplied by an 8300-V, three-phase, four-wire line. The plant load is estimated to be 300 kVA at 0.8 pf. Design a suitable measuring system.

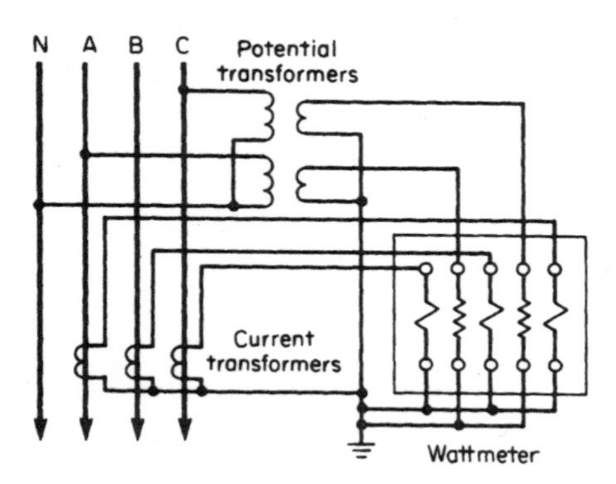

FIGURE 2.6 Three-phase, four-wire wattmeter circuit.

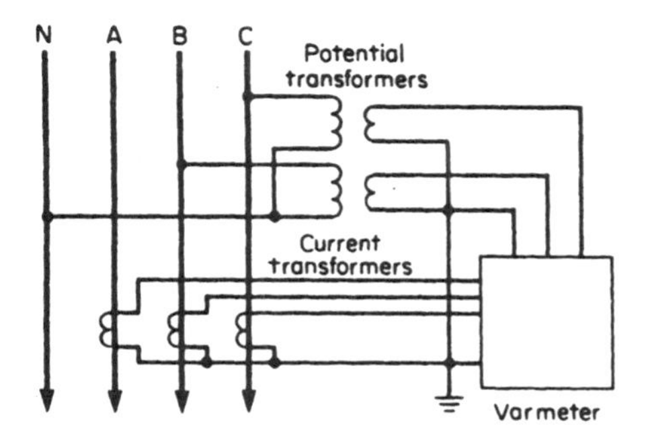

FIGURE 2.7 Varmeter connected to a three-phase, four-wire power line.

Calculation Procedure

1. Select Varmeter

The high-line voltage and large load dictate the use of a three-phase varmeter (rated at 120 V, 5 A) with current and potential transformers.

2. Select Current Transformer

I_{line} = 300,000/(1.73 × 8300) = 20.8 A. A 25:5-A current transformer is selected.

3. Select Potential Transformer

The potential transformer used with a three-phase, four-wire varmeter is connected line-to-neutral. The line-to-neutral voltage is 8300/1.73 = 4790 V; 4790/120 = 39.9. A 40:1 potential transformer is selected.

4. Determine a Suitable Scale for Meter

The phase angle θ is equal to $\cos^{-1} 0.8 = 36.9°$. The reactive power = 300,000 × sin 36.9° = 300,000 × 0.6 = 180,000 VARS. A scale providing a maximum reading of 200,000 VARS is selected.

5. Connect Varmeter to Line

The varmeter is connected to the power line by means of the current and potential transformers, as illustrated in Fig. 2.7.

Related Calculations. Varmeters are made for single- and three-phase, three- and four-wire systems. Center-zero scales are generally used for a varmeter. The instrument is designed to deflect to the right for a lagging power factor and to the left for a leading power factor. Many varmeters require an external compensator or phase-shifting transformer. For high-voltage systems with large loads, it is common practice to use current and potential transformers with varmeters designed for 120-V, 5-A operation.

POWER-FACTOR MEASUREMENT

The power factor of a group of four 30-hp electric motors in a manufacturing plant is to be measured. The motors are supplied by a 480/277-V, three-phase, three-wire line. The load is estimated to be 160 kVA, 0.85 pf. Determine how the power factor is to be measured.

Calculation Procedure

1. Select a Suitable Power-Factor Meter

A three-phase, three-wire power-factor meter is chosen for this application. Although self-contained power-factor meters are available, this application requires the use of current and potential transformers.

2. Select Current Transformer

$I_{line} = 160{,}000/(1.73 \times 480) = 192$ A. Select a 200:5-A current transformer.

3. Select Potential Transformer

For this meter, the potential-transformer primary winding is connected line-to-line. The potential transformer ratio, therefore, is 480/120 = 4:1. A 4:1 potential transformer is chosen.

4. Connect Power-Factor Meter-to-Line

Figure 2.8 shows how the current and potential tansformers are connected to the meter and line.

Related Calculations. Power-factor meters are made for single-phase as well as three-phase systems. Polyphase power-factor meters are designed on the basis of balanced loads.

Calculation Procedure

1. Select a Suitable Meter

For this application, a 240-V, three-wire, single-phase wattmeter without a demand register is selected. The meter is capable of carrying 200 A, which is the smallest current rating available. Neither current nor voltage instrument transformers are required for this application.

2. Connect the Watthour Meter to Line

A meter socket, or pan, is commonly used as a means of mounting the meter as well as connecting it to the line. The connection to the meter socket, and the internal connections of the meter, are provided in Fig. 2.9.

Related Calculations. Watthour meters are available for single- and three-phase loads, with clock-type and cyclometer (digital) readouts, with and without demand registers (Fig. 2.10). The single-phase, three-wire watthour meter contains two current coils and one potential coil. These act as the stator windings of a two-phase induction motor having a solid aluminum disk as its rotor.

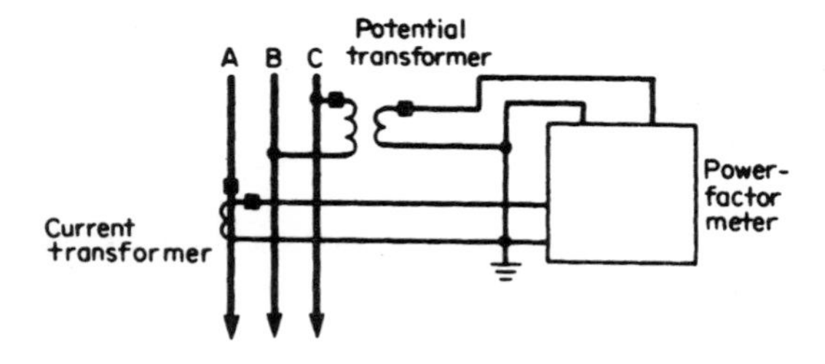

FIGURE 2.8 Power-factor meter connected to a three-phase, three-wire power line.

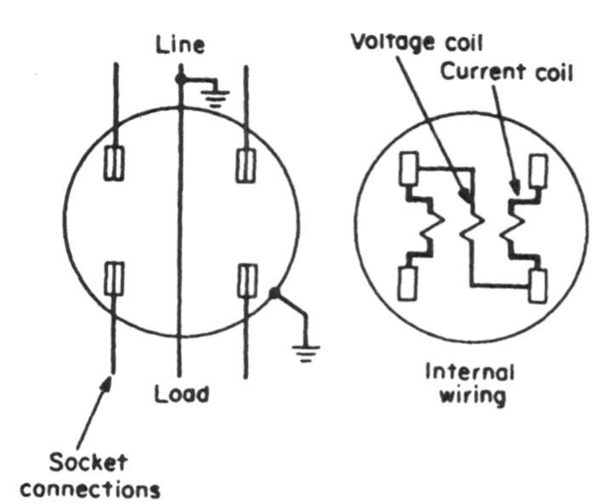

FIGURE 2.9 Single-phase watthour meter connections.

ELECTRIC PEAK-POWER DEMAND METERING

It is necessary to measure the peak demand, on a 15-min basis, of an industrial plant for which the peak demand is estimated to be 150 kW. The plant is supplied by a three-phase, four-wire, 7200-V line. The power factor is estimated to be 0.8 at peak demand. Specify how the measurement is to be made.

Calculation Procedure

1. Select Demand Meter

This application calls for a three-phase, four-wire meter containing a demand register with 15-min time intervals. The application requires use of both current and potential transformers.

2. Select Current Transformer

$I_{\text{line}} = 150{,}000/(1.73 \times 7200 \times 0.8) = 15.1$ A. Select a 15:5-A current transformer with a 15,000-V insulation rating.

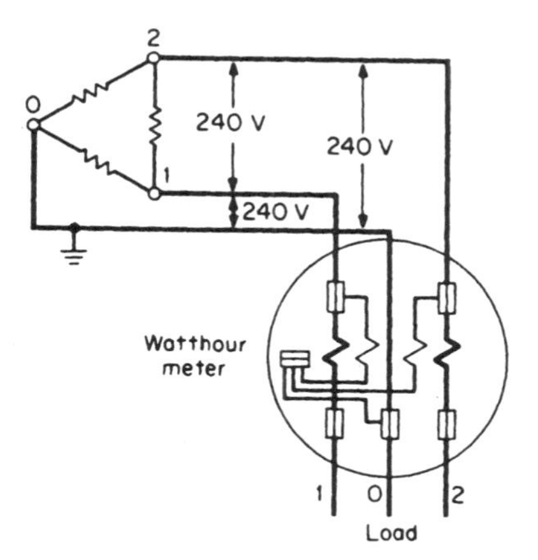

FIGURE 2.10 Three-phase, three-wire watthour meter connections.

3. Select Potential Transformer

The primary windings of these transformers are connected line-to-line and are therefore subject to 7200 V. The secondary winding provides 120 V to the meter. A transformer having a 7200:120, or 60:1, ratio is called for. An insulation rating of 15,000 V is required.

4. Connect Meter to Line

The meter connected to the line is shown in Fig. 2.11.

Related Calculations. Demand meters are incorporated into watthour meters for many commercial, institutional, and industrial plants that are subject to demand

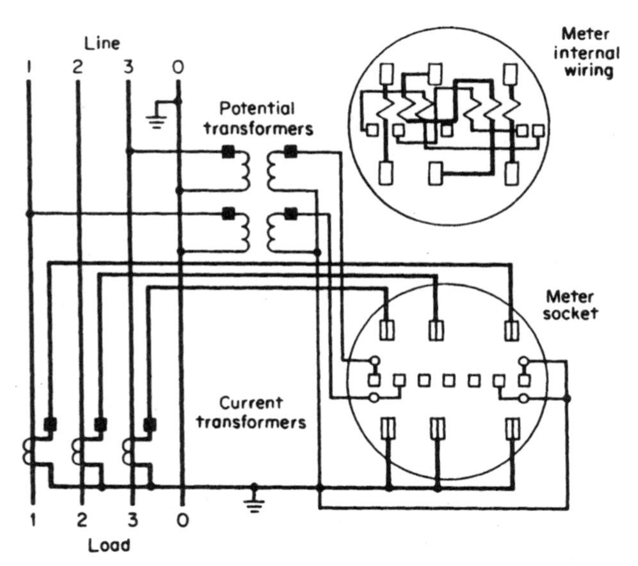

FIGURE 2.11 Connecting a three-phase watthour meter with a demand register to a power line.

charges by the electric utility. Demand meters operate as watthour meters over specified time intervals, usually 15- or 30-min periods. The highest 15- or 30-min consumption of energy is retained as an indication on the demand register until it is manually reset to zero. This is usually done on a monthly basis. The calibration of the demand register is in kilowatts and takes into account the period during which energy consumption is accumulated.

The selection of a demand meter is similar to the selection of a watthour meter. It is based on the type of service (single- or three-phase), the line voltage, anticipated current, and the time interval over which the definition of peak demand is based.

TEMPERATURE MEASUREMENT

The temperature of the cylinder head in a diesel engine is to be remotely indicated at a control-room panel. A temperature range of 0 to 200°C is to be accommodated. Design a system for making the measurement.

Calculation Procedure

1. Select System Type

Pneumatic and electrical analog systems are used for transmission of nonelectrical quantities. Availability of air supply and environmental and maintenance factors govern the choice of system. In this application, an electrical system is chosen.

2. Select Sensor

A copper-Constantan (type T) thermocouple is selected because of its small size, flexibility, and suitability for the temperature range to be measured.

3. Select Transmitter

Industrial instrumentation systems involve use of electronic or pneumatic instruments, called *transmitters,* to convert the weak signal produced by the sensor to a standard form. In electrical analog systems, this is usually a direct current, ranging from 4 to 20 mA. Transmitters are always used with *receivers,* or *indicators,* from which power is derived. A two-wire transmitter requires only two conductors to be run from transmitter to receiver. In this case a two-wire millivolt-to-current transmitter is selected.

The particular model is designed to accept the output of a type T thermocouple over a temperature range of 0 to 200°C (approximately 0 to 10 mV). The minimum temperature corresponds to 4 mA of output current while the maximum temperature corresponds to 20 mA of output current.

4. Select Indicator

Indicators, or receivers, for electrical analog instrumentation systems are designed to accommodate 4 to 20 mA dc. Calibrations may show 0 to 100 percent or the actual temperature. Recording indicators are used where automatic recording is required. Pointer-type instruments are traditional but digital displays are now available. In this case, a non-recording pointer-type instrument, with the temperature scale shown on the face of the instrument, is selected.

5. Connect System Components

The connections between thermocouple, transmitter, and indicator are illustrated in Fig. 2.12.

Related Calculations. Temperature-dependent resistors (TDRs), thermocouples, thermistors, and semiconductor sensors are used for industrial temperature measurement. The temperature range to be accommodated, ease of installation, and environmental factors affect the selection.

Question

Describe a system to monitor the temperature of a power component digitally, using a temperature-dependent current-source-type transducer (Fig. 2.13). The transducer is to give 1 μA per K, and the system should have a range of 0–100°C. The A/D converter has a range 0–5 V.

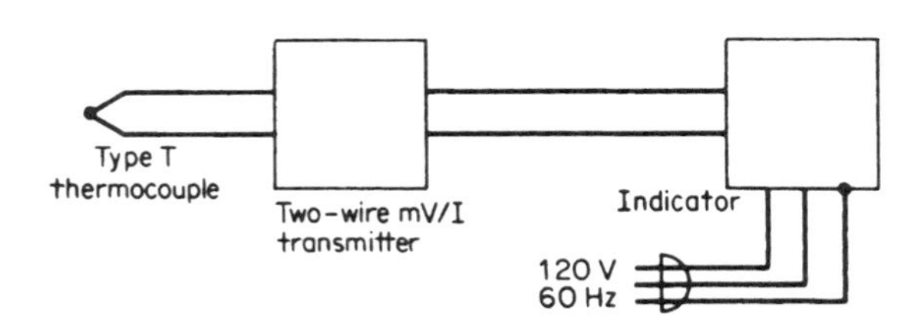

FIGURE 2.12 Temperature-measurement instrumentation.

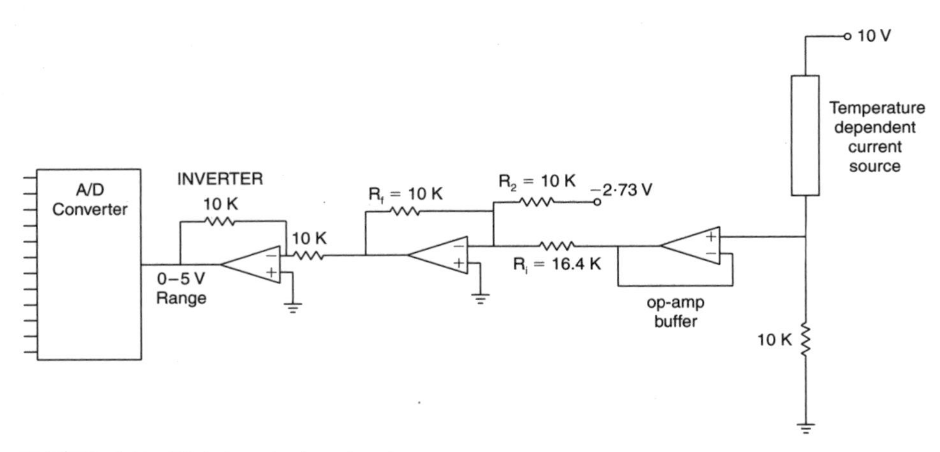

FIGURE 2.13 Digital monitoring of temperature.

Calculation Procedure

Commercial transducers of this type may operate with any supply between 5 and 15 V. Assuming the diode can accept source voltages of this range, the current should be dependent on temperature only. A resistance is placed so that a measurable voltage is obtained. 10K is chosen, as it will generate a voltage of appropriate magnitude. 0°C will produce 2.73 V and 100°C will produce 3.73 V. These will determine the values for the summing amplifier.

We now use the relationship for the summer,

$$[I_1/R_1 + I_2/R_2]R_f = -V_{\text{out}}$$

For the given conditions, the following resistances and reference voltage are satisfactory:

$$R_2 = 10\text{K}$$

$$R_f = 10\text{K}$$

$$R_1 = 16.4\text{K}$$

The op-amp buffer is placed so as not to load the transducer. The inverter ensures that the voltage is positive at all times.

The A/D converter will now give full output for 100°C and zero output for 0°C.

PRESSURE MEASUREMENT

The pressure in a retort of a chemical processing plant is to be remotely indicated and recorded. The corrosive and toxic nature of the liquid requires isolation of the pressure-sensing instrument. There is no air supply available for instrumentation purposes. The pressure to be measured is usually between 414 and 552 kPa (60 and 80 psi) but can vary from 0 to 690 kPa (0 to 100 psi). Design a remote-measuring system.

Calculation Procedure

1. Select Suitable System

The lack of an air supply is a major factor in choosing an electrical analog system. A pressure-to-current transmitter equipped with an isolation diaphragm and capillary tube is required to avoid contact with the hazardous material within the retort. The transmitter is a two-wire transmitter with an output of 4 to 20 mA corresponding to pressures ranging from 0 to 690 kPa (0 to 100 psi). The other major component of the system is the indicator–recorder, which is located in the control room.

2. Connect and Install System

The pressure-to-current transmitter is located at the retort and connected to an isolating diaphragm, as shown in Fig. 2.14. The capillary tube connecting the diaphragm to the transmitter may be prefilled by the instrument manufacturer or may be filled at the site during installation. Care must be exercised to avoid the inclusion of the air bubbles, which renders the system unduly temperature-sensitive.

A two-wire cable connects the transmitter to the indicator–recorder, located in the control room. This instrument contains the necessary dc power supply and associated electronic circuitry, deriving power from the 120-V, 60-Hz power line.

Related Calculations. Remote indication of pressure is usually accomplished by means of a pressure sensor coupled to a transmitter, which is connected by electrical conductors or pneumatic tubing to a remote indicating instrument. Typically, the systems are either pneumatic, with 20.7 to 103.4 kPa (3 to 15 psi) representing the full span of pressures to be measured, or electrical analog, with 4 to 20 mA dc representing the full span of pressures to be measured. Electrical analog systems use either two- or four-wire transmitters to convert the sensed pressure to an electrical current.

Question

Describe the instrumentation needed for monitoring incipient failures in 12-kV transformers and line insulators (Fig. 2.15).

Calculation Procedure

Insulation breakdown in transformers and insulators is often preceded by random low magnitudes of high-frequency currents (100s or 1000s of kHz). Since CTs may be unable

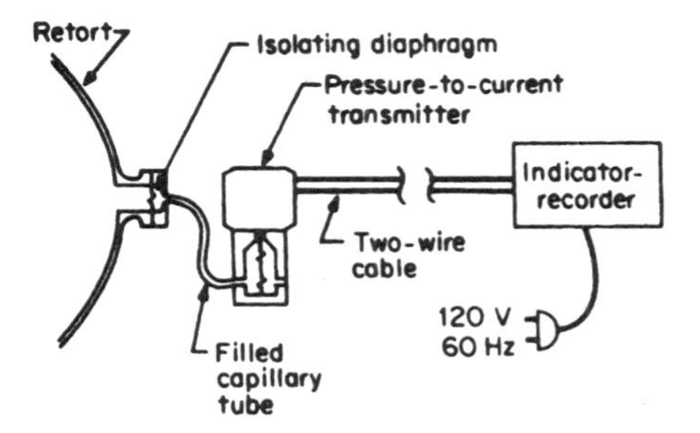

FIGURE 2.14 Pressure instrumentation system.

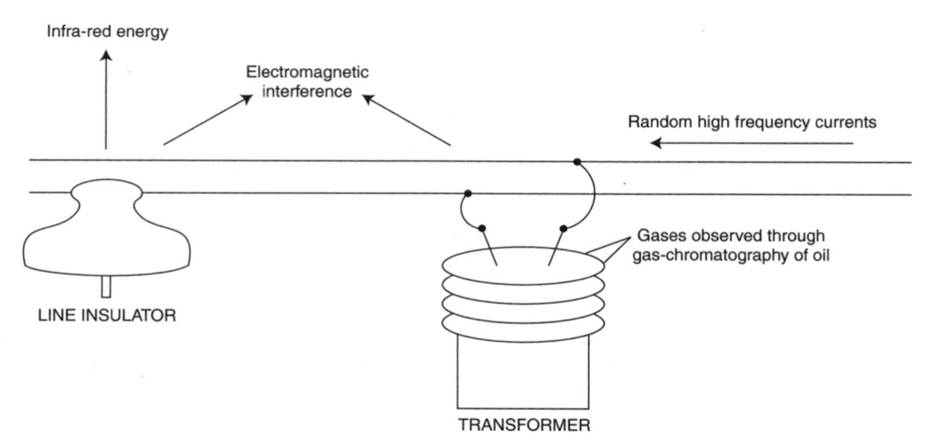

FIGURE 2.15 Instrumentation for early detection of incipient failure in insulators and transformers.

to respond to transients of these frequencies, a series resistance in the line may be used when appropriate. The voltage across the resistor can be monitored in real-time, or it can be recorded for processing later.

Failing insulation also gives off electromagnetic interference, which is picked up by appropriate detectors or simply by an AM radio. An infrared detector may also be used to pick up the heat energy given off by failing insulation.

Gas chromatography has been used successfully to detect the gases created in the oils of failing transformers.

BIBLIOGRAPHY

Anderson, P. M. 1999. *Power System Protection*. New York: McGraw-Hill.

Considine, Douglas M. 1974. *Process Instruments and Controls Handbook,* 2nd ed. New York: McGraw-Hill.

Cooper, W. D., and A. D. Helfrick. 1986. *Electronic Instrumentation and Measurement Techniques,* Englewood Cliffs, N.J.: Prentice-Hall.

Doebelin, Ernest. 1990. *Measurement Systems; Application and Design*, 4th ed. New York: McGraw-Hill.

Fink, Donald G., and John M. Carroll. 1978. *Standard Handbook for Electrical Engineers*, 11th ed. New York: McGraw-Hill.

Kantrowitz, Philip, Gabriel Kousourou, and Lawrence Zucker. 1979. *Electronic Measurements.* Englewood Cliffs, N.J.: Prentice-Hill.

Khan, Shahriar. 1997. *Failure Characterization of Distribution Insulators and New Methods of Predictive Maintenance,* Ph.D. Dissertation. Texas A&M University.

Khan, Shahriar, B. Don Russell, and K. Butler. 1994. *A Predictive Maintenance Approach to Power Distribution Systems*. Bozeman, M.T.: North American Power Symposium.

Kirk, Franklyn W., and Nicholas R. Rimboi. 1975. *Instrumentation*, 3rd ed. Chicago: American Technical Society.

Prensky, Sol D., and Richard S. Castellucis. 1982. *Electronic Instrumentation,* 3rd ed. Englewood Cliffs, N.J.: Prentice-Hall.

Prewitt, Roger W., and Stephen W. Fardo. 1979. *Instrumentation: Transducers; Experimentation and Applications*. Indianapolis: Sams Publishing Co.

Soisson, Harold E. 1975. *Instrumentation in Industry*. New York: Wiley.

SECTION 3

DC MOTORS AND GENERATORS

Lawrence J. Hollander, P.E.
Dean of Engineering Emeritus
Union College

DIRECT-CURRENT GENERATOR USED AS TACHOMETER, SPEED/VOLTAGE MEASUREMENT

A tachometer consists of a small dc machine having the following features: lap-wound armature, four poles, 780 conductors on the armature (rotor), field (stator) flux per pole = 0.32×10^{-3} Wb. Find the speed calibration for a voltmeter of very high impedance connected to the armature circuit.

Calculation Procedure

1. Determine the Number of Paths in the Armature Circuit

For a lap winding, the number of paths (in parallel), a, is always equal to the number of poles. For a wave winding, the number of paths is always equal to two. Therefore, for a four-pole, lap-wound machine $a = 4$.

2. Calculate the Machine Constant, **k**

Use the equation: machine constant $k = Np/a\pi$, where $N = 780/2 = 390$ turns (i.e., two conductors constitute one turn) on the armature winding, $p = 4$ poles, and $a = 4$ parallel paths. Thus, $(390)(4)/(4\pi) = 124.14$ and $k = 124.14$.

3. Calculate the Induced Voltage as a Function of Mechanical Speed

The average induced armature voltage, e_a (proportional to the rate of change of flux linkage in each coil) $= k\phi\omega_m$, where ϕ = flux per pole in webers, ω_m = mechanical speed of the rotor in radians per second, and e_a is in volts. Thus, $e_a/\omega_m = k\phi = (124.14)(0.32 \times 10^{-3}\ \text{Wb}) = 0.0397\ \text{V}\cdot\text{s/rad}$.

4. Calculate the Speed Calibration of the Voltmeter

Take the reciprocal of the factor e_a/ω_m, finding $1/0.0397 = 25.2$ rad/V·s. Thus, the calibration of the high-impedance voltmeter scale is such that each 1-V division is equivalent to a speed of 25.2 rads/s. With each revolution being equal to 2π rad, the calibration also is equivalent to a speed of $25.2/(2\pi)$ or 4 r/s for each 1-V division, or (4 r/s)(60 s/min) = 240 r/min for each 1-V division.

Related Calculations. The small dc generator, mechanically coupled to a motor, gives an output voltage that is directly proportional in magnitude to the speed of rotation. Not only may that voltage be read on a voltmeter-scale calibrated in units of speed but also it may be applied to a control circuit of the larger motor in such a way as to make desired corrections in speed. This type of tachometer affords a simple means for regulating the speed of a motor.

SEPARATELY EXCITED DC GENERATOR'S RATED CONDITIONS FROM NO-LOAD SATURATION CURVE

A separately excited dc generator has a no-load saturation curve as shown in Fig. 3.1 and an equivalent circuit as shown in Fig. 3.2. Its nameplate data are as follows: 5 kW, 125 V, 1150 r/min, and armature circuit resistance $R_a = 0.40\ \Omega$. Assume that the generator is driven at 1200 r/min and that the field current is adjusted by field rheostat to equal 2.0 A. If the load is the nameplate rating, find the terminal voltage, V_t. Armature reaction and brush losses may be neglected.

Calculation Procedure

1. Calculate the Armature Induced Voltage, $\mathbf{e_a}$

The no-load saturation curve is obtained at rated speed of 1150 r/min. In this case, however, the driven speed is 1200 r/min. For a given field flux (field current), the induced

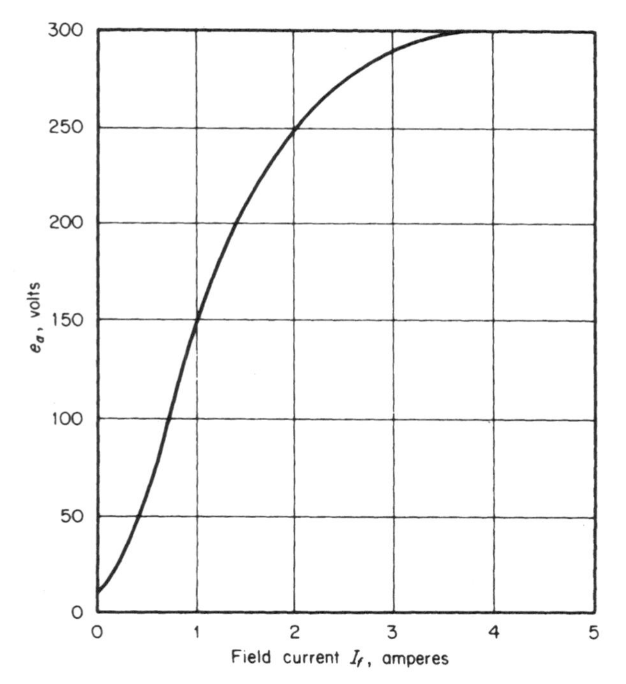

FIGURE 3.1 No-load saturation curve for a separately excited dc generator, at 1150 r/min.

voltage is proportional to speed. Thus, if I_f is 2.0 A, the induced voltage at 1150 r/min from the no-load saturation curve is equal to 250 V. The induced voltage at the higher speed of 1200 r/min is $e_a = (1200/1150)(250 \text{ V}) = 260.9$ V.

***2. Calculate the Rated Load Current,* I$_L$**

The rated load current is obtained from the equation: I_L = nameplate kilowatt rating/nameplate kilovolt rating = 5 kW/0.125 kV = 40 A. Note that the speed does not enter into this calculation; the current, voltage, and power ratings are limited by (or determined by) the thermal limitations (size of copper, insulation, heat dissipation, etc.).

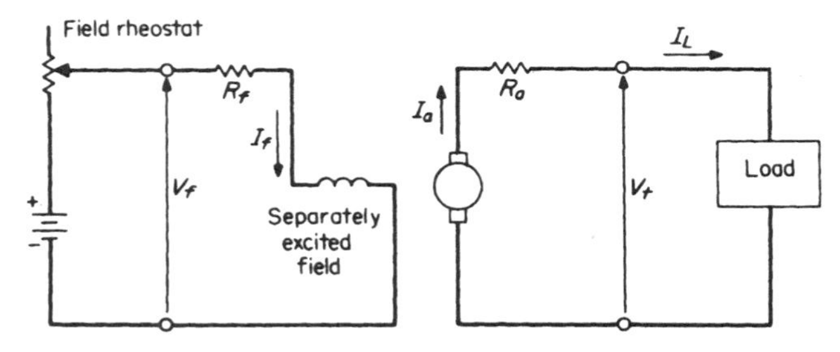

FIGURE 3.2 Equivalent circuit of a separately excited dc generator.

***3. Calculate the Terminal Voltage,* V_t**

The terminal voltage is determined from Kirchhoff's voltage law; or simply $V_t = e_a - I_L R_a = 260.9 \text{ V} - (40 \text{ A})(0.4 \ \Omega) = 244.9 \text{ V}$.

Related Calculations. This same procedure is used for calculating the terminal voltage for any variation in speed. If there were a series field or if there were a shunt field connected across the armature, essentially the same procedure is followed, with proper allowance for the additional circuitry and the variation in the total field flux, as in the next three problems.

TERMINAL CONDITIONS CALCULATED FOR DC COMPOUND GENERATOR

A compound dc generator is connected long-shunt. The data for the machine are as follows: 150 kW, 240 V, 625 A, series-field resistance $R_s = 0.0045 \ \Omega$, armature-circuit resistance $R_a = 0.023 \ \Omega$, shunt-field turns per pole = 1100, and series-field turns per pole = 5. Calculate the terminal voltage at rated load current when the shunt-field current is 5.0 A and the speed is 950 r/min. The no-load saturation curve of this machine at 1100 r/min is shown in Fig. 3.3.

Calculation Procedure

1. Draw the Equivalent Circuit Diagram

The equivalent circuit diagram is shown in Fig. 3.3.

***2. Calculate the Current in the Series Field,* I_s**

With reference to Fig. 3.3, the current in the series field is $I_s = I_a = I_L + I_f = 625 \text{ A} + 5.0 \text{ A} = 630 \text{ A}$.

3. Convert Series-Field Current to Equivalent Shunt-Field Current

So that the no-load saturation curve (Fig. 3.4) may be used, the series-field current is converted to equivalent shunt-field current; then, a total field flux is determined from the shunt-field current plus the equivalent shunt-field current of the series field. The equivalent shunt-field current of the series field is calculated from the relation: actual ampere-turns of series field/actual turns of shunt field = (630 A)(5 turns)/1100 turns = 2.86 A. The total flux in terms of field currents = shunt-field current + equivalent shunt-field current of series field = 5.0 A + 2.86 A = 7.86 A.

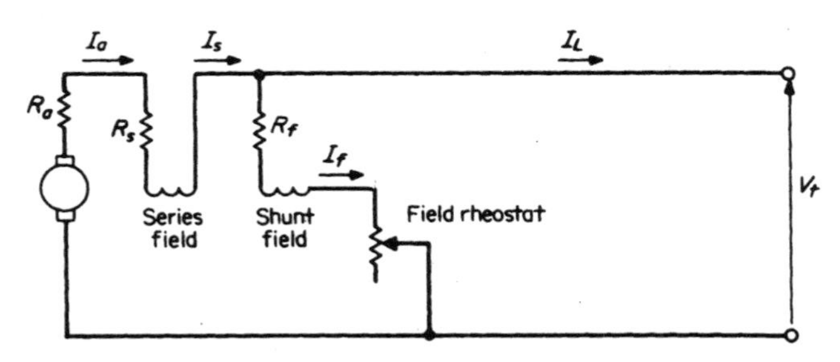

FIGURE 3.3 Equivalent circuit for a long-shunt compound dc generator.

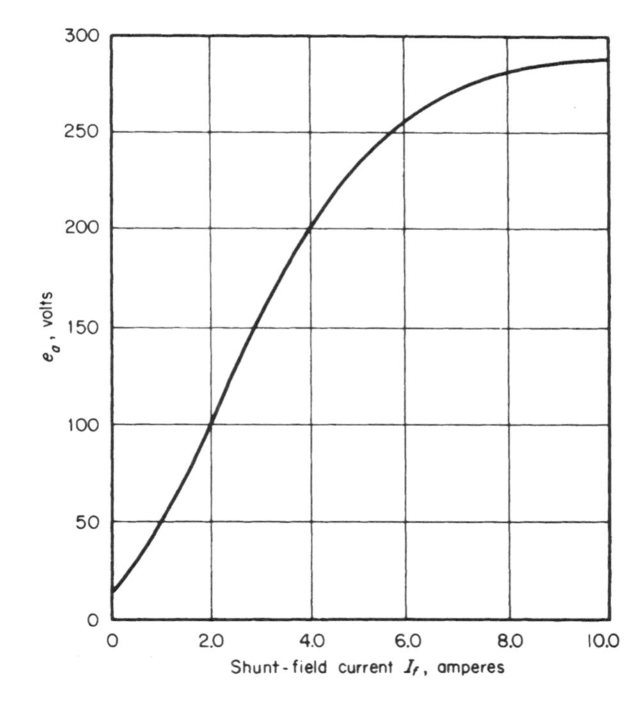

FIGURE 3.4 No-load saturation curve for a 240-V, 1100-r/min dc generator.

4. Calculate Armature Induced Voltage, e_a

With reference to the no-load saturation curve of Fig. 3.4, enter the graph at field current of 7.86 A. The no-load induced voltage $e_a = 280$ V at 1100 r/min. When the machine is driven at 950 r/min, the induced voltage $e_a = (280 \text{ V})(950/1100) = 242$ V.

5. Calculate the Terminal Voltage, V_t

The terminal voltage is calculated from Kirchhoff's voltage law: $V_t = e_a - I_a(R_a + R_s) = 242 \text{ V} - (630 \text{ A})(0.023 \ \Omega + 0.0045 \ \Omega) = 224.7$ V.

Related Calculations. In the long-shunt connection the series field is connected on the armature side of the shunt field; in the short-shunt connection the series field is connected on the terminal side of the shunt field. In the latter case, the series-field current is equal to I_L rather than I_a; Kirchhoff's voltage equation still is applicable. There is little difference between long-shunt and short-shunt compounding. In both cases, it is necessary to determine the net flux per pole; the series-field flux may be additive or subtractive. In this problem, it was assumed to be additive.

ADDED SERIES FIELD CALCULATED TO PRODUCE FLAT COMPOUNDING IN DC GENERATOR

A given dc generator is to be flat compounded. Its data are: 150 kW, 240 V, 625 A, 1100 r/min, armature-circuit resistance $R_a = 0.023 \ \Omega$, shunt-field turns per pole $= 1100$, series-field turns per pole $= 5$, and series-field resistance $R_s = 0.0045 \ \Omega$. It is driven at a constant speed of 1100 r/min, for which the no-load saturation curve is given in Fig. 3.4.

Determine the shunt-field circuit resistance for rated conditions (assuming long-shunt connections) and calculate the armature current to be diverted from the series field, for flat compounding at 250 V, and the resistance value of the diverter.

Calculation Procedure

1. Calculate the Value of the Shunt-Field Circuit Resistance

From the no-load saturation curve (Fig. 3.4), the 250-V point indicates a shunt-field current I_f of 5.6 A. A field-resistance line is drawn on the no-load saturation curve (a straight line from the origin to point 250 V, 5.6 A); this represents a shunt-field circuit resistance of (250 V)/(5.6 A) = 44.6 Ω. That is, for the no-load voltage to be 250 V, the shunt-field resistance must be 44.6 Ω.

2. Calculate the Armature Values at Rated Conditions

For an armature current of 625 A, I_aR_a = (625 A)(0.023 Ω) = 14.4 V. The armature current $I_a = I_f + I_L$ = 5.6 A + 625 A = 630.6 A. The induced armature voltage $e_a = V_t + I_a(R_a + R_s)$ = 250 V + (630.6 A)(0.023 Ω + 0.0045 Ω) = 267.3 V.

3. Calculate Needed Contribution of Series Field at Rated Conditions

At rated conditions, the current through the series field is 630.6 A (the same as I_a); refer to the long-shunt equivalent circuit shown in Fig. 3.3. For the induced voltage of 267.3 V (again referring to the no-load saturation curve), the indicated excitation is 6.6 A. Thus, the contribution from the series field = 6.6. A − 5.6 A = 1.0 A (in equivalent shunt-field amperes). The actual series-field current needed to produce the 1.0 A of equivalent shunt-field current is proportional to the number of turns in the respective fields = (1.0 A)(1100/5) = 220 A.

4. Calculate Diverter Resistance

Because the actual current through the series field would be 630.6 A and 220 A is sufficient to produce flat compounding (V_t = 250 V at no load and at full load), 630.6 A − 220 A = 410.6 A must be diverted through a resistor shunted across the series field. The value of that resistor = voltage across the series field/current to be diverted = (220 A × 0.0045 Ω)/410.6 A = 0.0024 Ω.

Related Calculations. In this problem, armature reaction was ignored because of lack of data. A better calculation would require saturation curves not only for no-load but also for rated full-load current. That full-load current curve would droop below the no-load curve in the region above the knee of the curve (above 250 V in this example). With these data taken into account, the needed series-field contribution to the magnetic field would be greater than what was calculated, and the current to be diverted would be less. Effectively, armature reaction causes a weakening of the magnetic field, in addition to distortion of that field. The actual calculations made in this problem would result in slight undercompounding, rather than flat compounding.

CALCULATION OF INTERPOLE WINDINGS FOR A DC GENERATOR

A dc lap-wound generator has the following data: 500 kW, 600 V, 4 poles, 4 interpoles, 464 armature conductors. The magnetomotive force (mmf) of the interpoles is 20 percent greater than that of the armature. Determine the number of turns on the interpoles.

Calculation Procedure

1. Calculate the Number of Armature Turns per Pole

If there are 464 armature conductors, then the number of turns is one-half of that amount; every two conductors constitute one turn. The machine has 4 poles; thus, armature turns per pole = (armature conductors/2)/number of poles = (464/2)/4 = 58.

2. Calculate the Current per Armature Path

For a lap winding, the number of paths (in parallel) is always equal to the number of poles; in this problem, it is 4. The armature current per path = $I_a/4$. This represents the current in each armature conductor or in each turn.

3. Calculate the Number of Turns on the Interpoles

The interpoles (also known as commutating poles) are for the purpose of providing mmf along the quadrature axis to reverse the flux in that space caused by the cross-magnetizing effect of armature reaction (see Fig. 3.5). They have a small number of turns of large cross section because the interpoles, being in series with the armature, carry armature current. The mmf of the armature per pole is: armature turns per pole × armature current in each conductor = $58(I_a/4) = 14.5I_a$ ampere-turns. If the interpoles have an mmf 20 percent greater, then each one is $(1.20)(14.5)I_a = 17.4I_a$ ampere-turns. The calculated number of turns on each interpole = 17.4; in actuality, use 17 turns.

Related Calculations. The function of interpoles is to improve commutation, that is, to eliminate sparking of the brushes as they slide over the commutator bars. The mmf of the interpoles usually is between 20 to 40 percent greater than the armature

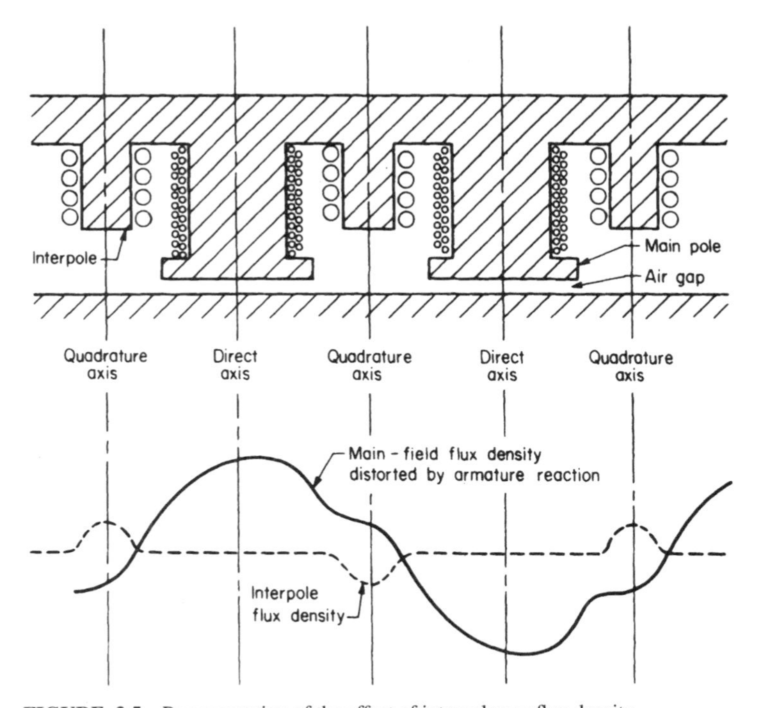

FIGURE 3.5 Representation of the effect of interpoles on flux density.

mmf, when the number of interpoles is equal to the number of main poles. If one-half of the number of interpoles is used, the mmf design is usually 40 to 60 percent greater than the armature mmf. These calculations are always on a per-pole basis. Because interpoles operate on armature current, their mmf is proportional to the effects of armature reaction.

DESIGN OF COMPENSATING WINDING FOR DC MACHINE

A lap-wound dc machine has the following data: four poles, ratio of pole face to pole span = 0.75, total number of armature slots = 33, number of conductors per slot = 12. Design a compensating winding for each pole face (i.e., find the number of conductors to be placed in each pole face and the number of slots on that face). See Fig. 3.6.

Calculation Procedure

1. Calculate Armature mmf in One Pole Span

The number of armature turns in a pole span is (total number of armature conductors/2)/number of poles = [(12 conductors per slot)(33 slots)/2]/4 poles = 49.5 armature turns. The current in each conductor is the current in each parallel path. For a lap-wound machine, the number of parallel paths is the same as the number of poles; in this problem, it is 4. The armature mmf in a pole span = (armature turns in a pole span) × (armature current per path) = $(49.5)(I_a/4) = 12.4\ I_a$ ampere-turns.

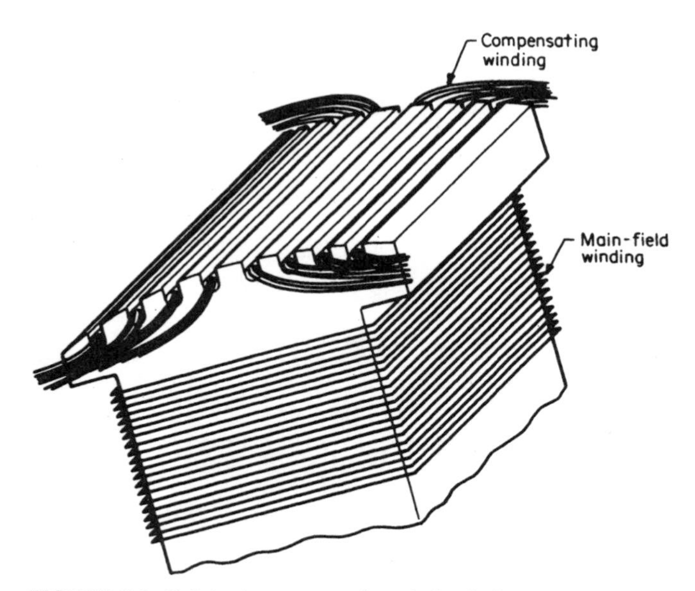

FIGURE 3.6 Pole having compensating winding in face.

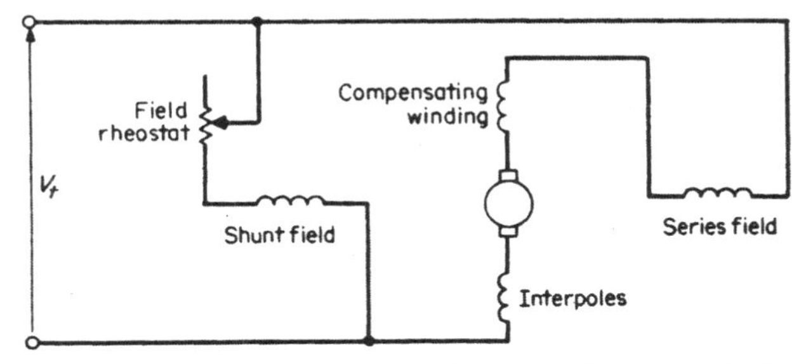

FIGURE 3.7 Connections showing dc compound machine with interpoles and compensating windings.

2. *Calculate the Armature Ampere-Turns per Pole Face*

The ratio of the pole face to pole span is given as 0.75. Therefore, the armature (mmf) ampere-turns per pole face = (0.75) × (12.4 I_a) = 9.3 ampere-turns.

3. *Calculate the Number of Conductors in the Compensating Winding*

The purpose of the compensating winding is to neutralize the armature mmf that distorts the flux density distribution. Under sudden changes in loading, there would be a flashover tendency across adjacent commutator bars; the compensating-winding effect overcomes this and is proportional to the armature current because it is connected in the armature circuit. Therefore, to design the compensating winding, it is necessary to have it produce essentially the same number of ampere-turns as the armature circuit under each pole face. In this problem, each pole face must produce 9.3 ampere-turns; this requires that each pole face have (2)(9.3 turns) = 18.6 conductors (say 18 or 19 conductors).

4. *Calculate the Number of Slots in the Pole Face*

The slots in the pole face should not have the same spacing as on the armature; otherwise, reluctance torques would occur as the opposite slots and teeth went in and out of alignment. The number of slots on the armature opposite a pole face is (33 slots)(0.75 ratio)/4 poles = 6.2 slots. For a different number of slots, 5 could be selected and 4 conductors could be placed in each slot (the total being 20 conductors). Another solution might be 8 slots with 2 conductors in each slot (the total being 16 conductors).

Related Calculations. In many cases, machines will have both interpoles (see previous problem) and compensating windings. The winding connections of the machine are shown in Fig. 3.7.

STATOR AND ARMATURE RESISTANCE CALCULATIONS IN DC SELF-EXCITED GENERATOR

A self-excited dc generator has the following data: 10 kW, 240 V. Under rated conditions, the voltage drop in the armature circuit is determined to be 6.1 percent of the terminal voltage. Also, the shunt-field current is determined to be 4.8 percent of the rated-load current. Calculate the armature circuit resistance and the shunt-field circuit resistance.

Calculation Procedure

1. Calculation the Shunt-Field Current

Use the equation: rated-load current I_L = rated load/rated voltage = 10,000 W/240 V = 41.7 A. The shunt-field current $I_f = 0.048\ I_L = (0.048)(41.7\ \text{A}) = 2.0\ \text{A}$.

2. Calculate the Resistance of the Shunt-Field Circuit

Use the equation: shunt-field resistance $R_f = V_t/I_f = 240\ \text{V}/2.0\ \text{A} = 120\ \Omega$.

3. Calculate the Resistance of the Armature Circuit

The armature current $I_a = I_L + I_f = 41.7\ \text{A} + 2.0\ \text{A} = 43.7\ \text{A}$. The voltage drop in the armature circuit, I_aR_a, is given as 6.1 percent of the terminal voltage, V_t. Thus, $I_aR_a = (0.061)(240\ \text{V}) = 14.64\ \text{V}$. But $I_a = 43.7$ A; therefore, $R_a = 14.64\text{V}/43.7\ \text{A} = 0.335\ \Omega$.

Related Calculations. Voltage and current relations for dc machines follow from the circuit diagrams as in Figs. 3.2, 3.3, and 3.7, as may be appropriate. It is a good practice to draw the equivalent circuit to better visualize the combinations of currents and voltages.

EFFICIENCY CALCULATION FOR DC SHUNT GENERATOR

A dc shunt generator has the following ratings: 5.0 kW, 240 V, 1100 r/min, armature-circuit resistance $R_a = 1.10\ \Omega$. The no-load armature current is 1.8 A when the machine is operated as a motor at rated speed and rated voltage. The no-load saturation curve is shown in Fig. 3.4. Find the efficiency of the generator at rated conditions.

Calculation Procedure

1. Calculate the Full-Load Current, $\mathbf{I_L}$

Use the equation: I_L = rated capacity in watts/rated terminal voltage = 5000 W/240 V = 20.8 A. This is the full-load current when the machine is operated as a generator.

2. Calculate the Armature Induced Voltage, $\mathbf{e_a}$***, at No Load***

When the machine is run as a motor, the no-load armature current is 1.8 A. Use the equation: $e_a = V_t - R_aI_a = 240\ \text{V} - (1.10\ \Omega)(1.8\ \text{A}) = 238\ \text{V}$.

3. Calculate the Rotational Losses, $\mathbf{P_{\mathit{rot(loss)}}}$

Use the equation: $P_{\text{rot(loss)}} = e_aI_a = (238\ \text{V})(1.8\ \text{A}) = 428.4\ \text{W}$.

4. Calculate the Field Current Required at Full Load

Under generator operation at full load, the armature current is the sum of the load current plus the shunt-field current, $I_L + I_f$. Use the equation: $V_t = 240\ \text{V} = e_a - R_a(I_L + I_f)$. Rearranging, find $e_a = 240\ \text{V} + (1.10\ \Omega)(20.8\ \text{A} + I_f) = 262.9 + 1.10\ I_f$. From observation of the no-load saturation curve (Fig. 3.4), assume that I_f is 7.0 A; the equation is satisfied. Then $e_a = 262.9\ \text{V} + (1.10\ \Omega)\ (7.0\ \text{A}) = 270.6\ \text{V}$. It may be necessary to make one or two trial attempts before the equation is satisfied.

5. Calculate the Copper Loss

The armature current, I_a, is $I_f + I_L = 7.0\text{ A} + 20.8\text{ A} = 27.8\text{ A}$. The armature-circuit copper loss is $I_a^2 R_a = (27.8\text{ A})^2(1.10\ \Omega) = 850\text{ W}$. The shunt-field copper loss is $V_t I_f = (240\text{ V})(7.0\text{ A}) = 1680\text{ W}$. The total copper loss is $850\text{ W} + 1680\text{ W} = 2530\text{ W}$.

6. Calculate the Efficiency

Use the equation: η = output/(output + losses) = (5000 W)(100 percent)/(5000 W + 2530 W + 428.4 W) = 62.8 percent.

Related Calculations. It should be noted in this problem that the generator was operated as a motor, at no load, in order to determine the armature current; from that current, the rotational losses were obtained. The rotational losses include friction, windage, and core losses. The efficiency could be improved by increasing the number of turns on the field winding, reducing I_f.

TORQUE AND EFFICIENCY CALCULATION FOR SEPARATELY EXCITED DC MOTOR

A separately excited dc motor has the following data: 1.5 hp, 240 V, 6.3 A, 1750 r/min, field current $I_f = 0.36$ A, armature-circuit resistance $R_a = 5.1\ \Omega$, field resistance $R_f = 667\ \Omega$. Laboratory tests indicate that at rated speed mechanical losses (friction and windage) = 35 W, at rated field current magnetic losses = 32 W, and at rated armature current and rated speed stray-load losses = 22 W. The no-load saturation curve is given in Fig. 3.8. Determine rated torque, input armature voltage, input power, and efficiency at rated torque and speed.

Calculation Procedure

1. Calculate the Rated Torque

Use the equation: P (in horsepower) = $2\pi nT/33{,}000$, where n = speed in r/min and T = torque in lb · ft. Rearranging, find $T = 33{,}000P/2\pi n = (33{,}000)(1.5\text{ hp})/(2\pi \times 1750\text{ r/min}) = 4.50$ lb · ft. Or (4.50 lb · ft)(N · m)/(0.738 lb · ft) = 6.10 N · m.

2. Calculate the Electromagnetic Power

The electromagnetic power, P_e, is the sum of the (a) output power, (b) mechanical power including the mechanical losses of friction and windage, (c) magnetic power losses, and (d) the stray-load losses. Thus, $P_e = (1.5\text{ hp} \times 746\text{ W/hp}) + 35\text{ W} + 32\text{ W} + 22\text{ W} = 1208\text{ W}$. See Fig. 3.9.

3. Calculate the Armature Current and Induced Voltage at Rated Load

From the no-load saturation curve, Fig. 3.8, the induced voltage for a field current of 0.36 A, is 225 V. Use the equation: electromagnetic power $P_e = e_a I_a$. Thus, $I_a = P_e/e_a = 1208\text{ W}/225\text{ V} = 5.37\text{ A}$. The armature-circuit voltage drop is $I_a R_a = (5.37\text{ A})(5.1\ \Omega) = 27.4\text{ V}$.

4. Calculate the Armature- and Field-Circuit Copper Loss

Use the equation: armature-circuit copper loss $= I_a^2 R_a = (5.37\text{ A})^2(5.1\ \Omega) = 147\text{ W}$. The field-circuit copper loss $= I_f^2 R_f = (0.36\text{ A})^2(667\ \Omega) = 86\text{ W}$.

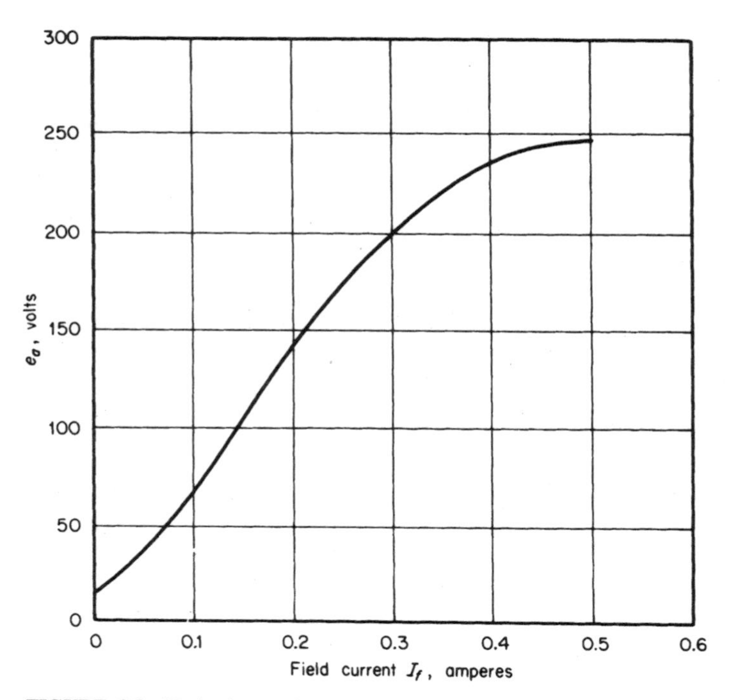

FIGURE 3.8 No-load saturation curve for dc motor, 1750 r/min.

5. *Calculate the Efficiency*

Power input = electromagnetic power P_e + armature-circuit copper loss + field-circuit copper loss = 1208 W + 147 W + 86 W = 1441 W. Thus, the efficiency η = (output × 100 percent)/input = (1.5 hp × 746 W/hp)(100 percent)/1441 W = 77.65 percent.

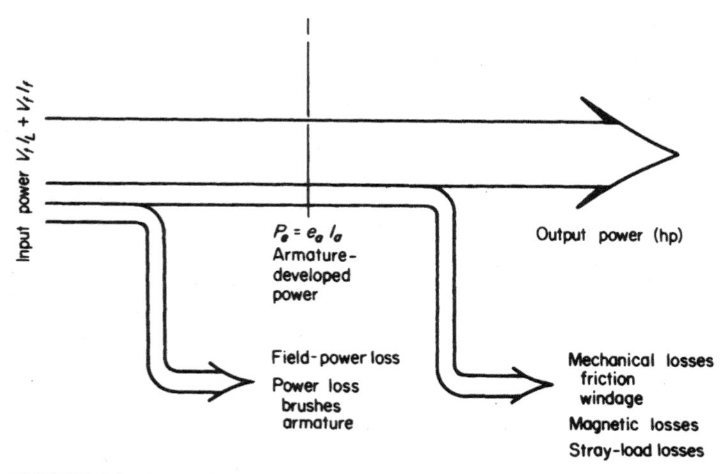

FIGURE 3.9 Power flow in a dc motor.

Related Calculations. The mechanical losses of friction and windage include bearing friction (a function of speed and the type of bearing), brush friction (a function of speed, brush pressure, and brush composition), and the windage resulting from air motion in the air spaces of the machine. The magnetic losses are the core losses in the magnetic material. Stray-load loss actually includes many of the previously mentioned losses but is a term accounting for increased losses caused by loading the motor. For example, the distortion of the magnetic field (armature reaction) increases with load and causes increased magnetic loss; it is included among the stray-load losses.

DESIGN OF A MANUAL STARTER FOR A DC SHUNT MOTOR

A dc shunt motor has the following data: 240 V, 18 A, 1100 r/min, and armature-circuit resistance $R_a = 0.33\ \Omega$. The no-load saturation curve for the machine is given in Fig. 3.4. The motor is to be started by a manual starter, shown in Fig. 3.10; the field current is set at 5.2 A. If the current during starting varies between 20 and 40 A, determine the starting resistors R_1, R_2, and R_3 needed in the manual starter.

Calculation Procedure

1. Calculate Armature Current at Standstill

At the first contact on the manual starter, 240 V dc is applied to the shunt field; the field rheostat has been set to allow $I_f = 5.2$ A. Thus, the armature current $I_a = I_L - I_f =$ 40 A $-$ 5.2 A $=$ 34.8 A. The 40 A is the maximum permissible load current.

2. Calculate the Value of the Starter Resistance, R_{ST}

Use the equation: $R_{ST} = (V_t/I_a) - R_a = (240\text{ V})/(34.8\text{ A}) - 0.330\ \Omega = 6.57\ \Omega$.

3. Calculate the Induced Armature Voltage, e_a, when the Machine Accelerates

As the machine accelerates, an induced armature voltage is created, causing the armature current to fall. When the armature current drops to the minimum allowed during

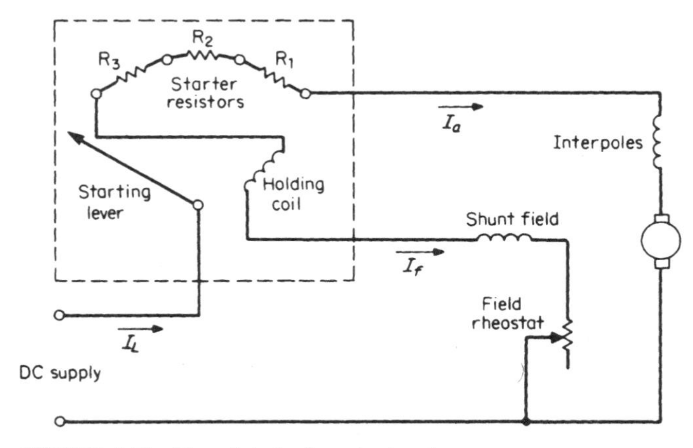

FIGURE 3.10 Manual starter for a shunt motor.

starting $I_a = I_L - I_f = 20\text{ A} - 5.2\text{ A} = 14.8\text{ A}$, the induced voltage is obtained from the equation: $e_a = V_t - I_a(R_a + R_{\text{ST}}) = 240\text{ V} - (14.8\text{ A})(0.330\ \Omega + 6.57\ \Omega) = 137.9\text{ V}$.

4. Calculate the Starter Resistance, $\mathbf{R}_{ST1}$

Use the equation: $R_{\text{ST1}} = (V_t - e_a)/I_a - R_a = (240\text{ V} - 137.9\text{ V})/34.8\text{ A} - 0.330\ \Omega = 2.60\ \Omega$.

5. Calculate the Induced Armature Voltage, $\mathbf{e_a}$, when Machine Again Accelerates

As in Step 3, $e_a = V_t - I_a(R_a + R_{\text{ST1}}) = 240\text{ V} - (14.8\text{ A})(0.330\ \Omega + 2.60\ \Omega) = 196.6\text{ V}$.

6. Calculate the Starter Resistance, $\mathbf{R}_{ST2}$

Use the equation: $R_{\text{ST2}} = (V_t - e_a)/I_a - R_a = (240\text{ V} - 196.6\text{ V})/34.8\text{ A} - 0.330\ \Omega = 0.917\ \Omega$.

7. Calculate the Induced Armature Voltage, $\mathbf{e_a}$, when Machine Again Accelerates

As in Steps 3 and 5, $e_a = V_t - I_a(R_a + R_{\text{ST2}}) = 240\text{ V} - (14.8\text{ A})(0.330\ \Omega + 0.917\ \Omega) = 221.5\text{ V}$.

8. Calculate the Three Starting Resistances, $\mathbf{R_1}$, $\mathbf{R_2}$, and $\mathbf{R_3}$

From the preceding steps, $R_1 = R_{\text{ST2}} = 0.917\ \Omega$, $R_2 = R_{\text{ST1}} - R_{\text{ST2}} = 2.60\ \Omega - 0.917\ \Omega = 1.683\ \Omega$, and $R_3 = R_{\text{ST}} - R_{\text{ST1}} = 6.57\ \Omega - 2.60\ \Omega = 3.97\ \Omega$.

Related Calculations. The manual starter is used commonly for starting dc motors. With appropriate resistance values in the starter, the starting current may be limited to any value. The same calculations may be used for automatic starters; electromagnetic relays are used to short-circuit sections of the starting resistance, sensing when the armature current drops to appropriate values.

CONSIDERATION OF DUTY CYCLE USED TO SELECT DC MOTOR

A dc motor is used to operate a small elevator in a factory assembly line. Each cycle of operation runs 5 min and includes four modes of operation as follows: (1) loading of the elevator, motor at standstill for 2 min; (2) ascent of the elevator, motor shaft horsepower required is 75 for 0.75 min; (3) unloading of the elevator, motor at standstill for 1.5 min; and (4) descent of the unloaded elevator, motor shaft horsepower (regenerative braking) required is -25 for 0.75 min. Determine the motor size required for this duty cycle.

Calculation Procedure

1. Draw a Chart to Illustrate the Duty Cycle

See Fig. 3.11.

2. Calculate the Proportional Heating of the Machine as a Function of Time

The losses (heating) in a machine are proportional to the square of the load current; in turn, the load current is proportional to the horsepower. Thus, the heating of the motor is proportional to the summation of ($P^2 \cdot$ time), where P is the power or horsepower expended (or absorbed during regenerative braking) during the time interval. In this case, the summation is $(75\text{ hp})^2(0.75\text{ min}) \cdot (-25\text{ hp})^2(0.75\text{ min}) = 4687.5\text{ hp}^2 \cdot \text{min}$.

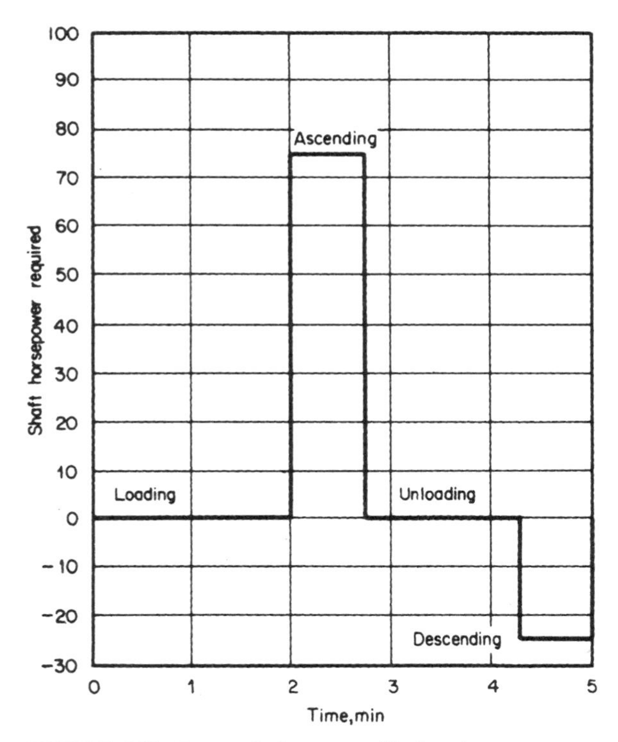

FIGURE 3.11 Duty cycle for an assembly-line elevator motor.

3. Consider the Variation in Cooling for Open (NEMA O Classification) or Closed (NEMA TE Classification) Motors

Allowance is made for the cooling differences between open or closed motors; this is particularly important because at standstill in open motors, without forced-air cooling, poorer cooling results. The running time and the stopped time are separated. At standstill, for open motors the period is divided by 4; for closed motors, the period is counted at full value. In this problem, assume a closed motor for which the effective total cycle time is 5 min. If the motor were of the open type, the effective cycle time would be 1.5 min running + 3.5/4 min at standstill = 2.375 min.

4. Calculate Root-Mean-Square Power

The root-mean-square power $P_{rms} = [(\Sigma P^2 \cdot \text{time})/(\text{cycle effective time})]^{0.5} = (4687.5\ \text{hp}^2 \cdot \text{min}/5\ \text{min})^{0.5} = 30.6$ hp for the case of the closed motor. $P_{rms} = (4687.5\ \text{hp}^2 \cdot \text{min}/2.375\ \text{min})^{0.5} = 44.4$ hp for the case of the open motor. Thus, the selection could be a 30-hp motor of the closed type or a 50-hp motor of the open type. It should be noted in this example that the peak power (75 hp used in ascending) is more than twice the 30 hp calculated for the closed motor; it may be desirable to opt for a 40-hp motor; the manufacturer's literature should be consulted for short-time ratings, etc. There is also the possibility of using separate motor-driven fans.

Related Calculations. The concept of the root-mean-square (rms) value used in this problem is similar to that used in the analysis of ac waves wherein the rms value of a current wave produces the same heating in a given resistance as would dc of the same ampere value. The heating effect of a current is proportional to I^2R. The rms value results from squaring the currents (or horsepower in this example) over intervals of time, finding the average value, and taking the square root of the whole.

CALCULATION OF ARMATURE REACTION IN DC MOTOR

A dc shunt motor has the following characteristics: 10.0 kW, 240 V, 1150 r/min, armature-circuit resistance $R_a = 0.72\ \Omega$. The motor is run at no load at 240 V, with an armature current of 1.78 A; the speed is 1225 r/min. If the armature current is allowed to increase to 50 A, because of applied load torque, the speed drops to 1105 r/min. Calculate the effect of armature reaction on the flux per pole.

Calculation Procedure

1. Draw the Equivalent Circuit and Calculate the Induced Voltage, e_a, at No Load

The equivalent circuit is shown in Fig. 3.12. The induced voltage is $e_a = V_t - I_aR_a = 240\ \text{V} - (1.78\ \text{A})(0.72\ \Omega) = 238.7\ \text{V}$.

2. Determine the No-Load Flux

Knowing the induced voltage at no load, one can derive an expression for the no-load flux from the equation: $e_a = k(\phi_{NL})\omega_m$, where k is the machine constant and ω_m is the mechanical speed in rad/s. The speed $\omega_m = (1225\ \text{r/min})(2\pi\ \text{rad/r})(1\ \text{min/60s}) = 128.3$ rad/s. Thus, $k\phi_{NL} = e_a/\omega_m = 238.7\ \text{V}/128.3\ \text{rad/s} = 1.86\ \text{V}\cdot\text{s}$.

3. Calculate the Load Value of Induced Voltage, e_a

The induced voltage under the loaded condition of $V_t = 240$ V and $I_a = 50$ A is $e_a = V_t - I_aR_a = 240\ \text{V} - (50\ \text{A})(0.72\ \Omega) = 204$ V. The speed is $\omega_m = (1105\ \text{r/min})(2\pi\ \text{rad/r})(1\ \text{min/60 s}) = 115.7$ rad/s. Thus, $k\phi_{load} = e_a/\omega_m = 204\ \text{V}/115.7\ \text{rad/s} = 1.76\ \text{V}\cdot\text{s}$.

4. Calculate the Effect of Armature Reaction on the Flux per Pole

The term, $k\phi$, is a function of the flux per pole; armature reaction reduces the main magnetic flux per pole. In this case, as the load increased from zero to 50 A (120 percent of rated full-load current of 10,000 W/240 V = 41.7 A), the percentage reduction of field flux $= (k\phi_{NL} - k\phi_{load})(100\ \text{percent})/k\phi_{NL} = (1.86 - 1.76)(100\ \text{percent})/1.86 = 5.4$ percent.

Related Calculations. It is interesting to note the units associated with the equation $e_a = k\phi\omega_m$ where e_a is in V, ϕ in Wb/pole, ω_m in rad/s. Thus, $k = e_a/\phi\omega_m = \text{V}/(\text{Wb/pole})(\text{rad/s}) = \text{V}\cdot\text{s/Wb}$. Because $\text{V} = d\phi/dt = \text{Wb/s}$, the machine constant, k, is unitless. Actually, the machine constant $k = Np/a\pi$, where N = total number of turns on

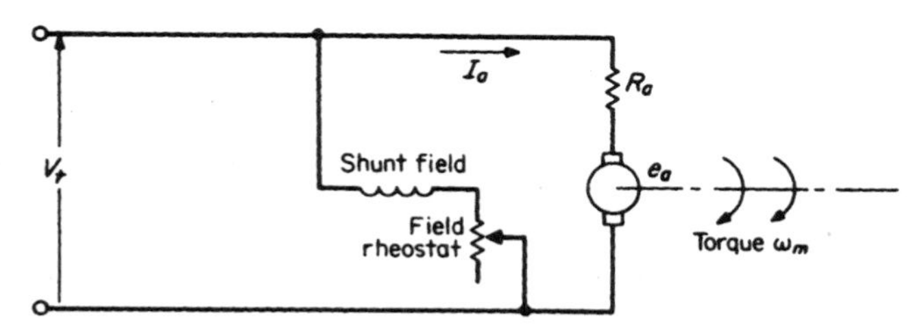

FIGURE 3.12 Equivalent circuit of a dc shunt motor.

the armature, p = the number of poles, and a = the number of parallel paths on the armature circuit. This combination gives the units of k as turns · poles/paths · radians.

DYNAMIC BRAKING FOR SEPARATELY EXCITED MOTOR

A separately excited dc motor has the following data: 7.5 hp, 240 V, 1750 r/min, no-load armature current I_a = 1.85 A, armature-circuit resistance R_a = 0.19 Ω. When the field and armature circuits are disconnected, the motor coasts to a speed of 400 r/min in 84 s. Calculate a resistance value to be used for dynamic braking, with the initial braking torque at twice the full-speed rated torque, and calculate the time required to brake the unloaded motor from 1750 r/min to 250 r/min.

Calculation Procedure

1. Calculate the Torque at Rated Conditions

The torque equation is $T = 33{,}000\ P/2\pi n$, where P = rating in hp and n = speed in r/min. T = (33,000)(7.5 hp)/(2π × 1750 r/min) = 22.5 lb · ft, or (22.5 lb · ft)(1 N · m/0.738 lb · ft) = 30.5 N · m.

2. Calculate Rotational Losses at Rated Speed

Allow for a 2-V drop across the brushes; the rotational losses are equal to (V_t − 2 V) $I_a - I_a^2 R_a$, where I_a = no-load current obtained from test. Rotational losses are (240 V − 2 V)(1.85 A) − (1.85 A)^2(0.19 Ω) = 439.6 W. This loss power can be equated to a torque: (439.6 W)(1 hp/746 W) = 0.589 hp, and (0.589 hp)(33,000)/(2π × 1750 r/min) = 1.77 lb · ft = (1.77 lb · ft)(1 N · m/0.738 lb · ft) = 2.40 N · m.

3. Calculate the Initial Braking Torque

The initial braking torque equals twice the full-speed rated torque; the electromagnetic torque, therefore, equals 2 × full-speed rated torque − rotational-losses torque = 2 × 30.5 N · m − 2.40 N · m = 58.6 N · m. This torque, which was determined at no load, is considered proportional to the armature current in the same manner as electromagnetic power equals $e_a I_a$. The electromagnetic *torque* = $k_1 I_a$. Hence, k_1 = no-load torque/I_a = 2.40 N · m/1.85 A = 1.3.

Under the condition of braking, the electromagnetic torque = 58.6 N · m = $k_1 I_a$, or I_a = 58.6 N · m/1.3 = 45 A.

4. Calculate the Value of the Braking Resistance

Use the equation $I_a = (V_t - 2\ \text{V})/(R_a + R_{\text{br}})$ = (240 V − 2 V)/(0.19 Ω + R_{br}) = 45 A. Solving the equation for the unknown, find R_{br} = 5.1 Ω.

5. Convert Rotational Speeds into Radians per Second

Use the relation (r/min)(1 min/60 s)(2π rad/r) = 1750 r/min, which multiplied by π/30 equals 183 rad/s. Likewise, 400 r/min = 42 rad/s, and 250 r/min = 26 rad/s.

6. Calculate Angular Moment of Inertia, J

The motor decelerated from 1750 r/min or 183 rad/s to 400 r/min or 42 rad/s in 84 s. Because $T = k_2\omega_0$, $k_2 = T/\omega_0$ = (2.40 N · m)/(183 rad/s) = 0.013. The acceleration torque is $T_{\text{acc}} = J(d\omega/dt)$. Thus, $k_2\omega_0 = J(d\omega/dt)$, or

$$\int_0^{84} dt = (J/k_2) \int_{183}^{42} (1/\omega) d\omega$$

Integrating we find

$$[t\ ]\,\big|_0^{84} = (J\ /0.013)\ (\ln \omega)\,\big|_{183}^{42}$$

Thus, $J = (84)(0.013)/(\ln 42 - \ln 183) = -0.742\ \text{kg} \cdot \text{m}^2$.

7. Calculate the Time for Dynamic Braking

From Step 3, the electromagnetic torque $= k_1 I_a$; from Step 4, $I_a = e_a/(R_a + R_{br})$. Thus, because $e_a = k_1\omega$, $I_a = k_1\omega/(R_a + R_{br}) = k_1\omega/5.29$, it follows that $J(d\omega/dt) = k_2\omega + k_1^2\,\omega/5.29$ or $dt = (5.29J\ d\omega)/(5.29k_2\omega + k_1^2\ \omega)$. This leads to

$$t = \frac{5.29 \times 0.742}{5.29 \times 0.013 + 1.3^2} \int_{183}^{26} \frac{1}{\omega}\, d\omega$$

The solution yields $t = 4.35$ s.

Related Calculations. This is an example of braking used on traction motors, whereby the motors act as generators to convert the kinetic energy into electrical energy, which is dissipated through resistance. Voltage generated by the machine is a function of speed and field current; the dissipating resistance is of fixed value. Therefore, the braking effect is related to the armature current and the field current.

A THREE-PHASE SCR DRIVE FOR A DC MOTOR

A three-phase, 60-Hz ac power source (440 V line-to-line) connected through silicon-controlled rectifiers (SCRs) is used to power the armature circuit of a separately excited dc motor of 10-hp rating (see Fig. 3.13). The data for the motor are: armature-circuit resistance $R_a = 0.42\ \Omega$ and I_f (field current) set so that at 1100 r/min when the machine is run as a generator, the full-load voltage is 208 V and the machine delivers the equivalent of 10 hp. Determine the average torque that can be achieved for a speed of 1100 r/min if the SCR's firing angle is set at 40° and at 45°.

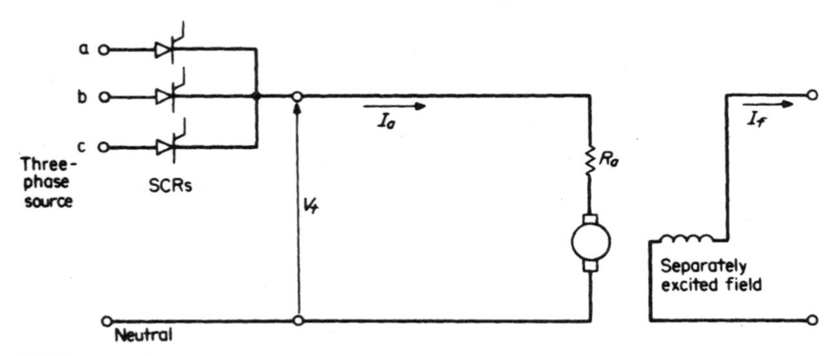

FIGURE 3.13 Three-phase SCR drive for a dc motor.

Calculation Procedure

1. Calculate the Average Value of SCR Output

The average value of the SCR output is determined from the waveforms shown in Fig. 3.14; *a*-phase voltage contributes one-third of the time, then *b*-phase voltage, then *c*-phase voltage. Let the firing angle be represented by δ, assuming that the reference axis is at 30°. Use the equation: $V_{avg} = (3\sqrt{3}/2\pi)\ V \cos\delta$, where V = the maximum value of one phase of the ac voltage (the equation assumes that only one SCR is operated at a time). For the case of 440 V line-to-line: $V = (440/\sqrt{3})\sqrt{2} = 359.3$ V.

For $\delta = 40°$, $V_{avg} = 0.827\ V \cos\delta = (0.827)(359.3\ \text{V})(\cos 40°) = 227.6$ V. For $\delta = 45°$, $V_{avg} = (0.827)(359.3\ \text{V})(\cos 45°) = 210.1$ V.

2. Convert the Speed to Radians per Second

The speed is given as 1100 r/min. Use the relation (1100 r/min)(2π rad/r)(1 min/60 s) = 115.2 rad/s.

3. Calculate the Torque Constant

From the equation, $T = kI_a$, the torque constant, k, can be calculated from the results of tests of the machine run as a generator. The armature current is determined from the test results also; thus, (10 hp)(746 W/hp)/208 V = 35.9 A. $T = (33{,}000)(\text{hp})/2\pi n$, where n = speed in r/min. $T = (33{,}000)(10)/(2\pi \times 1100) = 47.7$ lb · ft. Or (47.7 lb · ft)(1 N · m/0.738 lb · ft) = 64.7 N · m. Thus, $k = T/I_a = 64.7$ N · m/35.9 A) = 1.8 N · m/A.

4. Calculate the Armature Currents for the Two SCR Firing Angles

Use the equation $e_a = k\omega_m$. The units of k, the torque constant or the voltage constant, are N · m/A or V · s/rad. Thus, e_a = (1.8 V · s/rad)(115.2 rad/s) = 208 V. Use the equation $e_a = V_t - I_aR_a$, where the terminal voltage, V_t, is the average value of the SCR output. Solve for I_a; $I_a = (V_t - e_a)/R_a$. For the case of the 40° angle, I_a = (227.6 V − 208 V)/0.42 = 46.7 A. For the case of the 45° angle, I_a = (210.1 V − 208 V)/0.42 = 5.0 A.

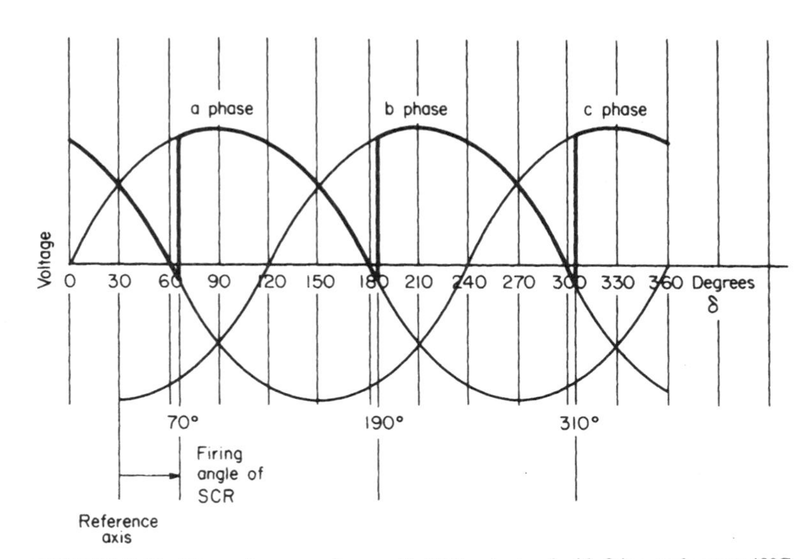

FIGURE 3.14 Three-phase waveforms with SCR output and with firing angle set at 40°C.

5. Calculate the Torques

Use the equation $T = kI_a$. For the 40° firing angle, $T = (1.8\ \text{N}\cdot\text{m/A})(46.7\ \text{A}) = 84.1\ \text{N}\cdot\text{m}$. For the 45° firing angle, $T = (1.8\ \text{N}\cdot\text{m/A})(5.0\ \text{A}) = 9\ \text{N}\cdot\text{m}$.

Related Calculations. The problem illustrates how a three-phase ac system may be used to power a dc motor. The system would be appropriate for motors larger than 5 hp. By varying the firing angle of the SCRs, the amount of current in the armature circuit may be controlled or limited. In this problem, the SCRs were connected for half-wave rectification. For much larger machines, full-wave rectification circuits may be used.

DIRECT-CURRENT SHUNT-MOTOR SPEED DETERMINED FROM ARMATURE CURRENT AND NO-LOAD SATURATION CURVE

The saturation curve for a dc shunt motor operating at 1150 r/min may be represented by the equation, $e_a = (250I_f)/(0.5 + I_f)$. See Fig. 3.15. The open-circuit voltage is e_a, in volts; the shunt-field current is I_f, in amperes. Data for the machine are as follows: armature-circuit resistance $R_a = 0.19\ \Omega$, R_f shunt-field-circuit resistance $= 146\ \Omega$, terminal voltage $V_t = 220$ V. For an armature current of 75 A, calculate the speed. The demagnetizing effect of armature reaction may be estimated as being 8 percent, in terms of shunt-field current at the given load.

Calculation Procedure

1. Calculate the Effective Shunt-Field Current

In a shunt motor, the terminal voltage is directly across the shunt field. Thus, $I_f = V_t/R_f = 220\ \text{V}/146\ \Omega = 1.5$ A. However, the demagnetizing effect of armature reaction is 8 percent, in terms of shunt-field current. Thus, $I_{f(\text{demag})} = (0.08)(1.5\ \text{A}) = 0.12$ A. This amount of demagnetization is subtracted from the shunt-field current to yield $I_{f(\text{effective})} = I_f - I_{f(\text{demag})} = 1.5\ \text{A} - 0.12\ \text{A} = 1.38$ A.

2. Calculate the Induced Voltage at 1150 r/min

The induced voltage, e_a, is now determined from the given saturation curve, of necessity at 1150 r/min. It may be calculated from the equation, with $I_{f(\text{effective})}$ substituted for I_f: $e_a = 250I_{f(\text{effective})}/(0.5 + I_{f(\text{effective})}) = (250)(1.38)/(0.5 + 1.38) = 183.5$ V. Alternatively, the voltage may be read directly from the graph of the saturation curve (see Fig. 3.15).

3. Calculate the Induced Voltage at the Unknown Speed

Use the equation: $e'_a = V_t - I_aR_a = 220\ \text{V} - (75\ \text{A})(0.19\ \Omega) = 205.8$ V.

4. Calculate the Unknown Speed

The induced voltage, e_a or e'_a, is proportional to the speed, provided the effective magnetization is constant. Thus, $e_a/e'_a = (1150\ \text{r/min})/n_x$. The unknown speed is $n_x = (1150\ \text{r/min})(e'_a/e_a) = (1150\ \text{r/min})(205.8\ \text{V})/183.5\ \text{V} = 1290$ r/min.

Related Calculations. DC motor speeds may be determined by referring to the saturation curve and setting proportions of induced voltage to speed, provided that for both sides of the proportion the effective magnetization is constant. The effective magnetization allows for the demagnetization of armature reaction. The procedure may be used with additional fields (e.g., series field) provided the effect of the additional fields can be related to the saturation curve; an iterative process may be necessary.

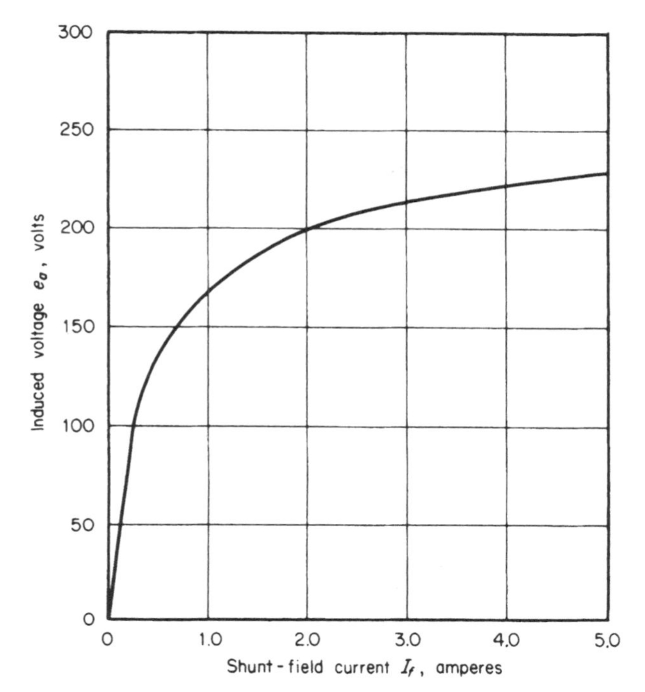

FIGURE 3.15 dc-motor saturation curve, $e_a = 250I_f/(0.5 + I_f)$; $n = 1150$ r/min.

CHOPPER DRIVE FOR DC MOTOR

A separately excited dc motor is powered through a chopper drive (see Fig. 3.16) from a 240-V battery supply. A square-wave terminal voltage, V_t, is developed that is 3 ms ON and 3 ms OFF. The motor is operating at 450 r/min. The machine's torque constant (or voltage constant) $k = 1.8$ N · m/A (or V · s/rad), and the armature-circuit resistance $R_a = 0.42\ \Omega$. Determine the electromagnetic torque that is developed.

Calculation Procedure

***1. Calculate the Terminal Voltage,* V_t**

The dc value of the pulsating terminal voltage (square wave) is its average value. Use the equation: $V_t = V(\text{time}_{ON})/(\text{time}_{ON} + \text{time}_{OFF}) = (240\text{ V})(3\text{ ms})/(3\text{ ms} + 3\text{ ms}) = 120\text{ V}$.

***2. Calculate the Induced Voltage,* e_a**

The average induced voltage, e_a, may be calculated from the equation, $e_a = k\omega$, where k = voltage constant in V · s/rad and ω = speed in rad/s. Thus, $e_a = (1.8\text{ N}\cdot\text{m/A})(450\text{ r/min})(2\pi\text{ rad/r})(1\text{ min}/60\text{ s}) = 84.8\text{ V}$.

***3. Calculate the Armature Current,* I_a**

The average armature current is determined from the general equation, $e_a = V_t - I_aR_a$. Rearranging, we find $I_a = (V_t - e_a)/R_a = (120\text{ V} - 84.8\text{ V})/0.42\ \Omega = 83.8\text{ A}$.

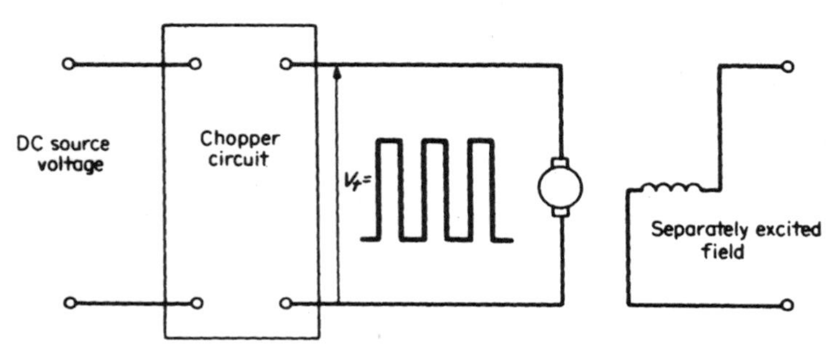

FIGURE 3.16 Chopper drive for a dc motor.

4. Calculate the Average Torque

The average torque is determined from the equation, $T = kI_a$, where k = the torque constant in N · m/A. Thus, T = (1.8 N · m/A)(83.8 A) = 150.84 N · m.

Related Calculations. The chopper is a switching circuit, usually electronic, that turns the output of the battery ON and OFF for preset intervals of time, generating a square wave for V_t'. The magnitude of the voltage at the terminals of the motor is varied by adjusting the average value of the square wave (either changing the widths of the pulses or the frequency of the pulses). Here we have a method for controlling motors that operate from batteries. Automotive devices operated with battery-driven motors would use chopper circuits for control.

DESIGN OF A COUNTER-EMF AUTOMATIC STARTER FOR A SHUNT MOTOR

A counter-emf automatic starter is to be designed for a dc shunt motor with the following characteristics: 115 V, armature-circuit resistance R_a = 0.23 Ω, armature current I_a = 42 A at full load. Determine the values of external resistance for a three-step starter, allowing starting currents of 180 to 70 percent of full load.

Calculation Procedure

***1. Calculate First-Step External Resistance Value,* R_1**

Use the equation, R_1 = [line voltage/(1.8 × I_a)] − R_a = [115 V/(1.8 × 42 A)] − 0.23 Ω = 1.29 Ω.

2. Calculate the Counter-emf Generated at First Step

With the controller set for the first step, the motor will attain a speed such that the counter-emf will reach a value to limit the starting current to 70 percent of full-load armature current. Thus, the counter-emf $E_a = V_t - (R_a + R_1)(0.70 \times I_a)$ = 115 V − (0.23 Ω + 1.29 Ω)(0.70 × 42 A) = 70.3 V. The drop across the resistance $R_a + R_1$ is 44.7 V.

***3. Calculate the Second-Step External Resistance Value,* R_2**

If the current is allowed to rise to 1.8 × 42 A through a voltage drop of 44.7 V, the new external resistance value is R_2 = [44.7 V/(1.8 × 42 A)] − 0.23 Ω = 0.36 Ω.

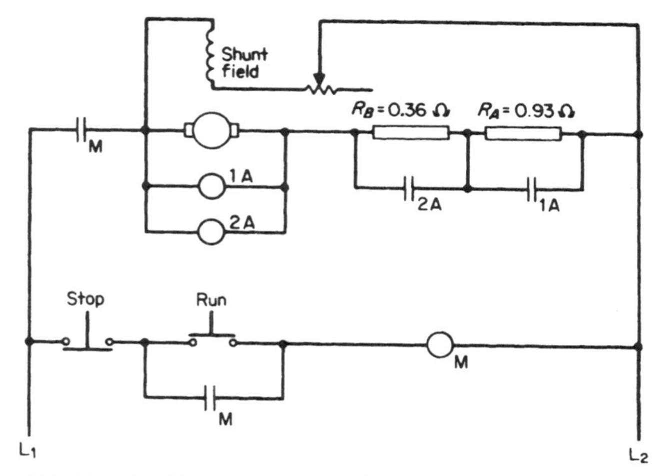

FIGURE 3.17 Three-step counter-emf automatic starter.

4. *Calculate the Counter-emf Generated at the Second Step*

As in the previous calculation, with the controller set for the second step, the motor will attain a speed such that the counter-emf $E_a = V_t - (R_a + R_2)(0.70 \times I_a) = 115\text{ V} - (0.23\ \Omega + 0.36\ \Omega)(0.70 \times 42\text{ A}) = 97.7\text{ V}$. The drop across the resistance $R_a + R_2 = 17.3$ V. This last step is sufficient to permit the accelerating motor to reach normal speed without exceeding 1.8×42 A.

5. *Draw the Circuit for the Three-Step Automatic Starter*

See Fig. 3.17. $R_A = R_1 - R_2 = 1.29\ \Omega - 0.36\ \Omega = 0.93\ \Omega$. $R_B = R_2 = 0.36\ \Omega$. When the run button is pressed, the closing coil M (main contactor) is energized, the shunt-field circuit is connected across the line, and the armature circuit is connected through the starting resistors to the line. In addition, a contact closes across the run button, serving as an interlock. The coil of contactor 1A is set to operate when the counter-emf = 70 V, thus short-circuiting R_A. The coil of contactor 2A is set to operate when the counter-emf = 98 V, thus short-circuiting R_B. In this manner, the external resistance has three steps: 1.29 Ω, 0.36 Ω, and 0.0 Ω. the starting current is limited to the range of 1.8×42 A and 0.7×42 A.

Related Calculations. Illustrated here is a common automatic starter employing the changing counter-emf developed across the armature during starting, for operating electromagnetic relays to remove resistances in a predetermined manner. Other starting mechanisms employ current-limiting relays or time-limit relays. Each scheme that is used is for the purpose of short-circuiting resistance steps automatically as the motor accelerates.

BIBLIOGRAPHY

Basak, Amitava. 1996. *Permanent-Magnet DC Linear Motors*. Oxford: Clarendon Press.

Beaty, H. Wayne, and James L. Kirtley, Jr. 1998. *Electric Motor Handbook*. New York: McGraw-Hill.

Hughes, Austin. 1993. *Electric Motors and Drives: Fundamentals, Types and Applications*. Oxford: Butterworth-Heinemann.

SECTION 4

TRANSFORMERS

Yilu (Ellen) Liu
Associate Professor
Electrical Engineering Department
Virginia Tech University

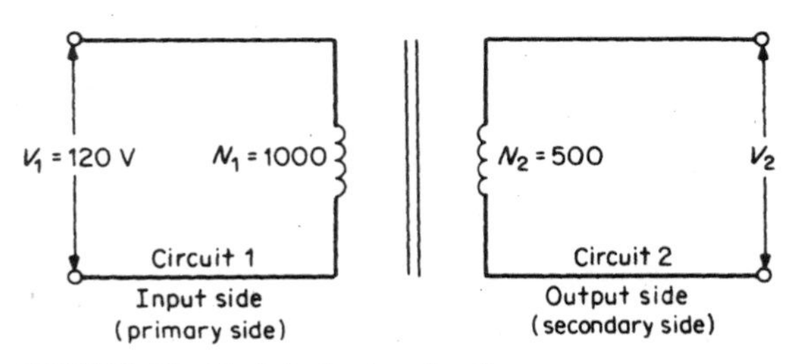

FIGURE 4.1 Ideal single-phase transformer.

ANALYSIS OF TRANSFORMER TURNS RATIO

Compute the value of the secondary voltage, the voltage per turn of the primary, and the voltage per turn of the secondary of the ideal single-phase transformer of Fig. 4.1.

Calculation Procedure

1. Compute the Value of the Secondary Voltage

The turns ratio, a, of a transformer is the ratio of the primary turns, N_1, to the secondary turns, N_2, or the ratio of the primary voltage, V_1, to the secondary voltage, V_2; $a = N_1/N_2 = V_1/V_2$. Substitution of the values indicated in Fig. 4.1 for the turns ratio yields $a = 1000/500 = 2$. Secondary voltage $V_2 = V_1/a = 120/2 = 60$ V.

2. Compute the Voltage per Turn of the Primary and Secondary Windings

Voltage per turn of the primary winding $= V_1/N_1 = 120/1000 = 0.12$ V/turn. Voltage per turn of the secondary winding $= V_2/N_2 = 60/500 = 0.12$ V/turn.

Related Calculations. This procedure illustrates a step-down transformer. A characteristic of both step-down and step-up transformers is that the voltage per turn of the primary is equal to the voltage per turn of the secondary.

ANALYSIS OF A STEP-UP TRANSFORMER

Calculate the turns ratio of the transformer in Fig. 4.1 when it is employed as a step-up transformer.

Calculation Procedure

1. Compute the Turns Ratio

In the step-up transformer, the low-voltage side is connected to the input, or primary, side. Therefore, $a = N_2/N_1 = V_2/V_1$; hence, $a = 500/1000 = 0.5$.

Related Calculations. For a particular application, the turns ratio, a, is fixed but is not a transformer constant. In this example, $a = 0.5$ when the transformer is used as a step-up transformer. In the previous example, $a = 2$ when the transformer is used as a step-down transformer. The two values of a are reciprocals of each other: $2 = 1/0.5$, and $0.5 = 1/2$.

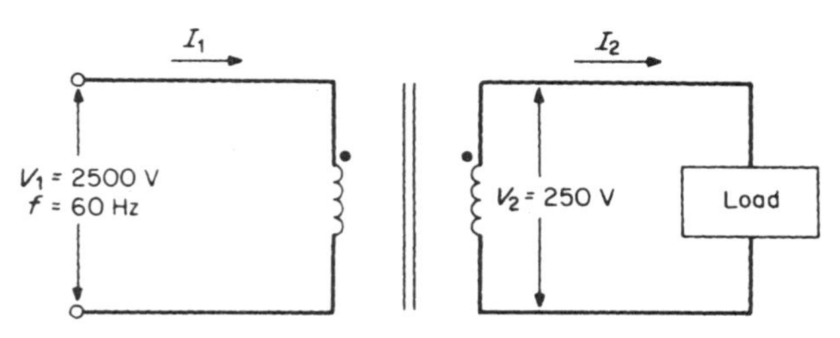

FIGURE 4.2 Single-phase transformer connected to load.

ANALYSIS OF A TRANSFORMER CONNECTED TO A LOAD

A 25-kVA single-phase transformer is designed to have an induced emf of 2.5 V/turn (Fig. 4.2). Calculate the number of primary and secondary turns and the full-load current of the primary and secondary windings.

Calculation Procedure

1. Compute the Number of Primary and Secondary Turns

$N_1 = V_1/(\text{V/turn}) = 2500\ \text{V}/(2.5\ \text{V/turn}) = 1000$ turns. Similarly, $N_2 = V_2/(\text{V/turn}) = 250/2.5 = 100$ turns. The turns ratio $a = N_1/N_2 = 1000/100 = 10{:}1$.

2. Compute the Full-Load Current of the Primary and Secondary Windings

Primary current $I_1 = (\text{VA})_1/V_1 = 25{,}000\ \text{VA}/2500\ \text{V} = 10$ A. Secondary current $I_2 = (\text{VA})_2/V_2 = 25{,}000\ \text{VA}/250\ \text{V} = 100$ A. The current transformation ratio is $1/a = I_1/I_2 = 10\ \text{A}/100\ \text{A} = 1{:}10$.

Related Calculations. One end of each winding is marked with a dot in Fig. 4.2. The dots indicate the terminals that have the same relative polarity. As a result of the dot convention, the following rules are established:

1. When current enters a primary that has a dot polarity marking, the current leaves the secondary at its dot polarity terminal.
2. When current leaves a primary terminal that has a dot polarity marking, the current enters the secondary at its dot polarity terminal.

Manufacturers usually mark the leads on the high-voltage sides as H_1, H_2, etc. The leads on the low-voltage side are marked X_1, X_2, etc. Marking H_1 has the same relative polarity as X_1, etc.

SELECTION OF A TRANSFORMER FOR IMPEDANCE MATCHING

Select a transformer with the correct turns ratio to match the 8-Ω resistive load in Fig. 4.3 to the Thevenin equivalent circuit of the source.

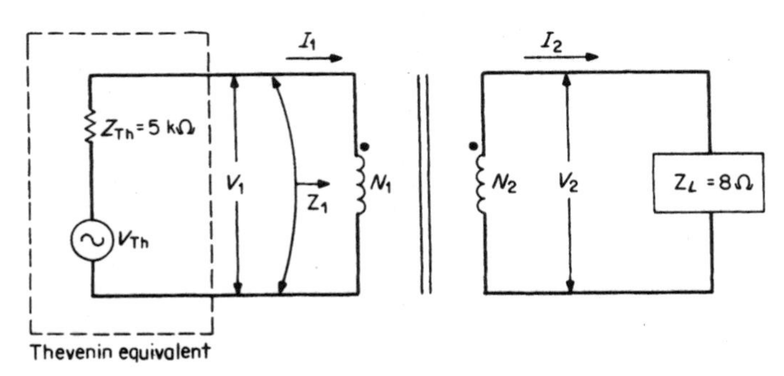

FIGURE 4.3 Transformer used for impedance matching.

Calculation Procedure

1. Determine the Turns Ratio

The impedance of the input circuit, Z_i, is 5000 Ω. This value represents the Thevenin impedance of the source. The load impedance, Z_L, is 8 Ω. To achieve an impedance match, the required turns ratio is

$$a = \sqrt{Z_i/Z_L} = \sqrt{5000\ \Omega/8\ \Omega} = 25$$

Therefore, the impedance-matching transformer must have a turns ratio of 25:1.

Related Calculations. The maximum power transfer theorem (Sec. 1) states that maximum power is delivered by a source to a load when the impedance of the load is equal to the internal impedance of the source. Because the load impedance does not always match the source impedance, transformers are used between source and load to ensure matching. When the load and source impedances are not resistive, maximum power is delivered to the load when the load impedance is the complex conjugate of the source impedance.

PERFORMANCE OF A TRANSFORMER WITH MULTIPLE SECONDARIES

Determine the turns ratio of each secondary circuit, the primary current, I_1, and rating in kVA of a transformer with multiple secondaries, illustrated in Fig. 4.4.

Calculation Procedure

1. Select the Turns Ratio of Each Secondary Circuit

Let the turns ratio of circuit 1 to circuit 2 be designated as a_2 and the turns ratio of circuit 1 to circuit 3 as a_3. Then, $a_2 = V_1/V_2 = 2000\ \text{V}/1000\ \text{V} = 2{:}1$ and $a_3 = V_1/V_3 = 2000\ \text{V}/500\ \text{V} = 4{:}1$.

2. Compute the Primary Current $\mathbf{I_1}$

Since $I_2 = V_2/Z_2$ and $I_3 = V_3/Z_3$, then $I_2 = 1000\ \text{V}/500\ \Omega = 20$ A and $I_3 = 500\ \text{V}/50\ \Omega = 10$ A. The ampere turns of the primary (N_1I_1) of a transformer equals the sum of the

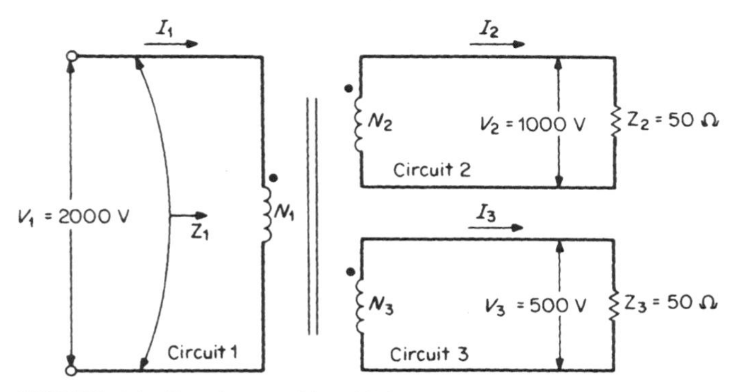

FIGURE 4.4 Transformer with multiple secondaries.

ampere turns of all the secondary circuits; therefore, $N_1I_1 = N_2I_2 + N_3I_3$. Solving for I_1, we have $I_1 = (N_2/N_1)I_2 + (N_3/N_1)I_3$. The turns ratio is $N_2/N_1 = 1/a_2$ and $N_3/N_1 = 1/a_3$. Hence, $I_1 = I_2/a_2 + I_3/a_3 = (20\text{ A})(\frac{1}{2}) + (10\text{ A})(\frac{1}{4}) = 12.5$ A.

3. Compute the Rating in kVA of the Transformer

The winding ratings are $\text{kVA}_1 = V_1I_1/1000 = (2000\text{ V})(12.5\text{ A})/1000 = 25$ kVA, $\text{kVA}_2 = V_2I_2/1000 = (1000\text{ V})(20\text{ A})/1000 = 20$ kVA, and $\text{kVA}_3 = V_3I_3/1000 = (500\text{ V})(10\text{ A})/1000 = 5$ kVA. The apparent power rating of the primary equals the sum of the apparent power ratings of the secondary. As a check, 25 kVA = (20 + 5) kVA = 25 kVA.

Related Calculations. When the secondary loads have different phase angles, the same equations still apply. The voltages and currents, however, must be expressed as phasor quantities.

IMPEDANCE TRANSFORMATION OF A THREE-WINDING TRANSFORMER

Calculate the impedance, Z_1, seen by the primary of the three-winding transformer of Fig. 4.4 using impedance-transformation concepts.

Calculation Procedure

1. Calculate Z_1

The equivalent reflected impedance of both secondaries or the total impedance "seen" by the primary, Z_1, is: $Z_1 = a_2^2 Z_2 \parallel a_3^2 Z_3$, where $a_2^2 Z_2$ is the reflected impedance of circuit 2 and $a_3^2 Z_3$ is the reflected impedance of circuit 3. Hence, $Z_1 = (2)^2(50\ \Omega) \parallel (4)^2(50\ \Omega) = 200 \parallel 800 = 160\ \Omega$.

2. Check the Value of Z_1 Found in Step 1

$Z_1 = V_1/I_1 = 2000\text{ V}/12.5\text{ A} = 160\ \Omega$.

Related Calculations. When the secondary loads are not resistive, the preceding equations still apply. The impedances are expressed by complex numbers, and the voltage and current are expressed as phasors.

SELECTION OF A TRANSFORMER WITH TAPPED SECONDARIES

Select the turns ratio of a transformer with tapped secondaries to supply the loads of Fig. 4.5.

Calculation Procedure

1. Calculate the Power Requirement of the Primary

When a transformer has multiple, or tapped, secondaries, $P_1 = P_2 + P_3 + \cdots$, where P_1 is the power requirement of the primary and P_2, P_3, . . . are the power requirements of each secondary circuit. Hence, $P_1 = 5 + 2 + 10 + 3 = 20$ W.

2. Compute the Primary Input Voltage, $\mathbf{V}_1$

From $P_1 = V_1^2/Z_1$, one obtains $V_1 = \sqrt{P_1 Z_1} = \sqrt{20 \times 2000} = 200$ V.

3. Compute the Secondary Voltages

$V_2 = \sqrt{5 \times 6} = 5.48$ V, $V_3 = \sqrt{2 \times 8} = 4$ V, $V_4 = \sqrt{10 \times 16} = 12.7$ V, and $V_5 = \sqrt{3 \times 500} = 38.7$ V.

4. Select the Turns Ratio

The ratios are $a_2 = V_1/V_2 = 200/5.48 = 36.5{:}1$, $a_3 = V_1/V_3 = 200/4 = 50{:}1$, $a_4 = V_1/V_4 = 200/12.7 = 15.7{:}1$, and $a_5 = V_1/V_5 = 200/38.7 = 5.17{:}1$.

Related Calculations. A basic transformer rule is that the total power requirements of all the secondaries must equal the power input to the primary of the transformer.

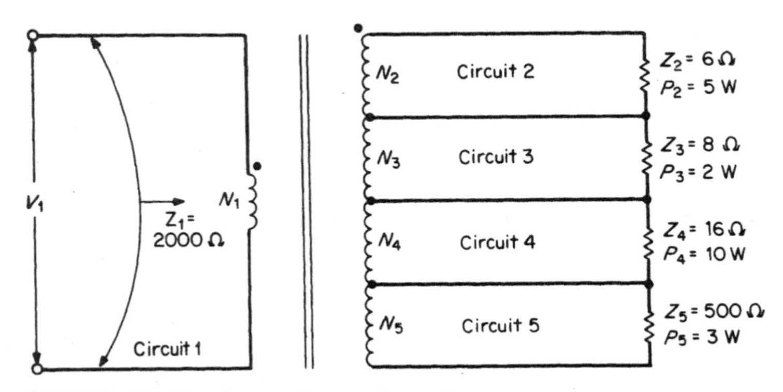

FIGURE 4.5 Transformer with tapped secondaries.

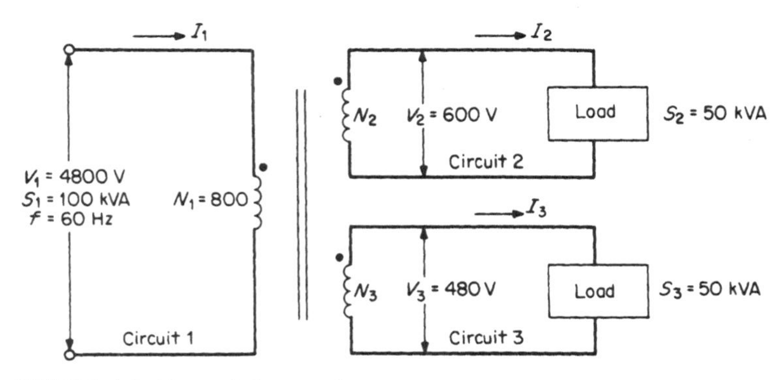

FIGURE 4.6 Three-winding transformer.

TRANSFORMER CHARACTERISTICS AND PERFORMANCE

Compute the number of turns in each secondary winding, the rated primary current at unity-power-factor loads, and the rated current in each secondary winding of the three-winding transformer of Fig. 4.6.

Calculation Procedure

1. Compute the Number of Turns in Each Secondary Winding

The turns ratio in winding 2 is $a_2 = V_1/V_2 = N_1/N_2 = 4800/600 = 8{:}1$; hence, $N_2 = N_1/a_2 = 800/8 = 100$ turns. Similarly, $a_3 = V_1/V_3 = N_1/N_3 = 4800/480 = 10{:}1$ from which $N_3 = N_1/a_3 = 800/10 = 80$ turns.

2. Compute the Rated Primary Current

$I_1 = (\text{VA})_1/V_1 = 100{,}000/4800 = 20.83$ A.

3. Compute the Rated Secondary Currents

$I_2 = (\text{VA})_2/V_2 = 50{,}000/600 = 83.8$ A and $I_3 = (\text{VA})_3/V_3 = 50{,}000/480 = 104.2$ A.

Related Calculations. This method may be used to analyze transformers with one or more secondary windings for power, distribution, residential, or commercial service. When the loads are not at unity power factor, complex algebra and phasors are used where applicable.

PERFORMANCE AND ANALYSIS OF A TRANSFORMER WITH A LAGGING POWER-FACTOR LOAD

Calculate the primary voltage required to produce rated voltage at the secondary terminals of a 100-kVA, 2400/240-V, single-phase transformer operating at full load. The power factor (pf) of the load is 80 percent lagging.

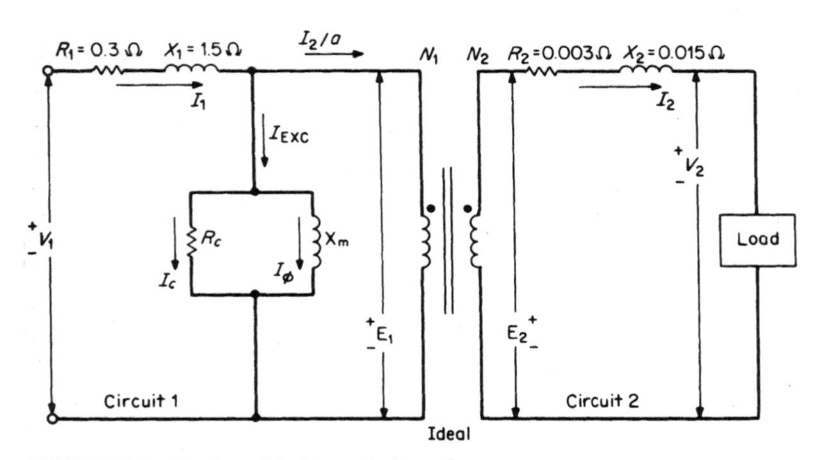

FIGURE 4.7 Circuit model of a practical transformer.

Calculation Procedure

1. Analyze the Circuit Model of Fig. 4.7

The model of Fig. 4.7 includes winding resistances, inductive reactances, and core and copper losses. The symbols are defined as follows:

V_1 = supply voltage applied to the primary circuit
R_1 = resistance of the primary circuit
X_1 = inductive reactance of the primary circuit
I_1 = current drawn by the primary from the power source
I_{EXC} = exciting current
I_c = core loss component of the exciting current; this component accounts for the hysteresis and eddy-current losses
I_ϕ = magnetizing component of exciting current
R_c = equivalent resistance representing core loss
X_m = primary self-inductance that accounts for magnetizing current
I_2/a = load component of primary current
E_1 = voltage induced in the primary coil by all the flux linking the coil
E_2 = voltage induced in the secondary coil by all the flux linking the coil
R_2 = resistance of the secondary circuit, excluding the load
X_2 = inductive reactance of the secondary circuit
I_2 = current delivered by the secondary circuit to the load
V_2 = voltage that appears at the terminals of the secondary winding, across the load

2. Draw Phasor Diagram

The phasor diagram for the model of Fig. 4.7 is provided in Fig. 4.8. The magnetizing current, I_ϕ, is about 5 percent of full-load primary current. The core loss component of the exciting current, I_c, is about 1 percent of full-load primary current. Current I_c is in phase

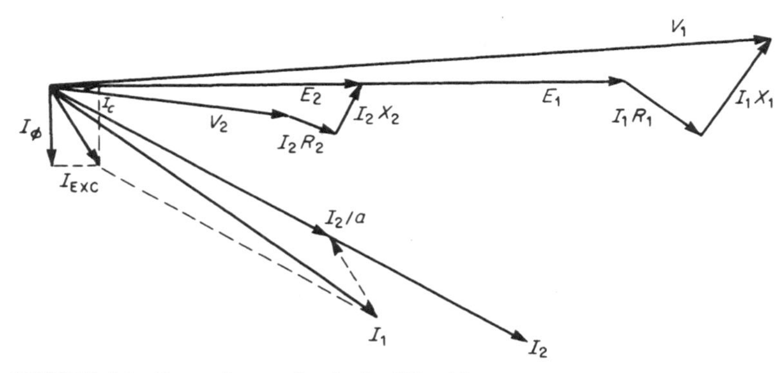

FIGURE 4.8 Phasor diagram for circuit of Fig. 4.7.

with E_1, and I_ϕ lags E_1 by 90°. The power factor of the exciting current, I_{EXC}, is quite low and lags E_1 by approximately 80°.

3. Simplify the Circuit of Fig. 4.7

The exciting current is neglected in the approximate transformer model of Fig. 4.9. The primary parameters R_1 and X_1 are referred to the secondary side as R_1/a^2 and X_1/a^2, respectively. Hence, $R_{EQ2} = R_1/a^2 + R_2$ and $X_{EQ2} = X_1/a^2 + X_2$. The equivalent impedance of the transformer referred to the secondary side is: $\mathbf{Z}_{\mathbf{EQ2}} = R_{EQ2} + jX_{EQ2}$.

4. Solve for $\mathbf{V}_1$

Transformer manufacturers agree that the ratio of rated primary and secondary voltages of a transformer is equal to the turns ratio: $a = V_1/V_2 = 2400/240 = 10{:}1$. $\mathbf{Z}_{\mathbf{EQ2}} = (R_1/a^2 + R_2) + j(X_1/a^2 + X_2) = (0.3/100 + 0.003) + j(1.5/100 + 0.015) = 0.03059\underline{/78.69°}$, but $|\mathbf{I}_2| = (VA)_2/V_2 = 100{,}000/240 = 416.67$ A. Using $\mathbf{V}_2$ as a reference, $\mathbf{I}_2 = 416.67\underline{/-36.87°}$ A, but $\mathbf{V}_1/a = \mathbf{V}_2 + \mathbf{I}_2\mathbf{Z}_{\mathbf{EQ2}} = 240\underline{/0°} + (416.67\underline{/-36.87°})(0.03059\underline{/78.69°}) = 249.65\underline{/1.952°}$. However, $|\mathbf{V}_1| = a|\mathbf{V}_1/a| = 10 \times 249.65 = 2496.5$ V.

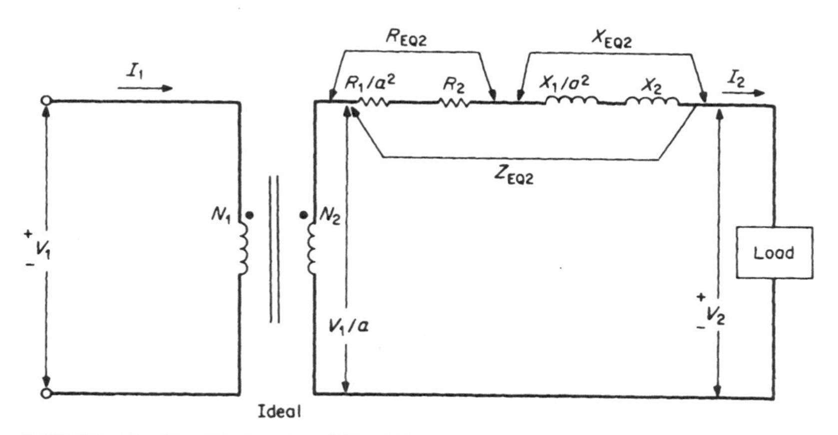

FIGURE 4.9 Simplified version of Fig. 4.7.

Related Calculations. For lagging power-factor loads, $\mathbf{V}_1$ must be greater than rated value (in this case 2400 V) in order to produce rated voltage across the secondary (240 V).

PERFORMANCE AND ANALYSIS OF A TRANSFORMER WITH A LEADING POWER-FACTOR LOAD

Calculate the primary voltage required to produce rated voltage at the secondary terminals of the transformer of Fig. 4.7. The transformer is operating at full load with an 80 percent leading power factor.

Calculation Procedure

1. Solve for V_1

With $\mathbf{V}_2$ as a reference, $\mathbf{I}_2 = 416.67\angle 36.87°$ A. Substituting, we have $\mathbf{V}_1/a = \mathbf{V}_2 + \mathbf{I}_2\mathbf{Z}_{EQ2} = 240\angle 0° + (416.67\angle 36.87°)(0.03059\angle 78.69°) = 234.78\angle 2.81°$. The magnitude $|\mathbf{V}_1| = a|\mathbf{V}_1/a| = (10)(234.78) = 2347.8$ V.

Related Calculations. When the load is sufficiently leading, as in this case, $|\mathbf{V}_1| =$ 2347.8 V. This value is less than the rated primary voltage of 2400 V to produce the rated secondary voltage.

CALCULATION OF TRANSFORMER VOLTAGE REGULATION

Calculate the full-load voltage regulation of the transformer shown in Fig. 4.7 for 80 percent leading and lagging power factors.

Calculation Procedure

1. Solve for the Full-Load Voltage Regulation at 80 Percent Lagging Power Factor

Transformer voltage regulation is defined as the difference between the full-load and the no-load secondary voltages (with the same impressed primary voltage for each case). Expressed as a percentage of the full-load secondary voltage, voltage regulation is: VR = $[(|\mathbf{V}_1/a| - |\mathbf{V}_2|)/|\mathbf{V}_2|] \times 100$ percent. From the results of the 80 percent lagging power factor example, VR = $(249.65 - 240)/240 \times 100$ percent = 4.02 percent.

The phasor diagram for a lagging power-factor condition, that is illustrated in Fig. 4.10, indicates positive voltage regulation.

2. Solve for the Full-Load Voltage Regulation at 80 Percent Leading Power Factor

From the 80 percent leading power-factor example, VR = $(234.78 - 240)/240 \times 100$ percent = -2.18 percent. The phasor diagram for a leading power-factor condition, shown in Fig. 4.11, illustrates negative voltage regulation.

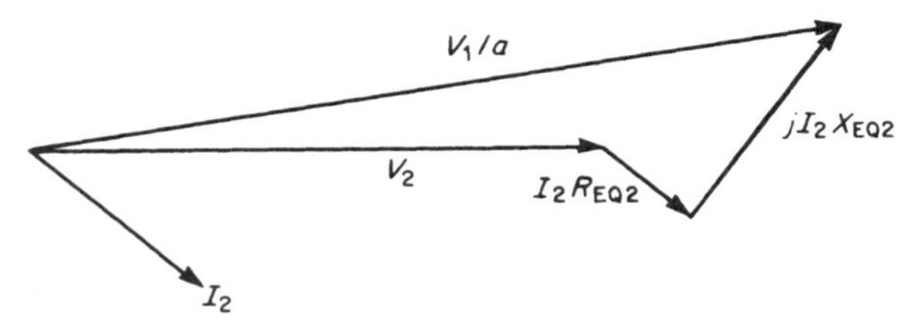

FIGURE 4.10 Phasor diagram of transformer for lagging power factor.

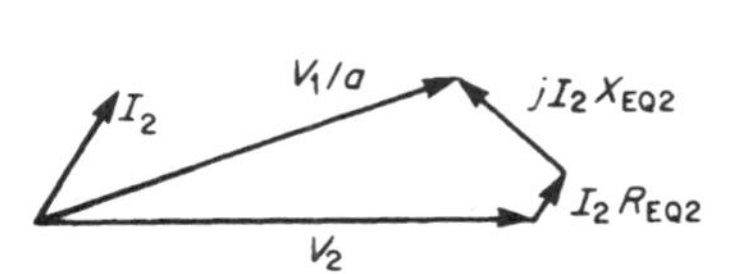

FIGURE 4.11 Phasor diagram of transformer for leading power factor.

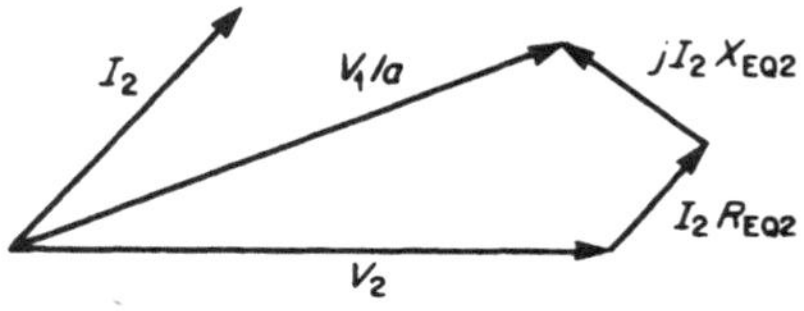

FIGURE 4.12 Phasor diagram for zero transformer voltage regulation.

Related Calculations. Negative regulation denotes that the secondary voltage increases when the transformer is loaded. This stems from a partial resonance condition between the capacitance of the load and the leakage inductance of the transformer. Zero transformer voltage regulation occurs when $V_1/a = V_2$. This condition, which occurs with a slightly leading power-factor load, is illustrated in Fig. 4.12.

CALCULATION OF EFFICIENCY

A 10-kVA transformer has 40-W iron loss at rated voltage and 160-W copper loss at full load. Calculate the efficiency for a 5-kVA, 80 percent power-factor load.

Calculation Procedure

1. Analyze the Losses

The sum of the hysteresis and eddy-current losses is called the core, or iron, loss of the transformer; this will be designated P_i. The core loss is a constant loss of the transformer.

The sum of the primary and secondary I^2R losses is called the copper loss, P_{cu}; $P_{cu} = I_1^2R_1 + I_2^2R_2$. This shows that copper losses vary with the square of the current.

2. Solve for Efficiency

Efficiency, η, can be found from $\eta = P_{out}/P_{in} = P_{out}/(P_{out} + \text{losses}) = P_{out}/(P_{out} + P_i + P_{cu}) = (\text{VA}_{load})(\text{pf})/[(\text{VA}_{load})(\text{pf}) + P_t + P_{cu}(\text{VA}_{load}/\text{VA}_{rating})^2]$. From this equation, $\eta = (5000 \times 0.8)/[5000 \times 0.8 + 40 + 160(5000/10{,}000)^2] = 0.98$ or 98 percent.

Related Calculations. The iron losses of a transformer are determined quite accurately by measuring the power input at rated voltage and frequency under no-load conditions.

Although it makes no difference which winding is energized, it is usually more convenient to energize the low-voltage side. (It is essential to use rated voltage for this test.)

The copper loss is measured by short-circuiting the transformer and measuring the power input at rated frequency and full-load current. It is usually convenient to perform the short-circuit test by shorting out the low-voltage side and energizing the high-voltage side; however, it does not matter if the procedure is reversed.

Because changing the power factor of the load does not change the losses, raising the load power factor will improve the efficiency of the transformer. The losses then become a smaller proportion of the total power input. The no-load efficiency of the transformer is zero. High loads increase the copper losses, which vary with the square of the current, thereby decreasing the efficiency. Maximum efficiency operation occurs at some intermediate value of load.

ANALYSIS OF TRANSFORMER OPERATION AT MAXIMUM EFFICIENCY

Calculate the load level at which maximum efficiency occurs for the transformer of the previous example. Find the value of maximum efficiency with a 100 percent power-factor load and a 50 percent power-factor load.

Calculation Procedure

1. Calculate the Load Level for Maximum Efficiency

Maximum efficiency occurs when the copper losses equal the iron losses. Then, $P_i = P_{cu}(\text{kVA}_{load}/\text{kVA}_{rating})^2$ or $40 = 160(\text{kVA}_{load}/10\text{ kVA})^2$. Solving, find $\text{kVA}_{load} = 5$ kVA.

2. Calculate Maximum Efficiency with 100 Percent Power-Factor Load

With the copper and core losses equal, $\eta = P_{out}(P_{out} + P_i + P_{cu}) = 5000(5000 + 40 + 40) = 0.9842$ or 98.42 percent.

3. Calculate Maximum Efficiency with 50 Percent Power-Factor Load

Maximum efficiency is $\eta = (5000 \times 0.5)/(5000 \times 0.5 + 40 + 40) = 0.969$ or 96.9 percent.

Related Calculations. Maximum efficiency occurs at approximately half load for most transformers. In this procedure, it occurs at exactly half load. Transformers maintain their high efficiency over a wide range of load values above and below the half-load value. Maximum efficiency for this transformer decreases with lower power factors for a value of 98.42 percent at 100 percent power factor to a value of 96.9 percent at 50 percent power factor. Efficiencies of transformers are higher than those of rotating machinery, for the same capacity, because rotating electrical machinery has additional losses, such as rotational and stray load losses.

CALCULATION OF ALL-DAY EFFICIENCY

A 50-kVA transformer has 180-W iron loss at rated voltage and 620-W copper loss at full load. Calculate the all-day efficiency of the transformer when it operates with the following unity-power-factor loads: full load, 8 h; half load, 5 h; one-quarter load, 7 h; no load, 4 h.

Calculation Procedure

1. Find the Total Energy Iron Losses

Since iron losses exist for the entire 24 h the transformer is energized, the total iron loss is $W_{i\,(\text{total})} = P_i t = (180 \times 24)/1000 = 4.32$ kWh.

2. Determine the Total Energy Copper Losses

Energy copper losses $W_{cu} = P_{cu}t$. Because copper losses vary with the square of the load, the total-energy copper losses are found as follows: $W_{cu\,(\text{total})} = (1^2 \times 620 \times 8 + 0.5^2 \times 620 \times 5 + 0.25^2 \times 620 \times 7)/1000 = 6.006$ kWh over the 20 h that the transformer supplies a load.

3. Calculate the Total Energy Loss

The total energy loss over the 24-h period is: $W_{\text{loss}\,(\text{total})} = W_{i\,(\text{total})} + W_{cu\,(\text{total})} = 4.32 + 6.006 = 10.326$ kWh.

4. Solve for the Total Energy Output

$W_{\text{out}\,(\text{total})} = 50 \times 8 + 50 \times \frac{1}{2} \times 5 + 50 \times \frac{1}{4} \times 7 = 612.5$ kWh.

5. Compute the All-Day Efficiency

All-day efficiency $= W_{\text{out(total)}}/[W_{\text{out(total)}} + W_{\text{loss(total)}}] \times 100$ percent $= 612.5/(612.5 + 10.236) \times 100$ percent $= 98.3$ percent.

Related Calculations. All-day efficiency is important when the transformer is connected to the supply for the entire 24 h, as is typical for ac distribution systems. It is usual to calculate this efficiency at unity power factor. At any other power factor, the all-day efficiency would be lower because the power output would be less for the same losses.

The overall energy efficiency of a distribution transformer over a 24-h period is high in spite of varying load and power-factor conditions. A low all-day efficiency exists only when there is a complete lack of use of the transformer, or during operation at extremely low power factors.

SELECTION OF TRANSFORMER TO SUPPLY A CYCLIC LOAD

Select a minimum-size transformer to supply a cyclic load that draws 100 kVA for 2 min, 50 kVA for 3 min, 25 kVA for 2 min, and no load for the balance of its 10-min cycle.

Calculation Procedure

1. Solve for the Apparent-Power Rating in kVA of the Transformer

When the load cycle is sufficiently short so that the temperature of the transformer does not change appreciably during the cycle, the minimum transformer size is the rms value of the load. Hence $S = \sqrt{(S_1^2 t_1 + S_2^2 t_2 + S_3^2 t_3)/[t(\text{cycle})]} = \sqrt{(100^2 \times 2 + 50^2 \times 3 + 25^2 \times 2)/10} = 53.62$ kVA.

Related Calculations. When selecting a transformer to supply a cyclic load, it is essential to verify that the voltage regulation is not excessive at peak load. The method used to select a transformer in this example is satisfactory provided the load cycle is short. If the load cycle is long (several hours), this method cannot be used. In that case the thermal time constant of the transformer must be considered.

ANALYSIS OF TRANSFORMER UNDER SHORT-CIRCUIT CONDITIONS

A transformer is designed to carry 30 times its rated current for 1 s. Determine the length of time that a current of 20 times the rating can be allowed to flow. Find the maximum amount of current that the transformer can carry for 2 s.

Calculation Procedure

1. Calculate the Time for 20 Times Rating

Transformers have a definite I^2t limitation because heat equals I^2R_{EQ} and R_{EQ} is constant for a particular transformer. (R_{EQ} represents the total resistance of the primary and secondary circuits.) Hence, $I^2_{\text{rating}} \cdot t_{\text{rating}} = I^2_{\text{new}} \cdot t_{\text{new}} = 30^2 \times 1 = 20^2 t_{\text{new}}$; solving, $t_{\text{new}} =$ 2.25 s.

2. Solve for the Maximum Permissible Current for 2 s

Since $30^2 \times 1 = I^2 \times 2, I = 21.21$ times full-load current.

Related Calculations. The thermal problem is basically a matter of how much heat can be stored in the transformer windings before an objectionable temperature is reached. The method followed in this example is valid for values of t below 10 s.

CALCULATION OF PARAMETERS IN THE EQUIVALENT CIRCUIT OF POWER TRANSFORMER BY USING THE OPEN-CIRCUIT AND SHORT-CIRCUIT TESTS

Both open-circuit and short-circuit tests are made on a 50-kVA, 5-kV/500-V, 60-Hz power transformer. Open-circuit test readings from the primary side are 5 kV, 250 W, and 0.4 A; short-circuit test readings from the primary side are 190 V, 450 W, and 9 A. Calculate the parameters of the equivalent circuit (R_c, X_m, R_1, X_2, R_2, X_2), referring to Fig. 4.7.

Calculation Procedure

1. Analyze the Equivalent Circuit Model in Fig. 4.7

The equivalent circuit symbols that have been used are defined on page 4.8. In order to determine all of the parameters of this equivalent circuit, we need to calculate the winding resistance, inductive reactance, the equivalent resistance representing the core loss, and the primary self-inductance that accounts for the magnetizing current.

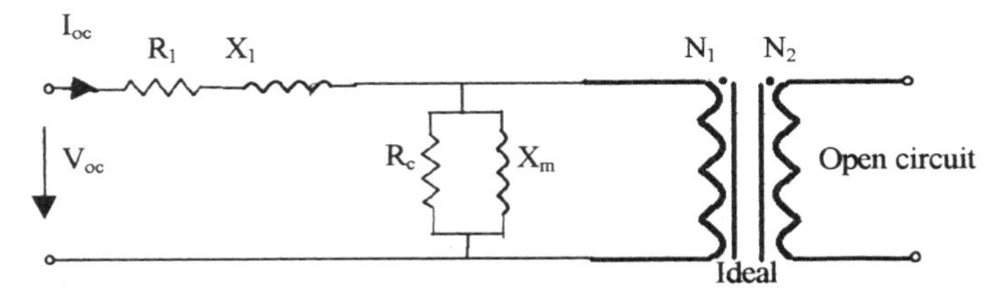

FIGURE 4.13 Equivalent circuit diagram of an open-circuit test for a power transformer.

2. Calculation of the Shunt Resister (R_c) *and the Magnetizing Reactance of the Transformer* (X_m) *by Open-Circuit Test*

The equivalent circuit diagram of the open-circuit test for power transformer is provided in Fig. 4.13.

Since R_c is much larger than R_1, we can omit the influence of R_1 when we calculate. Hence, $P_{oc} = V_{oc}^2/R_c$, $R_c = V_{oc}^2/P_{oc} = (5 \times 10^3)^2/250 = 100{,}000\ \Omega$. Also, since X_m is much larger than X_1, we can omit the influence of X_1 when we calculate. Then, $O_{oc} = V_{oc}^2/X_m$.

At first, we calculate the apparent power, $S_{oc} = V_{oc} \times I_{oc} = 5000 \times 0.4 = 2000$ VA, and the power factor, pf $= P_{oc}/S_{oc} = 250/2000 = 0.125$ (lagging). The power factor is the cosine of the angle between the voltage and current vector, so the pf angle $\theta = \cos^{-1} 0.125 = 82.8°$. Then, $Q_{oc} = V_{oc} \times I_{oc} \times \sin\theta = 5{,}000 \times 0.4 \times \sin 82.8 = 1984$ VAR. Finally, $X_m = V_{oc}^2/Q_{oc} = (5 \times 10^3)^2/1984 = 12{,}600\ \Omega$.

3. Calculation of the Coil Resistance (R_1, R_2) *and the Leakage Reactance* (X_1, X_2) *by Short-Circuit Test of Power Transformer*

The equivalent circuit diagram of short-circuit test of transformer is provided in Fig. 4.14.

Since the value of R_c, X_m is much larger than that of a^2R_2 and a^2X_2, we could get a simple equivalent circuit of the transformer for the short-circuit test. The equivalent circuit diagram referring to the primary side is shown in Fig. 4.15.

In most cases, R_1 is very close in value to a^2R_2, and X_1 is very close in value to a^2X_2, thus we could assume that $R_1 = a^2R_2$ and $X_1 = a^2X_2$. Then the equivalent circuit will be simplified further to the circuit shown in Fig. 4.16.

We define two parameters: $R_w = R_1 + a^2R_2 = 2R_1 = 2a^2R_2$ and $X_w = X_1 + a^2X_2 = 2X_2 = 2a^2X_2$. At first, we calculate the apparent power $S_{sc} = V_{sc} \times I_{sc} = 190 \times 9 = 1710$ VA, so we get $Q_{sc} = \sqrt{S_{sc}^2 - P_{sc}^2} = \sqrt{1710^2 - 450^2} = 1650$ VAR.

Then, $R_w = P_{sc}/I_{sc}^2 = 450/9^2 = 5.56\ \Omega$; $X_w = Q_{sc}/I_{sc}^2 = 1694/9^2 = 20.4\ \Omega$. Thus, $R_1 = R_w/2 = 5.56/2 = 2.78\ \Omega$. We know $a = 5000/500 = 10$. So, $R_2 = R_w/(2 \times a^2) = 5.56/(2 \times 10^2) = 27.8$ mΩ; and $X_2 = X_w/(2 \times a^2) = 20.4/(2 \times 10^2) = 102$ mΩ.

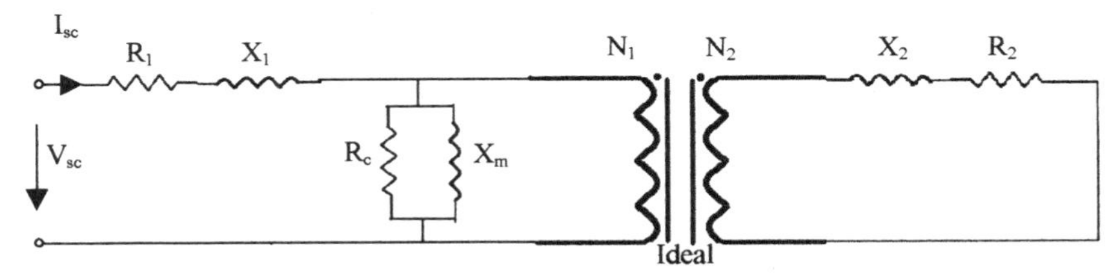

FIGURE 4.14 Equivalent circuit diagram of a short-circuit test for a power transformer.

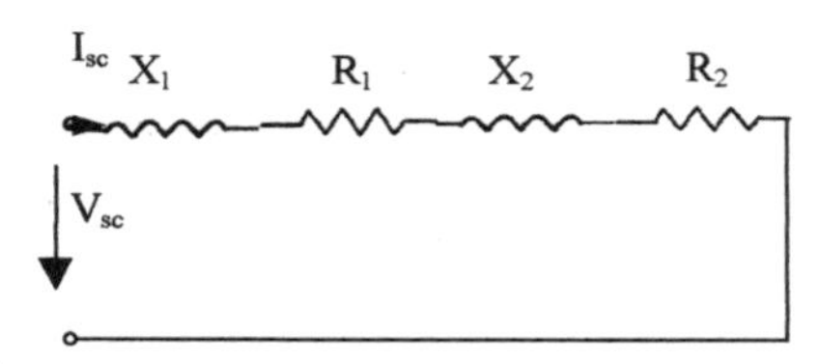

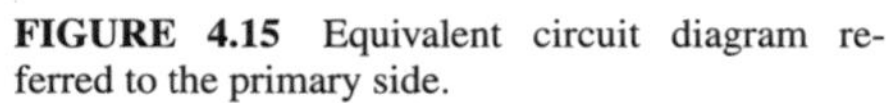

FIGURE 4.15 Equivalent circuit diagram referred to the primary side.

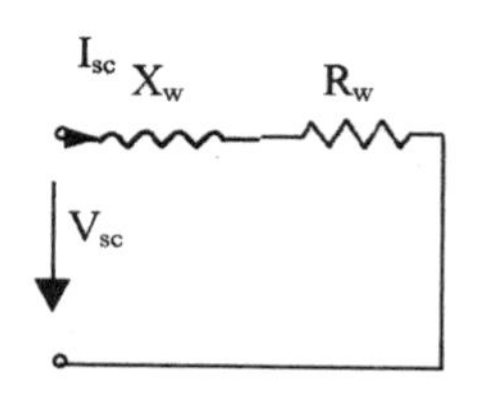

FIGURE 4.16 A more simplified equivalent circuit diagram.

Related Calculations. We could get the transformer exciting current from $I\varphi = I_{oc}$, the magnetizing current $I_m = V_{oc}/X_m$ and the current I_c supplying core losses when we conduct open-circuit and short-circuit tests on the transformer. Furthermore, the core loss of the transformer is approximately equal to P_{oc}, and the copper loss of the transformer is approximately equal to P_{sc}. This result could be used for the calculation of the efficiency of power transformer discussed on page 4.12.

PERFORMANCE OF A STEP-UP AUTOTRANSFORMER (BUCK/BOOST TRANSFORMER IN BOOST MODE)

Calculate the full-load currents of the 50-kVA, 2400-V, and 120-V windings of the 2400/120-V isolation transformer of Fig. 4.17*a*. When this transformer is connected as a step-up transformer in Fig. 4.17*b*, calculate its apparent power rating in kVA, the percent increase in apparent power capacity in kVA, and the full-load currents.

Calculation Procedure

1. Find the Full-Load Currents of the Isolation Transformer

$I_1 = (\text{VA})_1/V_1 = 50{,}000/2400 = 20.83$ A, and $I_2 = (\text{VA})_2/V_2 = 50{,}000/120 = 416.7$ A.

2. Determine the Power Rating in kVA of the Autotransformer

Because the 120-V winding is capable of carrying 416.7 A, the power rating in kVA of the autotransformer is $\text{VA}_2 = (2520 \times 416.7)/1000 = 1050$ kVA.

3. Calculate the Percent Increase in Apparent Power Using the Isolation Transformer as an Autotransformer

$\text{kVA}_{\text{auto}}/\text{kVA}_{\text{isolation}} = 1050/50 \times 100$ percent $= 2100$ percent.

4. Solve for the Full-Load Currents of the Autotransformer

Because the series winding (X_1 to X_2) has a full-load rating of 416.7 A, $I_2 = 416.7$ A. Current $I_1 = (\text{VA})_1/V_1 = (1050 \times 1000)/2400 = 437.5$ A. The current in the common winding is $I_c = I_1 - I_2 = 437.5 - 416.7 = 20.8$ A.

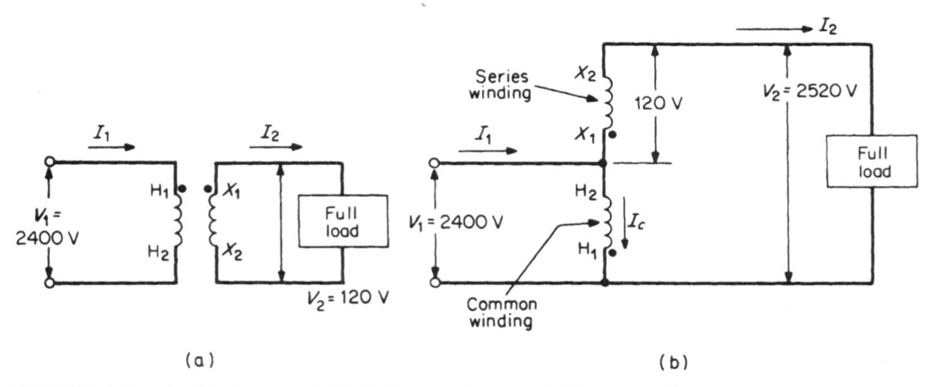

FIGURE 4.17 Application of an (*a*) isolation transformer and (*b*) autotransformer.

Related Calculations. With the circuit as an autotransformer, the power in kVA has increased to 2100 percent of its original value with the low-voltage winding operating at its rated capacity. The effect on the high-voltage winding is negligible because $I_c = 20.8$ A while I_1, with the circuit as an isolation transformer, is 20.83 A.

The increase in apparent-power capacity produced by connecting an isolation transformer as an autotransformer accounts for the smaller size in autotransformers of the same capacity in kVA compared with ordinary isolation transformers. However, this marked increase in capacity only occurs as the ratio of primary to secondary voltages in the autotransformer approaches unity.

ANALYSIS OF A DELTA-WYE THREE-PHASE TRANSFORMER BANK USED AS A GENERATOR-STEP-UP TRANSFORMER

Calculate the line current in the primary, the phase current in the primary, the phase-to-neutral voltage of the secondary, and the turns ratio of the 50-MVA, three-phase transformer bank of Fig. 4.18 when used as a generator-step-up transformer and operating at rated load.

Calculation Procedure

1. Find the Line Current in the Primary

$I_{1P} = S/(\sqrt{3} \times V_{\text{LP}}) = 50{,}000{,}000/(\sqrt{3} \times 13{,}000) = 2221$ A.

2. Determine the Value of the Phase Current in the Primary

$I_{\text{line}\Delta} = I_{\text{phase}}/\sqrt{3} = 2221/\sqrt{3} = 1282$ A.

3. Calculate the Phase-to-Neutral Voltage of the Secondary

$V_{1N} = V_{\text{LS}}/\sqrt{3} = 138{,}000/\sqrt{3} = 79{,}677$ V.

4. Solve for the Line Current in the Secondary

$I_{1S} = S/(\sqrt{3} \times V_{\text{LS}}) = 50{,}000{,}000/(\sqrt{3} \times 138{,}000) = 209$ A.

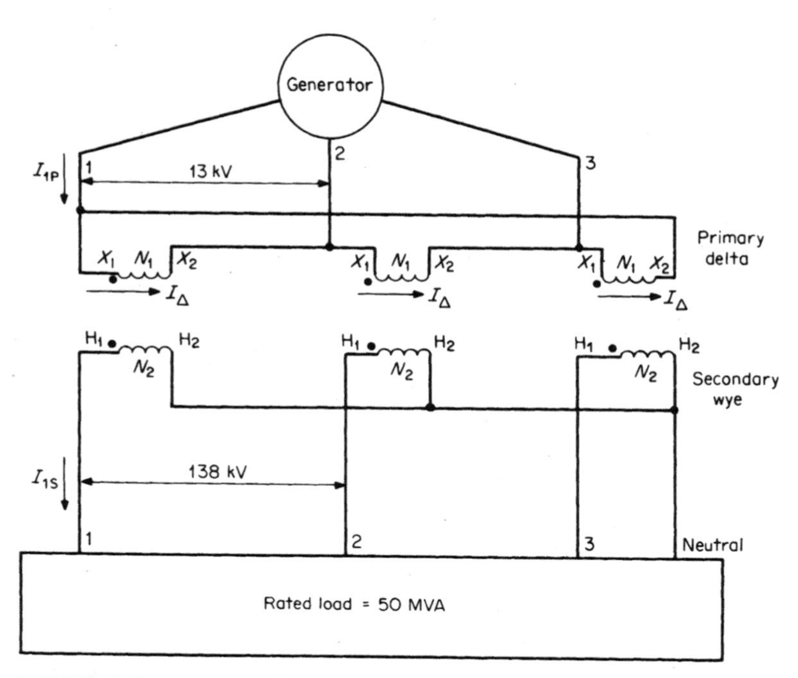

FIGURE 4.18 Delta-wye, three-phase transformer bank used as a generator step-up transformer.

5. Compute the Transformer Turns Ratio

The turns ratio is $a = N_1/N_2 = 13{,}000/79{,}677 = 0.163 = 1{:}6.13$.

Related Calculations. The line current in the secondary is related to the line current in the primary as follows: $I_{1S} = aI_{1P}/\sqrt{3} = (0.163 \times 2221)/\sqrt{3} = 209$ A, which checks the value of I_{1S} calculated in Step 4. The voltage phasor relations (Fig. 4.19) show that the secondary voltages of the wye side lead the primary voltages of the delta side by 30°.

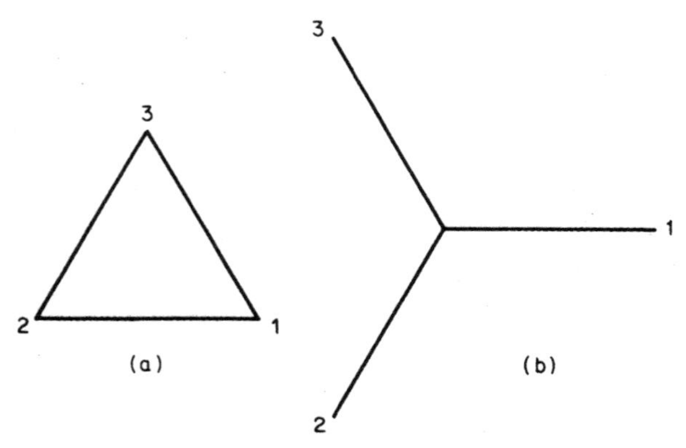

FIGURE 4.19 Voltage phasor relations for a step-up transformer with (*a*) delta primary at 13 kV and (*b*) wye secondary at 138 kV.

PERFORMANCE OF AN OPEN DELTA OR VEE-VEE SYSTEM

Each of the delta-delta transformers of Fig. 4.20 is rated at 40 kVA, 2400/240 V. The bank supplies a 80-kVA load at unity power factor. If transformer C is removed for repair, calculate, for the resulting vee-vee connection: load in kVA carried by each transformer, percent of rated load carried by each transformer, total apparent power rating in kVA of the transformer bank in vee-vee, ratio of vee-vee bank to delta-delta bank transformer ratings, and the percent increase in load on each transformer.

Calculation Procedure

1. Find the Load in kVA Carried by Each Transformer

Load per transformer = total kVA/$\sqrt{3}$ = 80 kVA/$\sqrt{3}$ = 46.2 kVA.

2. Determine the Percent of Rated Load Carried by Each Transformer

Percent transformer load = (load in kVA/transformer)/(rating in kVA per transformer) × 100 percent = 46.2 kVA/40 kVA × 100 percent = 115.5 percent.

3. Calculate the Total Rating in kVA of the Transformer Bank in Vee-Vee

Rating in kVA of vee-vee bank = $\sqrt{3}$ (rating in kVA per transformer) = $\sqrt{3} \times 40$ = 69.3 kVA.

4. Solve for the Ratio of the Vee-Vee Bank to the Delta-Delta Bank Transformer Ratings

Ratio of ratings = vee-vee bank/delta-delta bank = 69.3 kVA/120 kVA × 100 percent = 57.7 percent.

5. Compute the Percent Increase in Load on Each Transformer

Original load in delta-delta per transformer is 80 kVA/$\sqrt{3}$ = 26.67 kVA per transformer. Percent increase in load = (kVA per transformer in vee-vee)/(kVA per transformer in delta-delta) = 46.2 kVA/26.67 kVA × 100 percent = 173.2 percent.

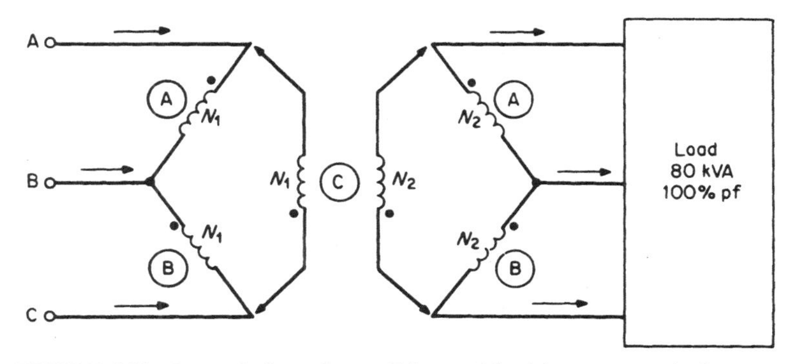

FIGURE 4.20 Removal of transformer C from a delta-delta system results in a vee-vee system.

Related Calculations. This example demonstrates that while the transformer load increases by 173.2 percent in a vee-vee system, each transformer is only slightly overloaded (115.5 percent). Because each transformer in a vee-vee system delivers line current and not phase current, each transformer in open delta supplies 57.7 percent of the total volt-amperes.

ANALYSIS OF A SCOTT-CONNECTED SYSTEM

A two-phase, 10-hp, 240-V, 60-Hz motor has an efficiency of 85 percent and a power factor of 80 percent. It is fed from the 600-V, three-phase system of Fig. 4.21 by a Scott-connected transformer bank. Calculate the apparent power drawn by the motor at full load, the current in each two-phase line, and the current in each three-phase line.

Calculation Procedure

1. Find the Apparent Power Drawn by the Motor

Rated output power of the motor is P_o = 746 W/hp × 10 hp = 7460 W. Active power, P, drawn by the motor at full load is P_o/η = 7460 W/0.85 = 8776 W, where η = efficiency. Apparent power drawn by the motor at full load is $S = P/\text{pf}$ = (8776 W)/0.8 = 10,970 VA.

2. Determine the Current in Each Two-Phase Line

Apparent power per phase = 10,970/2 = 5485 VA; hence, $I = S/V$ = 5485/240 = 22.85 A.

3. Calculate the Current in Each Three-Phase Line

$I = S/(\sqrt{3} \times V) = 10.970/(\sqrt{3} \times 600) = 10.56$ A.

Related Calculations. The Scott connection isolates the three-phase and two-phase systems and provides the desired voltage ratio. It can be used to change three-phase to two-

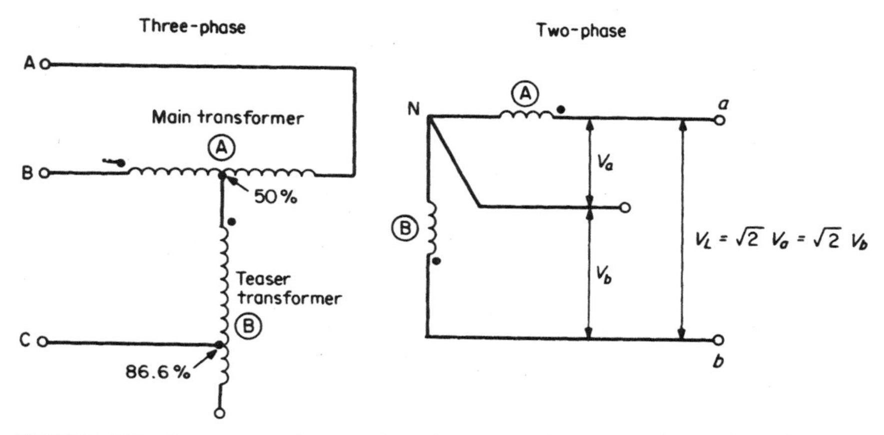

FIGURE 4.21 Scott-connected system, three-phase to two-phase or two-phase to three-phase.

phase or two-phase to three-phase. Because transformers are less costly than rotating machines, this connection is very useful when industrial concerns wish to retain their two-phase motors even through their line service is three-phase.

BIBLIOGRAPHY

Anderson, Edwin. 1985. *Electric Machines and Transformers*. Reston, Va.: Reston Publishing Co.

Elgerd, Olle Ingemar. 1982. *Electric Energy Systems Theory: An Introduction*, 2nd ed. New York: McGraw-Hill.

Gingrich, Harold W. 1979. *Electrical Machinery, Transformers, and Control*. Englewood Cliffs, N.J.: Prentice-Hall.

Henry, Tom. 1989. *Transformer Exam Calculations*. Orlando, Fla.: Code Electrical Classes & Bookstore.

Hicks, S. David. 1995. *Standard Handbook of Engineering Calculations*, 3rd ed. New York: McGraw-Hill.

Jackson, Herbert W. 1986. *Introduction to Electric Circuits*, 6th ed. Englewood Cliffs, N.J.: Prentice-Hall.

Johnson, Curtis B. 1994. *Handbook of Electrical and Electronics Technology*. Englewood Cliffs, N.J.: Prentice-Hall.

Kingsley, Charles, A. Ernest Fitzgerald, and Stephen Umans. 1990. *Electric Machinery,* 5th ed. New York: McGraw-Hill College Division.

Kosow, Irving L. 1991. *Electric Machinery and Transformers*, 2nd ed. Englewood Cliffs, N.J.: Prentice-Hall.

McPherson, George, and Robert D. Laramore. 1990. *An Introduction to Electrical Machines and Transformers*, 2nd ed. New York: John Wiley & Sons.

Richardson, David W. 1978. *Rotating Electric Machinery and Transformer Technology*. Reston, Va.: Reston Publishing Co.

Wildi, Theodore. 1981. *Electrical Power Technology*. New York: John Wiley & Sons.

SECTION 5

THREE-PHASE INDUCTION MOTORS

Hashem Oraee

Worcester Polytechnic Institute

INTRODUCTION

The induction motor is the most rugged and widely used machine in industry. Both stator winding and rotor winding carry alternating current. The alternating current is supplied to the stator winding directly and to the rotor winding by induction—hence the name *induction machine*.

The induction machine can operate both as a motor and as a generator. However, it is seldom used as a generator. The performance characteristics as a generator are not satisfactory for most applications.

The three-phase induction motor is used in various sizes. Large motors are used in pumps, fans, compressors, and paper mills.

EQUIVALENT CIRCUIT

The stator winding can be represented as shown in Fig. 5.1. Note that there is no difference in form between the stator equivalent circuit and that of the transformer primary winding. The only difference lies in the magnitude of the parameters. For example, the excitation current, I_ϕ, is considerably larger in the induction motor because of the air gap. The rotor winding can be represented as shown in Fig. 5.2.

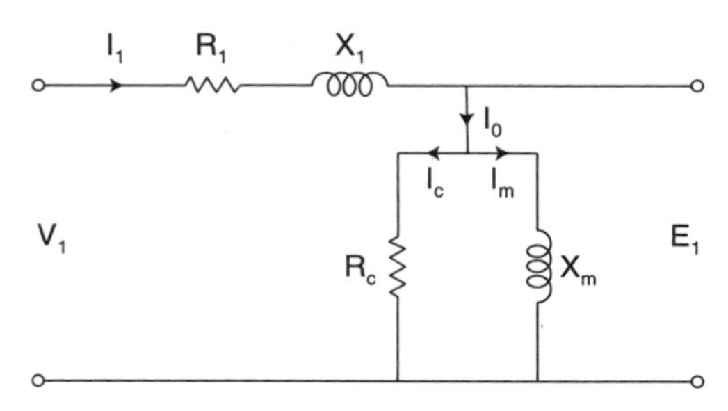

FIGURE 5.1 Stator equivalent circuit.

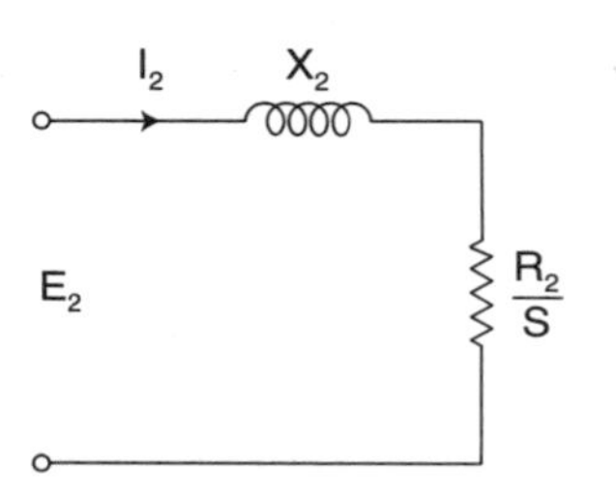

FIGURE 5.2 Rotor equivalent circuit.

The equivalent circuit of the induction motor is obtained by combining the stator and rotor equivalent circuit (Fig. 5.3). Note that the form of the equivalent circuit is identical to that of a two-winding transformer.

Example 1

A three-phase, 15-hp, 460-V, four-pole, 60-Hz, 1710-r/min induction motor delivers full output power to the load connected to its shaft. The friction and windage loss of the motor is 820 W. Determine the (a) mechanical power developed and (b) the rotor copper loss.

Solution

1. Full-load shaft power = 15 × 746 = 11,190 W. Mechanical power developed (P_m) = shaft power + friction and windage loss.

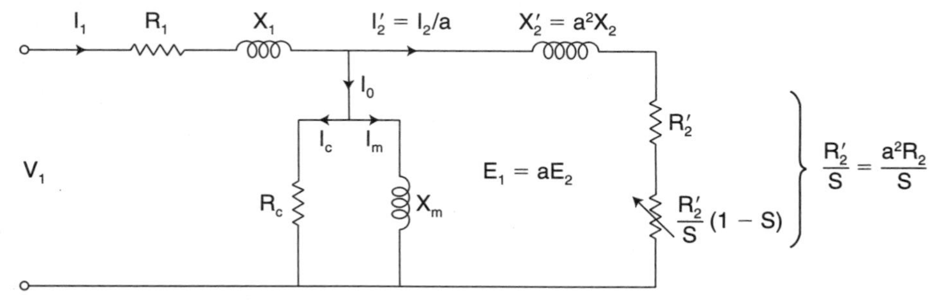

FIGURE 5.3 Complete equivalent circuit.

$$P_m = 11{,}190 + 820 = 12{,}010 \text{ W}$$

2. Synchronous speed $(n_s) = f/p$

$$n_s = 60/2 = 30 \times 60 = 1800 \text{ r/min}$$

Slip $(s) = (n_s - n)/n_s$

$$s = (1800 - 1710) \div 1800 = 0.05$$

Air gap power $(P_{ag}) = P_m \div (1 - s)$

$$P_{ag} = 12{,}010/(1 - 0.05) = 12{,}642.1 \text{ W}$$

Rotor copper loss $(P_{cu2}) = sP_{ag}$

$$P_{cu2} = 0.05 \times 12642.1 = 632.1 \text{ W}$$

DETERMINATION OF EQUIVALENT CIRCUIT PARAMETERS

The parameters of the equivalent circuit can be determined from the results of the no-load test, the locked-rotor test, and measurement of the dc resistance of the stator winding.

Example 2

Determine the equivalent circuit parameters of a 10-hp, 230-V, three-phase, four-pole wye-connected, double squirrel-cage induction motor (NEMA design C) from the following test data:

No-load test: $f = 60$ Hz; $V = 229.9$ V; $I = 6.36$ A; $P = 512$ W

Locked-rotor test at $f_1 = 15$ Hz: $V = 24$ V; $I = 24.06$ A; $P = 721$ W

Locked-rotor test at $f_1 = 60$ Hz: $V = 230$ V; $I = 110$ A; $P = 27{,}225$ W

Average dc stator winding resistance between stator terminals $= 0.42\ \Omega$

Solution

From the no-load test:

$$|Z_{NL}| = V_{NL} \div \sqrt{3} I_{NL} = 229.9/(1.732 \times 6.36) = 20.87 \text{ ohms}$$

$$R_{NL} = P_{NL}/3I_{NL}^2 = 512/(3 \times 6.36^2) = 4.22 \text{ ohms}$$

$$X_{NL} = \sqrt{|Z_{NL}|^2 - R_C^2} = \sqrt{20.87^2 - 4.22^2} = 20.44 \text{ ohms}$$

From the dc stator resistance measurement:

$$R_1 = 0.42 \div 2 = 0.21 \text{ ohms}$$

From the locked-rotor test:

$$Z_{IN} = V \div \sqrt{3}I = 24 \div (1.732 \times 24.06) = 0.576 \text{ ohms}$$

$$R_1 + R_2' = P \div 3I^2 = 721 \div (3 \times 24.06^2) = 0.415 \text{ ohms}$$

$$X_1 = X_2' = \sqrt{Z_{IN}^2 - (R_1 + R_2')^2} = \sqrt{0.576^2 - 0.415^2} = 0.404 \text{ ohms}$$

To determine leakage reactances at 60 Hz:

$$X_L = (60 \div 15)(X_1' + X_2') = 60/15 \times 0.404 = 1.616 \text{ ohms}$$

The IEEE test code for leakage reactance ratios are shown in Table 5.1. From Table 5.1, for a NEMA design C motor:

$$X_1 = 0.3\, X_L = 0.3 \times 1.616 = 0.485 \text{ ohms}$$

$$X_2 = 0.7\, X_L = 0.7 \times 1.616 = 1.131 \text{ ohms}$$

To find the magnetizing reactance:

$$X_m = X_{NL} - X_1 = 20.44 - 0.485 = 19.955 \text{ ohms}$$

$$P_g = P_{NL} - 3I^2R_1 = 27{,}225 - 3 \times 110^2 \times 0.21 = 19{,}602 \text{ W}$$

$$\omega_s = 2\pi f/p = (2\pi \times 60) \div 2 = 188.5 \text{ rad/s}$$

$$T_L = P_g \div \omega_s = 19{,}602 \div 188.5 = 104 \text{ N}\cdot\text{m}$$

To calculate rotor resistance and leakage reactance:

$$R_2' = (R_1 + R_2') - R_1 = 0.415 - 0.21 = 0.204 \text{ ohms}$$

From the locked-rotor test:

$$Z = \frac{V}{\sqrt{3}I} = 230 \div (1.732 \times 110) = 1.207 \text{ ohms}$$

$$R = \frac{P}{3I^2} = 27.225 \div (3 \times 110^2) = 0.75 \text{ ohms}$$

$$X = \sqrt{Z^2 - R^2} = \sqrt{1.207^2 - 0.75^2} = 0.95 \text{ ohms}$$

$$X_2' = X - X_1 = 0.95 - 0.485 = 0.465 \text{ ohms}$$

To calculate rotor resistance at start:

$$R_{fl} = R - R_1 = 0.75 - 0.21 = 0.54 \text{ ohms}$$

TABLE 5.1 IEEE Test Code for Empirical Ratios of Leakage Reactances

Reactance ratio	Squirrel-cage: Design class A	B	C	D	Wound rotor
X_1/X_L	0.5	0.4	0.3	0.5	0.5
X_2/X_L	0.5	0.6	0.7	0.5	0.5

$$R'_{2S} = R_{fl}\,[(X'_2 + X_m) \div X_m]^2 = 0.54 \times [(0.465 + 19.955) \div 19.955]^2 = 0.565\ \Omega$$

Note that both rotor resistance and leakage reactance at start have different values from those at running. Higher starting resistance and lower starting reactances are characteristics of a double-cage NEMA class C motor.

PERFORMANCE CHARACTERISTICS

The mechanical torque developed is given by

$$T_{mech} = 1/\omega_s \cdot \frac{3V_1^2}{[R_1 + (R'_2 \div s)]^2 + (X_1 + X'_2)^2} = R'_2 \div s \tag{5.1}$$

At low values of slip:

$$T_{mech} \approx 1/\omega_s \cdot \frac{3V_1^2}{R'_2} \cdot s \tag{5.2}$$

At high values of slip:

$$T_{mech} \approx 1/\omega_s \cdot \frac{3V_1^2}{(X_1 + X'_2)^2} = \frac{R'_2}{s} \tag{5.3}$$

The maximum (pullout, breakdown) torque developed by the motor is given by:

$$T_{max} = \frac{1}{2\omega_s} \cdot \frac{3V_1^2}{R_1 + \sqrt{[R_1^2 + (X_1 + X'_2)^2]}} \tag{5.4}$$

The maximum torque is independent of rotor resistance, but the value of rotor resistance determines the speed at which the maximum torque is developed. The losses in a three-phase induction motor are shown in the power flow diagram of Fig. 5.4.

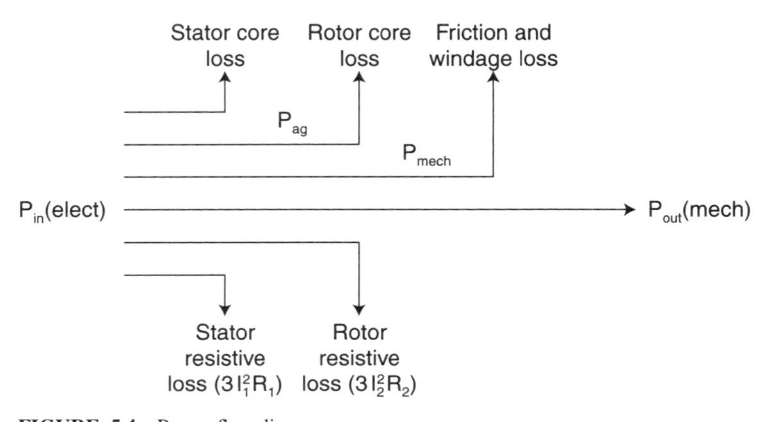

FIGURE 5.4 Power-flow diagram.

Example 3

A three-phase, 460-V, 1740-r/min, 60-Hz, four-pole, wound rotor induction motor has the following parameters per phase: $R_1 = 0.25\ \Omega$, $X_1 = X_2' = 0.5\ \Omega$, $R_2' = 0.2\ \Omega$, $X_m = 30\ \Omega$. The rotational losses are 1700 W. Find:

1. *a.* Starting current when started direct on full voltage
   b. Starting torque
2. *a.* Full-load slip
   b. Full-load current
   c. Ratio of starting current to full-load current
   d. Full-load power factor
   e. Full-load torque
   f. Ideal (internal) efficiency and motor efficiency at full load
3. *a.* Slip at which maximum torque is developed
   b. Maximum torque developed
4. How much external resistance per phase should be connected in the rotor circuit so that maximum torque occurs at start?

Solution

1. *a.* $V_1 = 460/\sqrt{3} = 265.6$ V. At start, $s = 1$. The input impedance

$$Z_1 = 0.25 + \frac{j30(0.2 + j0.5)}{0.2 + j30.5} = 1.08\,\underline{/66^\circ} \text{ ohms.}$$

The starting current

$$I_{st} = \frac{265.6}{1.08\underline{/66^\circ}} = 245.9\underline{/-66^\circ}\text{ A.}$$

b.

$$\omega_{syn} = \frac{1800}{60} \times 2\pi = 188.5 \text{ rad/s}$$

$$V_{Th} = \frac{265.6\,(j30.0)}{(0.25 + j30.5)} \approx 261.3 \text{ V}$$

$$Z_{Th} = \frac{j30(0.25 + j0.5}{0.25 + j30.5} = 0.55\,\underline{/63.9^\circ}$$

$$= 0.24 + j0.49$$

$$R_{Th} = 0.24\ \Omega$$

$$X_{Th} = 0.49 \simeq X_1$$

$$T_{st} = \frac{P_{ag}}{\omega_{syn}} = \frac{I_2'2R_2'/s}{\omega_{syn}}$$

$$= \frac{3}{188.5} \times \frac{261.3^2}{(0.24 + 0.2)^2 + (0.49 + 0.5)^2} \times \frac{0.2}{1}$$

$$= \frac{3}{188.5} \times (241.2)^2 \times \frac{0.2}{1}$$

$$= 185.2 \text{ N} \cdot \text{m}$$

2. ***a.***

$$s = \frac{1800 - 1740}{1800} = 0.0333$$

b.

$$\frac{R_2'}{s} = \frac{0.2}{0.0333} = 6.01 \ \Omega$$

$$Z_1 = (0.25 + j0.5) + \frac{(j30)(6.01 + j0.5)}{6.01 + j30.5}$$

$$= 0.25 + j0.5 + 5.598 + j1.596$$

$$= 6.2123 \underline{/19.7^\circ} \ \Omega$$

$$I_{FL} = \frac{265.6}{6.2123 \underline{/19.7}}$$

$$= 42.754 \underline{/-19.7^\circ} \text{ A}$$

c.

$$\frac{I_{\text{st}}}{I_{\text{FL}}} = \frac{245.9}{42.754} = 5.75$$

d.

$$\text{pf} = \cos(19.7^\circ) = 0.94 \text{ (lagging)}$$

e.

$$T = \frac{3}{188.5} \times \frac{(261.3)^2}{(0.24 + 6.01)^2 + (0.49 + 0.5)^2} \times 6.01$$

$$= \frac{3}{188.5} \times 41.29^2 \times 6.01$$

$$= 163.11 \text{ N} \cdot \text{m}$$

f. Air gap power:

$$P_{\text{ag}} = T\omega_{\text{syn}} = 163.11 \times 188.5 = 30{,}746.2 \text{ W}$$

Rotor copper loss:

$$P_2 = sP_{\text{ag}} = 0.0333 \times 30{,}746.2 = 1023.9 \text{ W}$$

$$P_{\text{mech}} = (1 - 0.0333)30{,}746.2 = 29{,}722.3 \text{ W}$$

$$P_{\text{out}} = P_{\text{mech}} - P_{\text{rot}} = 29{,}722.3 - 1700 = 28{,}022.3 \text{ W}$$

$$P_{\text{in}} = 3V_1I_1\cos\theta_1$$

$$= 3 \times 265.6 \times 42.754 \times 0.94 = 32{,}022.4 \text{ W}$$

$$\text{Eff}_{\text{motor}} = \frac{28{,}022.3}{32{,}022.4} \times 100 = 87.5\%$$

$$\text{Eff}_{\text{internal}} = (1 - s) = 1 - 0.0333 = 0.967 \rightarrow 96.7\%$$

3. ***a.***

$$s_{T_{\max}} = \frac{0.2}{[0.24^2 + (0.49 + 0.5)^2]^{1/2}} = \frac{0.2}{1.0187} = 0.1963$$

b.

$$T_{\max} = \frac{3}{2 \times 188.5}\left[\frac{261.3^2}{0.24 + [0.24^2 + (0.49 + 0.5)^2]^{1/2}}\right]$$

$$= 431.68 \text{ N} \cdot \text{m}$$

$$\frac{T_{\max}}{T_{\text{FL}}} = \frac{431.68}{163.11} = 2.65$$

4.

$$s_{T_{\max}} = 1 = \frac{R_2' + R_{\text{ext}}'}{[0.24^2 + (0.49 + 0.5)^2]^{1/2}} = \frac{R_2' + R_{\text{ext}}'}{1.0186}$$

$$R_{\text{ext}}' = 1.0186 - 0.2 = 0.8186\ \Omega/\text{phase}$$

PLUGGING

By interchanging two of the leads of a three-phase, three-wire distribution system, the phase sequence of the transmitted voltages is reversed. As a result, a backward-rotating field is generated. This field, of course, opposes the forward field that existed before the phase sequence of the applied voltages was reversed. On phase reversal, the motor will go through zero speed and then start in the reverse direction. When the motor approaches zero speed, however, the supply voltage is removed and the motor comes to a complete stop. The time that it takes for a motor to go from a speed ω_1 to a complete stop following a phase reversal of the applied voltages is given by

$$t = J\frac{\omega_s^2}{V^2 R_2}\left[R_2^2 \ln(2 - s) + 2R_1R_2(1 - s) + \frac{1}{2}(R_1^2 + X^2)(3 - 4s + s^2)\right] \qquad (5.5)$$

The value of the rotor resistance, as seen from the stator, that will result in minimum time while plugging a motor to stop is

$$R_2 = \sqrt{\frac{(R_1^2 + X^2)(3 - 4s + s^2)}{2 \ln(2 - s)}} \text{ ohms/phase} \qquad (5.6)$$

In driving a motor from rest up to synchronous speed, the energy dissipated in the rotor winding (W_{dr}) is given by

$$W_{dr} = J\frac{\omega_s^2}{2}\ \textit{Joules} \tag{5.7}$$

When a motor is brought to rest from synchronous speed by phase reversal, the energy dissipated in the rotor windings, $(W_{dr})_P$, is given by

$$(W_{dr})_P = 3\left(J\frac{\omega_s^2}{2}\right) \tag{5.8}$$

The energy dissipated in the rotor windings when the motor is subjected to a complete speed reversal $(W_{dr})_r$ is therefore given by

$$(W_{dr})_r = 4\left(\frac{1}{2}J\omega_s^2\right) \tag{5.9}$$

A three-phase induction motor will be overheated if the speed reversals are numerous enough to generate more heat than the machine is designed to dissipate. The nameplate efficiency of the machine specifically indicates the amount of power losses, and thus the maximum allowable rate of heat dissipation, under continuous operation. For example, the maximum rate of heat a 10-kW motor with an efficiency of 95 percent can safely dissipate is

$$\text{Losses} = P_{\text{out}}\left(\frac{1}{\eta} - 1\right) = 100\left(\frac{1}{0.94} - 1\right) = 6.38\ \text{kW}$$

Example 4

The per-phase equivalent circuit of a 440-V, three-phase, 60-Hz, 75-kW, 855-r/min, eight-pole induction motor is shown in Fig. 5.5. The motor drives a pure-inertia load of 6.5 kgm^2. Neglecting the effect of the magnetizing current and stator resistance, determine:

1. The time it takes to plug the motor to a complete stop and the associated energy dissipation in the rotor windings.
2. The time it takes to reverse the speed of the motor by plugging and associated energy dissipation in the rotor windings.

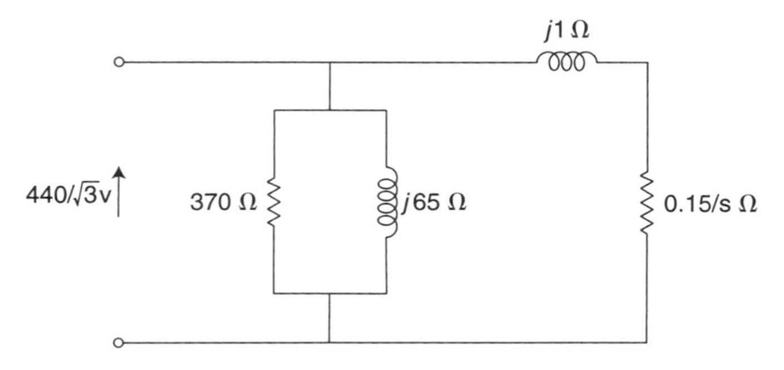

FIGURE 5.5 Equivalent circuit.

Solution

Synchronous speed $(n_s) = f/p$

$$n_s = 60/4 = 15 \times 60 = 900 \text{ r/min}$$

$$\text{Full-load operating slip } (s) = \frac{n_s - n}{n_s}$$

$$s = \frac{900 - 855}{900} = 0.05 \text{ p.u.}$$

1. Time to stop the motor by plugging is found from Eq. 5.5:

$$t = 6.5 \cdot \frac{[2\pi \cdot 900]^2}{60} \cdot \frac{1}{440^2 \cdot 0.15} \left[0.15^2 \ln (2 - 0.05) + \frac{1}{2}(1.0)^2 (3 - (4 \times j0.05 + 0.05^2)\right] = 2.82 \text{ s}$$

The energy dissipated in the rotor windings is calculated from Eq. 5.8:

$$(W_{\text{dr}})_p = 4\left[\frac{1}{2} \cdot 6.5 \left(\frac{2\pi \cdot 900}{60}\right)^2\right] = 86.61 \text{ kJ}$$

2. Time to reach full-load speed in the reverse direction:

$$t_2 = 6.5 \left(\frac{2\pi \cdot 900}{60}\right)^2 \cdot \frac{1}{440^2 \cdot 0.15}\left[(1 - 0.05)^2 \frac{1.0}{2} + 0.15^2 \ln\left(\frac{1}{0.05}\right)\right] = 1.13 \text{ s}$$

Thus, time taken to reverse motor speed by plugging is

$$t = t_1 + t_2 = 2.82 + 1.13 = 3.94 \text{ s}$$

The energy dissipated in the rotor windings is calculated from Eq. 5.9:

$$(W_{\text{dr}})_r = 4\left[\frac{1}{2} \cdot 6.5 \left(\frac{2\pi \cdot 900}{60}\right)^2\right] = 115.47 \text{ kJ}$$

BRAKING

To stop an induction motor, one can always disconnect it from the lines. The stopping time will vary depending upon the inertia, friction, and windage. Often the motor will be required to stop quickly and deceleration will then have to be controlled as required, for example, in a crane or hoist.

The function of braking is to provide means to dissipate the stored kinetic energy of the combined motor-system inertia during the period of deceleration from a higher to a lower, or zero, speed. The dissipation of kinetic energy may be external to the windings (external braking) or within them (internal braking). Sometimes a combination of internal and external braking may be employed to optimize braking performance.

External braking uses a form of mechanical brake that is coupled to the motor. It can be used for all types of motors and requires special mechanical coupling. Examples of external braking include friction (a shoe- or disk-type brake that can be activated mechanically, electrically, hydraulically, or pneumatically), eddy-current (a direct-connected magnetically coupled unit), hydraulic (coupled hydraulic pump), and magnetic-particle (a direct-connected magnetic-particle coupling) braking.

Methods for producing internal braking can be divided into two groups: countertorque (plugging) and generating (dynamic and regenerative braking). In the countertorque group, torque is developed to rotate the rotor in the opposite direction to that which existed before braking. In the generating group, torque is developed from the rotor speed. Dynamic torque is developed by motor speeds that are less than the synchronous speed, and energy is dissipated within the motor or in a connected load. Developed torque is zero at zero rotor speed.

Regenerative braking torque is developed by motor speeds that are higher than the synchronous speed. The motor is always connected to the power system and the generated power is returned to the power system.

For large motors, forced ventilation may be required during braking.

Example 5

In a three-phase induction motor, $|I_1| = 116.9$ A, $\omega_s = 188.5$ rad/s, $s = 0.0255$, $I_{\mathrm{fl}} = 425.6$ N · m, $Z_2 = 0.058 + j0.271\ \Omega$, and the saturated magnetizing reactance $X_m = 2.6\ \Omega$.

Calculate the dynamic braking torque T_{db} when the motor is running at rated load and a dc voltage is used to produce a rated equivalent ac in the stator (Fig. 5.6). Also, determine the maximum braking torque $T_{\mathrm{db(max)}}$ and the corresponding slip $s_{\max}$.

Solution

The dynamic braking torque (T_{db}) is calculated from

$$\begin{aligned} T_{\mathrm{db}} &= \frac{-3}{\omega_s} \cdot \frac{|I_1|^2 \cdot X_m^2}{[R_2/(1-s)]^2 + (X_m + X_2)^2} \cdot \frac{R_2}{(1-s)} \\ &= \frac{-3}{188.5} \cdot \frac{116.9^2 \cdot 2.6^2}{[0.058/(1-0.0255)]^2 + (2.6-0.271)^2} \cdot \frac{0.058}{(1\ 0.0255)} \\ &= -10.61 \mathrm{N} \cdot \mathrm{m}. \end{aligned}$$

Thus, the dynamic braking torque is

$$\left(\frac{10.61}{425.6}\right) \cdot 100 \text{ percent} = 2.5 \text{percent of full-load torque, that is, } T_{\mathrm{db}} = 2.5 \text{ percent of } T_{\mathrm{fl}}.$$

The maximum dynamic braking torque is calculated from

$$T_{\mathrm{db(max)}} = \frac{-3}{2\omega_s} \cdot \frac{|I_1|^2 \cdot X_m^2}{(X_m + X_2)} = \frac{-3}{2 \times 188.5} \cdot \frac{116.9^2 \cdot 2.6^2}{(2.6 + 0.271)} = -256.05\ \mathrm{N} \cdot \mathrm{m}$$

that is,

$$T_{\mathrm{db\,(max)}} = \frac{256.05}{425.6} \cdot (100 \text{ percent}) = (60.2 \text{ percent}) \cdot T_{\mathrm{fl}}.$$

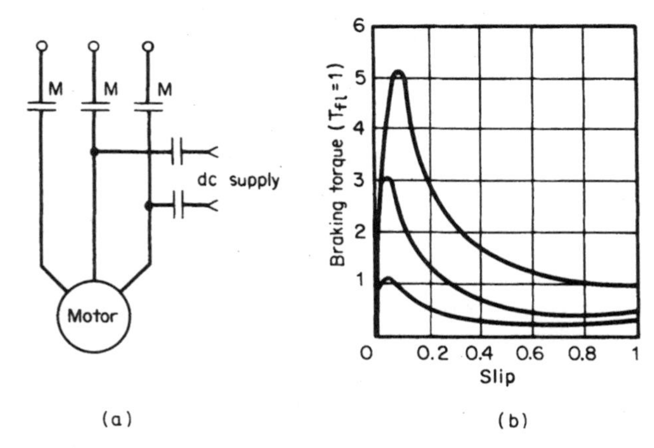

FIGURE 5.6 Dynamic braking using dc. (*a*) Circuit. (*b*) Torque-speed characteristics.

The corresponding slip

$$s_{\max} = 1 - \frac{R_2}{(X_m + X_2)} = 1 - \frac{0.058}{(2.6 + 0.271)} = 0.98$$

In the Examples, the braking torque was obtained by connecting a dc source to one phase of the motor after it is disconnected from the lines (Fig. 5.6*a*). With solid-state electronics, the problem of a separate dc source has been solved economically. The braking torque is low at the higher initial speed and increases to a high-peak value as the motor decelerates; however, it drops rapidly to zero speed (Fig. 5.6*b*). The losses are approximately the same as that for a single start. Higher braking torque can be obtained by using higher values of dc voltage and by inserting external resistance in the rotor circuit for a wound-rotor motor. Approximately 150 percent rated current (dc) is required to produce an average braking torque of 100 percent starting torque.

A form of dc braking using a capacitor-resistor-rectifier circuit where a variable dc voltage is applied as the capacitor discharges through the motor windings instead of a fixed dc voltage, as in dc braking, is illustrated in Fig. 5.7. The dc voltage and motor speed decrease together to provide a more nearly constant braking torque. The energy stored in the capacitor is all that is required for the braking power. This method, however, requires costly, large capacitors.

An induction motor once disconnected from the lines can produce braking torque by generator action if a bank of capacitors is connected to the motor terminals (Fig. 5.8). The braking torque may be increased by including loading resistors. If capacitors sized for power-factor correction are used, the braking torque will be small. To produce an initial peak torque of twice the rated torque, capacitance equal to about 3 times the no-load magnetizing apparent power in kVA is required. (This may produce a high transient voltage that could damage the winding insulation.) At a fairly high speed, the braking torque reduces to zero.

In ac dynamic braking (Fig. 5.9), "single-phasing" a three-phase induction motor will not stop it, but will generate a low braking torque with zero torque at zero speed. It is a very simple and comparatively inexpensive method. Losses within the motor winding may require a larger motor size to dissipate the heat. Braking torque in the case of a

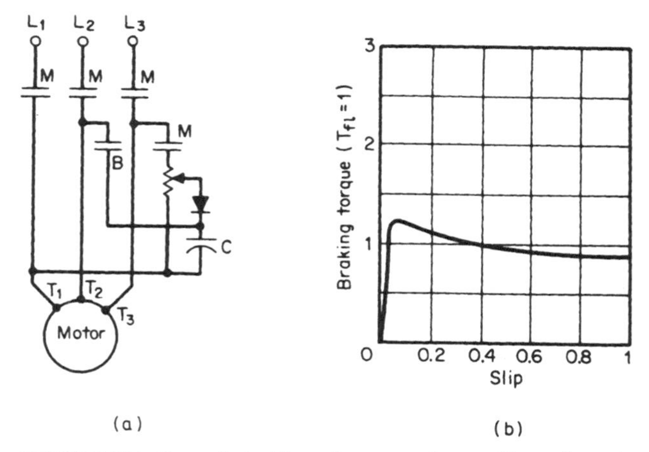

FIGURE 5.7 Dynamic braking using a capacitor-rectifier-resistor circuit. (*a*) Circuit. (*b*) Torque-speed characteristics.

wound rotor may be varied by the insertion of external resistance in the rotor circuit. Sometimes, a separate braking winding is provided.

In regenerative braking, torque is produced by running the motor as an induction generator and the braking power is returned to the lines. For example, regenerative braking can be applied to a two- or four-pole squirrel-cage motor initially running at, say, 1760 r/min by changing the number of poles from two to four. The synchronous speed now is 900 r/min and the machine runs as a generator. Regenerative braking is mainly used for squirrel-cage motors. The method becomes too involved when poles are changed for a wound-rotor motor.

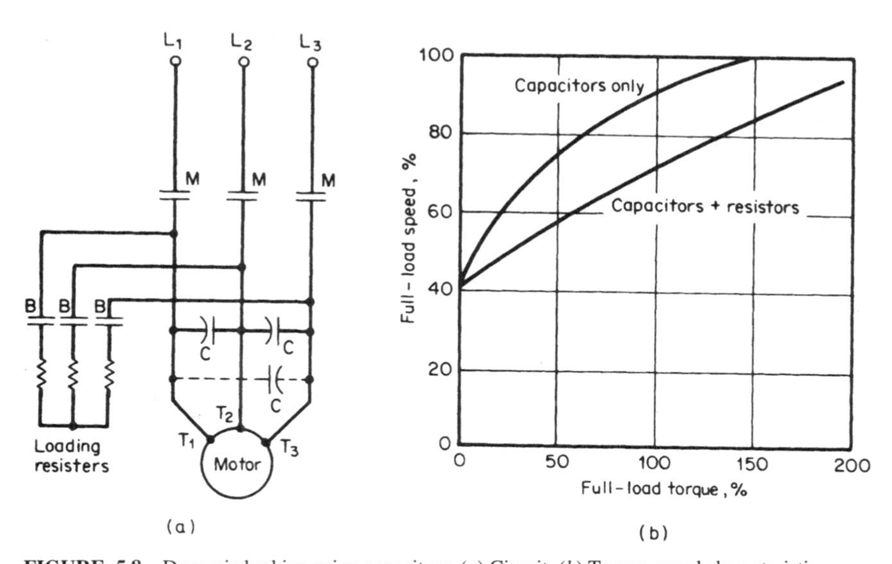

FIGURE 5.8 Dynamic braking using capacitors. (*a*) Circuit. (*b*) Torque-speed characteristics.

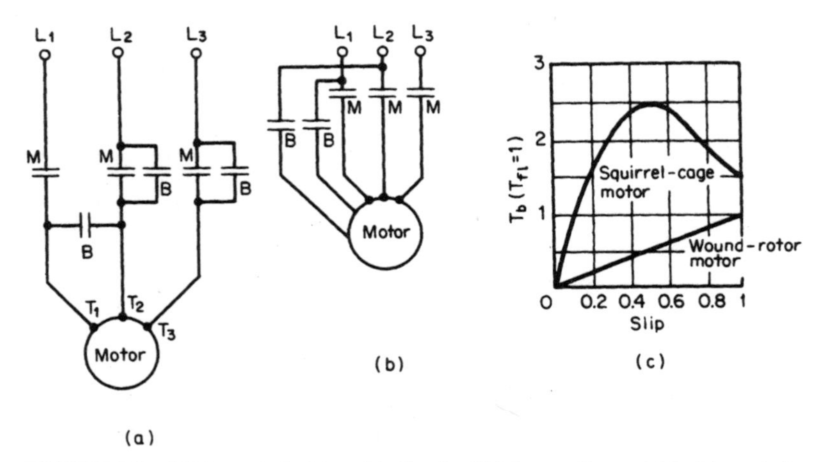

FIGURE 5.9 AC dynamic braking. (*a*) Circuit. (*b*) Motor with special braking winding. (*c*) Torque-speed characteristics.

AUTOTRANSFORMER STARTING

Autotransformer (or compensator) starting is the most commonly used reduced-voltage starter. A voltage equal to the tap voltage is maintained at the stator terminals during acceleration from the start to the transition. Usually, three taps, 0.5, 0.65, and 0.8, are provided.

Because of lower cost, two autotransformers in open-delta (vee connection) are employed. The result is a higher current in the third phase as well as a momentary open circuit during the transition. A very high transient current of short duration occurs during this momentary open circuit. The use of three autotransformers allows a closed-circuit transition, thereby eliminating the open-circuit transient and the unbalanced line current; this, however, adds to the cost of an additional transformer.

Example 6

Referring to Fig. 5.10, determine the line current I_{LL}, motor current I_{LA}, and starting torque T_{LA} for autotransformer starting at taps $\alpha = 0.5$, 0.65, and 0.8. At full voltage, the motor has I_L, the locked-rotor current in percent full-load current, equal to 600 percent and T_L, the locked-rotor torque in percent full-load torque, equal to 130 percent.

Select a transformer tap for the motor if the line current is not to exceed 300 percent of full-load current I_{fl} and starting torque is not to be less than 35 percent of full-load torque T_{fl} at start.

Solution

If three transformers are used, $I_{\mathrm{LL}} = \propto^2 I_L$ for $\propto = 0.5$, $I_{\mathrm{LL}} = 0.5^2 \times 600 = 150$ percent. If two transformers are used, $I_{\mathrm{LL}} = \propto^2 I_L + (15 \text{ percent}) = 165$ percent.

$$\text{Motor current } (I_{\mathrm{LA}}) = \propto I_L = 0.5 \times 600 = 300 \text{ percent}$$

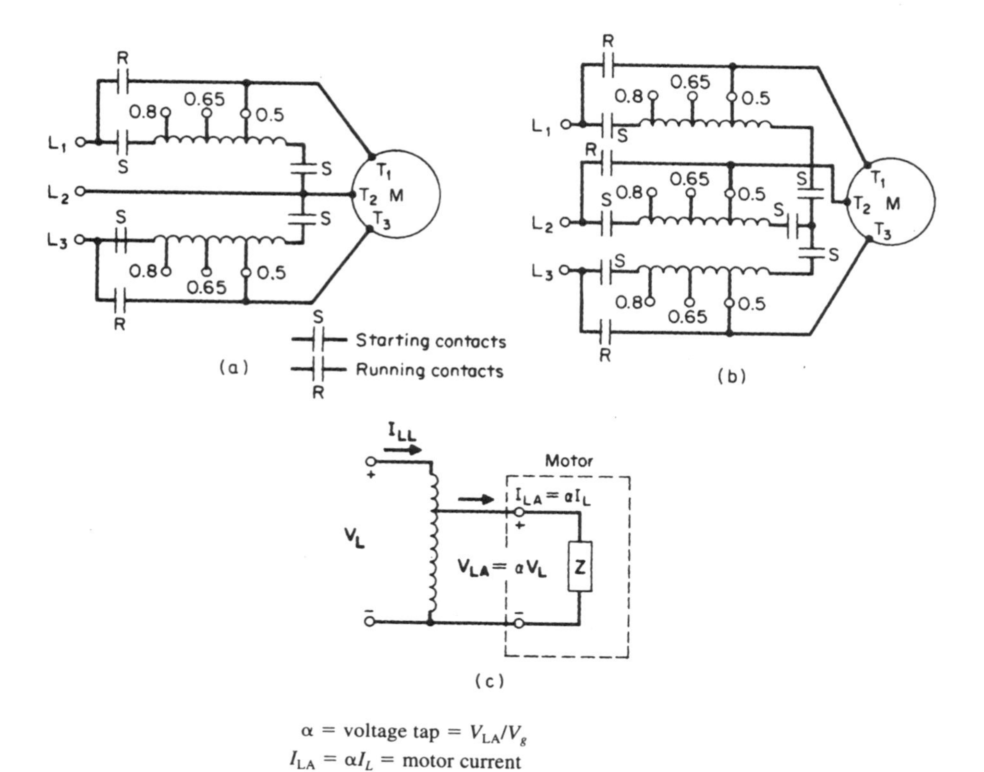

α = voltage tap = V_{LA}/V_g
$I_{\text{LA}} = \alpha I_L$ = motor current
$I_{\text{LL}} = \alpha I_{\text{LA}} = \alpha^2 I_L$ = line current for (*b*)
$= \alpha^2 I_1 + 0.15\, I_{\text{fl}}$ for (*a*)
$T_A = \alpha^2 T$ = motor torque with autotransformer

FIGURE 5.10 Autotransformer reduced-voltage starting. (*a*) Open-circuit transition. (*b*) Closed-circuit transition. (*c*) Equivalent circuit.

$$\text{Starting torque } (T_{\text{LA}}) = \alpha^2 I_L = 0.5^2 \times 130 = 32.5 \text{ percent}$$

Values for other taps are obtained in a similar way. The results are shown in Table 5.2.

To meet the current constraints that $I_{\text{LA}} \leq 300$ percent of I_{fl}, the tap may be either 0.5 or 0.65. To satisfy the torque constraints, $T_{\text{LA}} \geq 35$ percent of T_{fl}, the tap may be 0.65 or 0.8. Select tap 0.65.

TABLE 5.2 Values of Line Current

No.	Tap, α	Line current, I_{LL}: Two transformers		Line current, I_{LL}: Three transformers		Motor current, I_{LA}		Starting torque, T_{LA}	
1	0.5	27.5	165	25	150	50	300	25	32.5
2	0.65	44.8	269	42.3	254	65	390	42.3	55
3	0.8	66.5	399	64	384	80	480	64	83.2
		$\sigma_0 I_L$	$\sigma_0 I_{\text{fl}}$	$\sigma_0 I_L$	$\sigma_0 I_{\text{fl}}$	$\sigma_0 I_L$	$\sigma_0 I_{\text{fl}}$	$\sigma_0 T_L$	$\sigma_0 T_{\text{fl}}$

TABLE 5.3 Comparison of Resistance and Reactance Starting

No.	Feature	Resistance type	Reactor type
1	Starting pf	Higher	Lower
2	Developed torque around 75 to 85 percent n_s	Lower (15 to 20 percent T_{fl})	Higher (15 to 20 percent T_{fl})
3	Heat loss (I^2R)	Very high	Very small
4	Size	Larger	Smaller
5	Cost	Lower	Higher

RESISTANCE STARTING

During starting, the effective voltage at the motor terminals can be reduced through the use of a series resistance or reactance. With an increase in motor speed, the motor impedance increases and the starter impedance remains constant. This causes a greater voltage to appear at the motor terminals as the motor speed increases, resulting in higher developed torque. It provides "closed-circuit transition" and balanced line currents. These starters are designed for a single fixed-tap value of either 0.5, 0.65, or 0.8 of the line voltage.

Because of its flexibility and lower cost, the most widely used form of impedance starting is resistance starting. It could be designed to form a stepless starter through the aid of an automatically controlled variable resistance. For larger motors (greater than 200 hp) or for high voltages (greater than 2300 V), reactance starting is preferred. The features of both starting methods are summarized in Table 5.3.

Example 7

Referring to Fig. 5.11*a*, size the resistor R_S of a two-step resistance starter having taps T_1, T_2, and T_3 at 0.5, 0.65, and 0.8, respectively. Also, calculate the corresponding line current, pf, and torque. For the motor, $Z_L = 0.165 + j0.283\ \Omega$, $I_L = 539$ percent of I_{fl}, and $T_L = 128$ percent of T_{fl}.

Solution

In Table 5.4, the symbols V_{LR}, I_{LR}, and T_{LR} represent the motor locked-rotor quantities with the starter in. If I_{LR} is less than 400 percent, the tap may be 0.5 or 0.65 (see Table 5.4). If T_{LR} is greater than 35 percent, the tap may be 0.65 or 0.8. Hence, the selected tap is 0.65.

To determine R_S (Fig. 5.11*b*) for $\alpha = 0.65$, from Table 5.4 $R_S = 0.439\ \Omega$.

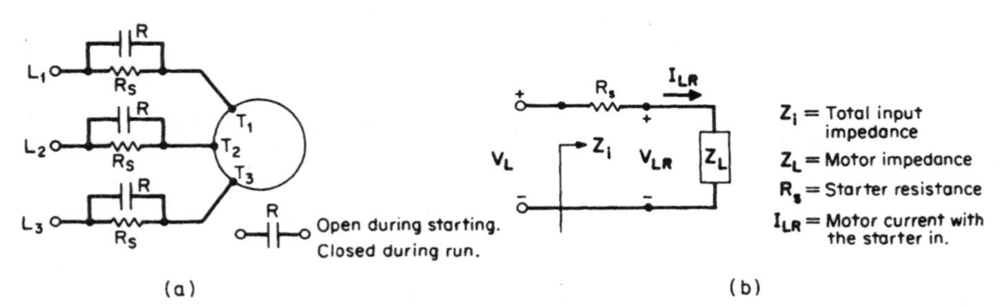

FIGURE 5.11 Resistance starting. (*a*) Connection diagram. (*b*) Equivalent circuit.

TABLE 5.4 Motor Locked-Rotor Resistance Quantities with the Starter In

No.	Equation		Tap α value 0.50	0.65	0.80
1	$I_{LR} = \dfrac{V_{LR}}{Z_L} = \dfrac{\alpha V_L}{Z_L} = \alpha I_L$	in percent I_{fl}	269.5	350.4	431.2
		percent I_L	50	65	80
2	$\lvert\mathbf{Z}_{IL}\rvert = \lvert R_L + jX_L\rvert = \dfrac{V_L}{I_{LR}} = \dfrac{V_L}{\alpha I_L}$				
	$= \left(\dfrac{Z}{\alpha}\right)\,\Omega$		0.844	0.649	0.528
3*	$R_{IL} = R_S + R_L = \sqrt{Z_{IL}^2 - X_L^2}\ \Omega$		0.810	0.604	0.471
4	$R_S = (R_{IL} - R_L)\ \Omega$		0.645	0.439	0.306
5	$\text{pf} = \cos(-\theta_{IL}) = R_{IL}/\lvert\mathbf{Z}_{IL}\rvert$		0.96	0.931	0.892
6	$\text{pf angle } \theta_{IL} = -\cos^{-1}(R_{IL}/\lvert\mathbf{Z}_{IL}\rvert)$		$-16.3°$	$-21.4°$	$-26.9°$
7	$T_{LR} = \alpha^2 T_L$	in percent T_{fl}	32	54.1	81.9
		percent T_L	25	42.3	64

*R_S = starter resistance (Fig. 5.9*b*).

REACTANCE STARTING

Example 8

Example 7 is now repeated using the reactance starter of Fig. 5.12.

Solution

In Table 5.5, the symbols V_{LX}, I_{LX}, and T_{LX} represent the motor locked-rotor quantities with the starter in. Since the starting current and torque are the same in both reactance and resistance starting, select the same tap of 0.65.

From the table, for $\alpha = 0.65$, $X_r = 0.390\ \Omega$. Hence, $L_S = X_r/(2\pi f_1) = 0.39/(2\pi \times 60) = 103$ mH.

SERIES-PARALLEL STARTING

In this method of starting (Fig. 5.13) dual-voltage windings are connected for higher voltage on a lower supply voltage during starting. Then, the connection is switched over to the normal lower-voltage connection.

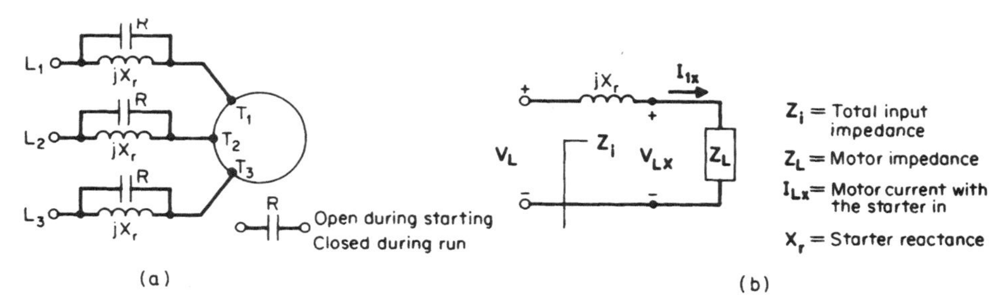

FIGURE 5.12 Reactance starting. (*a*) Connection diagram. (*b*) Equivalent circuit.

TABLE 5.5 Motor Locked-Rotor Resistance Quantities with the Starter In

No.	Equation		Tap α value 0.50	0.65	0.80
1	$I_{LX} = \frac{V_{LX}}{Z_L} = \frac{\alpha V_L}{Z_L} = \alpha I_L$	in percent I_{fl}	269.5	350.4	431.2
		percent I_L	50	65	80
2	$\lvert \mathbf{Z}_{IL} \rvert = \lvert R_L + jX_{IL} \rvert = V_L/I_{LX}$ $= V_L/\alpha I_L = (Z/\alpha)\ \Omega$		0.844	0.649	0.528
3	$X_{IL} = X_r + X_L = \sqrt{\lvert \mathbf{Z}_{IL} \rvert - \mathbf{R}_L^2}\ \Omega$		0.828	0.628	0.502
4	$X_r = (X_{IL} - X_L)\ \Omega$		0.590	0.390	0.264
5	$\text{pf} = \cos(-\theta_{IL}) = R_L/\lvert \mathbf{Z}_{IL} \rvert$		0.195	0.254	0.313
6	pf angle: $\theta_{IL} = -\cos^{-1}(R_L/Z_{IL})$		−78.8°	−75.3°	−71.8°
7	$T_{LX} = \alpha^2 T_L$	in percent T_{fl}	32	54.1	81.9
		percent T_L	25	42.3	64

Although it is occasionally employed for two-step starting, series-parallel starting is primarily used as the first step in a multistep increment starter because of its very low starting torque. Any 230/460-V motor can be connected for 230-V series-parallel starting.

Example 9

Calculate the starting line current I_{LS}, and starting torque, T_{LS}, of a 230/460-V motor to be connected for 460-V service with 230 V applied.

Solution

The starting line current is $I_{LS} = (230/460)\ I_L = 0.5\ I_L$. The starting torque is $T_{LS} = (230/460)^2\ T_L = 0.25\ T_L$.

MULTISTEP STARTING

Wound-rotor motors are universally started on full voltage. Reduction in the starting current is achieved through the addition of several steps of balanced resistors in the rotor circuit. As the motor accelerates, the rotor resistances can be lowered steadily in steps until

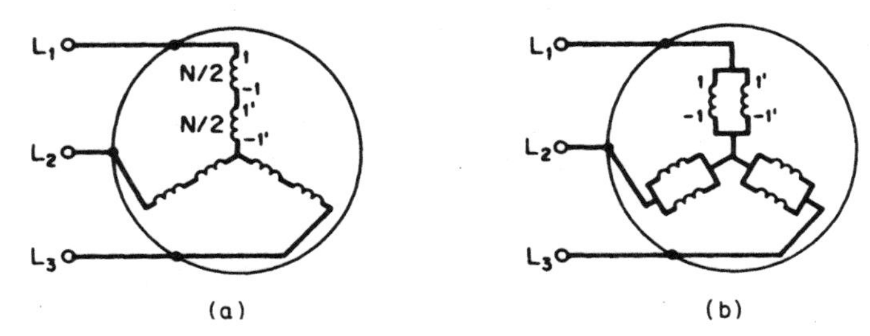

FIGURE 5.13 Series-parallel starting. (*a*) Series connection for start. (*b*) Parallel connection for run.

they are completely shorted out at the rated speed. Thus, the developed torque may be shaped to meet the requirements of the load within the limits on the current set by the power company. Reactors can be employed instead of resistors to reduce the starting current, though they are seldom used.

Example 10

Calculate the resistance of a five-step starter for a wound-rotor motor (Fig. 5.14*a*) with maximum starting torque; the inrushes shall be equal. Also, find the minimum motor current I_{min}. Motor data: $R_2/s_{maxT} = 1.192\ \Omega$, $s_{maxT} = 0.0795$, and $I_{maxT} = 760$ A.

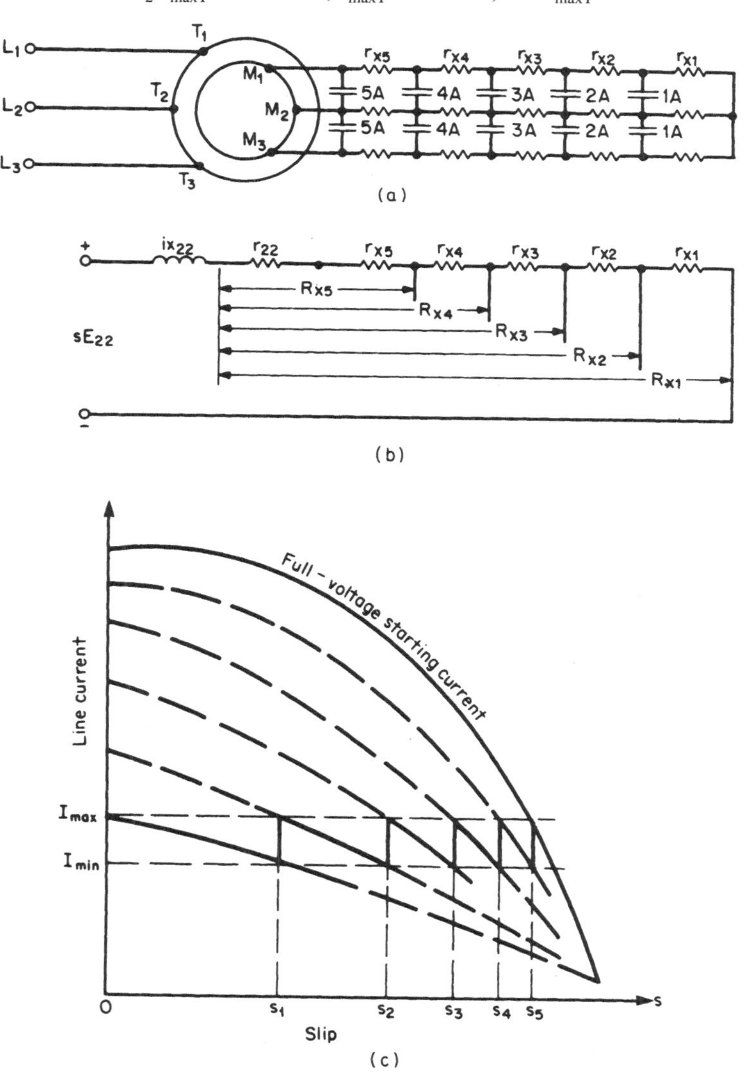

FIGURE 5.14 Five-step wound-rotor resistance starter. (*a*) Connection diagram. (*b*) Rotor equivalent circuit. (*c*) Acceleration curve.

Solution

From Fig. 5.14*b*, $R_{X1} = R_2/s_{maxT}$ (for $I_{max} = I_{maxT}$) $= 1.192$ as given. Slip $(s_1) = \sqrt[n]{s_{maxT}}$

$$s_1 = \sqrt[5]{0.0795} = 0.6027$$

$$R_{x2} = s_1 R_{x1} = 0.6027 \times 1.192 = 0.718 \text{ ohms}$$

Similarly,

$$R_{x3} = s_1 R_{x2} = 0.6027 \times 0.718 = 0.433 \text{ ohms}$$

$$R_{x4} = s_1 R_{x3} = 0.6027 \times 0.433 = 0.261 \text{ ohms}$$

And,

$$R_{x5} = s_1 R_{x4} = 0.6027 \times 0.261 = 0.157 \text{ ohms}$$

Also,

$$r_{x1} = R_{x1} - R_{x2} = 0.474 \text{ ohms}$$

$$r_{x2} = R_{x2} - R_{x3} = 0.285 \text{ ohms}$$

$$r_{x3} = R_{x3} - R_{x4} = 0.172 \text{ ohms}$$

$$r_{x4} = R_{x4} - R_{x5} = 0.104 \text{ ohms}$$

and,

$$r_{x5} = R_{x5} - R_2 = [0.157 - (1.192 \times 0.0795)] = 0.062 \text{ ohms.}$$

The acceleration curve is shown in Fig. 5.14*c*. The minimum motor current $I_{min} = I_{maxT}\, s_1 = 760 \times 0.6027 = 458$ A.

SPEED CONTROL

An induction motor is essentially a constant-speed motor when connected to a constant-voltage and constant-frequency power supply. The operating speed is very close to the synchronous speed. If the load torque increases, the speed drops by a very small amount. It is therefore suitable for use in substantially constant-speed drive systems. Many industrial applications, however, require several speeds or a continuously adjustable range of speeds. Traditionally, dc motors have been used in such adjustable-speed drive systems. However, dc motors are expensive, require frequent maintenance of commutators and brushes, and are prohibitive in hazardous atmospheres. Squirrel-cage induction motors, on the other hand, are cheap, rugged, have no commutators, and are suitable for high-speed applications. The availability of solid-state controllers, although more complex than those used for dc motors, has made it possible to use induction motors in variable-speed drive systems.

LINE-VOLTAGE CONTROL

The electromagnetic torque developed by an induction motor is proportional to the square of the impressed stator voltage. Line-voltage control is used mainly for small cage-rotor motors driving fan-type loads that are a function of speed. Disadvantages of this method include:

1. Torque is reduced with a reduction in stator voltage.
2. Range of speed control is limited.
3. Operation at higher than rated voltage is restricted by magnetic saturation.

Example 11

Calculate the speed and load torque as a percentage of the rated torque for a 10-hp, three-phase, four-pole squirrel-cage motor driving a fan when the supply voltage is reduced to one-half the normal value. The fan torque varies as the square of its speed. The slip at rated load is 4 percent, and the slip corresponding to maximum torque is 14 percent.

Solution

At two different operating speeds, n_1 and n_2, corresponding to two different supply voltages, V_1 and 0.5 V_1.

$$\frac{T_{m1}}{T_{m2}} = \frac{s_1}{s_2 \div 4}$$

and

$$\frac{T_{F1}}{T_{F2}} = \frac{n_1^2}{n_2^2} = \frac{(1 - s_1)^2}{(1 - s_2)^2}$$

Equating the motor and load torques, we get

$$\frac{s_1}{s_2 \div 4} = \frac{(1 - s_1)^2}{(1 - s_2)^2}$$

Therefore,

$$\frac{(1 - s_2)^2}{s_2} = \frac{(1 - s_1)^2}{4s_1} = \frac{(1 - 0.04)^2}{(4 \times 0.04)} = 5.76$$

and

$$s_2 = 0.131$$

Because s_2 is less than $s_{\text{maxT}} = 0.14$, the operation is in the stable region of the torque-speed characteristic (Fig. 5.15). Operating speed is:

$$n_2 = n_3(1 - s_2) = 120 \times 60 \times \frac{1 - 0.131}{4} = 1564 \text{ rpm}$$

Load torque is:

$$T_{F2} = T_{M2} = T_{r1}\left(\frac{s_2}{4s_1}\right) = T_{r1}\left(\frac{0.131}{0.16}\right) = 0.819\ T_{r1}$$

that is,

$$T_{F2} = 81.9 \text{ percent } T_{\text{rated}}$$

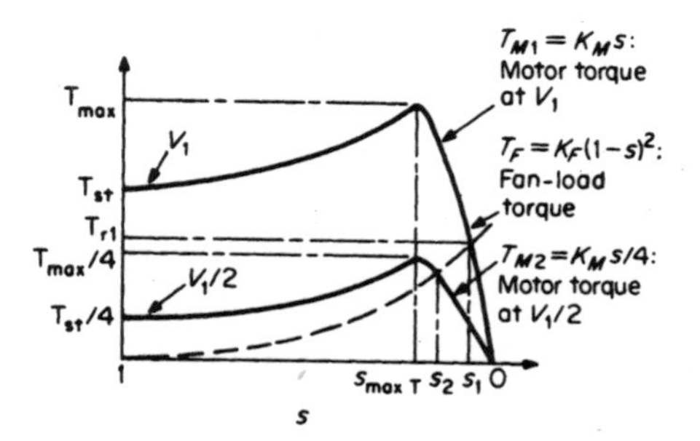

FIGURE 5.15 Using line-voltage control to vary speed of an induction motor.

FREQUENCY CONTROL

The synchronous speed of an induction motor is directly related to the frequency of the power to the motor. Hence, continuous speed control over a wide range for both squirrel-cage and wound-rotor motors may be realized by controlling the frequency. In addition to speed control, lower-frequency starting of an induction motor has the advantages of lower starting current and higher starting power factor. To maintain the air-gap flux density fairly constant, the motor input voltage per unit frequency is kept constant. This results in the maximum torque being very nearly constant over the high end of the speed range.

To overcome the resistive drops, the input voltage is not lowered below a value corresponding to a specified low frequency. A variable-frequency supply may be obtained from (1) a variable-frequency alternator, (2) a wound-rotor induction motor acting as a frequency changer, or (3) a solid-state static frequency converter. Method (3) is becoming more popular because of its lower cost and size.

Example 12

Calculate the maximum torque T_{max} and the corresponding slip s_{maxT}, of a 1000-hp, wye-connected, 2300-V, 16-pole, 60-Hz wound-rotor motor fed by a frequency converter (Fig. 5.16) at frequencies $f = 40$, 50, and 60 Hz. Motor data: $R_1 = 0.0721\ \Omega$, $X_1 = 0.605\ \Omega$, $R_2 = 0.0947\ \Omega$, and $X_2 = 0.605\ \Omega$. X_m (17.8 Ω) may be neglected.

Solution

The synchronous speed corresponding to the line frequency $f_l = 60$ Hz is

$$\omega_s = 2\pi f/p = 2\pi \times 60/8 = 47.12 \text{ rad/s}$$

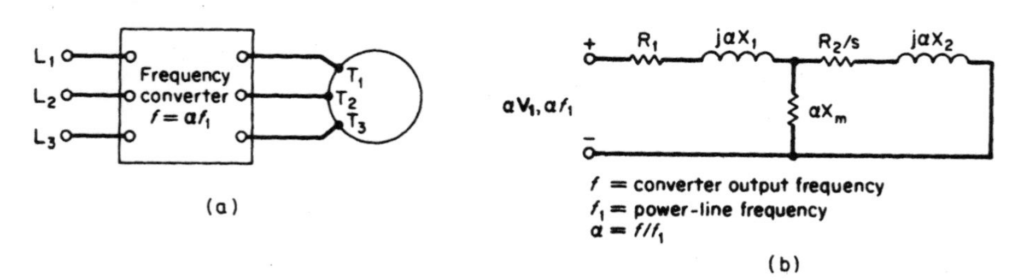

FIGURE 5.16 Employing a frequency converter for speed control of an induction motor. (*a*) Circuit diagram. (*b*) Equivalent circuit.

The slip corresponding to the maximum torque is given by

$$s_{\max T} \cong R_2 / \sqrt{R_1^2 + \alpha^2(X_1 + X_2)^2}$$

for $\alpha = f/f_l$. For $f_l = 60$ Hz, we obtain

$$s_{\max T} = 0.0947/\sqrt{0.0721^2 + 1^2(0.605 + 0.605)^2} = 0.078$$

Similarly, for $f_l = 50$ Hz, $s_{\max T} = 0.0937$ and for $f_l = 40$ Hz, $s_{\max T} = 0.117$

The maximum torque is given by

$$T_{\max} \cong \tfrac{3}{2}(V_1^2/\omega_s)\,\{\alpha/[R_1 + \sqrt{R_1^2 + \alpha^2(X_1 + X_2)^2}]\}$$

Substitution of values yields $T_{\max} = 43{,}709$ N · m for $f_l = 60$ Hz, 43,192 N · m for $f_l =$ 50 Hz, and 42,430 N · m for 40 Hz. The maximum torque changed very little for a 33 percent change in speed.

*SELECTION OF MOTOR STARTING AND SPEED CONTROL**

We want to choose a suitable starter and speed control for a 500-hp wound-rotor ac motor that must have a speed range of 2 to 1 with a capability for low-speed jogging. The motor is to operate at about 1800 r/min with current supplied at 4160 V, 60 Hz. An enclosed starter and controller is desirable from the standpoint of protection. What is the actual motor speed if the motor has four poles and a slip of 3 percent?

Selection of the Starter

Table 5.6 shows that a magnetic starter is suitable for wound-rotor motors in the 220- to 4500-V and 5- to 1000-hp range. Since the motor is in this range of voltage and horsepower, a magnetic starter will probably be suitable. Also, the magnetic starter is available in an enclosed cabinet, making it suitable for this installation.

Table 5.7 shows that a motor starting torque of approximately 200 percent of the full-load motor torque and current are obtained on the first point of acceleration. To calculate motor full-load speed at a slip of 3 percent, using $s = (n_s - n)/n$ gives,

$$n = n_s(1 - s) = 1800(1 - 0.03) = 1746 \text{ rpm}.$$

Selection of the Type of Speed Control

Table 5.7 summarizes the various types of adjustable-speed drives available today. This listing shows that power-operated contactors used with wound-rotor motors will give a 3 to 1 speed range with low-speed jogging. Since a 2 to 1 speed range is required, the proposed controller is suitable because it gives a wider speed range than needed.

Note from Table 5.7 that if a wider speed range were required, a thyratron control could produce a range up to 10 to 1 on a wound-rotor motor. Also, a wound-rotor direct-current motor set might be used too. In such an arrangement, an ac and dc motor are com-

*Adapted from Hicks. *Standard Handbook of Engineering Calculations*, McGraw-Hill, with permission of McGraw-Hill, Inc.

TABLE 5.6 Typical Alternating-Current Motor Starters*

		Typical range	
Motor type	Starter type	Voltage	Horsepower
Squirrel cage	Magnetic, full-voltage	110–550	1.5–600
	With fusible or nonfusible disconnect or circuit breaker	208–550	2–200
	Reversible	110–550	1.5–200
	Manual, full-voltage	110–550	1.5–7.5
	Manual, reduced-voltage, autotransformer	220–2500	5–150
	Magnetic, reduced-voltage, autotransformer	220–5000	5–1750
	Magnetic, reduced-voltage, resistor	220–550	5–600
Wound rotor	Magnetic, primary and secondary control	220–4500	5–1000
	Drums and resistors for secondary control	1000 max.	5–750
Synchronous	Reduced-voltage, magnetic	220–5000	25–3000
	Reduced-voltage, semimagnetic	220–2500	20–175
	Full-voltage, magnetic	220–5000	25–3000
High-capacity induction	Magnetic, full-voltage	2300–4600	Up to 2250
	Magnetic, reduced-voltage	2300–4600	Up to 2250
High-capacity synchronous	Magnetic, full-voltage	2300–4600	Up to 2500
	Magnetic, reduced-voltage	2300–4600	Up to 2500
High-capacity wound rotor	Magnetic primary and secondary	2300–4600	Up to 2250

*Based on Allis-Chalmers, General Electric, and Westinghouse units. This table is taken from *Standard Handbook of Engineering Calculations* by Hicks. Adapted from Hicks, *Standard Handbook of Engineering Calculations,* McGraw-Hill, with permission of McGraw-Hill, Inc.

bined on the same shaft. The rotor current is converted to dc by external silicon rectifiers and fed back to the dc armature through the commutator.

Tables 5.8 and 5.9 are used as a guide to the selection of starters and controls for alternating-current motors serving industrial, commercial, marine, portable, and residential applications.

To choose a direct-current motor starter, Table 5.8 is used as a guide.

Speed controls for dc motors can be chosen using Table 5.9 as a guide. DC motors are finding increasing use in industry. They are also popular in marine service.

MOTOR SELECTION FOR A CONSTANT LOAD

The horsepower rating and NEMA design class of an induction motor depend on the load-torque speed characteristic, duty cycle, inertia, temperature rise and heat dissipation, environmental conditions, and auxiliary drives. In the vast majority of applications, the selection process is relatively simple. But in a few applications, the selection process may involve criteria such as high starting torque, low starting current, intermittent duty, torque pulsations, high load inertia, or a duty cycle with frequent starting and stopping.

TABLE 5.7 Adjustable-Speed Drives*

	Drive types						
Drive features	Constant-voltage dc	Adjustable-voltage dc motor-generator set	Adjustable-voltage rectifier	Eddy-current clutch	Wound-rotor ac, standard	Wound-rotor thyratron	Wound-rotor dc-motor set
Power units required	Rectifier, dc motor,	AC motor, dc generator, dc motor	Rectifier, reactor,[a] dc motor	AC motor, eddy-current clutch	AC motor	AC motor thyratrons	AC dc motor, rectifier
Normal speed range	4–1	8–1 c-t +[b] 4–1 c-hp[c]	8–1 c-t + 4–1 c-hp[c]	34–1, 2 pole; 17–1, 4 pole	3–1	10–1[c]	3–1
Low speed for jogging	No[d]	Yes	Yes	Yes	Yes	Yes	Yes
Torque available	c-hp	c-t	c-t	c-t	c-t	c-t	c-t, c-hp
Speed regulation	10–15 percent	5 percent with regulator	5 percent with regulator	2 percent with regulator	Poor	±3 percent	5–7½ percent
Speed control	Field rheostat	Rheostats or pots	Rheostats or pots	Rheostats or pots	Steps, power contactors	Rheostats or pots	Rheostats or pots
Enclosures available	All	All	All	Open[e]	All	All	All
Braking: Regenerative	No	Yes	No	No	Yes	Yes	No
Dynamic	Yes	Yes	Yes	No[f]	Yes	Yes	Yes
Multiple operation	Yes	Yes	Yes	Yes	Yes	Yes	No
Parallel operation	Yes	Yes	Yes	Yes	No	Yes	Yes
Controlled acceleration, deceleration	Yes	Yes	Yes	Yes	No	Yes	No
Efficiency	80–85 percent	63–73 percent	70–80 percent	80–85 percent	80–85 percent	80–85 percent	80–85 percent
Top speed at maximum torque	83–87 percent	60–67 percent	60–70 percent	29 percent	29 percent	85–90 percent	73–78 percent

TABLE 5.7 Adjustable-Speed Drives (Continued)

Drive features	Drive types						
	Constant-voltage dc	Adjustable-voltage dc motor-generator set	Adjustable-voltage rectifier	Eddy-current clutch	Wound-rotor ac, standard	Wound-rotor thyratron	Wound-rotor dc-motor set
Rotor inertia[g]	100 percent[h]	100 percent	100 percent	75 percent	90 percent	90 percent	175 percent
Starting torque	200–300 percent	200–300 percent	200–300 percent	200–300 percent	200 percent	200–300 percent	200–300 percent
Number of comm., rings	1 comm.	2 comm.	1 comm.	None	1 set rings	1 set rings	1 comm., 1 set rings

[a]Used only in saturable-reactor designs.
[b]c-t—constant-torque, c-hp—constant horsepower.
[c]Units of 200 to 1 speed range are available.
[d]Low speed can be obtained using armature resistance.
[e]Totally enclosed units must be water- or oil-cooled.
[f]Eddy-current brake may be integral with unit.
[g]Based on standard dc motor.
[h]Normally is a larger dc motor since it has slower base speed.
*This table is taken from *Standard Handbook of Engineering Calculations* by Hicks.

TABLE 5.8 Direct-Current Motor Starters*

Type of starter	Typical users
Across-the-line	Limited to motors of less than 2 hp
Reduced-voltage, manual-control (face-piate type)	Used for motors up to 50 hp where starting is infrequent
Reduced-voltage, multiple-switch	Motors of more than 50 hp
Reduced-voltage, drum-switch	Large motors; frequent starting and stopping
Reduced-voltage, magnetic-switch	Frequent starting and stopping; large motors

*This table is taken from *Standard Handbook of Engineering Calculations* by Hicks.

TABLE 5.9 Direct-Current, Motor-Speed Controls*

Type of motor	Speed characteristic	Type of control
Series-wound	Varying; wide speed regulation	Armature shunt and series resistors
Shunt-wound	Constant at selected speed	Armature shunt and series resistors; field weakening; variable armature voltage
Compound-wound	Regulation is about 25 percent	Armature shunt and series resistors; field weakening; variable armature voltage

*This table is taken from *Standard Handbook of Engineering Calculations* by Hicks.

Selection of the proper horsepower rating will minimize initial and maintenance costs and improve operating efficiency and life expectancy. An undersized motor may result in overloads and a consequent reduction in life expectancy. Too large a motor horsepower rating will cause lower efficiency and power factor, greater space requirements, and higher initial cost.

Example 13

Select a motor driving a load requiring a torque of 55 N · m at 1764 r/min.

Solution

For a constant load, the motor horsepower is calculated from,

$$\text{hp} = \frac{\omega T}{746} = \frac{2\pi \times 1764/60 \times 55}{746} = 13.6 \text{ hp}.$$

A NEMA design B 15-hp motor for continuous duty is selected.

MOTOR SELECTION FOR A VARIABLE LOAD

A summary of various motor characteristics and applications is provided in Table 5.10.

TABLE 5.10 Summary of Motor Characteristics and Applications*

Polyphase motors				
Speed regulation	Speed control	Starting torque	Breakdown torque	Applications
General-purpose squirrel-cage (design B):				
Drops about 3 percent for large to 5 percent for small sizes	None, except multispeed types, designed for two to four fixed speeds	100 percent for large, 275 percent for 1-hp, four-pole unit	200 percent of full load	Constant-speed service where starting torque is not excessive. Fans, blowers, rotary compressors, and centrifugal pumps
High-torque squirrel-cage (design C):				
Drops about 3 percent for large to 6 percent for small sizes	None, except multispeed types, designed for two to four fixed speeds	250 percent of full load for high-speed to 200 percent for low-speed designs	200 percent of full load	Constant-speed service where fairly high starting torque is required infrequently with starting current about 550 percent of full load. Reciprocating pumps and compressors, crushers, etc.
High-slip squirrel-cage (design D):				
Drops about 10 to 15 percent from no load to full load	None, except multispeed types, designed for two to four fixed speeds	225 to 300 percent of full load, depending on speed with rotor resistance	200 percent. Will usually not stall until loaded to maximum torque, which occurs at standstill	Constant speed starting torque, if starting is not too frequent, and for high-peak loads with or without flywheels. Punch presses, shears, elevators, etc.
Low-torque squirrel-cage (design F):				
Drops about 3 percent for large to 5 percent for small sizes	None, except multispeed types, designed for two to four fixed speeds	50 percent of full load for high-speed to 90 percent for low-speed designs	135 to 170 percent of full load	Constant-speed service where starting duty is light. Fans, blowers, centrifugal pumps and similar loads

Wound-rotor:				
With rotor rings short-circuited drops about 3 percent for large to 5 percent for small sizes	Speed can be reduced to 50 percent by rotor resistance. Speed varies inversely as load	Up to 300 percent depending on external resistance in rotor circuit and how distributed	300 percent when rotor slip rings are short-circuited	Where high starting torque with low starting current or where limited speed control is required. Fans, centrifugal and plunger pumps, compressors, conveyors, hoists, cranes, etc.
Synchronous:				
Constant	None, except special motors designed for two fixed speeds	40 percent for slow- to 160 percent for medium-speed 80 percent pf. Specials develop higher	Unity-pf motors, 170 percent; 80 percent pf motors, 225 percent. Specials up to 300 percent	For constant-speed service, direct connection to slow-speed machines and where power-factor correction is required
Series:				
Varies inversely as load. Races on light loads and full voltage	Zero to maximum depending on control and load	High. Varies as square of voltage. Limited by commutation, heating, capacity	High. Limited by commutation, heating, and line capacity	Where high starting torque is required and speed can be regulated. Traction, bridges, hoists, gates, car dumpers, car retarders
Shunt:				
Drops 3 to 5 percent from no load to full load	Any desired range depending on design type and type of system	Good. With constant field, varies directly as voltage applied to armature	High. Limited by commutation, heating, and line capacity	Where constant or adjustable speed is required and starting conditions are not severe. Fans, blowers, centrifugal pumps, conveyors, wood- and metal-working, machines, and elevators
Compound:				
Drops 7 to 20 percent from no load to full load depending on amount of compounding	Any desired range, depending on design and type of control	Higher than for shunt, depending on amount of compounding	High. Limited by commutation, heating, and line capacity	Where high starting torque and fairly constant speed is required. Plunger pumps, punch presses, shears, bending rolls, geared elevators, conveyors, hoists

TABLE 5.10 Summary of Motor Characteristics and Applications* (Continued)

DC and single-phase motors				
Speed regulation	Speed control	Starting torque	Breakdown torque	Applications
Split-phase:				
Drops about 10 percent from no load to full load	None	75 percent for large to 175 percent for small sizes	150 percent for large to 200 percent for small sizes	Constant-speed service where starting is easy. Small fans, centrifugal pumps and light-running machines, where polyphase is not available
Capacitor:				
Drops about 5 percent for large to 10 percent for small sizes	None	150 percent to 350 percent of full load depending on design and size	150 percent for large to 200 percent for small sizes	Constant-speed service for any starting duty and quiet operation where polyphase current cannot be used
Commutator:				
Drops about 5 percent for large to 10 percent for small sizes	Repulsion induction, none. Brush-shifting types, 4–1 at full load	250 percent for large to 350 percent for small sizes	150 percent for large to 250 percent for small sizes	Constant-speed service for any starting duty where speed control is required and polyphase current cannot be used

*This table is from *Standard Handbook of Engineering Calculations* by Hicks.

Example 14

Select a motor for a load cycle lasting 3 min and having the following duty cycle: 1 hp for 35 s, no load for 50 s, and 4.4 hp for 95 s.

Solution

Using the relationship,

$$hp_{rms} = \sqrt{(\Sigma\, hp^2 \times \text{time})/(\text{running time} + \text{standstill time}/K)}.$$

where $K = 4$ for an enclosed motor and $K = 3$ for an open motor. Assume $K = 3$. Substituting the given values, we find:

$$hp_{rms} = \sqrt{(1^2 \times 35 + 0 + 4.4^2 \times 95)/(35 + 95 + 50/3)} = 3.57 \text{ hp}.$$

A 3-hp design B open-type motor is selected.

BIBLIOGRAPHY

Alger, Philip L. 1995. *Induction Machines: Their Behavior and Use*. Newark, N.J.: Gordon & Breach Science Publishers.

Beaty, H. Wayne and James L. Kirtley, Jr. 1998. *Electric Motor Handbook*. New York: McGraw-Hill.

Cochran, Paul L. 1989. *Polyphase Induction Motors: Analysis, Design, and Applications*. New York: Marcel Dekker.

SECTION 6

SINGLE-PHASE MOTORS

Lawrence J. Hollander, P.E.
Dean of Engineering Emeritus
Union College

EQUIVALENT CIRCUIT OF A SINGLE-PHASE INDUCTION MOTOR DETERMINED FROM NO-LOAD AND LOCKED-ROTOR TESTS

A single-phase induction motor has the following data: 1 hp, two pole, 240 V, 60 Hz, stator-winding resistance $R_s = 1.6\ \Omega$. The no-load test results are $V_{\rm NL} = 240$ V, $I_{\rm NL} = 3.8$ A, $P_{\rm NL} = 190$ W; the locked-rotor test results are $V_{\rm LR} = 88$ V, $I_{\rm LR} = 9.5$ A, $P_{\rm LR} = 418$ W. Establish the equivalent circuit of the motor.

Calculation Procedure

1. Calculate the Magnetizing Reactance, X_ϕ

The magnetizing reactance essentially is equal to the no-load reactance. From the no-load test, $X_\phi = V_{\rm NL}/I_{\rm NL} = 240\ {\rm V}/3.8\ {\rm A} = 63.2\ \Omega$. One-half of this (31.6 Ω) is assigned to the equivalent circuit portion representing the forward-rotating mmf waves and the other half to the portion representing the backward-rotating mmf waves. See Fig. 6.1. The no-load power, 190 W, represents the rotational losses.

2. Calculate the Impedance Values from Locked-Rotor Test

For the locked-rotor test in induction machines, the magnetizing branch of the equivalent circuit is considered to be an open circuit, because the ratio of impedances X_L/X_ϕ is very small. Considering $V_{\rm LR}$ as reference (phase angle = 0), use the equation $P_{\rm LR} = V_{\rm LR} I_{\rm LR} \cos\theta_{\rm LR}$. Thus, $\theta_{\rm LR} = \cos^{-1}(P_{\rm LR}/V_{\rm LR}I_{\rm LR}) = \cos^{-1}[418\ {\rm W}/(88\ {\rm V})(9.5\ {\rm A})] = \cos^{-1} 0.5 = 60°$. The effective stator current is $\mathbf{I}'_s = \mathbf{I}_{\rm LR} - \mathbf{I}_\phi = \mathbf{I}_{\rm LR} - \mathbf{V}_{\rm LR}/jX_\phi = 9.5\underline{/-60°} - 88\ {\rm V}/j63.2\ \Omega = 4.75 - j8.23 + j1.39 = 4.75 - j6.84 = 8.33\underline{/-55.2°}$ A. The rotor impedance referred to the stator is $\mathbf{V}_{\rm LR}/\mathbf{I}'_s = 88\underline{/0°}\ {\rm V}/8.33\underline{/-55.2°}\ {\rm A} = 10.56\underline{/55.2°}\ \Omega$.

Also, from the locked-rotor test, $P_{\rm LR} = I'^2_s(R_s + R'_r)$; $R_s + R'_r = P_{\rm LR}/I'^2_s = 418\ {\rm W}/(8.33$

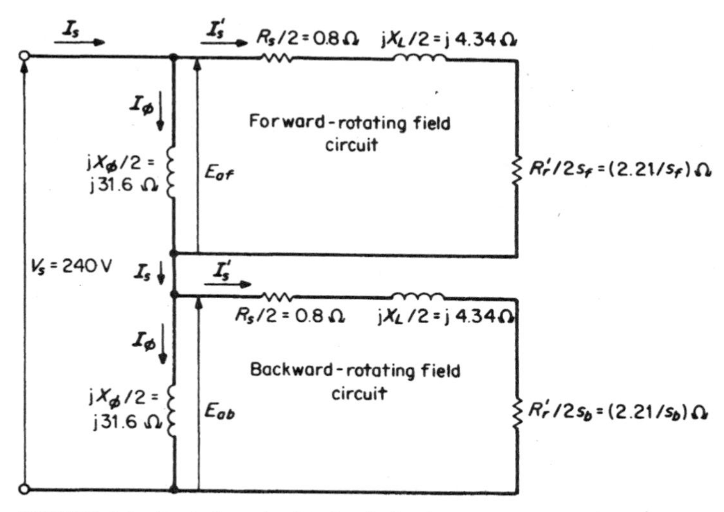

FIGURE 6.1 Equivalent circuit of a single-phase motor, referred to the stator.

A)^2 = 6.02 Ω. $R'_r = (R_s + R'_r) - R_s = 6.02 - 1.6 = 4.42$ Ω. Finally, leakage reactance is $\mathbf{X_L} = \mathbf{Z'_r} - (R_s + \mathbf{R'_r}) = 10.56\underline{/+55.2°} - 6.02 = 6.02 + j8.67 - 6.02 = j8.67$ Ω.

3. Draw the Equivalent Circuit

See Fig. 6.1, wherein $R_s/2 = 1.6/2 = 0.8$ Ω, $X_L/2 = 8.67/2 = 4.34$ Ω, and $R'_r/2 = 4.42/2 = 2.21$ Ω. Also:

R_s = resistance of stator winding, Ω
R'_r = resistance of rotor winding referred to the stator, Ω
X_ϕ = magnetizing reactance, Ω
X_L = leakage reactance, Ω
I_ϕ = magnetizing current, A
I_s = actual stator current, A
I'_s = effective stator current $(I_s - I_\phi)$, A
I_{LR} = locked-rotor current, A
V_{LR} = locked-rotor voltage, V
s_f = forward slip
s_b = backward slip $= 2 - s_f$
E_{af} = counter-emf developed by forward rotating fields of stator and rotor mmf's, V
E_{ab} = counter-emf developed by backward-rotating fields of stator and rotor mmf's, V
V_s = terminal voltage applied to the stator, V
Z'_r = impedance of rotor circuit referred to the stator, Ω

Related Calculations. The open-circuit (no-load) and the short-circuit (locked-rotor) tests are similar to those for a transformer or for a polyphase induction motor. The data are used in a similar fashion and the equivalent circuits have many similarities. In a single-phase machine, the single waveform is divided into two half-amplitude rotating fields: a forward-rotating wave and a backward-rotating wave. This device leads to the division of the equivalent circuit into a forward-designated section and a backward-designated section. The use of this equivalent circuit is demonstrated in subsequent problems.

TORQUE AND EFFICIENCY CALCULATIONS FOR SINGLE-PHASE INDUCTION MOTOR

For the 1-hp single-phase induction motor in the previous problem, calculate the shaft torque and efficiency when the motor is operating at a speed of 3470 r/min.

Calculation Procedure

1. Calculate the Forward and Backward Slips

The synchronous speed in r/min for a two-pole, 60-Hz machine is calculated from the equation $n_{sync} = 120f/p$ where f = frequency in Hz and p = number of poles; n_{sync} = (120)(60 Hz)/2 poles = 3600 r/min. To calculate the forward slip use the equation $s_f = (n_{sync} - n_{actual})/n_{sync}$ = (3600 r/min − 3470 r/min)/(3600 r/min) = 0.036. The backward slip is $s_b = 2 - 0.036 = 1.964$.

2. Calculate the Total Impedance of the Forward Equivalent Circuit, Z_f

The total impedance of the forward equivalent circuit is $\mathbf{Z_f} = R_s/2 + R'_r/2s_f + jX_L/2$ in parallel with $jX_\phi/2$. Thus, $R_s/2 + R'_r/2s_f + jX_L/2 = 0.8 + 2.21/0.036 + j4.34 = 62.2 + j4.34 = 62.4\angle 3.99°\ \Omega$.. $\mathbf{Z_f} = (62.4\angle 3.99°)(31.6\angle 90°)/(62.2 + j4.34 + j31.6) = 27.45\angle 63.97° = 12.05 + j24.67\ \Omega$.

3. Calculate the Total Impedance of the Backward Equivalent Circuit, Z_b

The total impedance of the backward equivalent circuit is $\mathbf{Z_b} = R_s/2 + R'_r/2s_b + jX_L/2$ in parallel with $jX_\phi/2$. Thus, $R_s/2 + R'_r/2s_b + jX_L/2 = 0.8 + 2.21/1.964 + j4.34 = 1.93 + j4.34 = 4.75\angle 66.03°\ \Omega$. $\mathbf{Z_b} = (4.75\angle 66.03°)(31.6\angle 90°)/(1.93 + j4.34 + j31.6) = 4.17\angle 69.1° = 1.49 + j3.89\ \Omega$.

4. Calculate the Total Circuit Impedance, Z

The total circuit impedance is $\mathbf{Z} = \mathbf{Z_f} + \mathbf{Z_b} = (12.05 + j24.67) + (1.49 + j3.89) = 13.54 + j28.56 = 31.61\angle 64.63°\ \Omega$.

5. Calculate the Power Factor, pf, and the Source Current, I_s

The power factor pf = cos 64.63° = 0.43. The source current (stator current) is $\mathbf{V_s}/\mathbf{Z} = 240\angle 0°\ \text{V}/31.61\angle 64.63°\ \Omega = 7.59\angle -64.63°\ \text{A}$.

6. Calculate the Forward and Backward Counter-emf's

The forward and backward components of the counter-emf's are each prorated on the source voltage, according to the ratio of the forward and backward equivalent impedances, $\mathbf{Z_f}$ and $\mathbf{Z_b}$, to the combined impedance, $\mathbf{Z}$. Thus, $\mathbf{E_{af}} = \mathbf{V_s}(\mathbf{Z_f}/\mathbf{Z}) = (240\angle 0°\ \text{V})(27.45\angle 63.97°/31.61\angle 64.63°) = 208.4\angle -0.66°\ \text{V}$. Similarly, $\mathbf{E_{ab}} = \mathbf{V_s}(\mathbf{Z_b}/\mathbf{Z}) = (240\angle 0°\ \text{V})(4.17\angle 69.1°/31.61\angle 64.63°) = 31.67\angle 4.47°\ \text{V}$.

7. Calculate the Forward- and Backward-Component Currents, I'_s

The forward-component current is $\mathbf{I'_{sf}} = \mathbf{E_{af}}/\mathbf{Z'_{rf}} = 208.4\angle -0.66°\ \text{V}/62.4\angle 3.99°\ \Omega = 3.34\angle -4.65°\ \text{A}$. The backward-component current is $\mathbf{I'_{sb}} = \mathbf{E_{ab}}/\mathbf{Z'_{rb}} = 31.67\angle 4.47°\ \text{V}/4.75\angle 66.03°\ \Omega = 6.67\angle -61.56°\ \text{A}$.

8. Calculate the Internally Developed Torque, T_{int}

The internally developed torque is T_{int} = forward torque − backward torque = $(30/\pi)(1/n_{\text{sync}})[(I'_{\text{sf}})^2(R'_r/2s_f) - (I'_{\text{sb}})^2(R'_r/2s_b)] = (30/\pi)(1/3600\text{r/min})[(3.34)^2(2.21/0.036) - (6.67)^2(2.21/1.964)] = 1.684\ \text{N}\cdot\text{m}$.

9. Calculate the Lost Torque in Rotational Losses, T_{rot}

The lost torque in rotational losses is $T_{\text{rot}} = P_{\text{NL}}/\omega_m$, where ω_m = rotor speed in rad/s = (3470 r/min)(π/30) = 363.4 rad/s. T_{rot} = 190 W/(363.4 rad/s) = 0.522 N · m.

10. Calculate the Shaft Torque, T_{shaft}

The shaft torque is $T_{\text{shaft}} = T_{\text{int}} - T_{\text{rot}}$ = 1.684 N · m − 0.522 N · m = 1.162 N · m.

11. Calculate Power Input and Power Output

To calculate the power input, use the equation $P_{\text{in}} = V_s I_s \cos\theta$ = (240 V)(7.59 A)(0.43) = 783.3 W. The power output is calculated from the equation $P_{\text{out}} = T_{\text{shaft}}(n_{\text{actual}}\ \text{r/min})(\pi/30)$ = (1.162 N · m)(3470 r/min)(π/30) = 422.2 W.

12. Calculate the Efficiency, η

Use the equation for efficiency: η = (output × 100 percent)/input = (422.2 W)(100 percent)/ 783.3 W = 53.9 percent.

Related Calculations. It should be observed from Step 8, where T_{int} = forward torque − backward torque, that if the machine is at standstill, both s_f and $s_b = 1$. By study of the equivalent circuit, it is seen that $R'_r/2s_f = R'_r/2s_b$, which leads to the conclusion that the net torque is 0. In subsequent problems, consideration will be given to methods used for starting, that is, the creation of a net starting torque.

DETERMINATION OF INPUT CONDITIONS AND INTERNALLY DEVELOPED POWER FROM THE EQUIVALENT CIRCUIT FOR SINGLE-PHASE INDUCTION MOTORS

A four-pole, 220-V, 60-Hz, ¼-hp, single-phase induction motor is operating at 13 percent slip. The equivalent circuit is shown in Fig. 6.2 (different from the previous two problems in that the stator resistance and leakage reactance are separated from the forward and backward circuits). Find the input current, power factor, and the internally developed power.

Calculation Procedure

1. Calculate the Forward- and Backward-Circuit Impedances, Z_f and Z_b

Use the equation $\mathbf{Z_f} = (R'_r/2s_f + jX'_r/2)(jX_\phi)/2)/(R'_r/2s_f + jX'_r/2 + jX_\phi/2) = (185 + j5.0)(j140)/(185 + j5.0 + j140) = 110.2\underline{/53.6^\circ} = 65.5 + j88.6\ \Omega$. Similarly, $\mathbf{Z_b} = (13 + j5.0)(j140)/(13 + j5.0 + j140) = 13.4\underline{/26.1^\circ} = 12.0 + j5.9\ \Omega$.

2. Calculate the Total Impedance of the Equivalent Circuit, Z

The total impedance of the equivalent circuit is $\mathbf{Z} = \mathbf{Z_s} + \mathbf{Z_f} + \mathbf{Z_b} = 12 + j10 + 65.5 + j88.6 + 12.0 + j5.9 = 89.5 + j104.5 = 137.6\underline{/49.4^\circ}\ \Omega$.

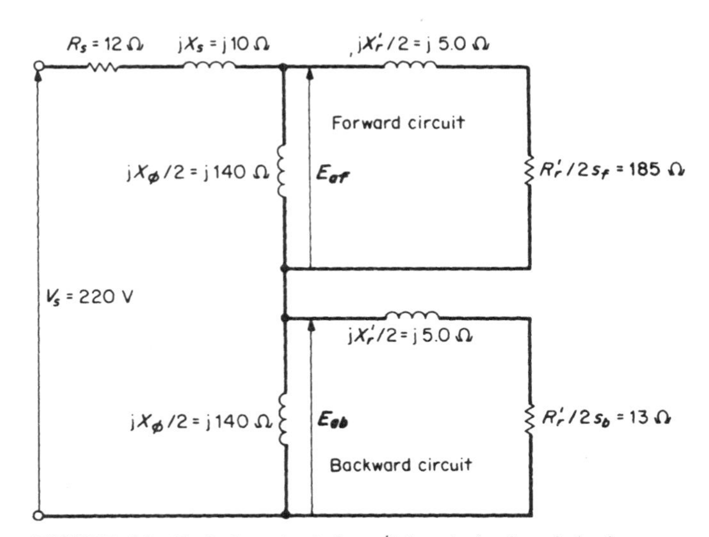

FIGURE 6.2 Equivalent circuit for a ¼-hp, single-phase induction motor.

3. Calculate the Power Factor, pf and Input Current, I_s

The power factor is pf = cos 49.4° = 0.65. The input current is $\mathbf{I_s} = \mathbf{V_s}/\mathbf{Z} = 220\angle 0°$ V/$137.6\angle 49.4°$ Ω = $1.6\angle -49.4°$ A.

4. Calculate the Forward and Backward Counter-emf's

As in the previous problem, $\mathbf{E_{af}} = V_s(Z_f/Z) = (220\angle 0°\text{ V})(110.2\angle 53.6°/137.6\angle 49.4°) = 176.2\angle 4.2°$ V. $\mathbf{E_{ab}} = \mathbf{V_s}(\mathbf{Z_b}/\mathbf{Z}) = (220\angle 0°\text{ V})(13.4\angle 26.1°/137.6\angle 49.4°) = 21.4\angle -23.3°$ V.

5. Calculate the Forward- and Backward-Component Currents, I'_s

The forward-component current is $\mathbf{I'_{sf}} = \mathbf{E_{af}}/\mathbf{Z'_{rf}} = 176.2\angle 4.2°/185\angle 1.55° = 0.95\angle 2.65°$ A. The backward-component current is $\mathbf{I'_{sb}} = \mathbf{E_{ab}}\mathbf{Z'_{rb}} = 21.4\angle -23.2°/13.9\angle 21° = 1.54\angle -44.3°$ A.

6. Calculate the Internally Developed Power, $\mathbf{P}_{int}$

Use the equation $P_{\text{int}} = [(R'_r/2s_f)I'^2_{\text{sf}} - (R'_r/2s_b)I'^2_{\text{sb}}]\,(1 - s) = [(185)(0.95)^2 - (13)(1.54)^2](1 - 0.13) = 118.4$ W; rotational losses must be subtracted from this in order to determine the power output, P_{out}. The power input is $P_{\text{in}} = V_sI_s \cos\theta = (220\text{ V})(1.6\text{ A})(0.65) = 228.8$ W.

Related Calculations. This problem, compared with the previous problem, illustrates a slightly different version of the equivalent circuit for a single-phase induction motor. It should be noted that in the backward portion of the circuit the magnetizing reactance, $X_\phi/2$, is so much larger than the values of X and R in the circuit that it may be neglected. Likewise, the forward portion of the circuit, $X'_r/2$, becomes insignificant. An approximate equivalent circuit is used in the next problem. These variations of equivalent circuits give similar results depending, of course, on what precise information is being sought.

DETERMINATION OF INPUT CONDITIONS AND INTERNALLY DEVELOPED POWER FROM THE APPROXIMATE EQUIVALENT CIRCUIT FOR SINGLE-PHASE INDUCTION MOTORS

An approximate equivalent circuit is given in Fig. 6.3, for the ¼-hp single-phase induction motor having the following data: 60 Hz, $V_s = 220$ V, $s_f = 13$ percent. Find the input current, power factor, and the internally developed power.

Calculation Procedure

1. Calculate the Forward- and Backward-Circuit Impedances, Z_f and Z_b

Notice that in the approximate equivalent circuit, the resistance element in the backward portion is $R'_r/(2)(2)$ rather than $R'_r/(2)(2 - s_f)$, where $(2 - s_f) = s_b$. $R'_r = 2s_f(185\ \Omega) = (2)(0.13)(185) = 48.1\ \Omega$. Thus, $R'_r/(2)(2) = 48.1\ \Omega/4 = 12.0\ \Omega$. This differs from the 13.0 Ω in the previous problem. Because of the magnitude of the difference between $jX_\phi/2$ and $jX'_r/2$, the latter will be neglected in calculating $\mathbf{Z_f}$. Thus, $\mathbf{Z_f} = (jX_\phi/2)(R'_r/2s_f)/(jX_\phi/2 + R'_r/2s_f) = (140\angle 90°)(185\angle 0°)/(j140 + 185\angle 0°) = 111.6\angle 52.9° = 67.3 + j89.0\ \Omega$. The calculation for $\mathbf{Z_b}$ simply is taken directly from the approximate equivalent circuit: $\mathbf{Z_b} = R'_r/4 + j(X'_r/2) = 12.0 + j5.0 = 13.0\angle 22.6°\ \Omega$.

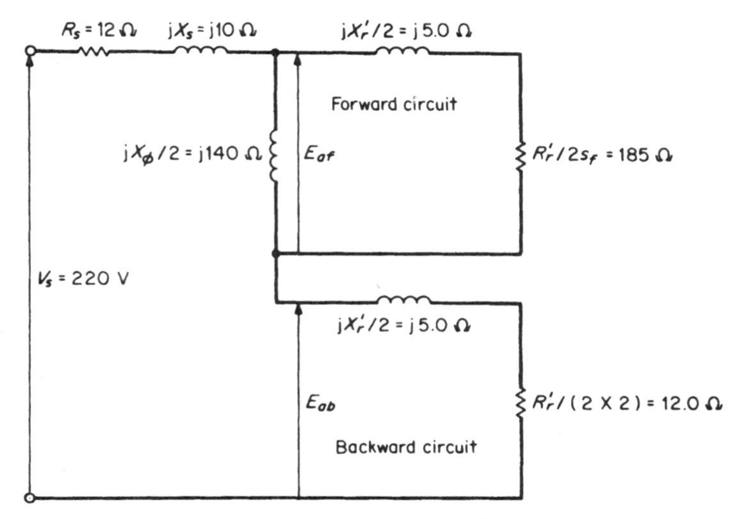

FIGURE 6.3 Approximate equivalent circuit for a $^1/_4$-hp, single-phase induction motor.

2. Calculate the Total Impedance of the Equivalent Circuit, Z

The total impedance of the equivalent circuit is $\mathbf{Z} = \mathbf{Z}_s + \mathbf{Z}_f + \mathbf{Z}_b = 12 + j10 + 67.3 + j89.0 + 12 + j5.0 = 91.3 + j104 = 138.4\angle 48.7°\ \Omega$.

3. Calculate the Power Factor, pf, and Input Current, I_s

The power factor is pf = cos 48.7° = 0.66. The input current is $\mathbf{I}_s = \mathbf{V}_s/\mathbf{Z} = 220\angle 0°\ \text{V}/138.4\angle 48.7°\ \Omega = 1.59\angle -48.7°$ A.

4. Calculate the Forward and Backward Counter-emf's

As in the previous two problems, $\mathbf{E}_{af} = \mathbf{V}_s(\mathbf{Z}_f/\mathbf{Z}) = (220\angle 0°\ \text{V})(111.6\angle 52.9°/138.4\angle 48.7°) = 177.4\angle 4.2°$ V. $\mathbf{E}_{ab} = \mathbf{V}_s(\mathbf{Z}_b/\mathbf{Z}) = (220\angle 0°\ \text{V})(13.0\angle 22.6°/138.4\angle 48.7°) = 20.7\angle -26.1°$ V.

5. Calculate the Forward- and Backward-Component Currents, I_s'

The forward-component current is $\mathbf{I}'_{sf} = \mathbf{E}_{af}/\mathbf{Z}'_{rf} = 177.4\angle 4.2°\ \text{V}/185\angle 0°\ \Omega = 0.96\angle 4.2°$ A. (*Note:* $jX_r'/2$ was neglected.) The backward-component current is $\mathbf{I}'_{sb} = \mathbf{E}_{ab}/\mathbf{Z}'_{rb} = 20.7\angle -26.1°\ \text{V}/13.0\angle 22.6°\ \Omega = 1.59\angle -48.7°$ A.

6. Calculate the Internally Developed Power, $\mathbf{P}_{int}$

Use the equation $P_{\text{int}} = [(R_r'/2s_f)I'^2_{sf} - (R_r'/4)\ I'^2_{sb}](1 - s) = [(185\ \Omega)(0.96\ \text{A})^2 - (12\ \Omega)(1.59\ \text{A})^2](1 - 0.13) = 121.9$ W. As previously, rotational losses must be subtracted from this value in order to determine the power output P_{out}. The power input is $P_{\text{in}} = V_sI_s \cos\theta = (220\ \text{V})(1.59\ \text{A}) \cos 48.7° = 230.9$ W.

7. Compare the Results of the General and Approximate Equivalent Circuits

The comparison follows. (The general equivalent circuit was used in the previous problem).

Related Calculations. A reasonable amount of calculations are simplified by using the approximate equivalent circuit for single-phase machine analysis, even in this example

Component	General equivalent circuit	Approximate equivalent circuit
$\mathbf{Z_f}$, Ω	$110.2\angle 53.6°$	$111.6\angle 52.9°$
$\mathbf{I_s}$, A	$1.6\angle -49.4°$	$1.59\angle -48.7°$
$\mathbf{Z_b}$, Ω	$13.4\angle 26.1°$	$13.0\angle 22.6°$
$\mathbf{Z}$, Ω	$137.6\angle 49.4°$	$138.4\angle 48.7°$
$\mathbf{E_{af}}$, V	$176.2\angle 4.2°$	$177.4\angle 4.2°$
$\mathbf{E_{ab}}$, V	$21.4\angle -23.3°$	$20.7\angle -26.1°$
$\mathbf{I'_{sf}}$, A	$0.95\angle 2.65°$	$0.96\angle 4.2°$
$\mathbf{I'_{sb}}$, A	$1.54\angle -44.3°$	$1.59\angle -48.7°$
P_{int}, W	118.4	121.9
P_{in}, W	228.8	230.9

for a slip of 13 percent. Lower slips will result in smaller differences between the two methods of analysis.

LOSS AND EFFICIENCY CALCULATIONS FROM THE EQUIVALENT CIRCUIT OF THE SINGLE-PHASE INDUCTION MOTOR

For the ¼-hp, single-phase induction motor of the previous two problems, calculate losses and efficiency from the general equivalent circuit. Assume that the core loss = 30 W and the friction and windage loss = 15 W.

Calculation Procedure

1. Calculate the Stator-Copper Loss

The stator-copper loss is calculated from the general equation for heat loss, $P = I^2R$. In this case, stator-copper loss is $P_{s(loss)} = I_s^2 R_s = (1.6 \text{ A})^2(12\ \Omega) = 30.7$ W.

2. Calculate Rotor-Copper Loss

The rotor-copper loss has two components: loss resulting from the forward-circuit current, $P_{f(loss)}$, and loss resulting from the backward-circuit current, $P_{b(loss)}$. Thus, $P_{f(loss)} = s_f I'^2_{sf}(R'_r/2s_f) = (0.13)(0.95 \text{ A})^2(185\ \Omega) = 21.7$ W, and $P_{b(loss)} = s_b I'^2_{sb}(R'_r/2s_b)$, where $s_b = 2 - s_f$. $P_{b(loss)} = (2 - 0.13)(1.54 \text{ A})^2(13\ \Omega) = 57.7$ W.

3. Calculate the Total Losses

Stator-copper loss	30.7 W
Rotor-copper loss (forward)	21.7 W
Rotor-copper loss (backward)	57.7 W
Core loss (given)	30.0 W
Friction and windage loss (given)	15.0 W
Total losses	155.1 W

4. Calculate the Efficiency

Use the equation η = [(input – losses)(100 percent)]/input = [(228.8 W − 155.1 W)(100 percent)]/228.8 W = 32 percent. This is a low efficiency indicative of the machine running at a slip of 13 percent, rather than a typical 3 to 5 percent. Alternatively, the output power may be calculated from: P_{int} − friction and windage losses − core loss = 118.4 W − 15.0 W − 30.0 W = 73.4 W. Here, efficiency = (output power)(100 percent)/input power = (73.4 W)(100 percent)/228.8 W = 32 percent.

Related Calculations. The establishment of an equivalent circuit makes it possible to make many different calculations. Also, once the equivalent circuit is created, these calculations may be made for any presumed value of input voltage and operating slip.

STARTING-TORQUE CALCULATION FOR A CAPACITOR MOTOR

A four-pole capacitor motor (induction motor) has the following data associated with it: stator main-winding resistance R_{sm} = 2.1 Ω, stator auxiliary-winding resistance R_{sa} = 7.2 Ω, stator main-winding leakage reactance X_{sm} = 2.6 Ω, stator auxiliary-winding leakage reactance X_{sa} = 3.0 Ω, reactance of capacitance inserted in series with the stator auxiliary winding X_{sc} = 65 Ω, rotor-circuit resistance referred to main winding R'_r = 3.9 Ω, rotor-circuit leakage reactance referred to stator main winding X'_r = 2.1 Ω, magnetizing reactance referred to stator main winding X_ϕ = 75.0 Ω. Calculate the starting torque; assume that the motor is rated ¼-hp at 115 V, 60 Hz, and that the effective turns ratio of the auxiliary winding to the main winding is 1.4. See Fig. 6.4.

Calculation Procedure

1. Calculate the Forward- and Backward-Circuit Impedances, Z_f and Z_b

A capacitor motor is similar to a two-phase induction motor, except that the two phases are unbalanced or unsymmetrical; their current relationship is usually not at 90°, but is determined by the capacitance in series with the auxiliary winding. At start (or standstill), s = 1, forward slip s_f = 1, and backward slip $s_b = (2 - s_f) = 1$.

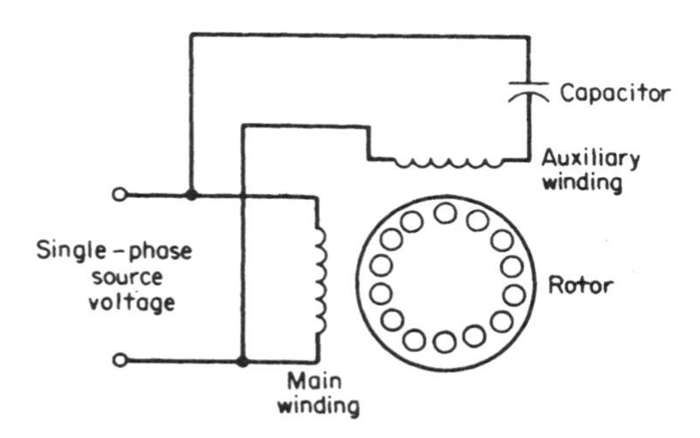

FIGURE 6.4 Single-phase, capacitor induction motor.

At standstill, the forward- and backward-circuit impedances are equal: $\mathbf{Z_f} = \mathbf{Z_b} = (jX_\phi)(R'_r + jX'_r)/(jX_\phi + R'_r + jX'_r) = (j75)(3.9 + j2.1)/(j75 + 3.9 + j2.1) = 4.28\angle 31.2° = 3.66 + j2.22\ \Omega$. The analysis departs from that for a single-phase machine and follows that for a two-phase machine; the rotor values are not divided by 2 as was done in the previous problems. The rotating fields of constant amplitude have one-half amplitude for single phase, unity amplitude for two phase, and three-halves amplitude for three phase.

2. Calculate the Total Impedance, Z_m, Referred to the Stator Main Winding

Use the equation $\mathbf{Z_m} = \mathbf{Z_{sm}} + \mathbf{Z_f} = 2.1 + j2.6 + 3.66 + j2.22 = 5.76 + j4.82 = 7.51\angle 39.9°\ \Omega$.

3. Calculate the Stator Main-Winding Current, I_{sm}

If the terminal voltage is the reference phasor, $\mathbf{I_{sm}} = \mathbf{V}_s/\mathbf{Z_m} = 115\angle 0°\ \text{V}/7.51\angle 39.9°\ \Omega = 15.3\angle -39.9°\ \text{A} = 11.74 - j9.81\ \text{A}$.

4. Calculate the Total Impedance, Z_a, Referred to the Stator Auxiliary Winding

The impedances used to calculate $\mathbf{Z_f}$ were referred to the main winding. In order to make the calculations relating to the auxiliary winding, the impedances now must be referred to *that* winding. This is done by multiplying $\mathbf{Z_f}$ by a^2, where a = effective turns ratio of the auxiliary winding to the main winding = 1.4. Thus, $\mathbf{Z_{fa}} = a^2\mathbf{Z_f} = (1.4)^2(3.66 + j2.22) = 7.17 + j4.35\ \Omega$.

The total impedance of the auxiliary stator winding is $\mathbf{Z_a} = \mathbf{Z_{sa}} + \mathbf{Z_{sc}} + \mathbf{Z_f} = 7.2 + j3.0 - j65 + 7.17 + j4.35 = 14.37 - j57.65 = 59.4\angle -76.0°\ \Omega$.

5. Calculate the Stator Auxiliary-Winding Current, I_{sa}

As in Step 3, $\mathbf{I_{sa}} = \mathbf{V}_t/\mathbf{Z_a} = 115\angle 0°\ \text{V}/59.4\angle -76.0°\ \Omega = 1.94\angle 76.0°\ \text{A} = 0.47 + j1.88\ \text{A}$.

6. Calculate the Starting Torque, $\mathbf{T}_{start}$

The starting torque is $T_{\text{start}} = (2/\omega_s)I_{\text{sm}}aI_{\text{sa}}R \sin\phi$, where $\omega_s = 4\pi f/p$ rad/s, f = frequency in Hz, p = number of poles, R = the resistance component of Z_f, and ϕ = angle separating I_m and I_a. Thus, $\omega_s = (4\pi)(60\ \text{Hz})/4 = 188.5$ rad/s. And $T_{\text{start}} = (2/188.5)(15.3\ \text{A})(1.4)(1.94\ \text{A})(3.66) \sin(39.9° + 76.0°) = 1.45\ \text{N}\cdot\text{m}$.

Related Calculations. It would appear that an unusually large capacitor has been used, resulting in an angle difference between I_{sm} and I_{sa} greater than 90°. However, in this case the capacitor remains in the circuit during running; when s_f and s_b are both different from unity, the angle would not be greater than 90°. This problem illustrates one method of handling currents from the two windings. Another method is with the use of symmetrical components of two unbalanced currents in two-phase windings.

STARTING TORQUE FOR A RESISTANCE-START SPLIT-PHASE MOTOR

A resistance-start split-phase motor has the following data associated with it: $^1/_2$ hp, 120 V, starting current in the auxiliary stator winding I_a = 8.4 A at a lagging angle of 14.5°, starting current in the main stator winding I_m = 12.65 A at a lagging angle of 40°. Determine the total starting current and the starting torque, assuming the machine constant is 0.185 V · s/A.

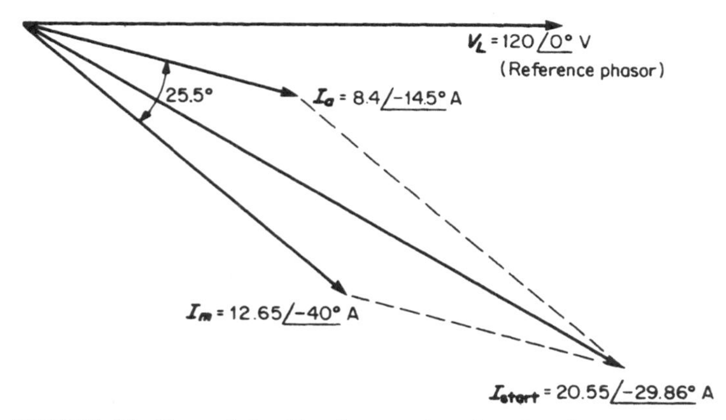

FIGURE 6.5 Phase relationships of currents in main and auxiliary windings of a resistance-start, split-phase motor.

Calculation Procedure

1. Divide the Starting Currents into In-Phase and Quadrature Components

The line voltage ($\mathbf{V}_L = 120\angle 0°$ V) is used as the reference phasor. Thus, at starting (as at locked-rotor) the current in the auxiliary stator winding is $\mathbf{I}_a = 8.4\angle -14.5°$ A $= 8.13 - j2.10$ A. Similarly, the current in the main stator winding is $\mathbf{I}_m = 12.65\angle -40.0°$ A $= 9.69 - j8.13$ A.

2. Calculate the Total In-Phase and Quadrature Starting Currents

The total in-phase starting current is $\mathbf{I}_{\text{in-phase}} = 8.13 + 9.69 = 17.82$ A. The total quadrature starting current is $\mathbf{I}_{\text{quad}} = -j2.10 - j8.13 = -j10.23$ A. Thus, the starting current is $\mathbf{I}_{\text{start}} = 17.82 - j10.23$ A $= 20.55\angle 29.86°$ A. The starting-current power factor is $\text{pf}_{\text{start}} = \cos 29.86° = 0.867$ lagging. See Fig. 6.5.

3. Calculate the Starting Torque

Use the equation $T_{\text{start}} = KI_mI_a \sin \theta_{ma}$, where K is the machine constant in V · s/A and θ_{ma} is the phase angle in degrees between $\mathbf{I}_m$ (main-winding current) and $\mathbf{I}_a$ (auxiliary-winding current). $T_{\text{start}} = (0.185 \text{ V} \cdot \text{s/A})(8.4 \text{ A})(12.65 \text{ A}) \sin 25.5° = 8.46$ V · A · s or 8.46 N · m.

Related Calculations. Split-phase motors achieve the higher resistance in the auxiliary windings by using fewer turns than in the main winding and using a smaller diameter wire. Therefore, the phase difference (usually less than 40°) does not allow for as high a starting torque as would be the case with a capacitor-start auxiliary winding. This type of starting mechanism is less expensive than the capacitor type for starting. Once started, the auxiliary winding is opened, the motor runs on the main winding only, and there is no difference in operation from the split-phase (resistance-start) or capacitor motor (capacitor-start).

SHADED-POLE MOTOR LOSSES AND EFFICIENCY

A four-pole shaded-pole motor (Fig. 6.6) has the following data associated with it: 120 V, full-load delivered power of 2.5 mhp (millihorsepower), 60 Hz, 350-mA full-load cur-

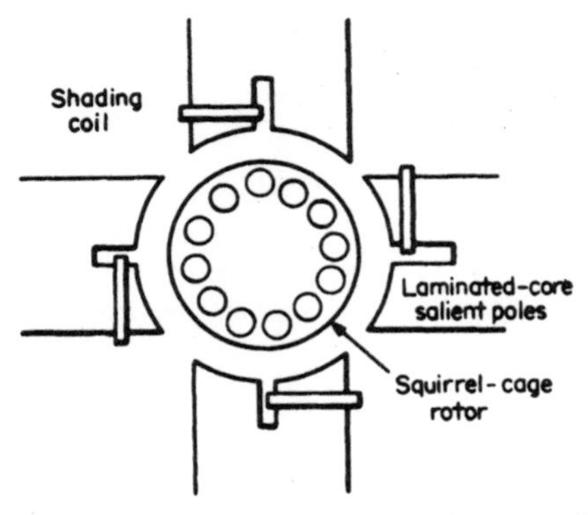

FIGURE 6.6 Four-pole, shaded-pole motor with a squirrel-cage rotor.

rent, 12-W full-load power input, 1525-r/min full-load speed, 1760-r/min no-load speed, 6.6 W no-load power input, 235-mA no-load current, and stator resistance measured with dc of 30 Ω. Calculate the losses and efficiency at full load.

Calculation Procedure

1. Calculate the Rotational Losses

From the no-load conditions, consider that the rotational losses of friction and windage are equal to the power input less the stator-copper loss. The stator resistance, measured with dc, was found to be 30 Ω; this is not the same as the effective ac value of resistance, which is influenced by nonuniform distribution of current over the cross section of the conductors (skin effect). The increase of resistance to ac as compared with dc may vary from 10 to 30 percent, the lower values being for small or stranded conductors and the higher values for large, solid conductors. Assume a value of 15 percent for this problem.

The rotational losses equal $P_{fw} = P_{NL} - I^2_{NL}(R_{dc})$(dc-to-ac resistance-correction factor) $= 6.6\text{ W} - (235 \times 10^{-3}\text{ A})^2(30\ \Omega)(1.15) = 6.6\text{ W} - 1.905\text{ W} = 4.69\text{ W}$.

2. Calculate Stator-Copper Loss at Full Load

At full load the stator-copper loss is $P_{scu} = I^2_{FL}(R_{dc})$(dc-to-ac resistance-correction factor) $= (350 \times 10^{-3}\text{ A})^2(30\ \Omega)(1.15) = 4.23\text{ W}$.

3. Calculate the Slip

The synchronous speed for a four-pole machine is obtained from the relation $n = 120\ f/p$, where $f =$ frequency in Hz and $p =$ the number of poles. Thus, $n = (120)(60)/4 = 1800$ r/min. Since the actual speed is 1525 r/min at full load, the slip speed is $1800 - 1525 = 275$ r/min, and the slip is (275 r/min)/(1800 r/min) $= 0.153$ or 15.3 percent.

4. Calculate Rotor-Copper Loss at Full Load

In induction machines, the rotor-copper loss is equal to the power transferred across the air gap multiplied by the slip. The power transferred across the air gap equals the input power minus the stator-copper loss. Thus, at full load, $P_{rcu} = (12\text{ W} - 4.23\text{ W})(0.153) = 1.2\text{ W}$.

5. Summarize the Full-Load Losses

Stator-copper loss	4.23 W
Rotor-copper loss	1.2 W
Friction and windage loss	4.69 W
Total losses	10.12 W

6. Calculate the Efficiency

The motor delivers 2.5 mhp; the input is 12 W or (12 W)(1 hp/746 W) = 16.1 mhp. Therefore, the efficiency is $\eta =$ (output)(100 percent)/input = (2.5 mhp)(100 percent)/16.1 mhp = 15.5 percent.

Alternatively, the developed mechanical power is equal to the power transferred across the air gap multiplied by $(1 - s)$, or (12 W − 4.23 W)(1 − 0.153) = 6.58 W. From this must be subtracted the friction and windage losses: 6.58 W − 4.69 W = 1.89 W, or (1.89 W)(1 hp/746 W) = 2 .53 mhp. The efficiency is η = (input − losses)(100 percent)/input = (12 W − 10.12 W)(100 percent)/12 W = 15.7 percent.

Related Calculations. The calculations in this problem are the same as for any induction motor. The rotor in this problem is of the squirrel-cage type, and part of each of the four salient stator poles is enclosed by heavy, short-circuited, single-turn copper coils.

SYNCHRONOUS SPEED AND DEVELOPED TORQUE FOR A RELUCTANCE MOTOR

A singly fed reluctance motor having an eight-pole rotor (Fig. 6.7) has a sinusoidal reluctance variation as shown in Fig. 6.8. The power source is 120 V at 60 Hz, and the 2000-turn coil has negligible resistance. The maximum reluctance is $\Re_q = 3 \times 10^7$ A/Wb, and the minimum reluctance is $\Re_d = 1 \times 10^7$ A/Wb. Calculate the speed of the rotor and the developed torque.

Calculation Procedure

1. Calculate the Speed

Use the equation for synchronous speed: $n = 120f/p$ r/min, where n = speed in r/min, f = frequency in Hz, and p = number of poles on the rotor. A reluctance motor operates at synchronous speed. Thus, n = (120)(60 Hz)/8 poles = 900 r/min.

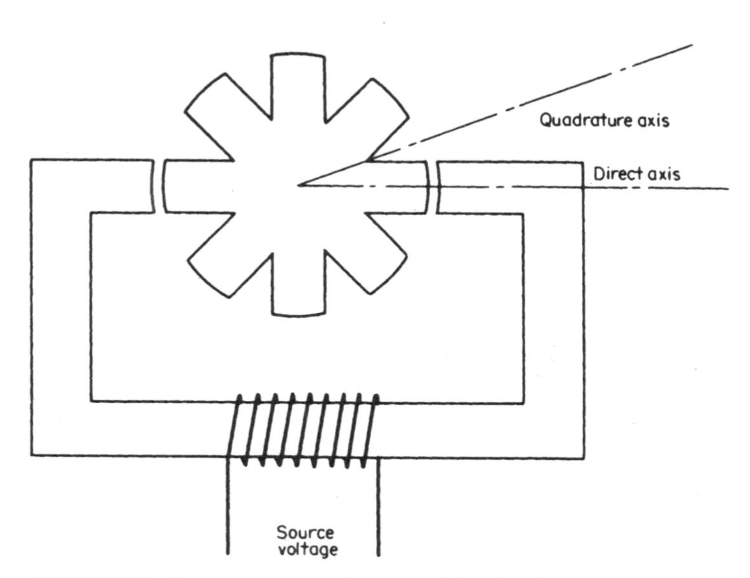

FIGURE 6.7 Reluctance motor with a singly fed eight-pole rotor.

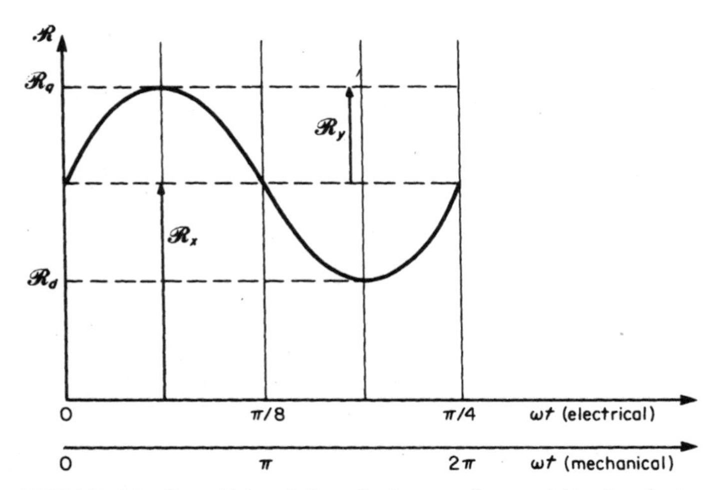

FIGURE 6.8 Sinusoidal variation of reluctance for an eight-pole reluctance motor.

2. Write the Equation for Variation of Reluctance

With reference to Fig. 6.8, $\Re_x = (\Re_q + \Re_d)/2 = (3 \times 10^7 \text{ A/Wb} + 1 \times 10^7 \text{ A/Wb})/2 = 2.0 \times 10^7$ A/Wb. $\Re_y = (\Re_q - \Re_d)/2 = (3 \times 10^7 \text{ A/Wb} - 1 \times 10^7 \text{ A/Wb})/2 = 1.0 \times 10^7$ A/Wb. Therefore, $\Re = \Re_x + \Re_v \sin (p\omega t)$ where p = number of poles on the rotor; $\Re = 2 \times 10^7 + 1 \times 10^7 \sin (8\omega t)$ A/Wb.

3. Calculate the Average Maximum Torque

Use the general equation for maximum time-average torque $T_{avg} = {}^1/_8\, p\phi^2_{max}\, \Re_v$ N · m, where p = number of poles on the rotor. The maximum flux is found from the equation $\phi_{max} = \sqrt{2}\, V_{source}/2\pi fn$ Wb, on the assumption that the n-turn excitation coil has negligible resistance, where f = frequency of the source voltage in Hz. Further, for synchronous speed to occur, and that there be an average torque, the electrical radians of the supply voltage, ω_e, must be related to the mechanical radians of the rotor, ω_m, such that $\omega_m = (2/p)\omega_e$.

The maximum flux is $\phi_{max} = (\sqrt{2})(120 \text{ V})/(2\pi 60)(2000 \text{ turns}) = 0.000225 \text{ Wb} = 2.25 \times 10^{-4}$ Wb. $T_{avg} = {}^1/_8(8 \text{ poles})(2.25 \times 10^{-4} \text{ Wb})^2 (1.0 \times 10^7 \text{ A/Wb}) = 5.066 \times 10^{-1}$ Wb · A or 0.507 N · m.

4. Calculate the Average Mechanical Power Developed

Use the equation $P_{mech} = T_{avg}\omega_m = (0.507 \text{ N} \cdot \text{m})(900 \text{ r/min})(2\pi \text{ rad/r})(1 \text{ min}/60 \text{ s}) =$ 47.8 N · m/s or 47.8 W.

Related Calculations. A necessary condition for reluctance motors, wherein the number of poles on the rotor differs from the number of stator poles, is that the electrical radians of the supply voltage relate to the mechanical radians of the rotor, as indicated in this problem. Also, it is fundamental that this solution is based on the variation of the reluctance being sinusoidal; this requirement is met approximately by the geometric shape of the iron. Also, it is assumed here that the electric-power source voltage is sinusoidal.

These calculations and those of the next problem make reference to average torque and power because in reluctance machines the instantaneous torque and power are not constant but vary. Only over a complete time cycle of the electrical frequency is there average or net torque.

MAXIMUM VALUE OF AVERAGE MECHANICAL POWER FOR A RELUCTANCE MOTOR

The singly fed reluctance motor shown in Fig. 6.9 has an iron stator and rotor with cross-sectional area 1 in. by 1 in. the length of the magnetic path in the rotor is 2 in. The length of each air gap is 0.2 in. The coil of 2800 turns is connected to a source of 120 V at 60 Hz; the resistance is negligible. The reluctance varies sinusoidally, the quadrature-axis reluctance $\Re_q$ is equal to 3.8 times direct-axis reluctance $\Re_d$. Determine the maximum value of average mechanical power.

Calculation Procedure

1. Calculate Direct-Axis Reluctance

Use the equation: $\Re_d = 2g/\mu_0 A$ A/Wb, where g = air gap length in meters, $\mu_0 = 4\pi \times 10^{-7}$ N · m/A^2, and A = cross section of iron in m^2. Converting the given dimensions of inches to meters, 1 in./(39.36 in./m) = 0.0254 m, 2 in. (39.36 in./m) = 0.0508 m, and 0.2 in. (39.36 in./m) = 0.00508 m. Thus, $\Re_d$ = (2)(0.00508 m)/[($4\pi \times 10^{-7}$)(0.0254 m)^2] = 1.253×10^7 A/Wb. And $\Re_q$, being 3.8 times $\Re_d$, is equal to (3.8)(1.253×10^7 A/Wb) or 4.76×10^7 A/Wb.

2. Write the Equation for the Variation of Reluctance

Refer to Fig. 6.10. $\Re_x = (\Re_d + \Re_q)/2$ = (1.253×10^7 A/Wb + 4.76×10^7 A/Wb)/2 = 3.0×10^7 A/Wb. $\Re_y = (\Re_q - \Re_d)/2$ = (4.76×10^7 A/Wb − 1.253×10^7 A/Wb)/2 = 1.75×10^7 A/Wb. Therefore, $\Re = \Re_x + \Re_y \sin(2\omega t) = 3.0 \times 10^7 + 1.75 \times 10^7 \sin(2\omega t)$ A/Wb.

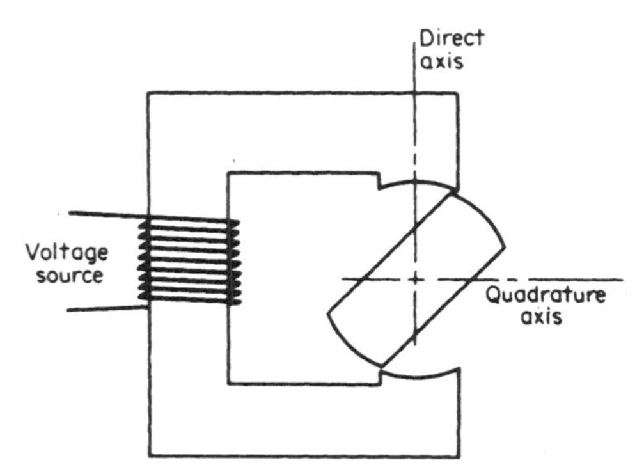

FIGURE 6.9 Singly fed reluctance motor.

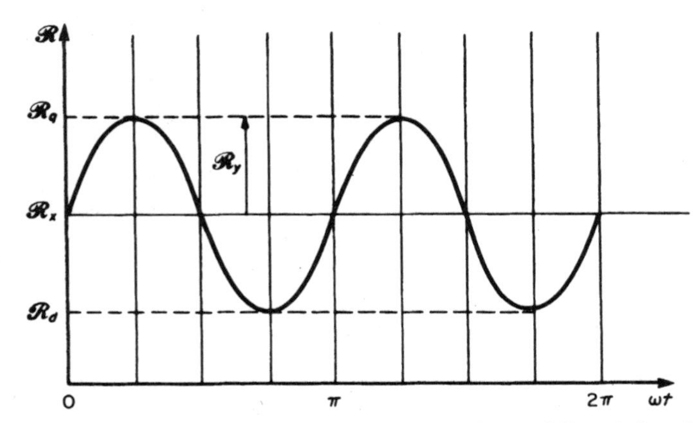

FIGURE 6.10 Sinusoidal variation of reluctance for one full revolution of rotor of a reluctance motor.

3. Calculate the Maximum Value of the Flux

The induced voltage in the coil is sinusoidal (resistance is negligible) and is equal to $Nd\phi/dt = -N\omega\phi_{max} \sin \omega t = V_{source}$. Thus, $\phi_{max} = V_{source}\sqrt{2}\ N\omega$, where N = number of turns, $\omega = 2\pi f$, and f = frequency in Hz; $\phi_{max} = (120\ \text{V})\sqrt{2}\ (2800\ \text{turns})\ (377\ \text{rad/s}) = 16 \times 10^{-5}$ Wb.

4. Calculate the Average Maximum Torque

Use the equation $T_{avg} = \Re_y \phi^2_{max}/4 = (1.75 \times 10^7\ \text{A/Wb})(16 \times 10^{-5}\ \text{Wb})^2/4 = 0.112$ Wb · A or 0.112 N · m.

5. Calculate the Mechanical Power Developed

Use the calculation $P_{mech} = T_{avg}\omega = (0.112\ \text{N} \cdot \text{m})(2\pi \times 60\ \text{rad/s}) = 42.4$ N · m/s or 42.2 W.

Related Calculations. The reluctance motor is a synchronous motor; average torque exists only at synchronous speed. This motor finds application in electric clocks and record players. It has the characteristic of providing accurate and constant speed. The starting conditions may be similar to induction starting; any method employing the principle of induction starting may be used.

BREAKDOWN TORQUE–SPEED RELATIONSHIP FOR A FRACTIONAL-HORSEPOWER MOTOR

A given shaded-pole, 60-Hz motor with a rating of 12.5 mhp has four poles. The breakdown torque is 10.5 oz · in. The shaded-pole speed–torque curve for this machine is given in Fig. 6.11. Determine (1) the speed and horsepower at breakdown, (2) the speed at rated horsepower, and (3) the rated horsepower converted to watts.

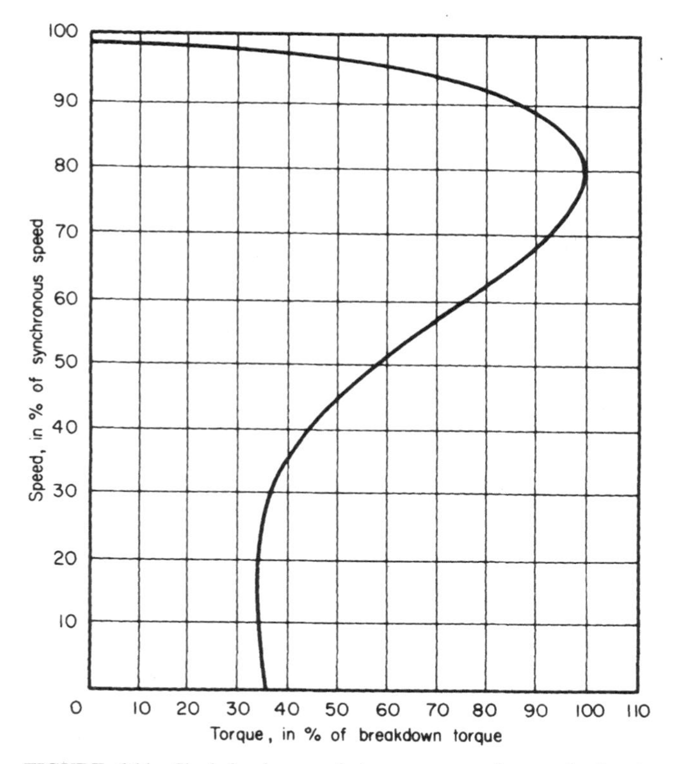

FIGURE 6.11 Shaded-pole speed–torque curve for a fractional-horsepower motor.

Calculation Procedure

1. Determine the Speed at Breakdown Torque

The synchronous speed is calculated from the equation: $n_{sync} = 120f/p$ r/min, where f = frequency in Hz and p = number of poles. Thus, $n_{sync} = (120)(60)/4 = 1800$ r/min. Refer to Fig. 6.11. At the point where the torque in percent of breakdown torque is 100, the speed in percent of synchronous speed is 80. Thus, the rotor speed is $n_{rot} = 0.80n_{sync} = (0.80)(1800) = 1440$ r/min.

2. Calculate the Horsepower at Breakdown

The general equation is horsepower = force × distance × speed. Thus, hp = (lb)(ft)(r/min)(2π rad/r)(hp·min)/33,000 ft·lb = (torque in lb·ft)(n_{rot} in r/min)/5252.1. Further, since 16 oz = 1 lb and 12 in. = 1 ft, hp = (torque in oz·in.)(n_{rot} in r/min)/1,008,403.2, or, approximately, hp = (torque in oz·in.)(n_{rot} in r/min) × 10^{-6}. At breakdown, hp = (10.5 oz·in.)(1440 r/min) × 10^{-6} = 15.12 × 10^{-3} hp = 15.12 mhp.

3. Calculate the Speed at Rated Horsepower

Convert the rated horsepower to torque, assuming a speed; referring to Fig. 6.11, use 95 percent of synchronous speed as a first approximation. Then torque in oz·in. =

hp $\times 10^6/(n_{rot}$ in r/min) = $(12.5 \times 10^{-3} \times 10^6)/(0.95)(1800$ r/min) = 7.31 oz · in. or 0.051 N · m. From this, (rated torque)(100 percent)/breakdown torque = (7.31)(100 percent)/10.5 = 69.6 percent. Refer to Fig. 6.11; read the percentage speed as 94 percent. Notice that in the range of torque from zero to 80 percent of breakdown torque, the speed variation is small, from about 98 to 92 percent of synchronous speed. Using the 94 percent value, the actual rotor speed at rated horsepower is (0.94)(1800 r/min) = 1692 r/min. Note that at this speed the rated horsepower converts to a torque of 7.39 oz · in. or 0.052 N · m. In these conversions, 1 ft · lb = 1.356 N · m, and a N · m/s = W.

4. Convert the Rated Horsepower to Watts

Use the relation: 746 W = 1 hp. The general equation is watts = (torque in oz · in.)(n_{rot} in r/min)/(1.352×10^3) = (7.39 oz · in.)(1692 r/min)/(1.352×10^3) = 9.26 W.

Related Calculations. Shaded-pole machines have low torque and horsepower values; the ounce-inch is a common unit of torque, and the millihorsepower (i.e., hp $\times 10^{-3}$) is a common unit of power. This problem illustrates the conversion of these units. These equations are applicable to fractional-horsepower machines.

FIELD- AND ARMATURE-WINDING DESIGN OF A REPULSION MOTOR

A ⅓-hp, 60-Hz, two-pole repulsion motor (Fig. 6.12) of the enclosed fan-cooled type operating at 3600 r/min has an efficiency of 65 percent. Other data for the machine are as follows: flux per pole = 3.5 mWb, terminal voltage to stator (field winding) = 240 V, stator-winding distributor factor $K_s = 0.91$, number of stator slots = 20 (with conductors distributed in eight slots per pole), number of rotor slots = 24 (with two coils per slot). Calculate (1) the number of conductors per slot in the stator (field winding), (2) the number of conductors and coils for the rotor (armature winding), and (3) the operating voltage of the armature winding.

Calculation Procedure

1. Calculate the Number of Conductors per Field Pole (Stator Winding)

Use the general equation $V = 4.44K_sZ_sf\phi_{pole}$ where V = terminal voltage, K_s = winding distribution factor (stator), Z_s = number of stator conductors, f = frequency in Hz, and ϕ_{pole} = flux per pole in Wb. The constant 4.44 represents $\sqrt{2}\pi$, where $\sqrt{2}$ is used to cause the voltage to be the rms value of ac, rather than the instantaneous value.

Rearranging the equation, find $Z_s = V/4.44K_sf\phi_{pole}$ = 240 V/(4.44)(0.91)(60 Hz)(3.5×10^{-3} Wb) = 283 conductors per pole.

2. Determine the Number of Stator Conductors per Slot

Although the stator has 20 slots, the conductors are distributed in eight slots per pole. If the 283 conductors per pole are divided by 8, the quotient is 35.375. Assume an even whole number such as 36 conductors per slot, yielding 288 conductors per pole (there being eight slots per pole wherein the conductors are to be distributed).

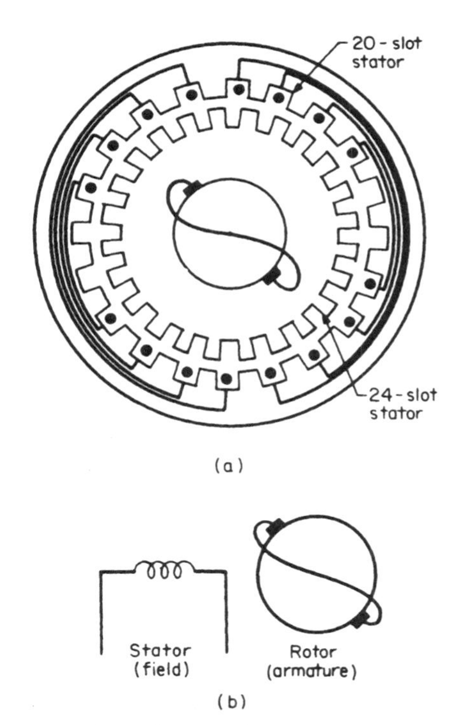

FIGURE 6.12 Repulsion motor showing stator and rotor arrangement. (*a*) Physical arrangement. (*b*) Simplified diagram.

3. Calculate the Number of Conductors per Armature Pole (Rotor Winding)

Use the general equation $E = \phi_{\text{pole}} Z_r np/\sqrt{2}fa$, where E = rotational emf in volts, Z_r = rotor (armature) conductors per pole, n = speed in r/min, p = number of poles, and a = number of parallel paths in the armature winding (for a lap winding, a = the number of poles; for a wave winding, $a = 2$).

Rearranging the equation, find $Z_r = \sqrt{2}\, Efa/\phi_{\text{pole}} np$. In this case, a voltage E is to be assumed as a first try; this voltage should be somewhat low, say 80 V, since the brushes are short-circuited in a repulsion motor. Thus, $Z_r = (\sqrt{2})(80 \text{ V})(60 \text{ Hz})(2)/(3.5 \times 10^{-3} \text{ Wb})(3600 \text{ r/min})(2) =$ 538 conductors per pole.

4. Calculate the Coils-per-Slot Arrangement

Because there are 24 slots on the armature, try an arrangement of two coils per slot = (24)(2) = 48 coils. The number of conductors must be a multiple of the number of coils. Try 12 conductors per coil (i.e., six turns per coil); this yields Z_r = (12 conductors per coil)(48 coils) = 576 conductors per pole.

Using this combination, calculate the rotational emf to compare with the first assumption. $E = (3.5 \times 10^{-3} \text{ Wb})(576 \text{ conductors})(3600 \text{ r/min})(2 \text{ poles})/(\sqrt{2})(60 \text{ Hz})(2 \text{ paths}) = 85.5 \text{ V}$. This is a reasonable comparison.

Related Calculations. The distribution factor for the stator winding in this problem accounts for the pitch factor and the breadth factor. A winding is of fractional pitch when the span from center to center of the coil sides that form a phase belt is less than the pole pitch. The breadth factor is less than unity when the coils forming a phase are distributed in two or more slots per pole.

The repulsion motor is similar in structure to the series motor: It has a nonsalient field (uniform air gap), and the commutated armature (rotor) has a short circuit across the brushes.

AC/DC TORQUE COMPARISON AND MECHANICAL POWER FOR A UNIVERSAL MOTOR

A universal motor operating on dc has the following conditions at starting: line current $I_L = 3.6$ A, starting torque $T_{\text{start}} = 2.3 \text{ N} \cdot \text{m}$. The motor is now connected to an ac source at 120 V/60 Hz. It may be assumed that the total resistance of the motor circuit is 2.7 Ω, the inductance is 36 mH, rotational losses are negligible, and the magnetic field strength varies linearly with the line current. Calculate for the ac condition: (1) the starting torque, (2) the mechanical power produced at 3.6 A, and (3) the operating power factor.

Calculation Procedure

1. Calculate the ac Impedance of the Motor Circuit

Refer to Fig. 6.13. The inductive reactance is calculated from the relation $X_L = \omega L = 2\pi f L$, where f = frequency in Hz and L = inductance in henrys. Thus $X_L = (2\pi)(60\ \text{Hz})(36 \times 10^{-3}\ \text{H}) = 13.6\ \Omega$. The impedance of the motor circuit is $R + jX_L = 2.7 + j13.6\ \Omega = 13.9\underline{/78.8}\ \ \Omega$.

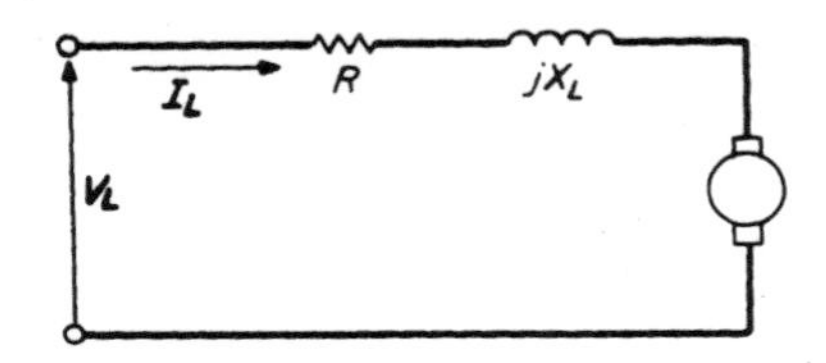

FIGURE 6.13 Equivalent circuit of a universal motor (ac operation).

2. Calculate the Flux

The flux in the magnetic circuit is proportional to the current in the circuit; the field circuit is in series with the armature. Use the general relation $T = k\phi I_L$, where T = torque in N · m and $k\phi$ = flux in Wb. For the dc condition at starting, $k\phi = T_{\text{start}}/I_L = 2.3\ \text{N}\cdot\text{m}/3.6\ \text{A} = 0.639\ \text{Wb}$.

For the ac condition at starting, $I_L = V_L/Z = 120\ \text{V}/13.9\ \Omega = 8.63\ \text{A}$, where Z = circuit impedance. Thus, $k\phi$, being proportional to current, becomes for the ac case (0.639 Wb)(8.63 A)/3.6 A = 1.53 Wb.

3. Calculate the ac Starting Torque

Use the equation again: $T = k\phi I_L$. $T_{\text{start(ac)}} = k\phi I_L = (1.53\ \text{Wb})(8.63\ \text{A}) = 13.2\ \text{Wb}\cdot\text{A}$, or 13.2 N · m.

4. Calculate the Counter-emf, e_a

Use the general equation: $\boldsymbol{e_a} = \mathbf{V_L} - \mathbf{I_L Z}$. Because the phasor relation of $\mathbf{I_L}$ to $\mathbf{V_L}$ is not known, except that it should be a small angle, and e_a should be approximately close to the value of $\mathbf{V_L}$ (say about 85 percent), a rough first calculation will assume $\mathbf{V_L}$ and $\mathbf{I_L}$ are in phase. Thus, $e_a = 120\underline{/0^\circ}\ \text{V} - (3.6\ \text{A})(13.9\underline{/78.8\Omega}) = 120 - 9.7 - j49.1 = 110.3 - j49.1 = 120.7\underline{/-24.0^\circ}\ \text{V}$. Now, repeat the calculation assuming I_L lags V_L by 24°. Thus, $e_a = 120\underline{/0^\circ} - (3.6\underline{/-24^\circ})(13.9\underline{/78.8^\circ}) = 120 - 50.04\underline{/54.8^\circ} = 120 - 28.8 - j40.9 = 91.2 - j40.9 = 99.95\underline{/-24.2^\circ}\ \text{V}$. Notice that the angle of −24° remains almost exactly the same. See Fig. 6.14.

5. Calculate the Mechanical Power, $\mathbf{P}_{mech}$

Use the equation $P_{\text{mech}} = e_a I_L = (99.95\ \text{V})(3.6\ \text{A}) = 359.8\ \text{W}$.

6. Calculate the Power Factor pf, of the Input

The power factor of the input is pf = cos θ, where θ is the angle by which I_L lags V_L; pf = cos 24.2° = 0.912. From this information $P_{\text{mech}} = V_L I_L \cos\theta$ − copper loss $= V_L I_L \cos\theta - I_L^2 R = (120\ \text{V})(3.6\ \text{A})(0.912) - (3.6\ \text{A})^2(2.7\ \Omega) = 393.98 - 34.99 = 359.0\ \text{W}$, which compares with 359.8 W calculated in the previous step (allowing for rounding off of numbers).

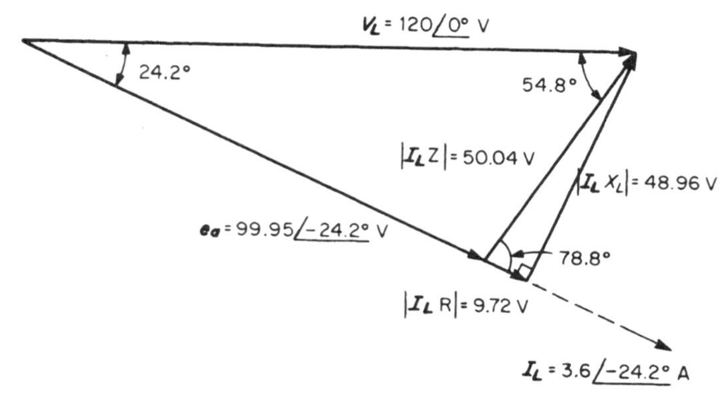

FIGURE 6.14 Phasor diagram of a universal motor circuit (ac operation).

Related Calculations. The universal motor is the same as a series-wound dc motor; it is capable of being operated on ac or dc. The equations used are similar in form, except for the consideration of inductive reactance and phase relations of voltage and current.

SINGLE-PHASE SERIES MOTOR (UNIVERSAL) EQUIVALENT CIRCUIT AND PHASOR DIAGRAM

A 400-hp ac series motor operating at full load has the following data associated with it: 240 V, 25 Hz, 1580 A, 350 kW, 1890 lb · ft (torque), 1111 r/min. The resistances and reactance of the elements are as follows: main series field, $R_{SF} = 0.0018\ \Omega$; armature and brushes, $R_{ab} = 0.0042\ \Omega$; interpole circuit, $R_{int} = 0.0036\ \Omega$; compensating field, $R_{comp} = 0.0081\ \Omega$; total series-circuit reactance (armature, main series field, interpole circuit, compensating field), $X_L = 0.046\ \Omega$. See Fig. 6.15. Calculate (1) the horsepower, (2) the efficiency, (3) the power factor, and (4) the counter-emf.

Calculation Procedure

1. Calculate the Horsepower Output

Use the equation: hp = (torque in lb · ft)(speed n in r/min)(2π rad/r)(hp · min/33,000 ft · lb) = (torque in lb · ft)(n in r/min)/5252 = (1890 lb · ft)(1111 r/min)/5252 = 399.8 hp.

2. Calculate the Efficiency

Use the equation for efficiency: η = (power output)(100 percent)/power input = (399.8 hp)(746 W/hp)(100 percent)/350 $\times$ 10^3 W) = 85.2 percent.

Alternatively, the efficiency may be approximated from the known losses; the copper losses = I^2R. The total circuit resistance is $R_{SF} + R_{ab} + R_{int} + R_{comp}$ = 0.0018 + 0.0042 + 0.0036 + 0.0081 = 0.0177 Ω. The copper losses = (1580 A)^2(0.0177 Ω) = 44,186 W. The efficiency is η = (output $\times$ 100 percent)/(output + losses) = (399.8 hp $\times$ 746 W/hp)(100 percent)/(399.8 hp $\times$ 746 W/hp + 44,186 W) =

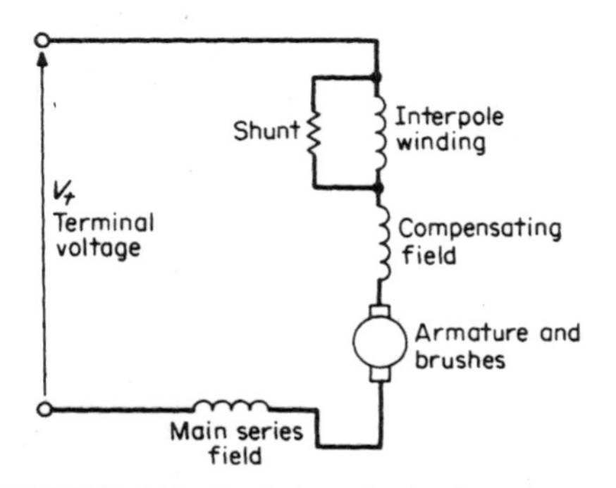

FIGURE 6.15 Equivalent circuit of a series motor.

87.1 percent. Of course, the alternative method does not account for the friction and windage losses, nor for the magnetic losses. These may be calculated from the difference in efficiencies of the two methods. Thus, 350×10^3 W − (399.8 hp)(746 W/hp) − 44,186 W = 7563 W, which may be taken as the friction, windage, and magnetic losses.

The calculation of efficiency becomes η = (399.8 hp)(746 W/hp)(100 percent)/[(399.8 hp)(746 W/hp) + 44,186 W + 7563 W] = 85.2 percent.

3. Calculate the Power Factor, pf

Use the equation for power: $W = VA \cos \theta$, where the power factor = $\cos \theta$. Thus, power factor is (hp)(746 W/hp)/VA = (399.8 hp)(746 W/hp)/(240 V)(1580 A) = 0.786. The power factor angle = $\cos^{-1} 0.786 = 38.1°$.

4. Calculate the Counter-emf

The counter-emf equals (terminal voltage V_t) − (resistive voltage drop) − (reactive voltage drop) = $240\angle 0°$ V − ($1580\angle -38.1°$ A)(0.0177 Ω) − j($1580\angle -38.1°$ A)(0.046 Ω) = 240 − 22.0 + j17.3 − 44.9 − j57.2 = 173.1 − j39.9 = $177.7\angle -13°$ V. See Fig. 6.16.

Related Calculations. Just as in the case of dc motors, series motors used for ac or dc usually have interpoles and compensating windings. The former are for the purpose of improving commutation, and the latter for neutralizing the field-distorting effects of armature reaction. Each of these elements contributes inductive reactance, which is important to consider in the case of ac operation.

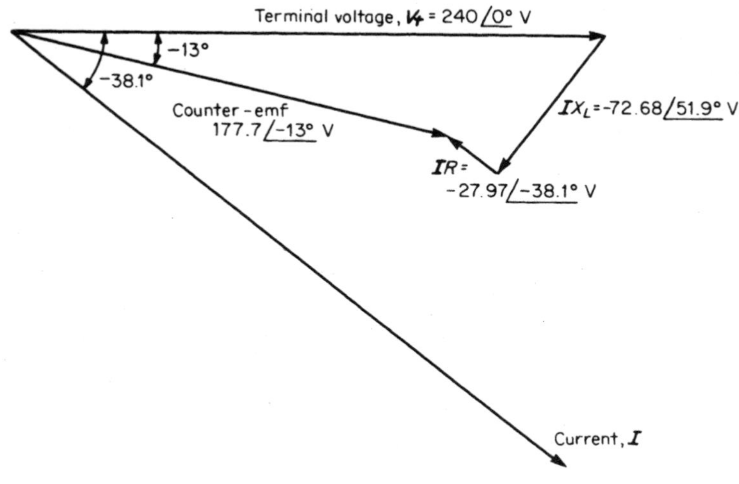

FIGURE 6.16 Phasor diagram of a single-phase series motor.

BIBLIOGRAPHY

Slemon, Gordon R., and A. Straughen. 1998. *Electric Machines*. Reading, Mass.: Addison-Wesley Publishing Co.

Matsch, Leander W. 1986. *Electromagnetic and Electromechanical Machines,* 3rd ed. New York: Harper & Row.

Stein, Robert. 1979. *Electric Power System Components*. New York: Van Nostrand-Reinhold.

Nasar, Syed A., and L. E. Unnewehr. 1979. *Electromechanics and Electric Machines,* 2nd ed. New York: Wiley.

Fitzgerald, Arthur E., Charles Kingsley, Jr., and Stephen D. Umans. 1990. *Electric Machinery*. New York: McGraw-Hill.

Koskow, Irving L. 1991. *Electric Machinery and Transformers*. Englewood Cliffs, N.J.: Prentice-Hall.

Fitzgerald, Arthur E., David E. Higginbotham, and Arvin Grabel. 1967. *Basic Electrical Engineering,* 3rd ed. New York: McGraw-Hill.

Smith, Ralph J., and Richard Dorf. 1992. *Circuits, Devices, and Systems*. New York: Wiley.

Morgan, Alan T. 1979. *General Theory of Electrical Machines*. Philadelphia: Heyden.

Say, Maurice C., and Herbert H. Woodson. 1959. *Electromechanical Energy Conversion*. New York: Wiley.

SECTION 7

SYNCHRONOUS MACHINES

Omar S. Mazzoni, Ph.D., P.E.
President
Systems Research International, Inc.

Marco W. Migliaro, P.E., Fellow IEEE
Chief Electrical and I&C Engineer
Florida Power & Light
Nuclear Division

PER-UNIT BASE QUANTITIES

Calculate the per-unit (p.u.) base quantities for a 150-MVA, 13.8-kV, 60-Hz, three-phase, two-pole synchronous, machine that has the following constants: d-axis mutual inductance between rotor and stator, $L_{ad} = 0.0056$ H; d-axis mutual inductance between stator winding a and rotor, $L_{afd} = 0.0138$ H; mutual inductance between stator winding a and d-axis amortisseur, $L_{akd} = 0.0054$ H; q-axis mutual inductance between rotor and stator, $L_{aq} = 0.0058$ H; and mutual inductance between stator winding a and q-axis amortisseur, $L_{akq} = 0.0063$ H. The per-unit system used should be the reciprocal mutual per-unit system. This denotes a per-unit system where the per-unit mutual inductances between the rotor and stator circuits are reciprocal. This also implies that $\overline{L}_{ad} = \overline{L}_{afd} = \overline{L}_{akd}$ and $\overline{L}_{aq} = \overline{L}_{akq}$. (*Note*: The bar over symbol denotes a per-unit value.)

Calculation Procedure

1. Select Base Values

Select VA_{base} = 150 MVA, V_{base} = 13.8 kV, and f_{base} = 60 Hz. From these values, other base quantities may be derived.

2. Calculate rms Stator Phase Current Base, $I_{s(base)}$

$I_{s(base)} = (MVA_{base} \times 1000)/(\sqrt{3} \times kV_{base}) = (150)(1000)/(\sqrt{3})(13.8) = 6276$ A.

3. Calculate Peak Stator Phase Current Base, $i_{s(base)}$

The current is $i_{s(base)} = \sqrt{2}I_{s(base)} = (\sqrt{2})(6276) = 8876$ A.

4. Calculate Stator Base Impedance, $Z_{s(base)}$

$Z_{s(base)} = kV^2_{base}/MVA_{base} = 13.8^2/150 = 1.270\ \Omega$.

5. Calculate Stator Base Inductance, $L_{s(base)}$

$L_{s(base)} = Z_{s(base)}/\omega_{base} = 1.270/377 = 3.37 \times 10^{-3}$ H.

6. Calculate Field Base Current, $i_{fd(base)}$

The current is $i_{fd(base)} = (L_{ad}/L_{afd})i_{s(base)} = (0.0056/0.0138)\ 8876 = 3602$ A.

7. Calculate Field Base Impedance, $Z_{fd(base)}$

$Z_{fd(base)} = (MVA_{base} \times 10^6)/i^2_{fd(base)} = (150 \times 10^6)/3602^2 = 11.56\ \Omega$.

8. Calculate Field Base Inductance, $L_{fd(base)}$

$L_{fd(base)} = Z_{fd(base)}/\omega_{base} = 11.56/377 = 30.66 \times 10^{-3}$ H.

9. Calculate Field Base Voltage, $e_{fd(base)}$

The voltage is $e_{fd(base)} = (MVA_{base} \times 10^6)/i_{fd(base)} = (150 \times 10^6)/3602 = 41{,}644$ V.

10. Calculate Direct-Axis Armortisseur Base Current, $i_{kd(base)}$

The current is $i_{kd(base)} = (L_{ad}/L_{akd})i_{s(base)} = (0.0056/0.0054)(8876) = 9204$ A.

11. Calculate Direct-Axis Amortisseur Base Impedance, $Z_{kd(base)}$

$Z_{kd(base)} = (MVA_{base} \times 10^6)/i^2_{kd(base)} = (150 \times 10^6)/9204^2 = 1.77\ \Omega$.

12. Calculate Direct-Axis Amortisseur Base Inductance, $L_{kd(base)}$

$L_{kd(base)} = Z_{kd(base)}/\omega_{base} = 1.77/377 = 4.70 \times 10^{-3}$ H.

13. Calculate Quadrature-Axis Amortisseur Base Current, $i_{kq(base)}$

The current is $i_{kq(base)} = (L_{aq}/L_{akq})i_{s(base)} = (0.0058/0.0063)(8876) = 8172$ A.

14. Calculate Quadrature-Axis Amortisseur Base Impedance, $Z_{kq(base)}$

$Z_{kq(base)} = (MVA_{base} \times 10^6)/i^2_{kd(base)} = (150 \times 10^6)/8172^2 = 2.246\ \Omega$.

15. Calculate Quadrature-Axis Amortisseur Base Inductance, $L_{kq(base)}$

$L_{kq(base)} = Z_{kq(base)}/\omega_{base} = 2.246/377 = 5.96 \times 10^{-3}$ H.

16. Calculate Base Mutual Inductance Between Amortisseur and Field, $L_{fkd(base)}$

$L_{fkd(base)} = (i_{fd(base)}/i_{kd(base)})L_{fd(base)} = (3602/9204)(30.66 \times 10^{-3}) = 12 \times 10^{-3}$ H.

17. Calculate Base Flux Linkage, $\phi_{s(base)}$

$\phi_{s(base)} = L_{s(base)}i_{s(base)} = (3.37 \times 10^{-3})\ 8876 = 29.9$ Wb·turns.

18. Calculate Base Rotation Speed in r/min

The base speed is $120f_{\text{base}}/P = (120)(60/2) = 3600$ r/min, where P is the number of poles.

19. Calculate Base Torque, $\mathbf{T}_{(base)}$

$T_{\text{base}} = (7.04\ \text{MVA}_{\text{base}} \times 10^6)/\text{r/min}_{\text{base}} = (7.04)(150)(10^6)/3600 = 293$ klb · ft (397.2 kN · m).

PER-UNIT DIRECT-AXIS REACTANCES

Calculate the synchronous, transient, and subtransient per-unit reactances for the direct axis of the machine in the previous example for which the field resistance $r_{\text{fd}} = 0.0072$ Ω, stator resistance $r_s = 0.0016$ Ω, stator leakage inductance $L_l = 0.4 \times 10^{-3}$ H, field self-inductance $L_{\text{ffd}} = 0.0535$ H, d-axis amortisseur self-inductance $L_{\text{kkd}} = 0.0087$ H, d-axis amortisseur resistance $r_{\text{kd}} = 0.028$ Ω, and the q-axis amortisseur resistance $r_{\text{kq}} = 0.031$ Ω. Assume the leakage inductances in the d- and q-axes are equal, which is generally a reasonable assumption for round-rotor machines.

Calculation Procedure

1. Calculate Per-Unit Values for $\mathbf{L}_{ad}$ and $\mathbf{L}_l$

From the previous example, $L_{\text{s(base)}} = 0.00337$ H. Therefore, $\bar{L}_{\text{ad}} = L_{\text{ad}}/L_{\text{s(base)}} = 0.0056/0.00337 = 1.66$ p.u. and $\bar{L}_l = L_l/L_{\text{s(base)}} = 0.0004/0.00337 = 0.12$ p.u. (*Note:* Bar over symbol designates per-unit value.)

2. Calculate Per-Unit Value of d-Axis Synchronous Inductance, $\bar{\mathbf{L}}_{\mathbf{d}}$

$\bar{L}_d = \bar{L}_{\text{ad}} + \bar{L}_l = 1.66 + 0.12 = 1.78$ p.u.

3. Calculate Per-Unit Value of d-axis Synchronous Reactance, $\bar{\mathbf{X}}_{\mathbf{d}}$

$\bar{X}_d = \bar{\omega}\bar{L}_d$. By choosing $f_{\text{base}} = 60$ Hz, the rated frequency of the machine is $\bar{f} = 60/f_{\text{base}} = 60/60 = 1$. It also follows that $\bar{\omega} = 1$ and $\bar{X}_d = \bar{L}_d = 1.78$ p.u.

4. Calculate $\bar{\mathbf{L}}_{ffd}$, $\bar{\mathbf{L}}_{kkd}$, $\bar{\mathbf{L}}_{afd}$, and $\bar{\mathbf{L}}_{akd}$

$\bar{L}_{\text{ffd}} = L_{\text{ffd}}/L_{\text{fd(base)}} = 0.0535/(30.66 \times 10^{-3}) = 1.74$ p.u. $\bar{L}_{\text{kkd}} = L_{\text{kkd}}/L_{\text{kd(base)}} = 0.0087/(4.7 \times 10^{-3}) = 1.85$ p.u. $\bar{L}_{\text{afd}} = L_{\text{afd}}/L_{\text{s(base)}}(i_{\text{s(base)}}/i_{\text{fd(base)}}) = 0.0138/(0.00337)(8876/3602) = 1.66$ p.u. $\bar{L}_{\text{adk}} = L_{\text{adk}}/[{}^2\!/_3(L_{\text{kd(base)}}i_{\text{kd(base)}}/i_{\text{s(base)}})] = 0.0054/({}^2\!/_3)(4.7 \times 10^{-3})(9204/8876) = 1.66$ p.u.

5. Calculate $\bar{\mathbf{L}}_{afd}$

$\bar{L}_{\text{fd}} = \bar{L}_{\text{ffd}} - \bar{L}_{\text{afd}} = 1.74 - 1.66 = 0.08$ p.u.

6. Calculate $\bar{\mathbf{L}}_{kd}$

$\bar{L}_{\text{kd}} = \bar{L}_{\text{kkd}} - \bar{L}_{\text{akd}} = 1.85 - 1.66 = 0.19$ p.u.

7. Calculate Per-Unit Values of d-axis Transient Inductance, $\bar{\mathbf{L}}'_{\mathbf{d}}$ and $\bar{\mathbf{X}}'_{\mathbf{d}}$

$\bar{L}'_d = \bar{X}'_d = \bar{L}_{\text{ad}}\bar{L}_{\text{fd}}/\bar{L}_{\text{ffd}} + \bar{L}_1 = (1.66)(0.08)/1.74 + 0.12 = 0.196$ p.u.

***8. Calculate Per-Unit Values of* d*-Axis Subtransient Inductance,* $\overline{L}''_d$ *and* $\overline{X}''_d$**

$\overline{L}''_d = \overline{X}''_d = (1/\overline{L}_{kd} + 1/\overline{L}_{ad} + 1/\overline{L}_{fd}) + \overline{L}_l = (1/0.19) + (1/1.66) + (1/0.08) + 0.12 = 0.174$ p.u.

PER-UNIT QUADRATURE-AXIS REACTANCES

Calculate the synchronous and subtransient per-unit reactances for the quadrature axis of the machine in the first example. Additional data for the machine are *q*-axis amortisseur self-inductance $L_{kkq} = 0.0107$ H, and the mutual inductance between stator winding *a*- and *q*-axis $L_{akq} = 6.3 \times 10^{-3}$ H.

Calculation Procedure

1. Calculate $\overline{L}_{aq}$

From values obtained in the two previous examples, $\overline{L}_{aq} = L_{aq}/L_{s(base)} = 0.0058/(3.37 \times 10^{-3}) = 1.72$ p.u.

2. Calculate $\overline{L}_q$

$\overline{L}_q = \overline{L}_{aq} + \overline{L}_l = 1.72 + 0.12 = 1.84$ p.u.

3. Calculate $\overline{L}_{kkq}$ ***and*** $\overline{L}_{akq}$

$\overline{L}_{kkq} = L_{kkq}/L_{kq(base)} = 0.0107/(5.96 \times 10^{-3}) = 1.80$ p.u. $\overline{L}_{akq} = L_{akq}/(L_{s(base)})(i_{s(base)}/i_{kq(base)}) = (6.3 \times 10^{-3})/(3.37 \times 10^{-3})(8876/8172) = 1.72$ p.u.

4. Calculate Per-Unit Value of* q*-axis Amortisseur Leakage Inductance, $\overline{L}_{kq}$

$\overline{L}_{kq} = \overline{L}_{kkq} - \overline{L}_{akq} = 1.80 - 1.72 = 0.08$ p.u.

5. Calculate Per-Unit Transient Inductance, $\overline{L}'_q$

$\overline{L}'_q = \overline{L}_{aq}\overline{L}_{kq}/(\overline{L}_{aq} + \overline{L}_{kq}) + \overline{L}_l = (1.72)(0.08)/(1.72 + 0.08) + 0.12 = 0.196$ p.u.

Related Calculations. Transient inductance $\overline{L}'_q$ is sometimes referred to as the *q*-axis subtransient inductance.

PER-UNIT OPEN-CIRCUIT TIME CONSTANTS

Calculate the per-unit field and subtransient open-circuit time constants for the direct axis of the machine in the first example. Use results obtained in previous examples.

Calculation Procedure

1. Calculate $\overline{r}_{fd}$ ***and*** $\overline{r}_{kd}$

The quantities are $\overline{r}_{fd} = r_{fd}/Z_{fd(base)} = 0.0072/11.56 = 6.23 \times 10^{-4}$ p.u. and $\overline{r}_{kd} = r_{kd}/Z_{kd(base)} = 0.028/1.77 = 0.0158$ p.u.

2. Calculate Field Open-Circuit Time Constant, $\overline{T}'_{do}$

$\overline{T}'_{do} = \overline{L}_{ffd}/\overline{r}_{fd} = 1.74/(6.23 \times 10^{-4}) = 2793$ p.u.

3. Calculate Subtransient Open-Circuit Time Constant, $\overline{T}''_{do}$

$\overline{T}''_{do} = (1/\overline{r}_{kd})(\overline{L}_{kd} + \overline{L}_{fd}\overline{L}_{ad}/\overline{L}_{ffd}) = (1/0.0158)[0.19 + (0.08)(1.66/1.74)] = 16.9$ p.u.

PER-UNIT SHORT-CIRCUIT TIME CONSTANTS

Calculate the per-unit transient and subtransient short-circuit time constants for the direct axis of the machine in the first example. Also, calculate the direct-axis amortisseur leakage time constant.

Calculation Procedure

1. Calculate Per-Unit Transient Short-Circuit Time Constant, $\overline{T}'_d$

$\overline{T}'_d = (1/\overline{r}_{fd})[\overline{L}_{fd} + \overline{L}_l\,\overline{L}_{ad}/(\overline{L}_l + \overline{L}_{ad})] = [1/(6.23 \times 10^{-4})][0.08 + (0.12)(1.66)/(0.12 + 1.66)] = 308$ p.u.

2. Calculate Per-Unit Subtransient Short-Circuit Time Constant, $\overline{T}''_{do}$

$$\frac{1}{\overline{r}_{kd}}\left[\overline{L}_{fd} + \frac{1}{(1/\overline{L}_{ad}) + (1/\overline{L}_{fd}) + (1/\overline{L}_l)}\right]$$

$$= \frac{1}{0.0158}\left[0.19 + \frac{1}{(1/1.66) + (1/0.08) + (1/0.12)}\right] = 15 \text{ p.u.}$$

3. Calculate Per-Unit Amortisseur Leakage Time Constant, $\overline{T}_{kd}$

$\overline{T}_{kd} = \overline{L}_{kd}/\overline{r}_{kd} = 0.19/0.158 = 12$ p.u.

Related Calculations. Per-unit open-circuit and short-circuit time constants for the quadrature axis of the machine may be calculated with procedures similar to those used in the two previous examples.

To calculate a time constant in seconds, multiply the per-unit quantity by its time base, $1/\omega_{(base)} = 1/377$ s; $T = \overline{T}/377$. For example, $T'_d = 308/377 = 0.817$ s.

STEADY-STATE PHASOR DIAGRAM

Calculate the per-unit values and plot the steady-state phasor diagram for a synchronous generator rated at 100 MVA, 0.8 pf lagging, 13.8 kV, 3600 r/min, 60 Hz, operating at rated load and power factor. Important machine constants are $\overline{X}_d = 1.84$ p.u., $\overline{X}_q = 1.84$ p.u., and $X'_d = 0.24$ p.u. The effects of saturation and machine resistance may be neglected.

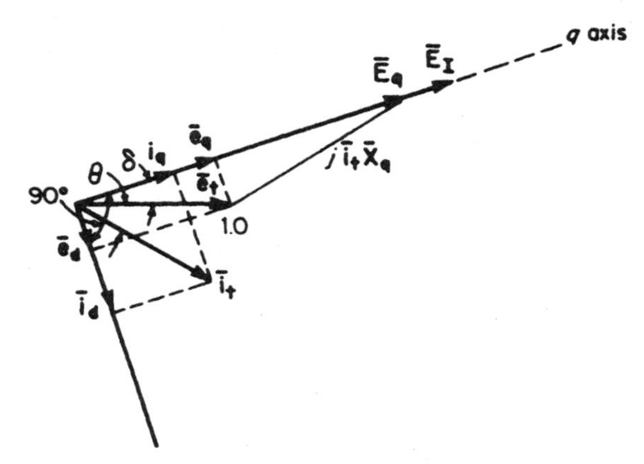

FIGURE 7.1 Phasor diagram for a synchronous generator.

Calculation Procedure

1. Determine Reference

If the VA and V base values are equal to the machine ratings, $I_{s(\text{base})} = (\text{MVA} \times 1000)/\sqrt{3}\ \text{kV} = (100)(1000)/(\sqrt{3})(13.8) = 4184$ A. Because the base voltage has been taken as 13.8 kV, the per-unit rms terminal voltage is $\overline{E}_t = 1.0$ p.u. From this, the per-unit peak voltage, $\bar{\mathbf{e}}_t = \overline{\mathbf{E}}_t = 1.0$ p.u. Quantity $\bar{\mathbf{e}}_t$ will be chosen as the reference phasor: $\bar{\mathbf{e}}_t = 1.0\underline{/0}$ p.u.

2. Locate q*-Axis*

Calculate a fictitious voltage $\overline{\mathbf{E}}_q = |\overline{\mathbf{E}}_q|/\delta$, where δ is the machine internal power angle. $\overline{\mathbf{E}}_q$ may be calculated by $\overline{\mathbf{E}}_q = \bar{\mathbf{e}}_t + \bar{\mathbf{i}}_t(\bar{r} + jX_q)$. But $\bar{\mathbf{i}}_t = 1.0\underline{/\theta°}$ p.u., where $\theta = \cos^{-1} 0.8 = -36.9°$. Therefore, $\overline{\mathbf{E}}_q = 1\underline{/0°} + 1.0\underline{/-36.9°} \times j1.84 = 2.38\underline{/35°}$ p.u. Power angle $\delta = 35°$.

3. Calculate the d*- and* q*-axis Components*

The d- and q-axis components may now be found by resolving $\bar{\mathbf{e}}_t$ and $\bar{\mathbf{i}}_t$ into components along the d- and q-axes, respectively: $\bar{\mathbf{e}}_q = |\bar{\mathbf{e}}_t| \cos\delta = (1.0)(0.819) = 0.819$ p.u. $\bar{\mathbf{e}}_d = |\bar{\mathbf{e}}_t| \sin\delta = (1.0)(0.574) = 0.574$ p.u. and $\bar{\mathbf{i}}_q = |\bar{\mathbf{i}}_t| \cos(\delta - \theta) = 1.0 \cos(35° + 36.9°) = 0.311$ p.u. $\bar{\mathbf{i}}_d = |\bar{\mathbf{i}}_t| \sin(\delta - \theta) = 0.951$ p.u.

4. Calculate $\overline{E}_I$

Voltage $\overline{\mathbf{E}}_I$ lies on the q-axis and represents the d-axis quantity, field current. $\overline{\mathbf{E}}_I = \overline{X}_{ad}\,\bar{\mathbf{i}}_{fd} = \bar{\mathbf{e}}_q + \overline{X}_d\bar{\mathbf{i}}_d + \bar{r}\bar{\mathbf{i}}_d = 0.819 + (1.84)(0.951) = 2.57$ p.u.

5. Draw Phasor Diagram

The phasor diagram is drawn in Fig. 7.1.

GENERATOR-CAPABILITY CURVE

The generator-capability curve, supplied by the manufacturer, is used to determine the ability of a generator to deliver real (MW) and reactive (Mvar) power to a network. Determine the capability curve, in per-unit values, of a generator with the following characteristics: 980 kVA, pf = 0.85, synchronous reactance $\overline{X}_d = 1.78$ p.u., maximum value of generator internal voltage $\overline{E}_{\max} = 1.85$ p.u., terminal voltage $\overline{V} = 1.0$ p.u., $\delta =$ load angle, and the system reactance, external to generator, is $\overline{X}_e = 0.4$ p.u. Consider steady-state stability as the limit of operation for a leading power factor.

Calculation Procedure

1. Calculate Stator-Limited Portion

The stator-limited portion is directly proportional to the full-power output, which is an arc of a circle with radius $\overline{R}_S = 1.0$ p.u. (curve *ABC* of Fig. 7.2).

2. Calculate Field-Limited Portion

The field-limited portion is obtained from the following expression: $\overline{P} = (3\overline{V}\overline{E}_{\max}/\overline{X}_d)\sin\delta + j[(\sqrt{3}\,\overline{V}\overline{E}_{\max}/\overline{X}_d\cos\delta - \sqrt{3}\,\overline{V}^2/\overline{X}_d]$, which is a circle with the center at 0,

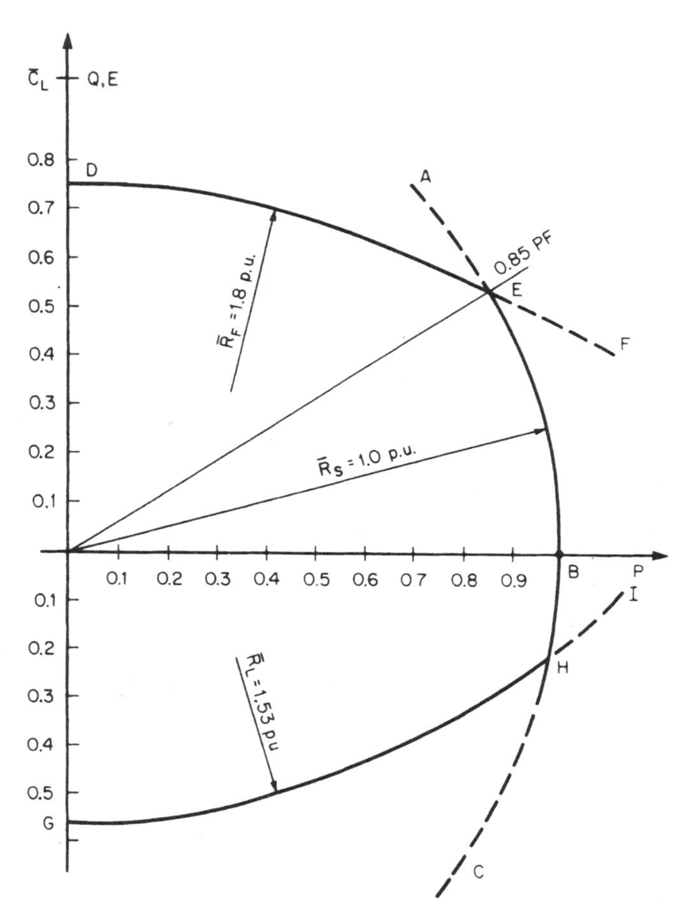

FIGURE 7.2 A generator-capability curve.

$-j\sqrt{3}\,\overline{V}^2/\overline{X}_d$, and whose radius is $\overline{R}_F = \sqrt{3}\,\overline{V}\overline{E}_{max}/\overline{X}_d = \sqrt{3}\,(1.0)(1.85)/1.78 = 1.8$ p.u. Because $\overline{V} = 1.0$ p.u., the center is located at $0, j(-1.0 \times \sqrt{3})/1.78 = 0, -j0.97$ (curve *DEF*, Fig. 7.2).

3. Calculate Steady-State Stability Curve

The steady-state stability curve is given by an arc of the circle defined by $\overline{C}_L$ = center $= j\overline{V}^2/2(1/\overline{X}_e - 1/\overline{X}_d) = j/2(1/0.4 - 1/1.78) = j0.97$. $\overline{R}_L$ = radius $= \overline{V}^2/2(1/\overline{X}_e + 1/\overline{X}_d) = 1/2(1/0.4 + 1.78) = 1.53$ p.u. (curve *GHI*, Fig. 7.2).

Related Calculations. Synchronous machines are capable of producing and consuming megawatts and megavars. When a generator is overexcited, it generates kilovars and delivers them to the system. When a generator is underexcited, negative megavars flow from the system into the machine. When the machine operates at unity power factor, it is just self-sufficient in its excitation.

Because operation of a generator is possible at any point within the area bounded by the capability curve, operators make use of the capability curve to control machine output within safe limits. The stator-limited portion relates to the current-carrying capacity of the stator-winding conductors. The field-limited portion relates to the area of operation under overexcited conditions, where field current will be higher than normal. The steady-state stability portion relates to the ability of the machine to remain stable.

The curves in Fig. 7.3 are typical capability characteristics for a generator. A family of curves corresponding to different hydrogen cooling pressures is shown. The rated power factor is 0.8; rated apparent power is taken as 1.0 p.u. on its own rated MVA base. This means that the machine will deliver rated apparent power down to 0.85 lagging power factor. For a lower power factor, the apparent power capability is lower than rated.

Many of the older, large steam-turbine generators in operation currently are being evaluated for the potential of increasing their power output. The industry refers to this as "power uprate." In many instances, the original designs of these machines included design margins that were used to account for uncertainties or were used for conservatism. Some machines also had generators with capabilities that exceeded the capabilities of the steam-turbine. With more sophisticated analysis tools available, increases in power output (and revenue) may potentially be obtained for a relatively small investment. Improved turbine blade designs and materials can also offer additional power output by replacement of the original turbine rotors. An uprate normally requires a study to determine if auxiliary systems (e.g., cooling water, feedwater, and condensate) can support the uprate. This study would include evaluation of system modifications required to support the uprate (e.g., motor horsepower increases and pump impeller replacements).

It is also possible to increase the generator power output by increased hydrogen pressure from the normal operating pressure to pressures as high as 75 psig. Increasing the hydrogen pressure results in increased cooling for the machine. With increased cooling, there is a greater machine rated power output available, since additional current can be safely carried by the conductors due to the increased cooling. However, a full evaluation of the generator design and its supporting systems must be conducted in order to determine the acceptability of the increased generator hydrogen pressure. (This study would normally involve the generator designer/manufacturer). These factors may include the effect on the hydrogen coolers and seal oil system, as well as, the ability of the excitation system to support the desired increase in power. Additionally, once the output power of the machine is increased, the capability of the generator transformer and switchyard equipment (e.g., circuit breakers) must be evaluated.

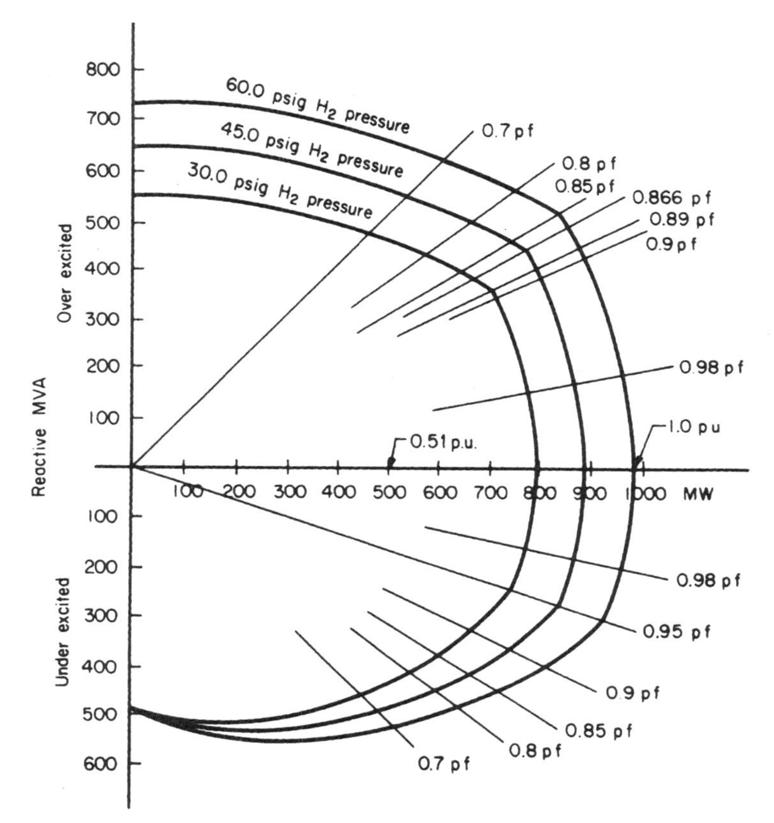

FIGURE 7.3 Capability curves for a hydrogen inner-cooled generator rated at 983 MVA, 0.85 pf, 22 kV.

GENERATOR REGULATION

Determine the regulation of a generator with the following characteristics: armature resistance $\overline{R}_a = 0.00219$ p.u.; power factor = 0.975; and open-circuit, zero power-factor and short-circuit saturation as in Fig. 7.4.

Calculation Procedure

1. Calculate Potier Reactance, $\overline{\mathbf{X}}_\mathbf{p}$

Use the relation $E_0 = E + \sqrt{3}I_aX_p$ from zero power-factor conditions, where E_0 = voltage at no load, I_a = armature current at full load, and E is the terminal voltage. In Fig. 7.4, RD = RE + DE; therefore, DE = $\sqrt{3}I_aX_p$. By definition, $\overline{X}_p = I_aX_p/(V_{LL}/\sqrt{3}) = \text{DE}/\overline{V}_{LL} = \text{DE/RE} = 0.43$ p.u.

2. Calculate Voltage behind Potier Reactance, $\overline{\mathbf{E}}_\mathbf{p}$

From Fig. 7.4, $\overline{E}_p = 1.175$ p.u.

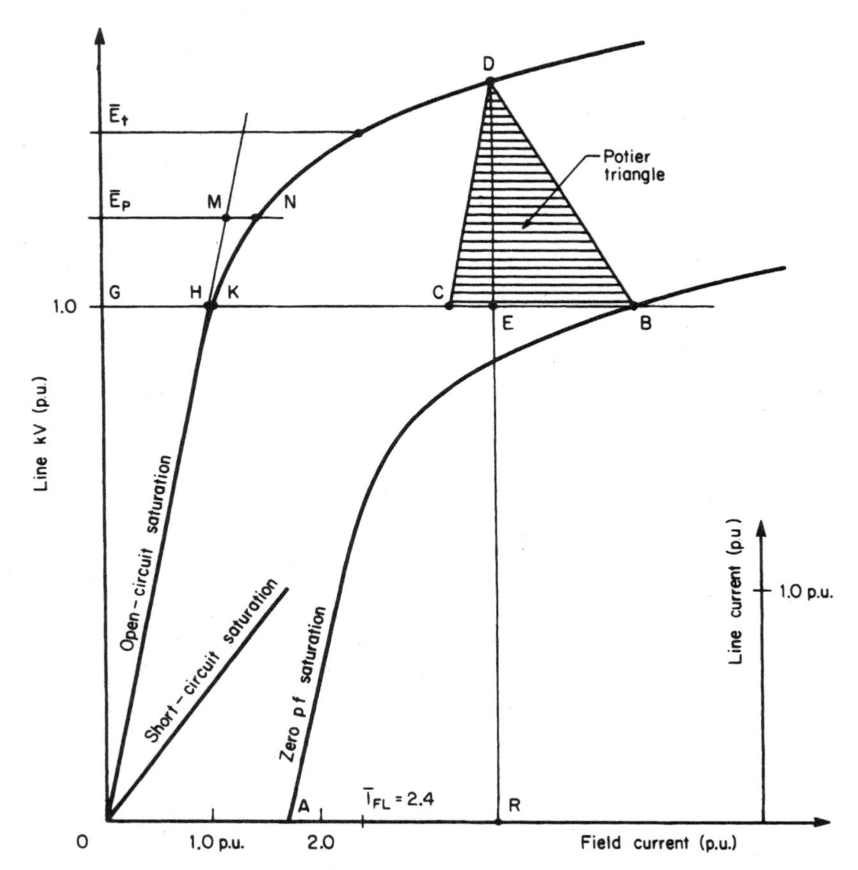

FIGURE 7.4 Determination of generator regulation.

3. Determine Excitation Required to Overcome Saturation, $\bar{\mathbf{I}}_{FS}$

From Fig. 7.4, $\bar{I}_{\text{FS}} = MN = 0.294$ p.u.

4. Determine Excitation for Air-Gap Line, $\bar{\mathbf{I}}_{FG}$

From Fig. 7.4, $\bar{I}_{\text{FG}} = GH = 1.0$ p.u.

5. Determine Excitation for Full-Load Current on Short-Circuit Saturation Curve, $\bar{\mathbf{I}}_{FSI}$

From Fig. 7.4, $\bar{I}_{\text{FSI}} = 0A = 1.75$ p.u.

6. Calculate Excitation for Full-Load Current, $\bar{\mathbf{I}}_{FL}$

From Fig. 7.5, $\bar{I}_{\text{FL}} + \bar{I}_{\text{FS}} + [(\bar{I}_{\text{FG}} + \bar{I}_{\text{FSI}} \sin \phi)^2 + (\bar{I}_{\text{FSI}} \cos \phi)^2]^{1/2} = 0.294 + [(1 + 1.75 \times 0.2223)^2 + (1.75 \times 0.975)^2]^{1/2} = 2.4$ p.u.

7. Determine $\bar{\mathbf{E}}_t$ ***on Open-Circuit Saturation Curve Corresponding to Full-Load Field Excitation,*** $\mathbf{I}_{FL}$

From Fig. 7.4, $\overline{E_t} = 1.33$ p.u.

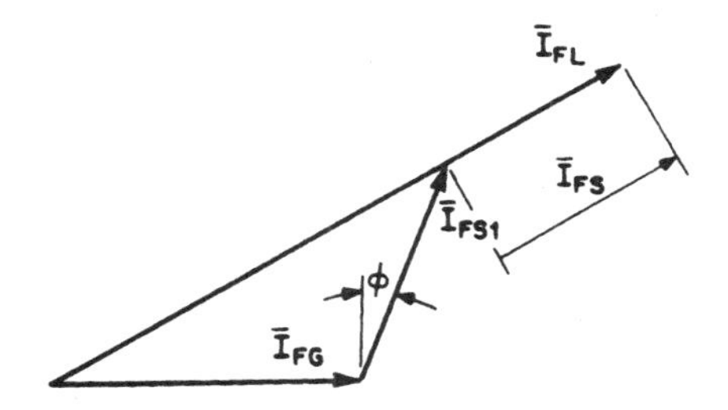

FIGURE 7.5 Generator excitation at full-load current.

8. Calculate Percent Regulation, % R

% R $= (\bar{E}_t - \bar{E})/\bar{E} = (1.33 - 1.0)/1.0 = 33\%$.

GENERATOR SHORT-CIRCUIT RATIO

Calculate the short-circuit ratio (SCR) for a generator having characteristic curves shown in Fig. 7.4.

Calculation Procedure

1. Determine Excitation Value for Full-Load Current, $\bar{\mathbf{I}}_{FSI}$, on the Short-Circuit Curve

$\bar{I}_{FSI} = 0A = 1.75$ p.u.

2. Determine Excitation, $\bar{\mathbf{I}}_{FV}$, Required to Produce Full Voltage on Open-Circuit Curve

$\bar{I}_{FV} = GK = 1.06$ p.u.

3. Calculate Short-Circuit Ratio (SCR)

SCR $= GK/0A = 1.06/1.75 = 0.6$.

POWER OUTPUT AND POWER FACTOR

Calculate the maximum output power for an excitation increase of 20 percent for a 13.8-kV wye-connected generator having a synchronous impedance of 3.8 Ω/phase. It is connected to an infinite bus and delivers 3900 A at unity power factor.

Calculation Procedure

1. Draw Phasor Diagrams

See Fig. 7.6. Subscript o indicates initial conditions. Voltage **V** is the line-to-neutral terminal voltage and **E** is the line-to-neutral voltage behind the synchronous reactance. Angle δ is the machine internal power angle and ϕ is the angle between the phase voltage and phase current.

2. Calculate Voltage behind Synchronous Reactance

$E = [(IX_s)^2 - V^2]^{1/2} = [(3900 \times 3.8/1000)^2 - (13.8/\sqrt{3})^2]^{1/2} = 14.54$ kV.

3. Calculate Maximum Power, $\mathbf{P}_{max}$

$P_{max} = 3EV/X_s \sin\delta$, where $\sin\delta = 1$ for maximum power. For a 20 percent higher excitation, $P_{max} = [(3)(1.2)(14.54)(13.8)/\sqrt{3}]/3.8 = 110$ MW.

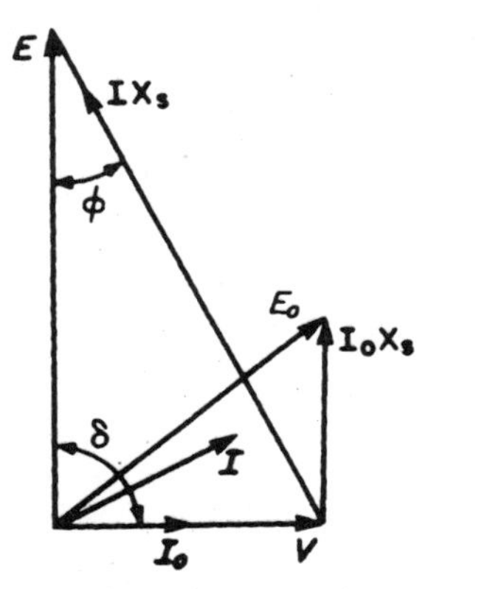

FIGURE 7.6 Generator phasor diagrams: power factor (pf) and power output.

4. Calculate Power Factor

The power factor is $\cos \phi = E/IX_s = 14.54/(3.9)(3.8) = 0.98$ lagging.

High-temperature super-conducting (HTS) generators are in the process of development and promise to offer many advantages over conventional generators. These advantages relate to lower losses and smaller, more compact designs. Related calculations for evaluating the performance of HTS machines will include specific issues in the area of their stability assessment and particular cooling characteristics.

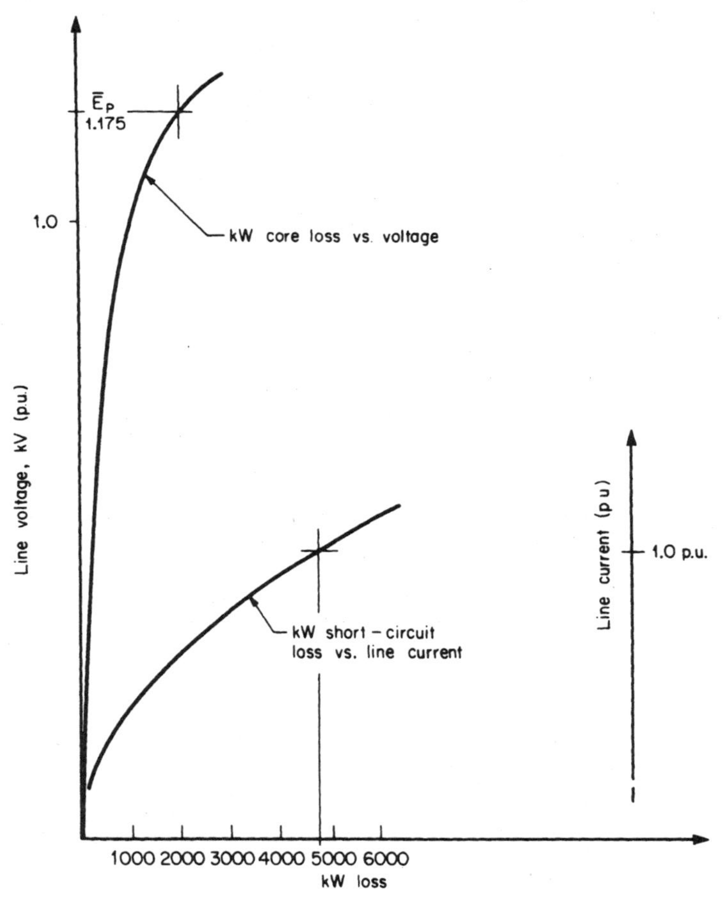

FIGURE 7.7 Generator loss curves.

GENERATOR EFFICIENCY

Determine the efficiency of a generator having the same basic characteristics of the generator in the "Generator Regulation" example. Additional data include armature full-load current $I_a = 28{,}000$ A; core and short-circuit losses as in Fig. 7.7; friction and windage loss, 500 kW (from drive motor input); armature resistance $R_a = 0.0011$ Ω/phase; excitation voltage at rated load, 470 V; excitation current for air-gap line, 3200 A; and output voltage 25 kV.

Calculation Procedure

1. Compute Core Loss

From Fig. 7.7 and the Potier voltage, E_p, (Fig. 7.4), for $\overline{E}_p = 1.175$ p.u., core loss = 2100 kW.

2. Determine Short-Circuit Loss

From Fig. 7.7, for 1.0 p.u. line current, short-circuit loss = 4700 kW.

3. Calculate Armature-Copper Loss

Armature-copper loss = $I_a^2 R_a = (28{,}000)^2(0.0011) = 862{,}000$ W = 862 kW.

4. Calculate Stray-Current Loss

Stray-current loss = short-circuit loss − armature-copper loss = 4700 − 862 = 3838 kW.

5. Calculate Power for Excitation

Power for excitation = field voltage × I_{FL} = (470)(2.4)(3200)/1000 = 3610 kW, where I_{FL} is from the "Generator Regulation" example.

6. Determine the Total Losses

Friction and windage	500 kW
Core loss	2,100
Armature copper loss	862
Stray current loss	3,838
Power for excitation	3,610
Total losses	10,910 kW

7. Calculate Generator Output, P_o

$P_o = \sqrt{3} \times \text{kV} \times \text{A} \times \text{pf} = (\sqrt{3})(25)(28{,}000)(0.975) = 1{,}182{,}125$ kW

8. Calculate Generator Efficiency

Efficiency = (power output)/(power output + total losses) = 1,182,125/(1,182,125 + 10,910) = 0.99 or 99 percent.

SYNCHRONIZING POWER COEFFICIENT

Calculate the synchronizing power coefficient at rated load for the following generator: 75,000 kW, terminal voltage $\overline{V} = 1.0$ p.u., armature current $\overline{I}_a = 1$ p.u., quadrature axis reactance $\overline{X}_q = 1.8$ p.u., and pf = 0.80 lagging. Neglect the resistive component of the armature.

Calculation Procedure

1. Calculate Rated Load Angle, δ

The angle is $\delta = \tan^{-1} [\overline{X}_q \cos \phi \, \overline{I}_a / (\overline{I}_a \, \overline{X}_q \sin \phi + \overline{V})] = \tan^{-1} \{(1.8)(0.80)(1.0)/[(1.0)(1.80)(0.6) + 1]\} = 35°$.

***2. Calculate Synchronizing Power Coefficient,* P$_r$**

P_r = (rated kW)/(rated load angle × 2π/360) = (75)(1000)/(35)(2π/360) = 122,780 kW/rad.

GENERATOR GROUNDING TRANSFORMER AND RESISTOR

Determine the size of a transformer and resistor required to adequately provide a high-resistance ground system for a wye-connected generator rated 1000 MVA, 26 kV, 60 Hz. In addition, generator capacitance = 1.27 μF, main transformer capacitance = 0.12 μF, generator lead capacitance = 0.01 μF, and auxiliary transformer capacitance = 0.024 μF.

Calculation Procedure

***1. Calculate Generator Line-to-Neutral Voltage,* V$_{L\text{-}N}$**

$V_{\text{L-N}} = (26 \text{ kV})/\sqrt{3} = 15$ kV.

***2. Calculate Total Capacitance,* C$_T$**

$C_T = (1.27 + 0.12 + 0.01 + 0.024)$ μF = 1.424 μF.

***3. Calculate Total Capacitive Reactance,* X$_{CT}$**

$X_{\text{CT}} = 1/2\pi f C_T = 1/(6.28)(60)(1.424 \times 10^{-6}) = 1864$ Ω.

4. Select* R = X$_{CT}$ *to Limit Transient Overvoltage during a Line-to-Ground Fault

Assume a 19.92/0.480-kV transformer. The resistance reflected to the primary is $R' = N^2R$, where R is the required resistor. Solve $R = R'N^2$, where $N = 19.92/0.480 = 41.5$, and find $R = 1864/41.5^2 = 1.08$ Ω.

5. Calculate Transformer Secondary Voltage,* V$_s$, *during a Line-to-Ground Fault

$V_s = V/N = 15{,}000/41.5 = 361$ V.

6. Calculate Current, I_s, through Grounding Resistor

$I_s = V_s/R = 361/1.08 = 334.3$ A.

7. Calculate Required Continuous Rating in kVA of Grounding Transformer

The rating is kVA $= I_sV_s = (334.3)(361) = 120.7$ kVA.

8. Select Short-Time Rated Transformer

From ANSI standards, a 50-kVA transformer may be used if a 9-min rating is adequate.

9. Calculate Generator Line-to-Ground Fault Current, I_f

$I_f = V/X_{CT} = 15{,}000/1864 = 8.05$ A.

POWER-FACTOR IMPROVEMENT

An industrial plant has a 5000-hp induction motor load at 4000 V with an average power factor of 0.8, lagging, and an average motor efficiency of 90 percent. A new synchronous motor rated at 3000 hp is installed to replace an equivalent load of induction motors. The synchronous motor efficiency is 90 percent. Determine the synchronous motor current and power factor for a system current of 80 percent of the original system and unity power factor.

Calculation Procedure

1. Calculate Initial System Rating, kVA_o

The rating is $kVA_o = (hp/\eta)(0.746/pf) = (5000/0.9)/(0.746/0.8) = 5181$ kVA, where η is efficiency.

2. Calculate Initial System Current, I_o

$I_o = kVA/\sqrt{3}\,V = 5181/(\sqrt{3})(4) = 148$ A.

3. Calculate New System Current, I

$I = 0.8I_o = (0.8)(746) = 598$ A (see Fig. 7.8).

4. Calculate New Induction-Motor Current, I_i

$I_i = 0.746\ hp/\sqrt{3}\,V\ \eta pf_i = (0.746)(2000)/(\sqrt{3})(4)(0.9)(0.8) = 299$ A.

5. Calculate Synchronous-Motor Current, I_s

$I_s = [(I_i \sin \phi_i)^2 + (I - I_i \cos \phi_i)^2]^{1/2} = \{[(299)(0.6)]^2 + [598 - (299)(0.8)]^2\}^{1/2} = (149.4^2 + 358.8^2)^{1/2} = 401.1$ A

6. Calculate Synchronous-Motor Power Factor, pf_s

The power factor is $pf_s = 358.8/401.1 = 0.895$.

Related Calculations. For verification, the synchronous motor horsepower, hp_s, should equal 3000 hp, where $hp_s = 3VI_s\eta pf_s/0.746 = (\sqrt{3})(4)(401.1)(0.9)(0.895)/0.746 = 3000$ hp.

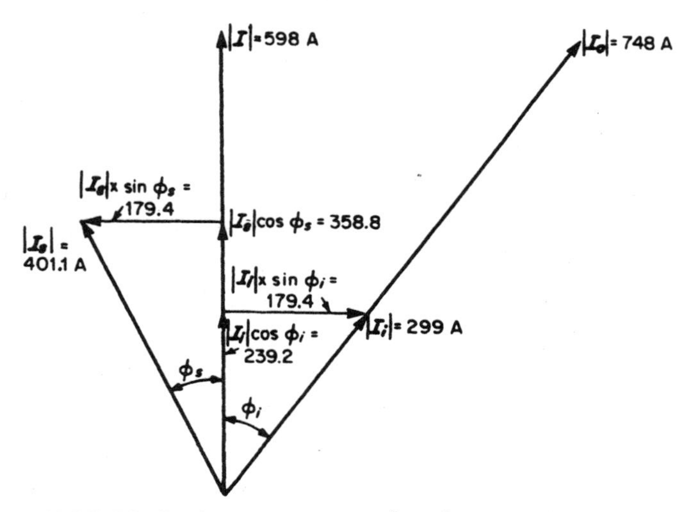

FIGURE 7.8 Synchronous-motor power-factor improvement.

BIBLIOGRAPHY

Fink, Donald G., and H. Wayne Beaty. 1999. *Standard Handbook for Electrical Engineers*, 14th ed. New York: McGraw-Hill.

Concordia, Charles. 1951. *Synchronous Machines*. Schenectady, N.Y.: General Electric Company.

Libby, Charles C. 1960. *Motor Selection and Application*. New York: McGraw-Hill.

Stevenson, William D., Jr. 1982. *Elements of Power System Analysis*, 4th ed. New York: McGraw-Hill.

Electrical Transmission and Distribution Reference Book. 1964. 4th ed. East Pittsburgh, P.A.: Westinghouse Electric Corporation.

ANSI C50.10. 1990. *Rotating Electrical Machinery—Synchronous Machines*. New York: American National Standards Institute.

ANSI C50.13. 1989. *Rotating Electrical Machinery—Cylindrical-Rotor Synchronous Generators*. New York: American National Standards Institute.

ANSI C50.14. 1977 (Reaffirmed 1989). *Requirements for Combustion Gas Turbine Driven Cylindrical Rotor Synchronous Generators*. New York: American National Standards Institute.

IEEE 115. 1995. *Test Procedure for Synchronous Machines Part I—Acceptance and Performance Testing*; *Part II—Test Procedures and Parameter Determination for Dynamic Analysis*. Piscataway, N.J.: The Institute of Electrical and Electronics Engineers.

NEMA MG 1. 1998. *Motors and Generators*. Arlington, V.A.: National Electrical Manufacturers Association.

SECTION 8

GENERATION OF ELECTRIC POWER

Hesham E. Shaalan
Assistant Professor
Georgia Southern University

MAJOR PARAMETER DECISIONS

The major parameter decisions that must be made for any new electric power-generating plant or unit include the choices of energy source (fuel), type of generation system, unit and plant rating, and plant site. These decisions must be based upon a number of technical, economic, and environmental factors that are to a large extent interrelated (see Table 8.1). Evaluate the parameters for a new power-generating plant or unit.

Calculation Procedure

1. Consider the Energy Source and Generating System

As indicated in Table 8.2, a single energy source or fuel (e.g., oil) is often capable of being used in a number of different types of generating systems. These include steam cycles, combined steam- and gas-turbine cycles (systems where the hot exhaust gases are delivered to a heat-recovery steam generator to produce steam that is used to drive a steam turbine), and a number of advanced technology processes such as fuel cells (i.e., systems having cathode and anode electrodes separated by a conducting electrolyte that convert liquid or gaseous fuels to electric energy without the efficiency limits of the Carnot cycle).

Similarly, at least in the planning stage, a single generic type of electric-power generating system (e.g., a steam cycle) can be designed to operate on any one of a number of fuels. Conversion from one fuel to another after plant construction does, however, generally entail significant capital costs and operational difficulties.

As Table 8.3 indicates, each combination of energy source and power-generating-system type has technical, economic, and environmental advantages and disadvantages that are unique. Often, however, in a particular situation there are other unique considerations that make the rankings of the various systems quite different from the typical values listed in Table 8.3. In order to make a determination of the best system, it is necessary to quantify and evaluate all factors in Table 8.3 (see page 8.4). Generally, this involves a complicated

TABLE 8.1 Major Parameter Decisions for New Plant

Parameter	Some alternatives
Energy source or fuel	Common fossil fuels (coal, oil, and natural gas) Nuclear fuels (uranium and thorium) Elevated water (hydroelectric) Geothermal steam Other renewable, advanced technology, or nonconventional sources
Generation system type	Steam-cycle (e.g., steam-turbine) systems (with or without cogeneration team for district heating and industrial steam loads) Hydroelectric systems Combustion-turbine (e.g., gas-turbine) systems Combined-cycle (i.e., combined steam- and gas-turbine) systems Internal-combustion engine (e.g., diesel) systems Advanced technology or nonconventional sources
Unit and plant rating	Capable of serving the current expected maximum electrical load and providing some spinning reserve for reliability and future load growth considerations Capable of serving only the expected maximum electrical load (e.g., peaking unit) Capable of serving most of the expected maximum load (e.g., using conservation or load management to eliminate the load that exceeds generation capacity)
Plant site	Near electrical load Near fuel source Near water source (water availability) Near existing electrical transmission system Near existing transportation system Near or on existing electrical-generation plant site

TABLE 8.2 Generic Types of Electric-Generating Systems

Energy source or fuel	Approximate percentage of total electric generation	Steam cycle 85 percent	Hydroelectric 13 percent	Combined steam- and gas-turbine cycle 1 percent	Combustion turbines 1 percent	Internal combustion engine (diesel) 1 percent	Photovoltaic	Wind turbine	Fuel cell	Magneto hydrodynamic	Thermoelectric	Thermionic	Open-steam or closed-ammonia cycle
Coal	44	X							X	X	X	X	
Oil	16	X		X	X	X			X	X	X	X	
Natural gas	14	X		X	X				X	X	X	X	
Elevated water supply	13		X										
Nuclear fission (uranium or thorium)	13	X											
Geothermal	0.15	X											
Refuse-derived fuels		X											
Shale oil		X		X	X				X	X	X	X	
Tar sands		X		X	X	X			X	X	X	X	
Coal-derived liquids and gases		X		X	X	X			X	X	X	X	
Wood		X											
Vegetation (biomass)		X											
Hydrogen		X				X			X	X	X	X	
Solar		X					X						
Wind								X					
Tides			X										
Waves			X										
Ocean thermal gradients													X
Nuclear fission			X										

TABLE 8.3 Comparison of Energy Source and Electrical-Generating Systems

Energy Source (fuel) and generation-system type	Fuel cost	System efficiency	Capital cost, $/kW	System operation and maintenance (excluding fuel) costs, $/MWh	Largest available unit ratings	System reliability and availability	System complexity	Fuel availability	Cooling water requirements	Major environmental impacts
Coal-fired steam cycle	Intermediate	High	Very high	Low to medium	Large	High	Very high	Best	Large	Particulars, SO_2 and oxides of nitrogen (NO_x) in stack gases; disposal of scrubber sludge and ashes
Oil-fired steam cycle	Highest	High	High	Lowest	Large	Very high	High	Fair	Large	SO_2 and NO_x in stack gases; disposal of scrubber sludge
Natural-gas–fired steam cycle	High	High	High	Lowest	Large	Very high	High	Fair	Large	NO_x in stack gases
Nuclear	Low	Intermediate	Highest	Medium	Largest	High	Highest	Good	Largest	Safety; radioactive waste disposal
Oil-fired combustion engine	Highest	Low	Lowest	Highest	Smallest	Lowest	Moderate	Fair	Smallest	SO_2 and NO_x in stack gases
Natural-gas–fired combustion turbine	High	Low	Lowest	Highest	Smallest	Lowest	Moderate	Fair	Smallest	NO_x in stack gases
Oil-fired combined cycle	Highest	Very high	Intermediate	Medium	Intermediate	Medium	Moderate	Fair	Moderate	SO_2 and NO_x in stack gases
Natural-gas–fired combined cycle	High	Very high	Intermediate	Medium	Intermediate	Medium	Moderate	Fair	Moderate	NO_x in stack gases
Hydroelectric	Lowest	Highest	Intermediate to highest	Low	Large	Highest if water is available	Lowest	Limited by area	Small	Generally requires construction of a dam
Geothermal steam	Low	Lowest	Intermediate	Medium	Intermediate	High	Low	Extremely limited by area	Low	H_2S emissions from system

tradeoff process and a considerable amount of experience and subjective judgment. Usually, there is no one system that is best on the basis of all the appropriate criteria.

For example, in a comparison between coal and nuclear energy, nuclear energy generally has much lower fuel costs but higher capital costs. This makes an economic choice dependent to a large extent on the expected capacity factor (or equivalent full-load hours of operation expected per year) for the unit. Coal and nuclear-energy systems have, however, significant but vastly different environmental impacts. It may well turn out that one system is chosen over another largely on the basis of a subjective perception of the risks or of the environmental impacts of the two systems.

Similarly, a seemingly desirable and economically justified hydroelectric project (which has the additional attractive features of using a renewable energy source and in general having high system availability and reliability) may not be undertaken because of the adverse environmental impacts associated with the construction of a dam required for the project. The adverse environmental impacts might include the effects that the dam would have on the aquatic life in the river, or the need to permanently flood land above the dam that is currently being farmed, or is inhabited by people who do not wish to be displaced.

2. Select the Plant, Unit Rating, and Site

The choice of plant, unit rating, and site is a similarly complex, interrelated process. As indicated in Table 8.3 (and 8.9), the range of unit ratings that are commercially available is quite different for each of the various systems. If, for example, a plant is needed with a capacity rating much above 100 MW, combustion turbine, diesel, and geothermal units could not be used unless multiple units were considered for the installation.

Similarly, the available plant sites can have an important impact upon the choice of fuel, power-generating system, and rating of the plant. Fossil-fuel or nuclear-energy steam-cycle units require tremendous quantities of cooling water [50.5 to 63.1 m^3/s (800,000 to 1,000,000 gal/min)] for a typical 1000-MW unit, whereas gas-turbine units require essentially no cooling water. Coal-fired units rated at 1000 MW would typically require over 2.7 million tonnes (3 million tons) of coal annually, whereas nuclear units rated at 1000 MW would typically require only 32.9 tonnes (36.2 tons) of enriched uranium dioxide (UO_2) fuel annually.

Coal-fired units require disposal of large quantities of ash and scrubber sludge, whereas natural-gas-fired units require no solid-waste disposal whatsoever. From each of these comparisons it is easy to see how the choice of energy source and power-generating system can have an impact on the appropriate criteria to be used in choosing a plant site. The location and physical characteristics of the available plant sites (such as proximity to and availability of water, proximity to fuel or fuel transportation, and soil characteristics) can have an impact on the choice of fuel and power-generating system.

3. Examine the Alternatives

Each of the more conventional electric-power generating systems indicated in Table 8.3 is available in a variety of ratings. In general, installed capital costs (on a dollars per kilowatt basis) and system efficiencies (heat rates) are quite different for the different ratings. Similarly, each of the more conventional systems is available in many variations of equipment types, equipment configurations, system parameters, and operating conditions.

For example, there are both pulverized-coal and cyclone boilers that are of either the drum or once-through type. Steam turbines used in steam cycles can be of either the tandem-compound or cross-compound type, with any number of feedwater heaters, and be either of the condensing, back-pressure, or extraction (cogeneration) type. Similarly, there are a number of standard inlet and reheat system conditions (i.e., temperatures and pressures). Units may be designed for base-load, intermediate-load, cycling, or peaking operation. Each particular combination of equipment type, equipment configuration, system parameters, and operating conditions has associated cost and operational advantages and disadvantages, which for a specific application must be evaluated and determined in somewhat the same manner that the fuel and electrical-generation system choice is made.

4. Consider the Electrical Load

The electrical load, on an electric-power system of any size generally fluctuates considerably on a daily basis, as shown by the shapes of typical daily load curves for the months April, August, and December in Fig. 8.1. In addition, on an annual basis, the system electrical load varies between a minimum load level, below which the electrical demand never falls, and a maximum or peak, load level which occurs for only a few hours per year. The annual load duration curve of Fig. 8.1*a* graphically shows the number of hours per year that the load on a particular power system exceeds a certain level.

For example, if the peak-power system load in the year (100 percent load) is 8100 MW, the load duration curve shows that one could expect the load to be above 70 percent of the peak (i.e., above 0.7×8100 MW $= 5760$ MW) about 40 percent of the year. The minimum load (i.e., load exceeded 100 percent of the time) is about 33 percent of the peak value.

Typically, for U.S. utility systems the minimum annual load is 27 to 33 percent of the peak annual load. Generally, the load level exceeds 90 percent of the peak value 1 to 5 percent of the time, exceeds 80 percent of the peak value 5 to 30 percent of the time, and exceeds 33 to 45 percent of the peak 95 percent of the time. Annual load factors [(average load/peak annual load) $\times$ 100 percent] typically range from 55 to 65 percent.

The frequency of a system fluctuates as the load varies, but the turbine governors always bring it back to 60 Hertz. The system gains or loses a few cycles throughout the day due to these fluctuations. When the accumulated loss or gain is about 180 cycles, the error is corrected by making all the generators turn either faster or slower for a brief period.

A major disturbance on a system, or contingency, creates a state of emergency. Immediate steps must be taken to prevent the contingency from spreading to other regions. The sudden loss of an important load or a permanent short-circuit on a transmission line constitutes a major contingency.

If a big load is suddenly lost, all the turbines begin to speed up and the frequency increases everywhere on the system. On the other hand, if a generator is disconnected, the speed of the remaining generators decreases because they suddenly have to carry the entire load. The frequency then starts to decrease at a rate that may reach 5 Hz/s, and no time must be lost under these conditions. Therefore, if conventional methods are unable to bring the frequency back to normal, some load must be dropped. Such load shedding is done by frequency-sensitive relays that open selected circuit breakers as the frequency falls.

Related Calculations. Generally, considerable economic savings can be obtained by using higher capital cost, lower operating cost units (such as steam-cycle units) to serve the base load (Fig. 8.1) and by using lower capital cost, higher operating cost units (such as combustion turbines) to serve the peaking portion of the load. The intermediate load range is generally best served by a combination of base-load, peaking, combined-cycle, and hydroelectric units that have intermediate capital and operating costs and have design provisions that reliably permit the required load fluctuations and hours per year of operation.

The optimum combination or mix of base-load, intermediate-load, and peaking power-generating units of various sizes involves use of planning procedures and production cost vs. capital cost tradeoff evaluation methods.

	April	August	December	Annual
Load factor	0.667	0.586	0.652	0.623
Min. load / peak load factor	0.371	0.378	0.386	0.330

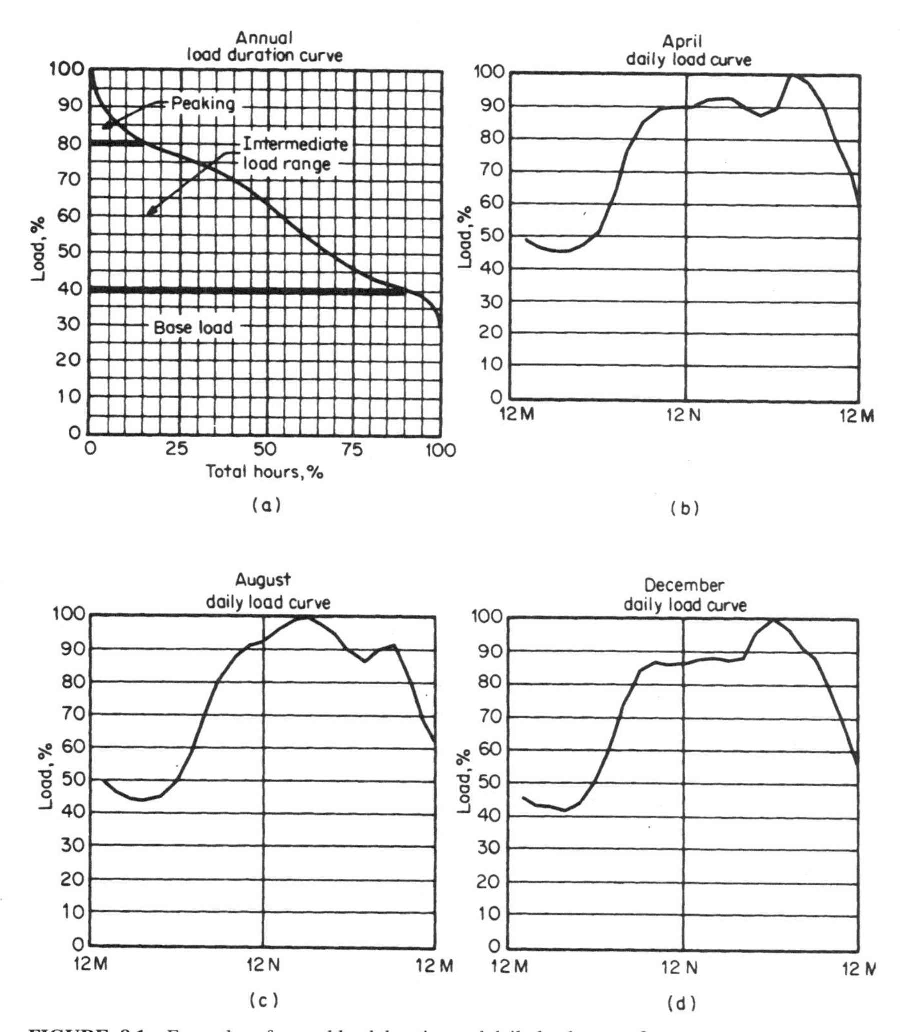

FIGURE 8.1 Examples of annual load duration and daily load curves for a power system.

OPTIMUM ELECTRIC-POWER GENERATING UNIT

Determine the qualities of an optimum new electric-power generating unit to be applied to an existing utility system. (Table 8.4 is a summary of all the necessary steps in making this kind of determination.)

TABLE 8.4 Steps to Determine the Optimum New Electric-Power Generating Unit

Step 1	Identify all possible energy source (fuel) and electric-generation-system combination alternatives.
Step 2	Eliminate alternatives that fail to meet system commercial-availability criteria.
Step 3	Eliminate alternatives that fail to meet energy source (fuel) commercial-availability criteria.
Step 4	Eliminate alternatives that fail to meet other functional or site-specific criteria.
Step 5	Eliminate alternatives that are always more costly than other feasible alternatives.
Step 5*a*	Calculate the appropriate annual fixed-charge rate.
Step 5*b*	Calculate fuel costs on a dollars per million Btu basis.
Step 5*c*	Calculate the average net generation unit heat rates.
Step 5*d*	Construct screening curves for each system.
Step 5*e*	Use screening-curve results to choose those alternatives to be evaluated further.
Step 5*f*	Construct screening curves for feasible renewable and alternative energy sources and generation systems and compare with alternatives in Step 5*e*.
Step 6	Determine coincident maximum predicted annual loads over the entire planning period.
Step 7	Determine the required planning reserve margin.
Step 8	Evaluate the advantages and disadvantages of smaller and larger generation-unit and plant ratings.
Step 8*a*	Consider the economy-of-scale savings associated with larger unit and plant ratings.
Step 8*b*	Consider the operational difficulties associated with unit ratings that are too large.
Step 8*c*	Take into account the range of ratings commercially available for each generation-system type.
Step 8*d*	Consider the possibility of jointly owned units.
Step 8*e*	Consider the forecast load growth.
Step 8*f*	Determine the largest unit and plant ratings that can be used in generation expansion plans.
Step 9	Develop alternative generation expansion plans.
Step 10	Compare generation expansion plans on a consistent basis.
Step 11	Determine the optimum generation expansion plan by using an iterative process.
Step 12	Use the optimum generation expansion plan to determine the next new generation units or plants to be installed.
Step 13	Determine the generator ratings for the new generation units to be installed.
Step 14	Determine the optimum plant design.
Step 15	Evaluate tradeoff of annual operation and maintenance costs vs. installed capital costs.
Step 16	Evaluate tradeoffs of thermal efficiency vs. capital costs and/or operation and maintenance costs.
Step 17	Evaluate tradeoff of unit availability (reliability) and installed capital costs and/or operation and maintenance costs.
Step 18	Evaluate tradeoff of unit rating vs. installed capital costs.

Calculation Procedure

1. Identify Alternatives

As indicated in Table 8.2, there are over 60 possible combinations of fuel and electric-power generating systems that either have been developed or are in some stage of development. In the development of power-generating expansion plans (to be considered later), it is necessary to evaluate a number of installation sequences with the various combinations of fuel and electric-power generating systems with various ratings. Even with the use of large computer programs, the number of possible alternatives is too large to reasonably evaluate. For this reason, it is necessary to reduce this large number of alternatives to a reasonable and workable number early in the planning process.

2. Eliminate Alternatives That Fail to Meet System Commercial Availability Criteria

Reducing the number of alternatives for further consideration generally begins with elimination of all of those systems that are simply not developed to the stage where they can be considered to be available for installation on a utility system in the required time period. The alternatives that might typically be eliminated for this reason are indicated in Table 8.5.

3. Eliminate Alternatives That Fail to Meet Energy-Source Fuel Commercial Availability Criteria

At this point, the number of alternatives is further reduced by elimination of all of those systems that require fuels that are generally not commercially available in the required quantities. Alternatives that might typically be eliminated for this reason are also indicated in Table 8.5.

4. Eliminate Alternatives That Fail to Meet Other Functional or Site-Specific Criteria

In this step those alternatives from Table 8.2 are eliminated that, for one reason or another, are not feasible for the particular existing utility power system involved. Such systems might include wind (unless 100- to 200-kW units with a fluctuating and interruptible power output can suffice), geothermal (unless the utility is located in the geyser regions of northern California), conventional hydroelectric (unless the utility is located in a region where elevated water is either available or can feasibly be made available by the construction of a river dam), and tidal hydroelectric (unless the utility is located near one of the few feasible oceanic coastal basin sites).

Table 8.5 is intended to be somewhat representative of the current technology. It is by no means, however, intended to be all-inclusive or representative for all electric-power

TABLE 8.5 Systems That Might Be Eliminated

Reason for elimination	Systems eliminated
Systems are not commercially available for installation on a utility system.	Fuel-cell systems Magnetohydrodynamic systems Thermoelectric systems Thermionic systems Solar photovoltaic or thermal-cycle systems Ocean thermal gradient open-steam or closed-ammonia cycle systems Ocean-wave hydraulic systems Nuclear-fusion systems
Energy source (fuel) is not commercially available in the required quantities.	Shale oil Tar sands Coal-derived liquids and gases Wood Vegetation Hydrogen Refuse-derived fuels
Systems typically do not satisfy other functional, feasibility, or site-specific criteria.	Wind Geothermal Conventional hydroelectric Tidal hydroelectric

generating installation situations. For example, in certain situations the electric-power generating systems that are used extensively today (such as coal and oil) may be similarly eliminated for such reasons as inability to meet government clean-air and/or disposal standards (coal), fuel unavailability for a variety of reasons including government policy (oil or natural gas), lack of a site where a dam can be constructed without excessive ecological and socioeconomic impacts (hydroelectric), or inability to obtain the necessary permits and licenses for a variety of environmental and political reasons (nuclear).

Similarly, even now, it is conceivable that in a specific situation a number of those alternatives that were eliminated such as wind and wood, might be feasible. Also, in the future, several of the generating systems such as solar photovoltaic might become available in the required ratings or might be eliminated because of some other feasibility criterion.

It should be emphasized that for each specific electric-generating-system installation it is necessary to identify those energy alternatives that must be eliminated from further consideration on the basis of criteria that are appropriate for the specific situation under consideration.

5. Eliminate Alternatives That Are Always More Costly Than Other Feasible Alternatives

In this step those remaining fuel and electric-power generating system alternatives from Table 8.2 are eliminated from further consideration that will not, under any reasonable foreseeable operational criteria, be less costly than other feasible alternatives.

Typically, the elimination of alternatives in this stage is based on a comparison of the total power-generating costs of the various systems, considering both the fixed costs (i.e., capital plus fixed operation and maintenance costs) and the production costs (fuel costs plus variable operation and maintenance costs) for the various systems.

This comparison is generally made by means of *screening curves,* such as those in Figs. 8.2 through 8.4. For each combination of fuel and electric-power generating system

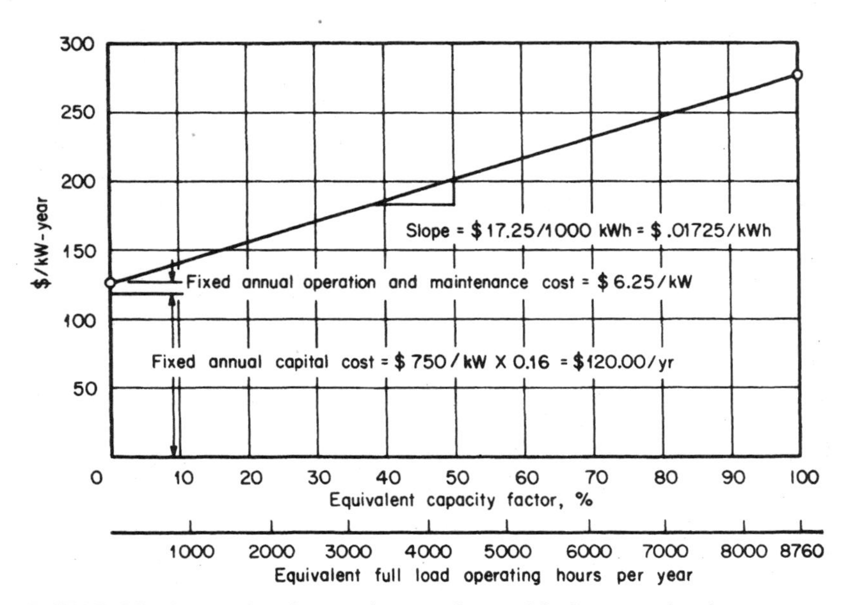

FIGURE 8.2 Construction of a screening curve for a coal-fired steam-cycle unit.

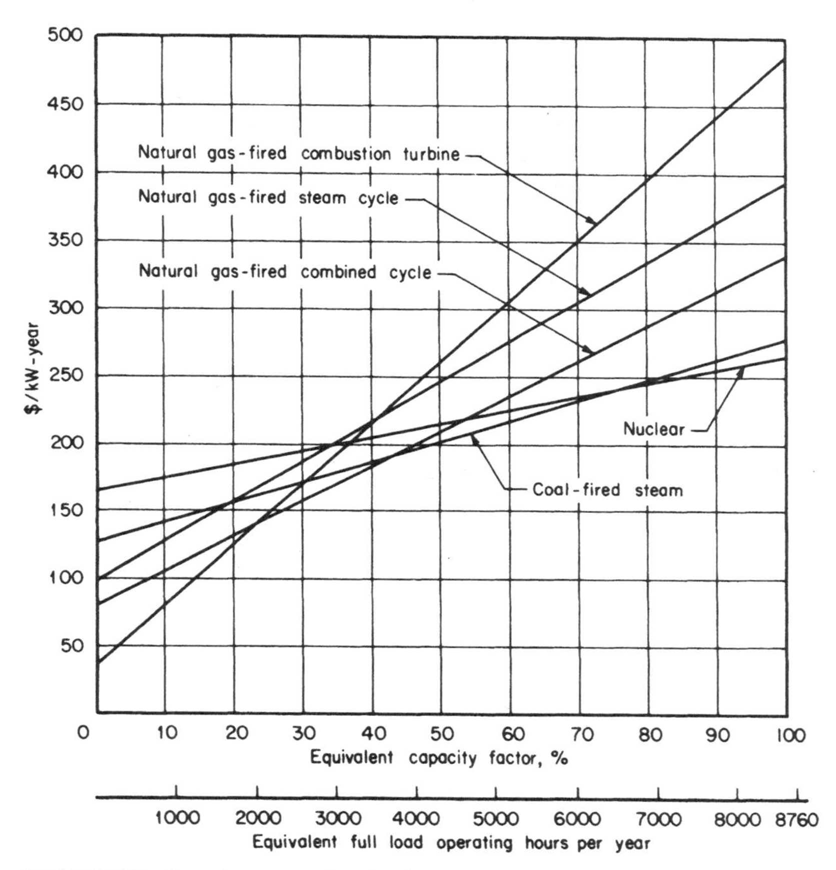

FIGURE 8.3 Screening curves for electric-generation-system alternatives based on assumption of availability of natural gas.

still under consideration, the annual operation costs per installed kilowatt (dollars per year per kilowatt) is plotted as a function of capacity factor (or equivalent full-load operation hours per year).

ANNUAL CAPACITY FACTOR

Determine the annual capacity factor of a unit rated at 100 MW that produces 550,000 MWh per year.

Calculation Procedure

1. Compute Annual Capacity Factor as a Percentage

The factor is

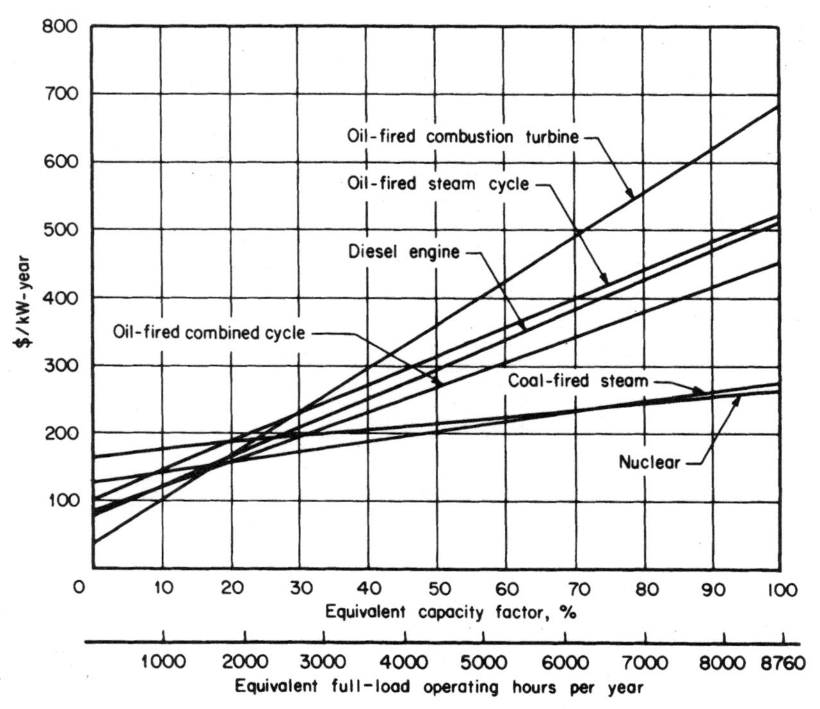

FIGURE 8.4 Screening curves for electric-generation-system alternatives based on assumption that natural gas is unavailable.

$$\left(\frac{550{,}000 \text{ MWh}/100 \text{ MW}}{8760 \text{ h/yr}}\right) 100 \text{ percent} = 68.2 \text{ percent}$$

2. Compute Annual Capacity Factor in Hours per Year

The factor is (68.2/100)(8760 h/yr) = 5550 h/yr.

ANNUAL FIXED-CHARGE RATE

Estimate the annual fixed rate for an investor-owned electric utility.

Calculation Procedure

1. Examine the Appropriate Factors

As shown in Table 8.6, the annual fixed-charge rate represents the average, or "levelized," annual carrying charges including interest or return on the installed capital, depreciation or return of the capital, tax expense, and insurance expense associated with the installation of a particular generating unit for the particular utility or company involved.

TABLE 8.6 Typical Fixed-Charge Rate for Investor-Owned Electric Utility

Charge	Rate, percent
Return	7.7
Depreciation	1.4
Taxes	6.5
Insurance	0.4
Total	16.0

Related Calculations. Fixed-charge rates for investor-owned utilities generally range from 15 to 20 percent; fixed-charge rates for publicly owned utilities are generally about 5 percent lower.

FUEL COSTS

Calculate fuel costs on a dollars per megajoule (and million Btu) basis.

Calculation Procedure

1. Compute Cost of Coal

On a dollars per megajoule (dollars per million Btu) basis, the cost of coal at \$39.68/tonne (\$36/ton) with a heating value of 27.915 MJ/kg (12,000 Btu/lb) is (\$39.68/tonne)/[(1000 kg/tonne)(27.915 MJ/kg)] = \$0.001421/MJ = \$1.50/million Btu.

2. Compute the Costs of Oil

On a dollars per megajoule basis, the cost of oil at \$28 per standard 42-gal barrel (\$0.17612/L) with a heating value of 43.733 MJ/kg (18,800 Btu/lb) and a specific gravity of 0.91 is (\$0.17612/L)/[(43.733 MJ/kg)(0.91 kg/L)] = \$0.004425/MJ = \$4.67/million Btu.

3. Compute the Cost of Natural Gas

On a dollars per megajoule basis, natural gas at \$0.1201/m^3 (\$3.40 per thousand standard cubic feet) with a heating value of 39.115 MJ/m^3 = 1050 Btu/1000 ft^3 costs (\$0.1201/m^3)/(39.115 MJ/m^3) = \$0.00307/MJ = \$3.24/million Btu.

4. Compute Cost of Nuclear Fuel

On a dollars per megajoule basis, nuclear fuel at \$75.36/MWday costs (\$75.36/MWday)/[(1.0 J/MWs)(3600 s/h)(24 h/day)] = \$0.00087/MJ = \$0.92/million Btu.

AVERAGE NET HEAT RATES

A unit requires 158,759 kg/h (350,000 lb/h) of coal with a heating value of 27.915 MJ/kg (12,000 Btu/lb) to produce 420,000 kW output from the generator. In addition, the unit has electric power loads of 20,000 kW from required power-plant auxiliaries, such as boiler feed pumps. Calculate the average net generation unit rate.

Calculation Procedure

1. Define the Net Heat Rate

The average net heat rate (in Btu/kWh or J/kWh) of an electric-power generating unit is calculated by dividing the total heat input to the system (in units of Btu/h or MJ/h) by the net electric power generated by the plant (in kilowatts), taking into account the boiler, turbine, and generator efficiencies and any auxiliary power requirements.

2. Compute the Total Heat Input to Boiler

Total heat input to boiler equals (158,759 kg/h)(27.915 MJ/kg) = 4.43×10^6 MJ/h = 4200×10^6 Btu/h.

3. Compute the Net Power Output of the Generating Unit

Net generating-unit power output is 420,000 kW − 20,000 kW = 400,000 kW.

4. Determine the Net Heat Rate of the Generating Unit

The net generating unit heat rate is (4.43×10^6 MJ/h)/400,000 kW = 11.075 MJ/kWh = 10,500 Btu/kWh.

CONSTRUCTION OF SCREENING CURVE

A screening curve provides a plot of cost per kilowattyear as a function of capacity factor or operating load. An example is the screening curve of Fig. 8.2 for a coal-fired steam-cycle system, based on the data in Table 8.7 (see pages 8.16 and 8.17). Assume the total installed capital cost for a 600-MW system is $450 million and the fixed-charge rate is 16 percent. In addition, assume the total fixed operation and maintenance cost is $3,750,000 per year for the unit. Verify the figures given in Fig. 8.2.

Calculation Procedure

1. Determine the Fixed Annual Capital Cost

The installed cost per kilowatt is ($\$450 \times 10^6$)/600,000 kW = \$750/kW. Multiplying by the fixed-charge rate, we obtain the fixed annual cost: (\$750/kW)(0.16) = \$120/kWyr.

2. Compute Fixed Operation and Maintenance Costs

Fixed operation and maintenance cost on a per-kilowatt basis is (\$3,750,000/yr)/600,000 kW = \$6.25/kWyr.

3. Compute Cost per Year at a Capacity Factor of Zero

The cost in dollars per year per kilowatt at a capacity factor of zero is \$126.25/kWyr, which is the sum of the annual fixed capital cost, \$120/kWyr, plus the annual fixed operation and maintenance cost of \$6.25/kWyr.

4. Determine the Fuel Cost

Coal at \$39.68/tonne with a heating value of 27.915 MJ/kg costs \$0.001421/MJ, as determined in a previous example. With an average unit heat rate of 11.075 MJ/kWh, the fuel cost for the unit on a dollars per kilowatthour basis is (\$0.001421/MJ)(11.075 MJ/kWh) = \$0.01575/kWh.

With a levelized variable operation and maintenance cost for the system of \$0.00150/kWh (Table 8.7), the total variable production cost for the coal-fired steam-cycle unit is \$0.01725/kWh (i.e., the fuel cost, \$0.01575/kWh, plus the variable operation and maintenance cost of \$0.00150/kWh, or \$0.01725/kWh). Hence, the total annual fixed and variable costs on a per-kilowatt basis to own and operate a coal-fired steam-cycle system 8760 h per year (100 percent capacity factor) would be \$126.25/kWyr + (\$0.01725/kWh)(8760 h/yr) = 126.26/kWh + \$151.11/kWyr = \$277.36/kWyr.

Related Calculations. As indicated in Fig. 8.2, the screening curve is linear. The y intercept is the sum of the annual fixed capital, operation, and maintenance costs and is a function of the capital cost, fixed-charge rate, and fixed operation and maintenance cost. The slope of the screening curve is the total variable fuel, operation, and maintenance cost for the system (i.e., \$0.01725/kWh), and is a function of the fuel cost, heat rate, and variable operation and maintenance costs.

Table 8.7 shows typical data and screening curve parameters for all those combinations of energy source and electric-power generating system (listed in Table 8.2) that were not eliminated on the basis of some criterion in Table 8.5. Table 8.7 represents those systems that would generally be available today as options for an installation of an electric-power–generating unit.

Figure 8.3 illustrates screening curves plotted for all non-oil-fired systems in Table 8.7. For capacity factors below 23.3 percent (2039 equivalent full-load operating hours per year), the natural-gas–fired combustion turbine is the least costly alternative. At capacity factors from 23.3 to 42.7 percent, the natural-gas–fired combined-cycle system is the most economical. At capacity factors from 42.7 to 77.4 percent, the coal-fired steam-cycle system provides the lowest total cost, and at capacity factors above 77.4 percent, the nuclear plant offers the most economic advantages. From this it can be concluded that the optimum generating plan for a utility electric power system would consist of some rating and installation sequence combination of those four systems.

Steam-cycle systems, combined-cycle systems, and combustion-turbine systems can generally be fired by either natural gas or oil. As indicated in Table 8.7, each system firing with oil rather than natural gas generally results in higher annual fixed capital cost and higher fixed and variable operation, maintenance, and fuel costs. Consequently, oil-fired systems usually have both higher total fixed costs and higher total variable costs than natural-gas–fired systems. For this reason, firing with oil instead of natural gas results in higher total costs at all capacity factors.

In addition, as indicated in Table 8.3, the fact that oil-fired systems generally have more environmental impact than natural-gas systems means that if adequate supplies of natural gas are available, oil-fired steam cycles, combined cycles, and combustion-turbine systems would be eliminated from further consideration. They would never provide any benefits relative to natural-gas–fired systems.

If natural gas were not available, the alternatives would be limited to the non-natural-gas–fired systems of Table 8.7. The screening curves for these systems are plotted in Fig. 8.4. From the figure, if natural gas is unavailable, at capacity factors below 16.3 percent oil-fired combustion turbines are the least costly alternative. At capacity factors from 16.3 to 20.0 percent, oil-fired combined-cycle systems are the most economical. At capacity factors from 20.0 to 77.4 percent a coal-fired steam-cycle system provides the lowest total cost, and at capacity factors above 77.4 percent nuclear is the best.

If any of those systems in Table 8.5 (which were initially eliminated from further consideration on the basis of commercial availability or functional or site-specific criteria) are indeed possibilities for a particular application, a screening curve should also be constructed for those systems. The systems should then be evaluated in the same manner as the systems indicated in Figs. 8.3 and 8.4. The screening curves for renewable energy

TABLE 8.7 Data Used for Screening Curves of Figs. 8.2 through 8.4

System	Total installed capital cost, millions of dollars	Unit rating, MW	Installed capital cost, $/kW	Annual levelized fixed-charge rate, percent	Annual fixed capital cost, $/kWyr	Total annual fixed O&M cost, millions of dollars per year	Annual fixed O&M costs, $/kWyr	Annual fixed capital, O&M costs, $/kWyr
Coal-fired steam cycle	450.0	600	750	16	120.00	3.75	6.25	126.25
Oil-fired steam-cycle	360.0	600	600	16	96.00	3.30	5.50	101.50
Natural-gas–fired steam cycle	348.0	600	580	16	92.80	3.00	5.00	97.80
Nuclear	900.0	900	1000	16	160.00	5.13	5.70	165.07
Oil-fired combined cycle	130.5	300	435	18	78.30	1.275	4.25	82.55
Natural-gas–fired combined cycle	126.0	300	420	18	75.60	1.20	4.00	79.60
Oil-fired combustion turbine	8.5	50	170	20	34.00	0.175	3.50	37.50
Natural-gas–fired combustion turbine	8.0	50	160	20	32.00	0.162	3.25	35.25
Diesel engine	3.0	8	375	20	75.00	0.024	3.00	78.00

*$/t = $/ton × 1.1023
$/L = ($/42-gal barrel) × 0.00629
$/m^3 = ($/MCF) × 0.0353
†MJ/kg = (Btu/lb) × 0.002326
MJ/m^3 = (Btu/SCF) × 0.037252
‡MJ/t = (million Btu/ton) × 1163
MJ/m^3 = (million Btu/bbl) × 6636
MJ/m^3 = (million Btu/MCF) × 37.257
MJ/MWday = (million Btu/MWday) × 1055
§$/MJ = ($/million Btu) × 0.000948
¶MJ/kWh = (Btu/kWh) × 0.001055
J/kWh = (Btu/kWh) × 1055

Fuel costs* $/standard unit	Fuel energy content†	Energy content per standard unit,‡ millions Btu per standard unit	Fuel cost,§ dollars per million Btu	Average net heat rate for unit,¶ Btu/kWh	Fuel cost, $/kWh	Levelized variable O&M costs, $/kWh	Total variable costs (fuel + variable O&M cost), $/kWh
$36/ton	12,000 Btu/lb	24 million Btu/ton	1.50	10,500	0.1575	0.00150	0.01725
$28/bbl	18,800 Btu/lb with 0.91 specific gravity	6 million Btu/barrel	4.67	10,050	0.04693	0.00130	0.04823
$3.40/1000 ft^3 (MCF)	1050 Btu per standard cubic foot (SCF)	1.05 million Btu/MCF	3.24	10,050	0.03256	0.00120	0.03376
$75.36/MWday		81.912 million Btu/MWday	92	11,500	0.01058	0.00085	0.01143
$28/bbl	18,800 Btu/lb with 0.91 specific gravity	6 million Btu/barrel	4.67	8,300	0.03876	0.00350	0.04226
$3.40/MCF	1050 Btu/SCF	1.05 million Btu/MCF	3.24	8,250	0.02673	0.00300	0.02973
$28/bbl	18,800 Btu/lb with 0.91 specific gravity	6 million Btu/barrel	4.67	14,700	0.06865	0.00500	0.07365
$3.40/MCF	1050 Btu/SCF	1.05 million Btu/MCF	3.24	14,500	0.04698	0.00450	0.05148
$28/bbl	18,800 Btu/lb with 0.91 specific gravity	6 million Btu/bbl	4.67	10,000	0.04670	0.00300	0.04970

sources such as hydroelectric, solar, wind, etc. are essentially horizontal lines because the fuel, variable operation, and maintenance costs for such systems are negligible.

NONCOINCIDENT AND COINCIDENT MAXIMUM PREDICTED ANNUAL LOADS

For a group of utilities that are developing generating-system expansion plans in common, the combined maximum predicted annual peak loads used in generating-system expansion studies should be the coincident maximum loads (demands) expected in the year under consideration. Any diversity (or noncoincidence) in the peaks of the various utilities in the group should be considered. Such diversity, or noncoincidence, will in general be most significant if all the various utilities in the planning group do not experience a peak demand in the same season.

Assume the planning group consists of four utilities that have expected summer and winter peak loads in the year under consideration, as indicted in Table 8.8. Determine the noncoincident and coincident annual loads.

Calculation Procedure

1. Analyze the Data in Table 8.8

Utilities A and D experience the highest annual peak demands in the summer season, and utilities B and C experience the highest annual peak demands in the winter season.

2. Compute Noncoincident Demands

Total noncoincident summer maximum demand for the group of utilities is less than the annual noncoincident maximum demand by [1 − (8530 MW/8840 MW)](100 percent) = 3.51 percent. The total noncoincident winter maximum demand is less than the annual noncoincident maximum demand by [1 − (8240 MW/8840 MW)](100 percent) = 6.79 percent.

3. Compute Coincident Demands

If the seasonal diversity for the group averages 0.9496 in the summer and 0.9648 in the winter, the total coincident maximum demand values used for generation expansion

TABLE 8.8 Calculation of Coincident Maximum Demand for a Group of Four Utilities

	Maximum demand, MW		
	Summer	Winter	Annual
Utility A	3630	3150	3630
Utility B	2590	2780	2780
Utility C	1780	1900	1900
Utility D	530	410	530
Total noncoincident maximum demand	8530	8240	8840
Seasonal diversity factor	0.9496	0.9648	
Total coincident maximum demand	8100 MW	7950 MW	8100 MW*

*Maximum of summer and winter.

planning in that year would be (8530 MW)(0.9496) = 8100 MW in the summer and (8240 MW)(0.9648) = 7950 MW in the winter.

REQUIRED PLANNING RESERVE MARGIN

All utilities must plan to have a certain amount of reserve generation capacity to supply the needs of their power customers in the event that a portion of the installed generating capacity is unavailable.

Reserve generating capability is also needed to supply any expected growth in the peak needs of electric utility customers that might exceed the forecast peak demands. In generating-system expansion planning such reserves are generally identified as a percentage of the predicted maximum annual hourly demand for energy.

Compute the reserve capacity for a group of utilities (Table 8.8) having a predicted maximum hourly demand of 8100 MW.

Calculation Procedure

1. Determine What Percentage Increases Are Adequate

Lower loss-of-load probabilities are closely related to higher planning reserve margins. Experience and judgment of most utilities and regulators associated with predominantly thermal power systems (as contrasted to hydroelectric systems) has shown that planning reserves of 15 to 25 percent of the predicted annual peak hourly demand are adequate.

2. Calculate Reserve and Installed Capacity

The range of additional reserve capacity is (0.15)(8100) = 1215 MW to (0.25)(8100) = 2025 MW. The total installed capacity is, therefore, 8100 + 1215 = 9315 MW to 8100 + 2025 = 10,125 MW.

Related Calculations. The reliability level of a particular generation expansion plan for a specific utility or group of utilities is generally determined from a loss-of-load probability (LOLP) analysis. Such an analysis determines the probability that the utility, or group of utilities, will lack sufficient installed generation capacity on-line to meet the electrical demand on the power system. This analysis takes into account the typical unavailability, because of both planned (maintenance) outages and unplanned (forced) outages, of the various types of electric-power generating units that comprise the utility system.

For generating-system expansion planning a maximum loss-of-load probability value of 1 day in 10 years has traditionally been used as an acceptable level of reliability for an electric-power system. Owing primarily to the rapid escalation of the costs of power-generating-system equipment and to limits in an electrical utility's ability to charge rates that provide for the financing of large construction projects, the trend recently has been to consider higher loss-of-load probabilities as possibly being acceptable.

Planning for generating-system expansion on a group basis generally results in significantly lower installed capacity requirements than individual planning by utilities. For example, collectively the utilities in Table 8.8 would satisfy a 15 percent reserve requirement with the installation of 9315 MW, whereas individually, on the basis of total annual noncoincident maximum loads, utilities would install a total of 10,166 MW to retain the same reserve margin of 15 percent.

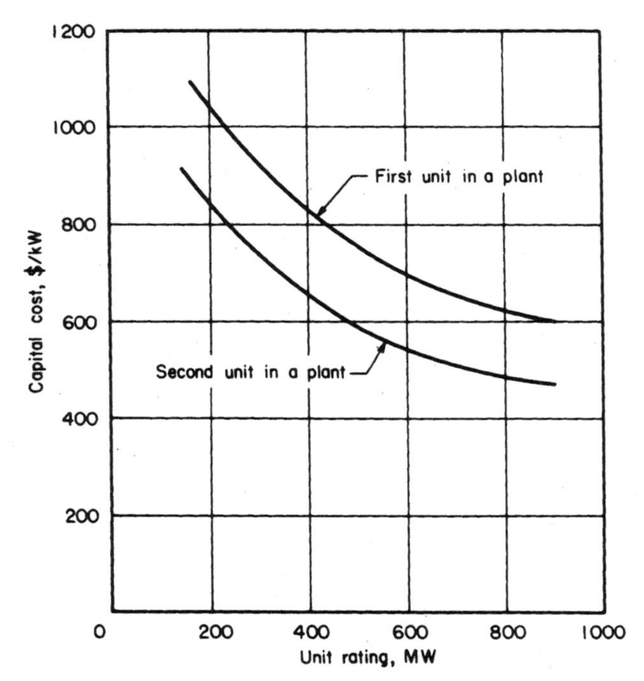

FIGURE 8.5 Typical capital costs vs. unit-rating trend for first and second coal-fired steam-cycle units.

Generating-expansion planning by a group also requires a certain amount of joint planning of the electrical transmission system to ensure that the interconnections between the various utilities in the planning group have sufficient capacity to facilitate the seasonal transfer of power between the utilities. This enables each utility in the planning group to satisfy the applicable reserve criteria at all times of the year.

For all types of electric-power generating units, it is generally the case that smaller unit ratings have higher installed capital costs (on a dollar per kilowatt basis) and higher annual fixed operation and maintenance costs (on a dollar per kilowatthour basis), with the increase in the costs becoming dramatic at the lower range of ratings that are commercially available for that type of electric-power generating unit. Figs. 8.5 and 8.6 illustrate this for coal-fired steam units. Smaller generating units also generally have somewhat poorer efficiencies (higher heat rates) than the larger units.

The installation of generating units that are too large, however, can cause a utility to experience a number of operational difficulties. These may stem from excessive operation of units at partial loads (where unit heat rates are poorer) or inability to schedule unit maintenance in a manner such that the system will always have enough spinning reserve capacity on-line to supply the required load in the event of an unexpected (forced) outage of the largest generation unit.

In addition, it is not uncommon for large units to have somewhat higher forced-outage rates than smaller units, which implies that with larger units a somewhat larger planning reserve margin might be needed to maintain the same loss-of-load probability.

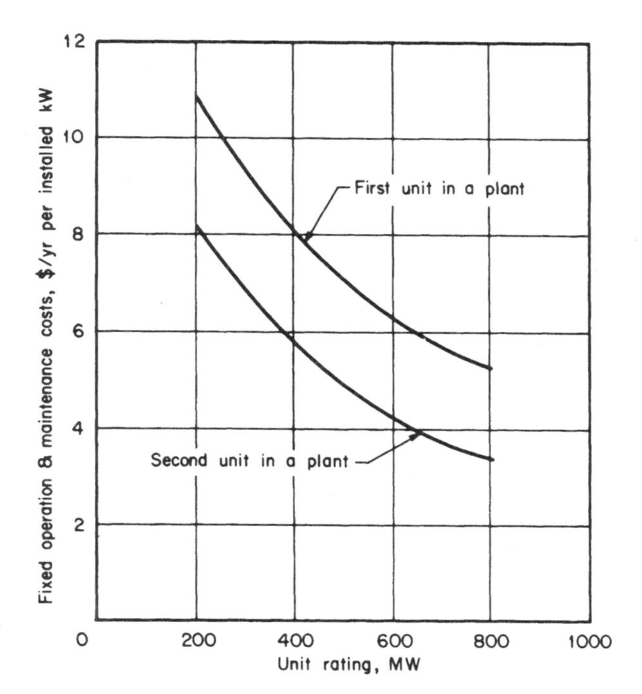

FIGURE 8.6 Typical fixed operation and maintenance costs vs. unit-rating trend for first and second coal-fired steam-cycle units.

RATINGS OF COMMERCIALLY AVAILABLE SYSTEMS

The ratings of the various system types indicated in Table 8.7 are typical of those that are commonly available today. Actually, each of the various systems is commercially available in the range of ratings indicated in Table 8.9. Evaluate the different systems.

Calculation Procedure

1. Consider Nuclear Units

Nuclear units are available in tandem-compound turbine configurations (Fig. 8.7*a*) where a high-pressure (HP) turbine and one to three low-pressure (LP) turbines are on one shaft system driving one generator at 1800 r/min in ratings from 500 to 1300 MW.

2. Consider Fossil-Fuel Units

Steam units fired by fossil fuel (coal, oil, or natural gas) are available in tandem-compound 3600-r/min configurations (Fig. 8.7*b*) up to 800 MW and in cross-compound turbine configurations (Fig. 8.7*c*) where an HP and intermediate-pressure (IP) turbine are on one shaft driving a 3600-r/min electric generator. One or more LP turbines on a second shaft system drive an 1800-r/min generator in ratings from 500 to 1300 MW.

TABLE 8.9 Commercially Available Unit Ratings

Type	Configuration*	Rating range
Fossil-fired (coal, oil, natural gas) steam turbines	Tandem-compound	20–800 MW
	Cross-compound	500–1300 MW
Nuclear steam turbine	Tandem-compound	500–1300 MW
Combined-cycle systems	Two- or three-shaft	100–300 MW
Combustion-turbine systems	Single-shaft	<1–110 MW
Hydroelectric systems	Single-shaft	<1–800 MW
Geothermal systems	Single-shaft	<20–135 MW
Diesel systems	Single-shaft	<1–20 MW

*See Fig. 8.7.

Related Calculations. Combined-cycle systems are generally commercially available in ratings from 100 to 300 MW. Although a number of system configurations are available, it is generally the case that the gas- and steam-turbine units in combined-cycle systems drive separate generators, as shown in Fig. 8.7*d*.

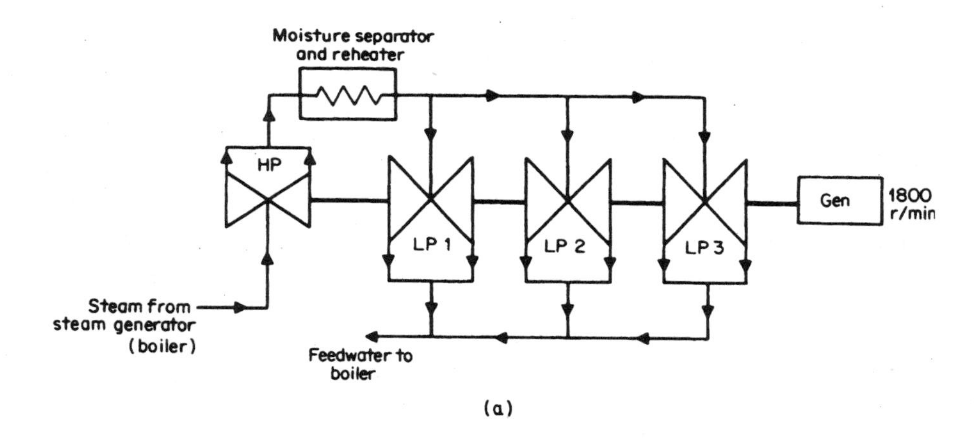

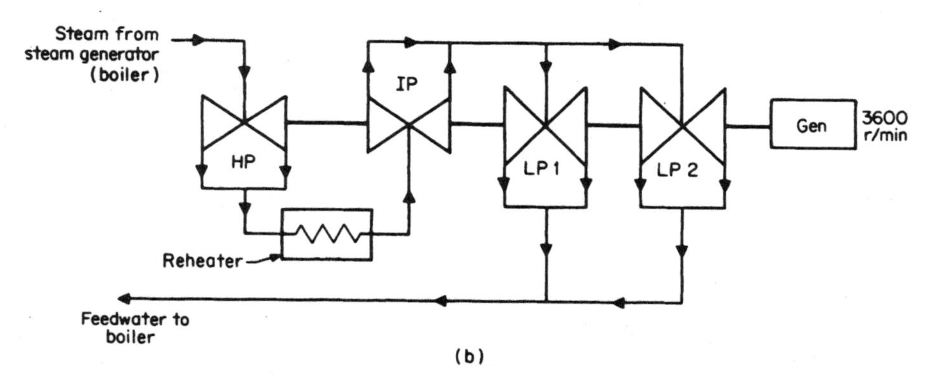

FIGURE 8.7 Various turbine configurations.

Because nuclear and larger fossil-fired unit ratings are often too large for a particular company or electric utility to assimilate in a single installation, it has become common for the smaller and medium-size utilities to install and operate these types of units on a joint, or pool, basis. In this arrangement, one utility has the responsibility for installation and operation of the units for all of the partners. Each of the utilities pays a percentage of all capital and operation costs associated with the unit in accordance with the ownership splits.

Such an arrangement enables all of the owners of the unit to reap the benefits of the lower installed capital costs (dollars per kilowatt) and lower operation costs (dollars per kilowatthour) typical of larger-size units. The operational difficulties associated with having too much of an individual company's total installed capacity in a single generating unit are minimized.

For a number of technical and financial reasons (including excessive fluctuations in reserve margins, uneven cash flows, etc.) utilities find it beneficial to provide for load growth with capacity additions every 1 to 3 years. Hence, the rating of units used in a generating-system expansion plan is to a certain extent related to the forecast growth in the period.

HYDROPOWER GENERATING STATIONS

Hydropower generating stations convert the energy of moving water into electrical energy by means of a hydraulic turbine coupled to a synchronous generator. The power that can be extracted from a waterfall depends upon its height and rate of flow. Therefore, the size and physical location of a hydropower station depends on these two factors.

The available hydropower can be calculated by the following equation:

$$P = 9.8 \times q \times h$$

where

P = available water power (kW)

q = water rate of flow (m^3/s)

h = head of water (m)

9.8 = coefficient used to take care of units

The mechanical power output of the turbine is actually less than the value calculated by the preceding equation. This is due to friction losses in the water conduits, turbine casing, and the turbine itself. However, the efficiency of large hydraulic turbines is between 90 and 94 percent. The generator efficiency is even higher, ranging from 97 to 99 percent, depending on the size of the generator.

Hydropower stations can be divided into three groups based on the head of water:

1. High-head development
2. Medium-head development
3. Low-head development

High-head developments have heads in excess of 300 m, and high-speed turbines are used. Such generating stations can be found in mountainous regions, and the amount of impounded water is usually small. Medium-head developments have heads between 30 m and 300 m, and medium-speed turbines are used. The generating station is typically fed by a large reservoir of water retained by dikes and a dam. A large amount of water is usually impounded behind the dam. Low-head developments have heads under 30 m, and low-speed turbines are used. These generating stations often extract the energy from flowing rivers, and no reservoir is provided. The turbines are designed to handle large volumes of water at low pressure.

LARGEST UNITS AND PLANT RATINGS USED IN GENERATING-SYSTEM EXPANSION PLANS

The group of utilities in Table 8.8 is experiencing load growth as shown in Table 8.10. Determine the largest nuclear and fossil unit ratings allowable through year 15 and beyond.

Calculation Procedure

1. Select Unit Ratings

Generally the largest unit installed should be 7 to 15 percent of the peak load of the utility group. For this reason, in Columns 5 and 7 of Table 8.10, 900-MW nuclear and 600-MW fossil units are selected through year 15. Beyond year 15, 1100-MW nuclear and 800-MW fossil were chosen.

2. Determine the Ratings for 1 Percent Growth per Year

For lower annual load growth of 1 percent per year instead of 2.1 to 3.2 percent as indicated, financial considerations would probably encourage the utility to install 600-MW nuclear and 300- to 400-MW fossil units instead of the 600- to 1100-MW units.

ALTERNATIVE GENERATING-SYSTEM EXPANSION PLANS

At this point, it is necessary to develop numerous different generating-system expansion plans or strategies. The development of two such plans or strategies is indicated in Table 8.10. The plans should be based upon the forecast maximum (peak) coincident hourly electrical demand (load) for each year in the planning period for the group of utilities that are planning together. The planning period for such studies is commonly 20 to 40 yr.

If the installed capacity for the group of utilities is initially 9700 MW, Columns 5 and 7 might be representative of two of the many generating-system expansion plans or strategies that a planner might develop to provide the required capacity for each year in the planning period. Determine the total installed capacity and percentage reserve for Plan B.

Calculation Procedure

1. Compute the Installed Capacity in Year 6

The total installed capacity in year 6 is 9700 MW initially, plus a 300-MW combined-cycle unit in year 2, plus a 900-MW nuclear unit in year 3, plus a 50-MW natural-gas–fired combustion-turbine unit in year 5, plus a 600-MW coal-fired steam unit in year 6 = 11,550 MW. This exceeds the required installed capacity of 11,209 MW.

2. Compute the Reserve Percentage

The percentage reserve is [(11,550 MW/9,747 MW) − 1.0](100 percent) = 18.5 percent, which exceeds the targeted planning reserve level of 15 percent.

Related Calculations. The excess of the actual reserves in a given generating-system expansion plan over the targeted planning reserve level increases the total cost of the plan but also to some degree improves the overall reliability level. It therefore needs to be considered in the comparison of generating-system plans.

The generating-system expansion plans developed usually contain only those types of electric-power generating systems, that were found in the screening curve analysis to yield minimum total annual cost in some capacity factor range. For example, if natural gas is available in sufficient quantities over the planning period, the types of electric-power generating systems used in the alternative generating-system expansion plans would be limited to nuclear units, coal-fired steam-cycle units, and natural-gas–fired combined-cycle and combustion-turbine units, as indicated in Table 8.10.

After a number of different generating-system expansion strategies are developed, the plans must be compared on a consistent basis so that the best plan to meet a given reliability index can be determined. The comparison between the various generating-system expansion plans is generally performed by calculating for each plan the production and investment costs over the life of the plan (20 to 40 yr) and then evaluating those costs using discounted revenue requirements (i.e., present worth, present value) techniques (see Sec. 19).

The production costs for each generating-system expansion plan are generally calculated by large computer programs that simulate the dispatching (or loading) of all the units on the entire power system, hourly or weekly, over the entire planning period. These programs generally employ a probabilistic technique to simulate the occasional unavailability of the various units on the power system. In addition, load forecast, economic, and technical data for each existing and new unit on the power system and for the power system as a whole (as indicated in the first four columns of Table 8.11) are required.

The investment costs for each plan are generally calculated by computer programs that simulate the net cash flows due to the investments in the various plans. Annual book depreciation, taxes, insurance, etc., appropriate for the particular utility involved, are considered. These programs generally require the economic data and corporate financial model data indicated in the last column of Table 8.11.

To determine the optimum generating-system plan over the planning period, sufficient generating-system expansion plans similar to those indicated in Table 8.10 must be developed and evaluated so that all of the reasonable combinations of electric-power–generating-system types, ratings, and installation timing sequences are represented. Even with the use of large computer programs, the number of possible alternative plans based on all combinations of plant types, ratings, etc. becomes too cumbersome to evaluate in detail. It is generally the case, therefore, that generating-system planners use an iterative process to determine the optimum plan.

For example, early in the evaluation process, a smaller number of alternative plans is evaluated. On the basis of a preliminary evaluation, one or more of those plans are modified in one or more ways and reevaluated on a basis consistent with the initial plan to determine if the modifications make the plan less than optimum.

As indicated in Table 8.12, it takes a number of years to license and construct a new power plant. The initial years of the various alternative generating-system expansion plans represent new power-generating facilities for which the utility is already committed. For this reason, the initial years of all of the alternative generating-system plans are generally the same.

TABLE 8.10 Two Alternative Generation Expansion Plans Developed for a Utility

Column 1	Column 2	Column 3	Column 4	Column 5	Column 6	Column 7	Column 8
				Generating-system expansion Plan A		Generating-system expansion Plan B	
Year	Forecast annual growth in peak load coincident, percent yr	Forecast maximum or peak coincident demand or load, MW	Required installed capacity with 15 percent minimum reserve margin, MW	Capacity installation, MW	Total installed capacity, MW	Capacity installation, MW	Total installed capacity, MW
0 (current year)		8,100		9700	9,700	9700	9,700
1	3.2	8,359	9,613	—	9,700	—	9,700
2	3.2	8,627	9,921	600 C	10,300	300 CC	10,000
3	3.2	8,903	10,238	50 CT	10,350	900 N	10,900
4	3.2	9,188	10,566	900 N	11,250	—	—
5	3.2	9,482	10,904	—	11,250	50 CT	10,950
6	2.8	9,747	11,209	50 CT	11,300	600 C	11,550
7	2.8	10,020	11,523	300 CC	11,600	50 CT	11,600
8	2.8	10,301	11,846	600 C	12,220	300 CC	11,900
9	2.8	10,589	12,177	50 CT	12,250	900 N	12,800
10	2.1	10,811	12,433	900 N	13,150	—	—
11	2.1	11,038	12,694	—	13,150	50 CT	12,850
12	2.1	11,270	12,961	—	13,150	600 C	13,450

13	2.1	11,507	13,233	300 CT	13,450	—	—
14	2.5	11,795	13,564	600 C	14,050	300 CC	13,750
15	2.5	12,089	13,903	—	—	900 N	14,650
16	2.5	12,392	14,250	1100 N	15,150	—	—
17	2.5	12,701	14,607	—	15,150	50 CT	14,700
18	2.7	13,044	15,001	50 CT	15,200	800 C	15,500
19	2.7	13,397	15,406	300 CC	15,500	300 CC	15,800
20	2.7	13,758	15,822	800 C	16,300	50 CT	15,850
21	2.7	14,130	16,249	50 CT	16,350	1100 N	16,950
22	2.7	14,511	16,688	800 C	17,150	—	—
23	3.0	14,947	17,189	50 CT	17,200	800 C	17,750
24	3.0	15,395	17,704	1100 N	18,300	—	—
25	3.0	15,857	18,235	—	—	1100 N	18,850
26	3.0	16,333	18,782	800 C	19,100	—	—
27	3.0	16,823	19,346	1100 N	20,200	800 C	19,650
28	3.0	17,327	19,926	—	—	300 CC	19,950
29	3.0	17,847	20,524	800 C	21,000	1100 N	21,050
30	3.0	18,382	21,140	800 C	21,800	800 C	21,850

Key: N = nuclear steam-cycle unit, C = coal-fired steam-cycle unit, CC = natural-gas–fired combined-cycle unit, and CT = natural-gas–fired combustion turbine.

TABLE 8.11 Data Generally Required for Computer Programs to Evaluate Alternative Expansion Plans

Load-forecast data	Data for each existing unit	Data for each new unit	General technical data regarding power system	Economic data and corporate financial model data
Generally determined from an analysis of historical load, energy requirement, and weather-sensitivity data using probabilistic mathematics	Fuel type	Capital cost and/or levelized carrying charges	Units required in service at all times for system area protection and system integrity	Capital fuel and O&M costs and inflation rates for various units
	Fuel cost	Fuel type		
	Unit incremental heat rates (unit efficiency)	Fuel costs	Hydroelectric unit type and data—run of river, pondage, or pumped storage	Carrying charge or fixed-charge rates for various units
Future annual load (MW) and system energy requirements (MWh) on a seasonal, monthly, or weekly basis	Unit fuel and startup	Unit incremental heat rates		
	Unit maximum and minimum rated capacities	Unit fuel and startup costs	Minimum fuel allocations (if any)	Discount rate (weighted cost or capital)
	Unit availability and reliability data such as partial and full forced outage rates	Unit availability and reliability data such as mature and immature full and partial forced-outage rates and scheduled outages	Future system load data	Interest rate during construction
Seasonal load variations				
Load-peak variance			Data for interconnected company's power system or power pool	Planning period (20–50 yr)
Load diversity (for multiple or interconnected power systems)	Scheduled outage rates and maintenance schedules	Unit commercial operation dates		Book life, tax life, depreciation rate and method, and salvage value or decommissioning cost for each unit
Sales and purchases to other utilities	O&M (fixed, variable, and average)		Reliability criteria such as a spinning reserve or loss-of-load probability (LOLP) operational requirements	
Seasonally representative load-duration curve shapes		Unit maximum and minimum capacities		Property and income tax rates
	Seasonal derating (if any) and seasonal derating period	Sequence of unit additions	Limitations on power system ties and interconnections with the pool and/or other companies	
	Sequence of unit retirements (if any)	Operation and maintenance costs (fixed and variable)		Investment tax credits
				Insurance rates
	Minimum downtime and/or dispatching sequence (priority) of unit use)	Time required for licensing and construction of each type of unit	Load management	
			Required licensing and construction lead time for each type of generation unit	

TABLE 8.12 Time Required to License and Construct Power Plants in the United States

Type	Years
Nuclear	8–14
Fossil-fired steam	6–10
Combined-cycle units	4–8
Combustion turbine	3–5

The resulting optimum generating-system expansion plan is generally used, therefore, to determine the nature of the next one or two power-generating facilities after the committed units. For example, if Plan A of Table 8.10 is optimum, the utility would already have to be committed to the construction of the 600-MW coal-fired unit in year 2, the 50-MW combustion turbine in year 3, and the 900-MW nuclear unit in year 4. Because of the required lead times for the units in the plan, the optimum plan, therefore, would in essence have determined that licensing and construction must begin shortly for the 50-MW combustion turbine in year 6, the 300-MW turbine combined-cycle unit in year 7, the 600-MW coal-fired unit in year 8, and the 900-MW nuclear unit in year 10.

GENERATOR RATINGS FOR INSTALLED UNITS

After determining the power ratings in MW of the next new generating units, it is necessary to determine the apparent power ratings in MVA of the electric generator for each of those units. For a 0.90 power factor and 600-MW turbine, determine the generator rating.

Calculation Procedure

1. Compute the Rating

Generator rating in MVA = turbine rating in MW/power factor. Hence, the generator for a 600-MW turbine would be rated at 600 MW/0.90 = 677 MVA.

Related Calculations. The turbine rating in MW used in the preceding expression may be the rated or guaranteed value, the 5 percent over pressure value (approximately 105 percent of rated), or the maximum calculated value [i.e., 5 percent over rated pressure and valves wide open (109 to 110 percent of rated)] with or without one or more steam-cycle feedwater heaters out of service. This depends upon the manner in which an individual utility operates its plants.

OPTIMUM PLANT DESIGN

At this point, it is necessary to specify the detailed design and configuration of each of the power-generating facilities. Describe a procedure for realizing an optimum plant design.

Calculation Procedure

1. Choose Design

Consider for example, the 600-MW coal-fired plant required in year 8; a single design (i.e., a single physical configuration and set of rated conditions) must be chosen for each component of the plant. Such components may include coal-handling equipment, boiler, stack-gas cleanup systems, turbine, condenser, boiler feed pump, feedwater heaters, cooling systems, etc. for the plant as a whole.

2. Perform Economic Analyses

In order to determine and specify the optimum plant design for many alternatives, it is necessary for the power-plant designer to repeatedly perform a number of basic economic analyses as the power-plant design is being developed. These analyses, almost without exception, involve one or more of the following tradeoffs:

a. Operation and maintenance cost vs. capital costs
b. Thermal efficiency vs. capital costs and/or operation and maintenance (O&M) costs
c. Unit availability (reliability) vs. capital costs and/or O&M costs
d. Unit rating vs. capital cost

ANNUAL OPERATION AND MAINTENANCE COSTS VS. INSTALLED CAPITAL COSTS

Evaluate the tradeoffs of annual O&M costs vs. installed capital for Units A and B in Table 8.13.

TABLE 8.13 Evaluation of Annual O&M Costs vs. Installed Capital Costs

Cost component	Unit A	Unit B
Net unit heat rate	10.55 MJ/kWh (10,000 Btu/kWh)	10.55 MJ/kWh (10,000 Btu/kWh)
Unit availability	95 percent	95 percent
Unit rating	600 MW	600 MW
Installed capital costs	$\$450 \times 10^6$	$\$455 \times 10^6$
Levelized or average fixed-charge rate	18.0 percent	18.0 percent
Levelized or average annual O&M cost (excluding fuel)	$\$11.2 \times 10^6$/yr	$\$9.7 \times 10^6$/yr
For Unit A:		
Annual fixed capital charges = $(\$450 \times 10^6)(18/100)$	=	$\$81.00 \times 10^6$/yr
Annual O&M cost (excluding fuel)	=	$\$11.20 \times 10^6$/yr
Total annual cost used for comparison with Unit B	=	$\$92.20 \times 10^6$/yr
For Unit B:		
Annual fixed capital charges = $(\$455 \times 10^6)(18/100)$	=	$\$81.90 \times 10^6$/yr
Annual O&M cost (excluding fuel)	=	$\$9.70 \times 10^6$/yr
Total annual cost used for comparison with Unit A	=	$\$91.60 \times 10^6$/yr

Calculation Procedure

1. Examine Initial Capital Costs

Units A and B have the same heat rate [10.550 MJ/kWh (10,000 Btu/kWh)], plant availability (95 percent), and plant rating (600 MW). As a result, the two alternatives would also be expected to have the same capacity factors and annual fuel expense.

Unit B, however, has initial capital costs that are $5 million higher than those of Unit A but has annual O&M costs (excluding fuel) that are $1.5 million less than those of Unit A. An example of such a case would occur if Unit B had a more durable, higher capital-cost cooling tower filler material [e.g., polyvinyl chloride (PVC) or concrete] or condenser tubing material (stainless steel or titanium), whereas Unit A had lower capital-cost wood cooling-tower filler or carbon steel condenser tubing.

2. Analyze Fixed and Annual Costs

For an 18 percent fixed-charge rate for both alternatives, the annual fixed charges are $900,000 higher for Unit B ($81.9 million per year vs. $81.0 million per year). Unit B, however, has annual O&M costs that are $1.5 million lower than those of Unit A. Unit B, therefore, would be chosen over Unit A because the resulting total annual fixed capital, operation, and maintenance costs (excluding fuel, which is assumed to be the same for both alternatives) are lower for Unit B by $600,000 per year. In this case, the economic benefits associated with the lower annual operation and maintenance costs for Unit B are high enough to offset the higher capital costs.

THERMAL EFFICIENCY VS. INSTALLED CAPITAL AND/OR ANNUAL OPERATION AND MAINTENANCE COSTS

Table 8.14 describes two alternative units that have different thermal performance levels but have the same plant availability (reliability) and rating. Unit D has a net heat rate (thermal performance level) that is 0.211 MJ/kWh (200 Btu/kWh) higher (i.e., 2 percent poorer) than that of Unit C but has both installed capital costs and levelized annual O&M costs that are somewhat lower than those of Unit C. Determine which unit is a better choice.

Calculation Procedure

1. Compute the Annual Fixed Charges and Fuel Cost

If the two units have the same capacity factors, Unit D with a higher heat rate requires more fuel than Unit C. Because Unit D has both installed capital costs and levelized annual operation and maintenance costs that are lower than those of Unit C, the evaluation problem becomes one of determining whether the cost of the additional fuel required each year for Unit D is more or less than the reductions in the annual capital and O&M costs associated with Unit D.

The simplest method of determining the best alternative is to calculate the total annual fixed charges and fuel costs using the following expressions:

Annual fixed charges (dollars per year) = TICC × FCR/100, where TICC = total installed capital cost for Unit C or D (dollars) and FCR = average annual fixed-charge rate (percent/yr).

TABLE 8.14 Evaluation of Thermal Efficiency vs. Installed Capital Costs and Annual Costs

Cost components	Unit C	Unit D
Net unit heat rate	10.550 MJ/kWh (10,000 Btu/kWh)	10.761 MJ/kWh (10,200 Btu/kWh)
Unit availability	95 percent	95 percent
Unit rating	600 MW	600 MW
Installed capital cost	$\$450 \times 10^6$	$\$445 \times 10^6$
Levelized or average fixed-charge rate	18.0 percent	18.0 percent
Levelized or average annual O&M costs (excluding fuel)	$\$11.2 \times 10^6$/yr	$\$11.1 \times 10^6$/yr
Levelized or average capacity factor	70 percent	70 percent
Levelized or average fuel cost over the unit lifetime	$1.50/million Btu ($0.001422/MJ)	$1.50/million Btu ($0.001422/MJ)
For Unit C:		
Annual fixed capital charges = ($\$450 \times 10^6$)(18/100)		= $\$81.00 \times 10^6$/yr
Annual O&M cost (excluding fuel)		= $\$11.20 \times 10^6$/yr
Annual fuel expense		
= (10.550 MJ/kWh)(600,000 kW)(8760 h/yr)(70/100)($0.001422/MJ)		= $\$55.19 \times 10^6$/yr
[= (10,000 Btu/kWh)(600,000 kW)(8760 h/yr)(70/100)($1.50/$10^6$ Btu)]		
Total annual cost used for comparison with Unit D		= $\$147.39 \times 10^6$/yr
For Unit D:		
Annual fixed capital charges = ($\$445 \times 10^6$)(18/100)		= $\$80.10 \times 10^6$/yr
Annual O&M cost (excluding fuel)		= $\$11.10 \times 10^6$/yr
Annual fuel expense		
= (10.761 MJ/kWh)(600,000 kW)(8760 h/yr)(70/100)($0.001422/MJ)		= $\$56.29 \times 10^6$/yr
[= (10,200 Btu/kWh)(600,000 kW)(8760 h/yr)(70/100)($1.50/$10^6$ Btu)]		
Total annual cost used for comparison with Unit C		= $\$147.49 \times 10^6$/yr

Annual fuel expense (dollars per year) = HR × rating × 8760 × CF/100 × FC/10^6, where HR = average net heat rate in J/kWh (Btu/kWh), rating = plant rating in kW, CF = average or levelized unit capacity factor in percent, and FC = average or levelized fuel costs over the unit lifetime in dollars per megajoule (dollars per million Btu). The calculated values for these parameters are provided in Table 8.14.

2. *Make a Comparison*

As shown in Table 8.14, even though the annual fixed charges for Unit C are $900,000 per year higher ($81.0 million per year vs. $80.10 million per year) and the annual O&M costs for Unit C are $100,000 per year higher ($11.2 million per year vs. $11.1 million per year), the resulting total annual costs are $100,000 per year lower for Unit C ($147.39 million per year vs. $147.49 million per year). This stems from the annual fuel expense for Unit C being $1.10 million per year lower than for Unit D ($55.19 million per year vs. $56.29 million per year) because the heat rate of Unit C is 0.2110 MJ/kWh (200 Btu/kWh) better. In this case, the Unit C design should be chosen over that of Unit D because the economic benefits associated with the 0.2110 MJ/kWh (200 Btu/kWh) heat-rate improvement more than offsets the higher capital and O&M costs associated with Unit C.

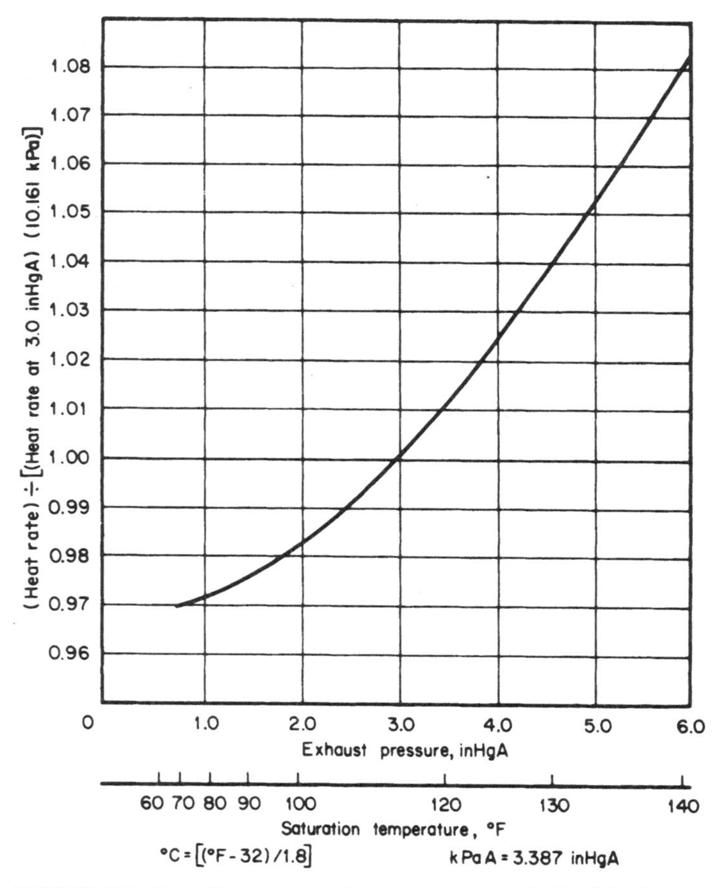

FIGURE 8.8 Typical heat rate vs. exhaust pressure curve for fossil-fired steam-cycle units.

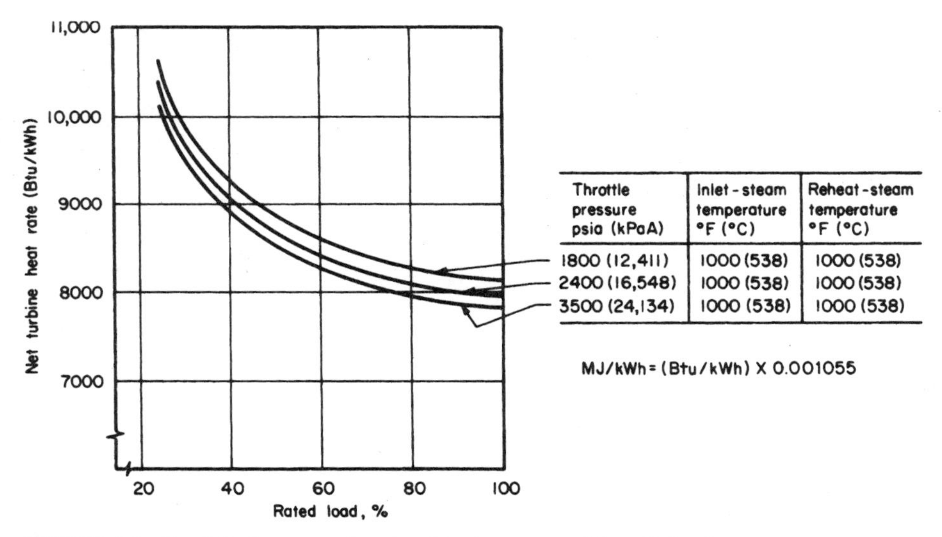

FIGURE 8.9 Fossil-fired steam-cycle unit turbine heat rates at 7.6-cmHgA exhaust pressure vs. percent-rated load.

Related Calculations. The heat rate for a steam-cycle unit changes significantly with the turbine exhaust pressure (the saturation pressure and temperature provided by the cooling system) as shown in Fig. 8.8; with the percentage of rated load (amount of partial load operation of the unit) as shown in Fig. 8.9; with the choice of throttle (gauge) pressure of 12,411 kPa (1800 psig) vs. a gauge pressure measured at 16,548 kPa (2400 psig) vs. 24,132 kPa (3500 psig) as shown in Fig. 8.9 and Table 8.15; with throttle and reheater temperature and reheater pressure drop; and with a number of steam-cycle configuration and component performance changes as shown in Table 8.16. Therefore it is necessary for the power-plant designer to investigate carefully the choice of each of these parameters.

For example, as shown in Fig. 8.8, a change in cooling-tower performance (such as a change in the cooling-tower dimensions or a change in the rated circulating water flow) that increases the turbine-exhaust saturation temperature from 44.79°C to 48.62°C would cause the turbine-exhaust saturation pressure to rise from an absolute pressure of

TABLE 8.15 Effect of Steam-Condition Changes on Net Turbine Heat Rates

Steam condition	Percent change in net heat rate			
Throttle pressure, psi	1800	2400	3500	3500
Number of reheats	1	1	1	1
Throttle pressure (change from preceding column)		1.9–2.1	1.8–2.0	1.6–2.0
50°F Δ throttle temperature	0.7	0.7–0.8	0.8–0.9	0.7
50°F Δ first reheat temperature	0.8	0.8	0.8	0.4
50°F Δ second reheat temperature				0.6
One point in percent Δ reheated pressure drop	0.1	0.1	0.1	0.1
Heater above reheat point	0.7	0.6	0.5–0.6	

kPa = psi × 6.895
Δ°C = ΔF/1.8

TABLE 8.16 Effect of Steam-Cycle Changes on Net Turbine Heat Rates

Cycle configuration	Change in net heat rate*	
	Percent	Btu/kWh†
1. Extraction line pressure drops of 3 percent rather than 5 percent (constant throttle flow)	−0.14	−11
2. Bottom heater drains flashed to condenser through 15°F drain cooler rather than 10°F‡	+0.01−0.02	+1−2
3. Change deaerator heater to closed-cascading type with a 5°F temperature difference (TD) and a 10°F drain cooler	+0.24	+19
4. Make all drain coolers 15°F rather than 10°F	+0.01	+1
5. Reduce demineralized condenser makeup from 3 percent to 1 percent	−0.43	−35
6. Make top heater 0°F TD rather than −3°TD (constant throttle flow)	+0.01	+1
7. Make low-pressure heater TDs 3°F rather than 5°F	−0.11	−9
8. Eliminate drain cooler on heater 7	+0.08	+6.1

*+, is poorer
†MJ/kWh = (Btu/kWh) × 0.001055
‡Δ°C = Δ°F/1.8

9.48 kPa (2.8 inHg) to 11.52 kPa (3.4 inHg), which, as indicated in Fig. 8.8, would increase the heat-rate factor from 0.9960 to 1.0085 (i.e., a change of 0.0125 or 1.25 percent). For a net turbine heat rate of 8.440 MJ/kWh (8000 Btu/kWh), this results in an increase of 0.106 MJ/kWh (100 Btu/kWh); that is, (8.440 MJ/kWh)(0.0125) = 0.106 MJ/kWh.

From Fig. 8.9, operation of a unit at a gauge pressure of 16,548 kPa/538°C/538°C (2400 psig/1000°F/1000°F) at 70 percent of rated load instead of 90 percent would increase the net turbine heat rate by 0.264 MJ/kWh (250 Btu/kWh) from 8.440 MJ/kWh (8000 Btu/kWh) to 8.704 MJ/kWh (8250 Btu/kWh).

From Table 8.15 and Fig. 8.9, a change in the throttle gauge pressure from 16,548 kPa (2400 psig) to 12,411 kPa (1800 psig) would increase the heat rate from 0.160 to 0.179 MJ/kWh (152 to 168 Btu/kWh), that is, from 1.9 to 2.1 percent of 8.440 MJ/kWh (8000 Btu/kWh).

REPLACEMENT FUEL COST

Table 8.17 describes the pertinent data for two alternatives that have the same net unit heat rate and rating. Unit F, however, has an average unit availability about 3 percent lower than Unit E (92 vs. 95 percent). The capacity factor for each unit is 70 percent. Determine the replacement fuel cost.

Calculation Procedure

1. Analyze the Problem

Because the ratings and heat rates are the same, it is convenient, for evaluation purposes, to assume that a utility would attempt to produce the same amount of electric power with either unit throughout the year. However, because of its lower plant availabil-

TABLE 8.17 Evaluation of Reliability vs. Installed Capital and O&M Costs

	Unit E	Unit F
Net unit heat rate	10.550 MJ/kWh (10,000 Btu/kWh)	10.550 MJ/kWh (10,000 Btu/kWh)
Unit availability	95 percent	92 percent
Unit rating	600 MW	600 MW
Installed capital cost	$\$450 \times 10^6$	$\$440 \times 10^6$
Levelized or average fixed charge rate	18 percent	18 percent
Levelized or average annual O&M cost (excluding fuel)	$\$11.2 \times 10^6$/yr	$\$12.0 \times 10^6$/yr
Desired levelized or average capacity factor	70 percent	70 percent
Actual levelized or average capacity factor	70 percent	67.8 percent
For Unit E:		
Annual fixed capital charges = ($\$450 \times 10^6$)(18/100)		= $\$81.00 \times 10^6$/yr
Annual O&M cost (excluding fuel)		= $\$11.20 \times 10^6$/yr
Total annual cost used for comparison with Unit F		= $\$92.20 \times 10^6$/yr
For Unit F:		
Annual fixed capital charges = ($\$440 \times 10^6$)(18/100)		= $\$79.20 \times 10^6$/yr
Annual O&M cost (excluding fuel)		= $\$12.00 \times 10^6$/yr
Replacement energy required for Unit F as compared with Unit E = (600 MW)(8760 h/yr)(70/100) [1 − (92 percent/95 percent)] = (600 MW)(8760 h/yr)(70 − 67.789)/100 = 116,210 MWh/yr		
Replacement energy cost penalty for Unit F as compared with Unit E = (116,210 MWh/yr)($15/MWh)		= $\$\ 1.74 \times 10^6$/yr
Total annual cost used for comparison with Unit E		= $\$92.94 \times 10^6$/yr

ity. Unit F would in general be expected to produce about 3 percent less electric power than Unit E. As a result, during a total of 3 percent of the year when Unit F would not be available, as compared with Unit E, the utility would have to either generate additional power or purchase power from a neighboring utility to replace the energy that Unit F was unable to produce because of its unavailability.

The difference between the cost of either the purchased or generated replacement power and the cost to generate that power on the unit with the higher plant availability represents a replacement energy cost penalty that must be assessed to the unit with the lower power availability (in the case, Unit F).

2. *Calculate the Replacement Energy Cost*

The replacement energy cost penalty is generally used to quantify the economic costs associated with changes in plant availability, reliability, or forced outage rates. The replacement energy cost penalty is calculated as follows: replacement energy cost penalty = RE × RECD, in dollars per hour, where RE = replacement energy required in MWh/yr and RECD = replacement energy cost differential in dollars per megawatthour.

The value of RECD is determined by RECD = REC − AGC_{ha}, where REC = cost to either purchase replacement energy or generate replacement energy on a less efficient or more costly unit, in dollars per megawatthour, and AGC_{ha} = average generation cost of the unit under consideration with the highest (best) availability, in dollars per megawatthour. The average generation cost is calculated as $AGC_{ha} = HR_{ha} \times FC_{ha}/10^6$, where HR_{ha} = heat rate of the highest availability unit under consideration in J/kWh (Btu/kWh) and FC_{ha} = the average or levelized fuel cost of the highest availability unit under consideration in dollars per megajoule (dollars per million Btu).

The replacement energy, RE, is calculated as follows: $RE = \text{rating} \times 8760 \times [(DCF/100)(1 - PA_{la}/PA_{ha})]$, where rating = the capacity rating of the unit in MW, DCF = desired average or levelized capacity factor for the units in percent, PA_{la} = availability of unit under consideration with lower availability in percent, and PA_{ha} = availability of unit under consideration with higher availability in percent.

Implied in this equation is the assumption that the actual capacity factor for the unit with lower availability (ACF_{la}) will be lower than for the unit with higher availability as follows: $ACF_{la} = DCF \times (PA_{la}/PA_{ha})$.

As shown in Table 8.17, even though the annual fixed charges for Unit E were $800,000 per year higher ($81.00 million vs. $79.20 million per year), the total resulting annual costs for Unit E were $740,000 lower ($92.20 million vs. $94.94 million per year) because Unit F had a $1.74 million per year replacement energy cost penalty and operation and maintenance costs that were $800,000 per year higher than Unit E.

Related Calculations. It generally can be assumed that the replacement energy (either purchased from a neighboring utility or generated on an alternate unit) would cost about $10 to $20 per megawatthour more than energy generated on a new large coal-fired unit. In the example in Table 8.17, a value of replacement energy cost differential of $15/MWh was used.

CAPABILITY PENALTY

Compare the capability (capacity) penalty for Units G and H in Table 8.18. Unit G has a rated capacity that is 10 MW higher than that of Unit H.

Calculation Procedure

1. Analyze the Problem

To achieve an equal reliability level the utility would, in principle, have to replace the 10 MW of capacity not provided by Unit H with additional capacity on some other new unit. Therefore, for evaluation purposes, the unit with the smaller rating must be assessed what is called a capability (capacity) penalty to account for the capacity difference. The capability penalty, CP, is calculated by: $CP = (\text{rating}_l - \text{rating}_s) \times CPR$, where rating_l and rating_s are the ratings of the larger and smaller units, respectively, in kW, and CPR = capability penalty rate in dollars per kilowatt.

2. Calculate the Capability Penalty

For example, if the units have capital costs of approximately $500/kW, or if the capacity differential between the units is provided by additional capacity on a unit that would cost $500/kW, the capability penalty assessed against Unit H (as shown in Table 8.18) is $5 million total. For an 18 percent fixed-charge rate, this corresponds to $900,000 per year.

Note that the annual operation and maintenance costs are the same for both units. In this case, those costs were not included in the total annual costs used for comparison purposes.

As shown in Table 8.18, even though alternative Unit H had a capital cost $2 million

TABLE 8.18 Evaluation of Unit Rating vs. Installed Capital Costs

	Unit G	Unit H
Unit rating	610 MW	600 MW
Net unit heat rate	10.550 MJ/kWh (10,000 Btu/kWh)	10.550 MJ/kWh (10,000 Btu/kWh)
Unit availability	95 percent	95 percent
Installed capital cost	$\$450 \times 10^6$	$\$448 \times 10^6$
Levelized or average fixed-charge rate	18 percent	18 percent
Levelized or average annual O&M cost (excluding fuel)	$\$11.2 \times 10^6$/yr	$\$11.2 \times 10^6$/yr
Capability penalty rate	$500/kW	$500/kW
For Unit G:		
Annual fixed capital charges $= (\$450 \times 10^6)(18/100)$		$= \$81.00 \times 10^6$/yr
Total annual cost used for comparison with Unit H		$= \$81.00 \times 10^6$/yr
For Unit H:		
Annual fixed capital charges $= (\$448 \times 10^6)(18/100)$		$= \$80.64 \times 10^6$/yr
O&M costs same as alternate Unit G		
Total capability penalty for Unit H as compared with Unit G $= (610\ \text{MW} - 600\ \text{MW})(1000\ \text{kW/MW})(\$500/\text{kW})$ $= \$5.00 \times 10^6$/yr		
Annual capability penalty $= (\$5.0 \times 10^6)(18/100)$		$= \$\ 0.90 \times 10^6$/yr
Total annual costs used for comparison with Unit G		$= \$81.54 \times 10^6$/yr

lower than alternative Unit G, when the capability penalty is taken into account, alternative Unit G would be the economic choice.

Related Calculations. Applying the evaluation techniques summarized in Tables 8.13 through 8.18 sequentially makes it possible to evaluate units that fall into more than one (or all) of the categories considered and thereby to determine the optimum plant design.

BIBLIOGRAPHY

Decher, Reiner. 1996. *Direct Energy Conversion: Fundamentals of Electric Power Production*. Oxford University Press.

Elliot, Thomas C. 1997. *Stanford Handbook of Power Plant Engineering*. New York: McGraw-Hill.

Lausterer, G. K., H. Weber, and E. Welfonder. 1993. *Control of Power Plants and Power Systems*. New York and London: Pergamon Press.

Li, Kam W., and A. Paul Priddey. 1985. *Power Plant System Design*. New York: Wiley.

Marks' Standard Handbook for Mechanical Engineers. 1996. New York: McGraw-Hill.

Van Der Puije, Patrick. 1997. *Electric Power Generation,* 2nd ed. Boca Rotan, Fla.: Chapman & Hall Publishers.

Wildi, Theodore. 2000. *Electrical Machines, Drives, and Power Systems*, 4th ed. Englewood Cliffs, N.J.: Prentice-Hall.

Wood, Allen J., and Bruce Wollenberg. 1996. *Power Generation, Operation and Control.* New York: Wiley.

SECTION 9

OVERHEAD TRANSMISSION LINES AND UNDERGROUND CABLES

Richard A. Rivas*
Electrical Engineer, M.Sc.
Assistant Professor
Universidad Simón Bolívar
Caracas, Venezuela
Ph.D. Candidate
University of British Columbia
British Columbia, Canada

*Original author of this section was John S. Wade, Jr., Ph.D., The Pennsylvania State University.

INTRODUCTION

Overhead transmission lines are composed of aluminium conductors that, even in modest capacities, are stranded in spiral fashion for flexibility. These are primarily classified as:

AAC: all-aluminum conductors

AAAC: all-aluminum alloy conductors

ACSR: aluminum conductors, steel reinforced

ACAR: aluminum conductors, alloy reinforced

Aluminum conductors are compared in conductivity with the International Annealed Copper Standard (IACS) in Table 9.1. The table lists the percent conductivity, as well as the temperature coefficient of resistance, α, expressed per °C above 20°C.

CONDUCTOR RESISTANCE

Calculate the resistance of 1000 ft (304.8 m) of solid round aluminum conductor, type EC-H19 (AWG No.1) at 20 and 50°C. The diameter = 0.2893 in. (7.35 mm).

TABLE 9.1 Comparison of Aluminum and Copper Conductors

Material	Percent conductivity	α
Aluminum:		
EC-H19	61.0	0.00403
5005-H19	53.5	0.00354
6201-T81	52.5	0.00347
Copper:		
Hard-drawn	97.0	0.00381

Calculation Procedure

1. Calculate Resistance at 20°C

Use $R = \rho l/A$, where R is the resistance in ohms, ρ is the resistivity in ohm-circular mils per foot ($\Omega \cdot$ cmil/ft), l is the length of conductor in feet, and A is the area in circular mils (cmil), which is equal to the square of the conductor diameter given in mils. From Table 9.1, the percentage of conductivity for EC-H19 is 61 percent that of copper. For IACS copper, $\rho = 10.4\ \Omega \cdot$ cmil/ft. Therefore, for aluminum, the resistivity is $10.4/0.61 = 17.05\ \Omega \cdot$ cmil/ft. The resistance of the conductor at 20°C is $R = (17.05)(1000)/289.3^2 = 0.204\ \Omega$.

2. Calculate Resistance at 50°C

Use $R_T = R_{20°}\ [1 + \alpha(T - 20°)]$, where R_T is the resistance at the new temperature, $R_{20°}$ is the resistance at 20°C, and α is the temperature coefficient of resistance. From Table 9.1, $\alpha = 0.00403$. Hence, $R_{50°} = 0.204[1 + 0.00403(50 - 20)] = 0.229\ \Omega$.

INDUCTANCE OF SINGLE TRANSMISSION LINE

Using flux linkages, determine the self-inductance, L, of a single transmission line.

Calculation Procedure

1. Select Appropriate Equation

Use $L = \lambda/i$, where L is the self-inductance in H/m, λ is the magnetic flux linkage in Wb $\cdot$ turns/m, and i is the current in amperes.

2. Consider Magnetic Flux Linkage

Assume a long, isolated, round conductor with uniform current density. The existing flux linkages include those internal to the conductor partially linking the current and those external to the conductor, which links all of the current. These will be calculated separately and then summed, yielding total inductance, $L_T = L_{\text{int}} + L_{\text{ext}}$.

3. Apply Ampere's Law

The magnitude of the magnetic flux density, **B,** in Wb/m^2 produced by a long current filament is $B = \mu i/2\pi r$, where $\mu =$ the permeability of the flux medium ($4\pi \times 10^{-7}$ H/m for free space and nonferrous material) and $r =$ radius to **B** from the current center in meters. The direction of **B** is tangential to encirclements of the enclosed current, and clockwise if positive current is directed into this page (right-hand rule). The differential flux linkages per meter external to a conductor of radius a meters are $d\lambda = \mu i\ dr/2\pi r$ Wb $\cdot$ turns/m and $r > a$.

4. Consider Flux Inside Conductor

The calculation of differential flux linkages is complicated by **B,** which is a function of only that part of the current residing inside the circle passing through the measuring point of **B.** The complication is compounded by the reduction in current directly affecting $d\lambda$. Thus, if $\mathbf{B} = (\mu i/2\pi r)(\pi r^2/\pi a^2)$ Wb/m^2, then $d\lambda = (\mu i/2\pi r)(\pi r^2/\pi a^2)^2 = (\mu i/2\pi)(r^3/a^4)$ Wb $\cdot$ turns/m for $r < a$.

5. Calculate the Internal Inductance

Integrating the expression in (4) yields: $\lambda = (\mu i/2\pi)(\frac{1}{4})$, from which:

$$L_{\text{int}} = (10^{-7}/2) \text{ H/m.}$$

Related Calculations. Calculation of the inductance due to the flux external to a long isolated conductor yields an infinite value, since r varies from a to infinity, but then such an isolated conductor is not possible.

INDUCTANCE OF TWO-WIRE TRANSMISSION LINE

Consider a transmission line consisting of two straight, round conductors (radius a meters), uniformly spaced D meters apart, where $D \gg a$ (Fig. 9.1). Calculate the inductance of the line.

Calculation Procedure

1. Consider Flux Linkages

It is realistic to assume uniform and equal but opposite current density in each conductor. The oppositely directed currents, therefore, produce a net total flux linkage of zero because the net current in any cross section of both conductors is zero. This is true for any multiconductor system whose cross-sectional currents add to zero (for example, in a balanced three-phase system).

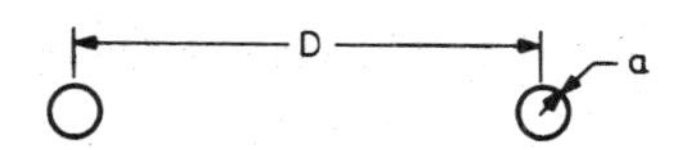

FIGURE 9.1 A two-wire transmission line.

2. Calculate Inductance of One Conductor

Use $\lambda = (\mu i/2\pi)\ [\frac{1}{4} + \ln(D/a)]$ Wb · turns/m. Because $L = \lambda/i$, $L = (2 \times 10^{-7})\ [\frac{1}{4} + \ln(D/a)]$ H/m. Inductance L may be expressed in more compact form by $L = (2 \times 10^{-7}) \ln(D/r')$, where $r' = a \exp(-\frac{1}{4})$ is the *geometric mean radius,* GMR. The value of r' is $0.788a$.

3. Calculate Total Inductance, L_T

$L_T = 2L = (4 \times 10^{-7})\ln(D/r')$ H/m. In more conventional units, $L_T = 1.482 \times \log(D/r')$ mH/mi.

INDUCTIVE REACTANCE OF TWO-WIRE TRANSMISSION LINE

Calculate the inductive reactance of 10 mi (16.1 km) of a two-conductor transmission line (Fig. 9.1), where $D = 8$ ft (2.44 m) and $a = 0.1$ in. (2.54 mm) at a frequency of 60 Hz (377 rad/s).

Calculation Procedure

1. Calculate the Geometric Mean Radius

The GMR is $r' = (0.7788)(2.54 \times 10^{-3}) = 0.001978$ m.

2. Calculate $\mathbf{L_T}$

$L_T = (4 \times 10^{-7})\ln(D/r') = (4 \times 10^{-7})\ln(2.44/0.001978) = 28.5 \times 10^{-7}$ H/m.

3. Calculate Inductive Reactance $\mathbf{X_L}$

$X_L = (377)(28.5 \times 10^{-7}\text{ H/m})(16.1 \times 10^3\text{ m}) = 17.3\ \Omega$.

Related Calculations. A larger conductor and/or smaller conductor spacing reduces the inductive reactance.

INDUCTANCE OF STRANDED-CONDUCTOR TRANSMISSION LINE

Determine the inductance of a transmission line having six identical round conductors (Fig. 9.2) arranged so that the currents on each side occupy the conductors equally with uniform current density.

Calculation Procedure

1. Calculate Flux of One Conductor

The flux linkages of conductor 1, which carries one-third of the current, can be deduced from the method established for the two-conductor line by: $\lambda_1 = (\mu/2\pi)(i/3)\ [\frac{1}{4} + \ln(D_{11'}/a) + \ln(D_{12'}/a) + \ln(D_{13'}/a) - \ln(D_{12}/a) - \ln(D_{13}/a)]$, where $D_{11'}$ is the distance between conductors 1 and 1′ and so on. When terms are collected, the equation becomes: $\lambda_1 = (\mu i/2\pi)\ \{\ln[(D_{11'}D_{12'}D_{13'})^{1/3}/(r'D_{12}D_{13})^{1/3}]\}$ Wb · turns/m. The flux linkages of the other two conductors carrying current in the same direction are found in a similar manner.

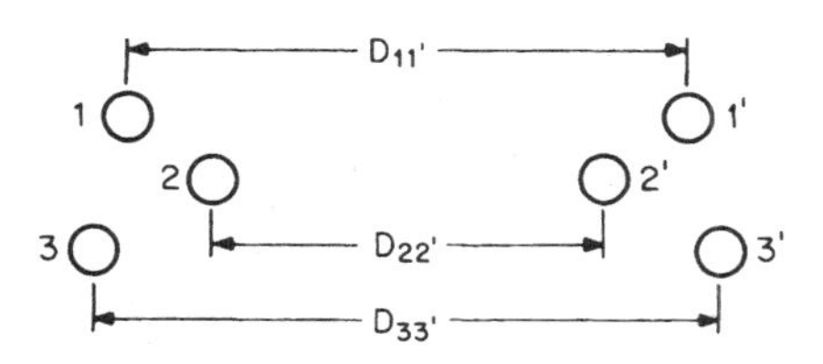

FIGURE 9.2 Stranded-conductor transmission line.

2. Calculate the Inductances

The inductances of conductors one, two, and three then become: $L_1 = \lambda_1/(i/3) = (3 \times 2 \times 10^{-7})[\ln(D_{11'}D_{12'}D_{13'})^{1/3}/(r'D_{12}D_{13})^{1/3}]$ H/m, $L_2 = \lambda_2/(i/3) = (3 \times 2 \times 10^{-7})\ [\ln(D_{21'}D_{22'}D_{23'})^{1/3}/(r'D_{12}D_{23})^{1/3}]$ H/m, and $L_3 = \lambda_3/(i/3) = (3 \times 2 \times 10^{-7})\ [\ln(D_{31'}D_{32'}D_{33'})^{1/3}/(r'D_{23}D_{13})^{1/3}]$ H/m.

The inductance of the unprimed side (Fig. 9.2) of the line is one-third the average value because the inductances are in parallel. Then $L_{avg} = (L_1 + L_2 + L_3)/3$ H/m and $L = (L_1 + L_2 + L_3)/9$ H/m.

***3. Determine the Total Inductance,* L_T**

Combining the sets of equations in Step 2, we obtain: $L_T = (2 \times 10^{-7})\{\ln[(D_{11'}D_{12'}D_{13'})(D_{21'}D_{22'}D_{23'})(D_{31'}D_{32'}D_{33'})]^{1/9}/(r'^3 D_{12}^2 D_{13}^2 D_{23}^2)^{1/9}\}$ H/m.

Related Calculations. To find the inductance of the primed side of the line in Fig. 9.2, follow the same procedure as above. Again, summing produces the total inductance.

The root of the product in the denominator for the expression of L_T is a GMR. (The root of the product in the numerator is called the geometric mean distance, GMD.) Although tabulated values are usually available, one may calculate GMR $= (r'^n D_{12}^2 D_{13}^2 D_{23}^2 \ldots D_{(n-1)n}{}^2)^{1/n}$ m, where $1/n$ is the reciprocal of the number of strands.

INDUCTANCE OF THREE-PHASE TRANSMISSION LINES

Determine the inductance per phase for a three-phase transmission line consisting of single conductors arranged unsymmetrically (Fig. 9.3).

Calculation Procedure

1. Use Flux-Linkage Method

The principle of obtaining inductance per phase by using the flux linkages of one conductor is utilized once again. If the line is unsymmetrical and it remains untransposed, the inductance for each phase will not be equal (transposition of a line occurs when phases *a*, *b*, and *c* swap positions periodically). Transposing a transmission line will equalize the inductance per phase. Inductance, however, varies only slightly when untransposed, and it is common practice, in hand calculations, to assume transposition as is done in what follows.

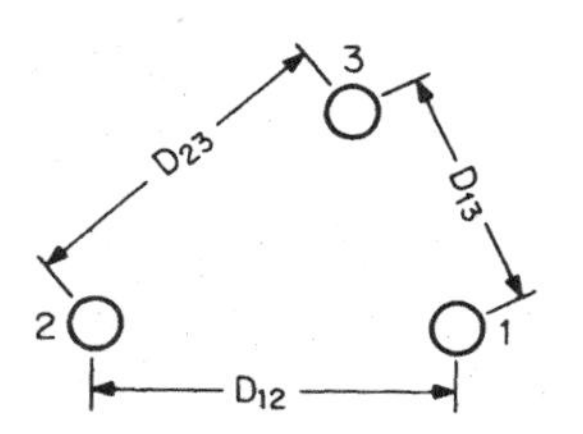

FIGURE 9.3 A three-phase transmission line.

Assume phase *a* shifts from position 1 to 2, and then to 3; phases *b* and *c* also move in the rotation cycle. The average flux linkages for phase *a* are then given by: $\lambda_a = (2 \times 10^{-7})/(3)[(\frac{1}{4}\mathbf{I_a} + \mathbf{I_a} \ln D_{12}/a + \mathbf{I_a} \ln D_{13}/a + \mathbf{I_b} \ln D_{21}/a + \mathbf{I_c} \ln D_{31}/a) + (\frac{1}{4}\mathbf{I_a} + \mathbf{I_a} \ln D_{21}/a + \mathbf{I_a} \ln D_{23}/a + \mathbf{I_b} \ln D_{32}/a + \mathbf{I_c} \ln D_{12}/a) + (\frac{1}{4}\mathbf{I_a} + \mathbf{I_a} \ln D_{32}/a + \mathbf{I_a} \ln D_{31}/a + \mathbf{I_b} \ln D_{13}/a + \mathbf{I_c} \ln D_{23}/a)]$, where $\mathbf{I_a}$, $\mathbf{I_b}$, and $\mathbf{I_c}$ are the rms phase currents. Then $\mathbf{I_a} + \mathbf{I_b} + \mathbf{I_c} = 0$. Also, $D_{12} = D_{21}$, $D_{23} = D_{32}$, and $D_{13} = D_{31}$. After terms are combined, the average flux linkage becomes: $\lambda_a = (2 \times 10^{-7})(\mathbf{I_a})/(3)[\ln(D_{12}D_{13}D_{23})/a^3 + \frac{3}{4}]$ Wb · turns/m.

2. Calculate **L**

Using r'^3 and dividing by $\mathbf{I_a}$, we find the inductance per phase is $L_\phi = (2 \times 10^{-7})[\ln(D_{12}D_{13}D_{23})^{1/3}/r']$ H/m.

Related Calculations. If the conductor for each phase is concentrically stranded, the distance between conductors remains the same, but r' is replaced by a tabulated GMR. The inductance in conventional form is given by: $L = 0.7411 \log[(D_{12}D_{13}D_{23})^{1/3}/\text{GMR}]$ mH/mi; D and GMR are given in feet.

PER-PHASE INDUCTIVE REACTANCE

Calculate the per-phase inductive reactance per mile (1600 m) for a three-phase line at 377 rad/s. The conductors are aluminum conductors, steel-reinforced (ACSR) Redwing (Table 9.2) arranged in a plane as shown in Fig. 9.4.

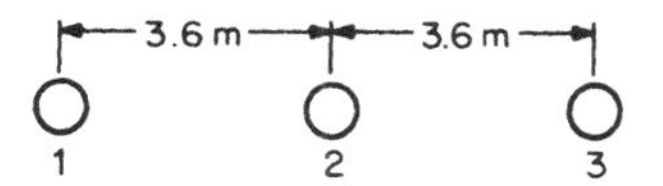

FIGURE 9.4 A plane-spaced transmission line.

Calculation Procedure

1. Calculate $\mathbf{L_\phi}$

From Table 9.2, GMR = 0.0373 ft (0.01 m). Substituting in the equation for per-phase inductance, we find $L_\phi = (2 \times 10^{-7})\ln(3.6 \times 7.2 \times 3.6)^{1/3}/0.01 = 12.2 \times 10^{-7}$ H/m.

2. Calculate Inductive Reactance, $\mathbf{X_L}$

$X_L = 377 \times 12.2 \times 10^{-7}$ H/m $\times$ 1600 m = 0.74 Ω/mi.

INDUCTANCE OF SIX-CONDUCTOR LINE

Calculate the per-phase inductance of the transmission line of Fig. 9.5 where the conductors are arranged in a double-circuit configuration.

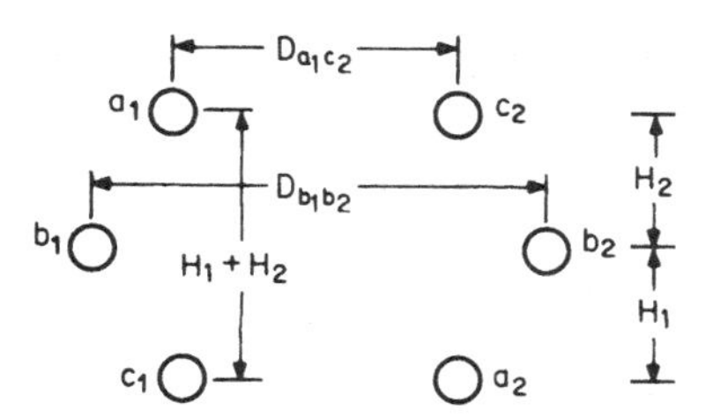

FIGURE 9.5 Line conductors arranged in a double circuit configuration.

Calculation Procedure

1. Use Suitable Expression for $\mathbf{L_\phi}$

Use $L_\phi = (2 \times 10^{-7})\ln(\text{GMD}/\text{GMR})$ H/m, where GMR is the GMR of a conductor.

2. Calculate GMD

The GMD includes the distances between all the phase combinations. However, the expression for GMD can be reduced to one-half the distances that are represented in the original expression and the root becomes $\frac{1}{6}$ rather than $\frac{1}{12}$. Thus, $\text{GMD} = (D_{a1b1}D_{a1b2}D_{a1c1}D_{a1c2}D_{b1c1}D_{b1c2})^{1/6}$ m.

TABLE 9.2 Aluminum Conductors, Steel Reinforced (ACSR)

Code word	Size, Mcmil	Stranding aluminum /steel	Outside diameter, in	Resistance		GMR, ft	Phase-to-neutral, 60 Hz, reactance at 1-ft spacing	
				DC, Ω/ 1000 ft at 20°C	AC, 60 Hz, Ω/mi at 25°C		Inductive Ω/mi, X_a	Capacitive Ω/mi, X_a
Waxwing	266.8	18/1	0.609	0.0646	0.3448	0.0198	0.476	0.1090
Partridge	266.8	2/76	0.642	0.0640	0.3452	0.0217	0.465	0.1074
Ostrich	300	26/7	0.680	0.0569	0.3070	0.0229	0.458	0.1057
Merlin	336.4	18/1	0.684	0.0512	0.2767	0.0222	0.462	0.1055
Linnet	336.4	26/7	0.721	0.0507	0.2737	0.0243	0.451	0.1040
Oriole	336.4	30/7	0.741	0.0504	0.2719	0.0255	0.445	0.1032
Chickadee	397.5	18/1	0.743	0.0433	0.2342	0.0241	0.452	0.1031
Ibis	397.5	26/7	0.783	0.0430	0.2323	0.0264	0.441	0.1015
Lark	397.5	30/7	0.806	0.0427	0.2306	0.0277	0.435	0.1007
Pelican	477	18/1	0.814	0.0361	0.1947	0.0264	0.441	0.1004
Flicker	477	24/7	0.846	0.0359	0.1943	0.0284	0.432	0.0992
Hawk	477	26/7	0.858	0.0357	0.1931	0.0289	0.430	0.0988
Hen	477	30/7	0.883	0.0355	0.1919	0.0304	0.424	0.0980
Osprey	556.5	18/1	0.879	0.0309	0.1679	0.0284	0.432	0.0981
Parakeet	556.5	24/7	0.914	0.0308	0.1669	0.0306	0.423	0.0969
Dove	556.5	26/7	0.927	0.0307	0.1663	0.0314	0.420	0.0965
Eagle	556.5	30/7	0.953	0.0305	0.1651	0.0327	0.415	0.0957
Peacock	605	24/7	0.953	0.0283	0.1536	0.0319	0.418	0.0957
Squab	605	26/7	0.966	0.0282	0.1529	0.0327	0.415	0.0953
Teal	605	30/19	0.994	0.0280	0.1517	0.0341	0.410	0.0944
Rook	636	24/7	0.977	0.0269	0.1461	0.0327	0.415	0.0950
Grosbeak	636	26/7	0.990	0.0268	0.1454	0.0335	0.412	0.0946
Egret	636	30/19	1.019	0.0267	0.1447	0.0352	0.406	0.0937
Flamingo	666.6	24/7	1.000	0.0257	0.1397	0.0335	0.412	0.0943
Crow	715.5	54/7	1.051	0.0240	0.1304	0.0349	0.407	0.0932
Starling	715.5	26/7	1.081	0.0238	0.1294	0.0355	0.405	0.0948
Redwing	715.5	30/19	1.092	0.0237	0.1287	0.0373	0.399	0.0920

3. Calculate GMR

Use GMR = $(\text{GMR}_c^3 D_{a1a2} D_{b1b2} D_{c1c2})^{1/6}$ m.

INDUCTIVE REACTANCE OF SIX-CONDUCTOR LINE

Calculate the per-phase inductive reactance of a six-conductor, three-phase line at 377 rad/s consisting of Teal ACSR conductors (Fig. 9.5). Distance $D_{a1c2} = 4.8$ m, $H_1 = H_2 = 2.4$ m, and $D_{b1b2} = 5.4$ m.

Calculation Procedure

1. Determine GMD

The necessary dimensions for calculating GMD are: $D_{a1b2} = D_{b1c2} = (2.4^2 + 5.1^2)^{1/2} = 5.64$ m, $D_{a1b1} = D_{b1c1} = (2.4^2 + 0.3^2)^{1/2} = 2.42$ m, $D_{a1c1} = 4.8$ m, and $D_{a1c2} = 4.8$ m. Hence, GMD $= [(2.42^2)(5.64^2)(4.8^2)]^{1/6} = 4.03$ m.

2. Determine GMR$_c$

From Table 9.2, for Teal, GMR$_c$ = 0.0341 ft (0.01 m). Then, $D_{a1a2} = D_{c1c2} = (4.8^2 + 4.8^2)^{1/2} = 6.78$ m, $D_{b1b2} = 5.4$ m, and GMR $= [(0.01^3)(6.78^2)(5.4)]^{1/6} = 0.252$ m.

3. Calculate Inductive Reactance per Phase

$X_L = (377)(2 \times 10^{-7})[\ln(4.03/0.252)] = 0.209 \times 10^{-3}$ Ω/m (0.336 Ω/mi).

INDUCTIVE REACTANCE OF BUNDLED TRANSMISSION LINE

Calculate the inductive reactance per phase at 377 rad/s for the bundled transmission line whose conductors are arranged in a plane shown in Fig. 9.6. Assume conductors are ACSR Crow.

Calculation Procedure

1. Determine GMD

Assume distances are between bundle centers and transposition of phases. Then, GMD $= [(9^2)(18)]^{1/3} = 11.07$ m. From Table 9.2, GMR$_c$ = 0.034 ft (0.01 m). The GMR

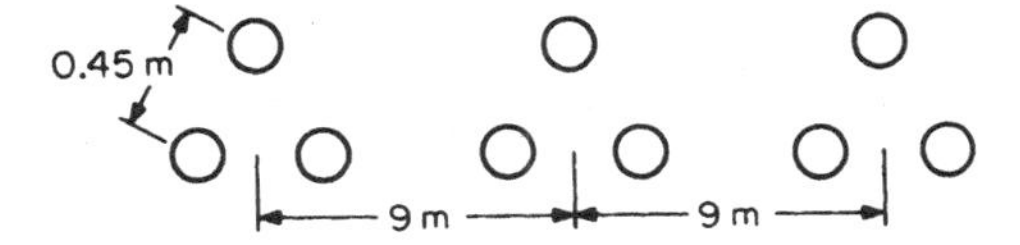

FIGURE 9.6 A bundled transmission line.

should include all conductor spacings from each other in the usual product form with, in this case, three values of GMR_c. Because of redundancy, GMR $= (0.01 \times 0.45^2)^{1/3} =$ 0.127 m.

2. *Calculate Inductive Reactance per Phase*

$X_L = 377 \times 2 \times 10^{-7} \times \ln(11.07/0.127) = 0.337 \times 10^{-3}\ \Omega/\text{m}\ (0.544\ \Omega/\text{mi})$.

Related Calculations. For a two-conductor bundle, GMR $= (GMR_c D)^{1/2}$ and for a four-conductor bundle, GMR $= (GMR_c D^3 2^{1/2})^{1/4}$. In each case, D is the distance between adjacent conductors.

For a two-conductor bundle, GMR $= (GMR_c D)^{1/2}$, and for a four-conductor bundle, GMR $= (GMR_c D^3 2^{1/2})^{1/4}$. In each case, D is the distance between adjacent conductors. For an n-conductor bundle as depicted in Fig. 9.7, GMR $= (n \cdot GMR_c \cdot A^{n-1})^{1/n}$, where A is the radius of the bundle. Similarly, for an n-conductor bundle, $r_{\text{equiv}} = (n \cdot r \cdot A^{n-1})^{1/n}$, where r is the external radius of the conductor and r_{equiv} is the equivalent external radius of the bundle (Dommel, 1992).

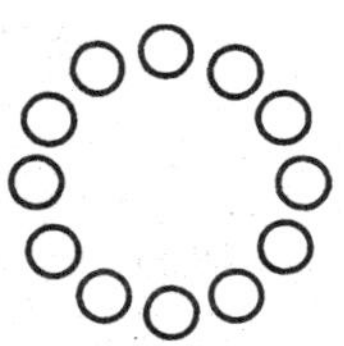

FIGURE 9.7 A multiconductor configuration replacing a large conductor.

At voltage levels above 230 kV, the corona loss surrounding single conductors, even though they are expanded by nonconducting central cores, becomes excessive. Therefore, to reduce the concentration of electric-field intensity, which affects the level of ionization, the radius of a single conductor is artificially increased by arranging several smaller conductors, in what approximates a circular configuration. This idea is depicted in Fig. 9.7. Other arrangements that prove satisfactory, depending on the voltage level, are shown in Fig. 9.8.

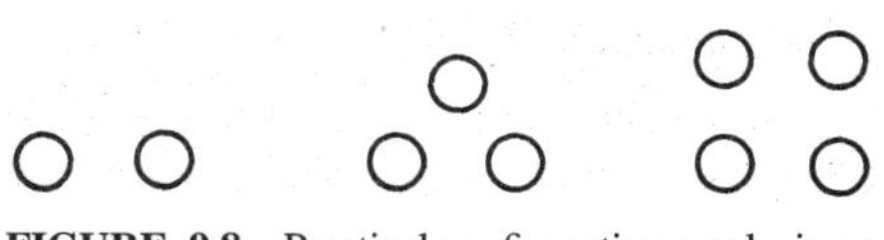

FIGURE 9.8 Practical configurations replacing a large conductor.

A benefit that accrues from bundling conductors of one phase of a line is an increase in GMR. Also, the inductance per phase is reduced, as is corona ionization loss.

INDUCTIVE REACTANCE DETERMINED BY USING TABLES

Determine the inductive reactance per phase using data in Tables 9.2 and 9.3 for ACSR Redwing with the spacing given in Fig. 9.9.

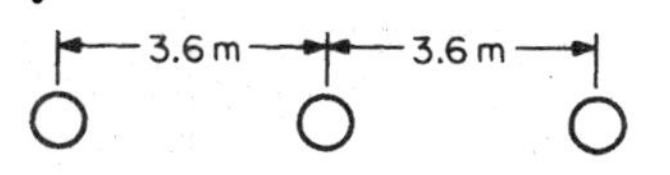

FIGURE 9.9 A three-phase line where conductors are separated by 3.6 m (12 ft).

Calculation Procedure

1. *Use Appropriate Tables*

The *Aluminum Electrical Conductor Handbook* provides tabulated data that reduce the amount of calculation necessary to find the inductive reactance for either a single- or three-phase line where circuits are neither paralleled nor bundled. To determine the reactance by this

TABLE 9.3 Separation Component X_d of Inductive Reactance at 60 Hz,* Ohms per Conductor per Mile†

	Separation of conductors											
	Inches											
Feet	0	1	2	3	4	5	6	7	8	9	10	11
0	—	−0.3015	−0.2174	−0.1682	−0.1333	−0.1062	−0.0841	−0.0654	−0.0492	−0.0349	−0.0221	−0.0106
1	0	0.0097	0.0187	0.0271	0.0349	0.0423	0.0492	0.0558	0.0620	0.0679	0.0735	0.0789
2	0.0841	0.0891	0.0938	0.0984	0.1028	0.1071	0.1112	0.1152	0.1190	0.1227	0.1264	0.1299
3	0.1333	0.1366	0.1399	0.1430	0.1461	0.1491	0.1520	0.1549	0.1577	0.1604	0.1631	0.1657
4	0.1682	0.1707	0.1732	0.1756	0.1779	0.1802	0.1825	0.1847	0.1869	0.1891	0.1912	0.1933
5	0.1953	0.1973	0.1993	0.2012	0.2031	0.2050	0.2069	0.2087	0.2105	0.2123	0.2140	0.2157
6	0.2174	0.2191	0.2207	0.2224	0.2240	0.2256	0.2271	0.2287	0.2302	0.2317	0.2332	0.2347
7	0.2361	0.2376	0.2390	0.2404	0.2418	0.2431	0.2445	0.2458	0.2472	0.2485	0.2498	0.2511
	Separation of conductors, ft											
8	0.2523	15	0.3286	22	0.3751	29	0.4086	36	0.4348	43	0.4564	
9	0.2666	16	0.3364	23	0.3805	30	0.4127	37	0.4382	44	0.4592	
10	0.2794	17	0.3438	24	0.3856	31	0.4167	38	0.4414	45	0.4619	
11	0.2910	18	0.3507	25	0.3906	32	0.4205	39	0.4445	46	0.4646	
12	0.3015	19	0.3573	26	0.3953	33	0.4243	40	0.4476	47	0.4672	
13	0.3112	20	0.3635	27	0.3999	34	0.4279	41	0.4506	48	0.4697	
14	0.3202	21	0.3694	28	0.4043	35	0.4314	42	0.4535	49	0.4722	

*From formula at 60 Hz, $X_d = 0.2794 \log_{10} d$, d = separation in feet.
†From *Electrical Transmission and Distribution Reference Book*, Westinghouse Electric Corporation, 1950.

method, it is convenient to express the inductive reactance by: $X_L = 2\pi f(2 \times 10^{-7})\ln(1/\mathrm{GMR}_c) + 2\pi f(2 \times 10^{-7})\ln$ GMD or $X_L = 2\pi f(0.7411 \times 10^{-3})\log(1/\mathrm{GMR}_c) + 2\pi f(0.7411 \times 10^{-3})\log$ GMD. the first term in the latter expression is the reactance at *1-ft spacing*, which is tabulated in Table 9.2 for 60 Hz. The second term is the *spacing component* of inductive reactance tabulated in Table 9.3 for a frequency of 60 Hz. Conversion from English to SI units, and vice versa, may be required.

2. Determine GMD

A distance of 3.6 m = 12 ft. Hence, GMD = $[(12^2)(24)]^{1/3}$ = 15.1 ft (4.53 m).

***3. Determine X*~L~**

From Tables 9.2 and 9.3, $X_L = 0.399 + 0.329 = 0.728$ Ω/mi. In SI units, $X_L = (0.728$ Ω/mi)(1 mi/1.6 km) = 0.45 Ω/km.

EFFECT OF MUTUAL FLUX LINKAGE

Referring to Fig. 9.10, find the voltage, in V/m, induced in a nearby two-conductor line by the adjacent three-phase transmission line carrying balanced currents having a magnitude of 50 A.

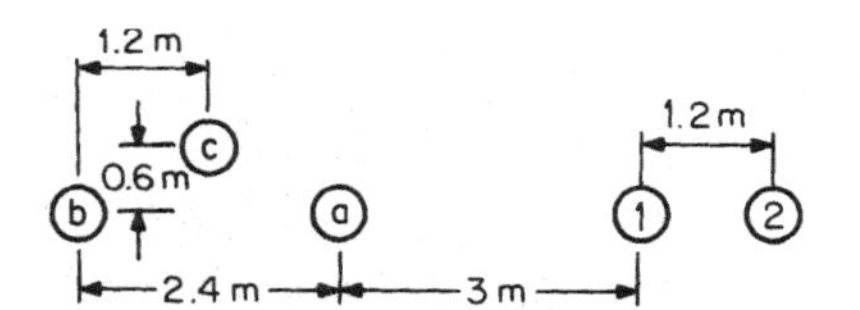

FIGURE 9.10 A three-phase line in proximity to a two-wire line.

Calculation Procedure

1. Define Approach

Flux linkages within the 1.2-m-wide plane of the two-conductor line from each phase of the transmission line shall be summed. Then, from Faraday's law, the derivative of this result will yield the answer.

2. Calculate Distances from Each Involved Conductor

$D_{b1} = 2.4 + 3 = 5.4$ m, $D_{b2} = 1.2 + 2.4 + 3 = 6.6$ m, $D_{a1} = 3$ m, $D_{a2} = 3 + 1.2 = 4.2$ m, $D_{c1} = (0.6^2 + 4.2^2)^{1/2} = 4.24$ m, and $D_{c2} = (0.6^2 + 5.4^2)^{1/2} = 5.43$ m.

3. Calculate Flux Linkages

$\lambda = \mu(i_a/2\pi)\ln(D_{a2}/D_{a1}) + \mu(i_b/2\pi)\ln(D_{b2}/D_{b1}) + \mu(i_c/2\pi)\ln(D_{c2}/D_{c1})$. This equation is a function of time. Substituting values in the expression and combining terms, we find $\lambda = \sqrt{2}[33.364 \sin \omega t + 20.07 \sin(\omega t - 120°) + 24.74 \sin(\omega t + 120°)] \times 10^{-7}$ Wb · turns/m, where $\omega = 2\pi f$.

4. Apply Faraday's Law

The voltage per unit length is $V = d\lambda/dt = \sqrt{2}[33.64\omega \cos \omega t + 20.07\omega \cos(\omega t - 120°) + 24.74\omega \cos(\omega t + 120°)] \times 10^{-7}$ V/m.

***5. Determine* V**

Transforming to the frequency domain and rms values, one obtains $V = (0.424 + j0.143) \times 10^{-3}$ V/m = $(0.68 + j0.23)$ V/mi.

INDUCTIVE IMPEDANCES OF MULTICONDUCTOR TRANSMISSION LINES INCLUDING GROUND-RETURN CORRECTIONS

Consider a transmission line consisting of n straight conductors. For the sake of simplicity, let us depict only the conductors i and k with their respective images, and assume that the per-phase ac resistance has been obtained from Table 9.2. Determine the self and mutual impedances of such a line taking into account the ground returns.

Calculation Procedure

1. Calculate the Complex Penetration Depth

Traditionally, ground-return corrections have been carried out by using series-based asymptotic approximation of Carson's infinite integrals (Carson, 1926). However, truncation errors might be unacceptable for the impedances in cases with wide separation among conductors, frequency higher than power frequency, or low earth resistivity (Dommel, 1985).

To represent current return through homogeneous ground, the ground can be replaced by an ideal plane placed below the ground surface at a distance $\overline{p}$ equal to the complex penetration depth for plane waves (Fig. 9.11). Such an approach has been proposed in Dunbanton (1969), Gary (1976), and Deri et al. (1981) and produces results that match those obtained from Carson's correction terms. The main advantage of this method is that it allows use of simple formulae for self and mutual impedances—formulae derived from using the images of the conductors—and, therefore, obtaining accurate results through the use of electronic calculators.

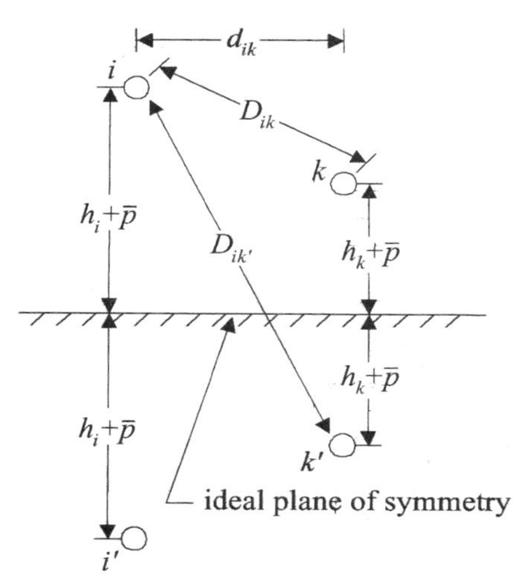

FIGURE 9.11 Transmission line geometry for impedance calculations including complex penetration depth.

The value of the complex penetration depth in m is given by $\overline{p} = \sqrt{\rho/(j\omega\mu)}$, where ρ is the ground resistivity in $\Omega \cdot$ m, j is the imaginary number, ω is the angular frequency in rad/s—equal to 377 rad/s for 60 Hz—, and μ is the ground permeability in H/m. Assuming that the ground permeability is equal to the permeability in free space, μ_o, $\mu = 4\pi \times 10^{-7}$ H/m.

2. Calculate the Self-Impedance $\overline{Z}_{ii}$

The self-impedance, $\overline{Z}_{ii}$, which is one of the diagonal elements of the series impedance matrix, represents the impedance in Ω/m of the loop "conductor i/ground return." Such an impedance can be determined from $\overline{Z}_{ii} = R_i + j\omega \dfrac{\mu_o}{2\pi} \ln \dfrac{2(h_i + \overline{p})}{\text{GMR}_i}$, where R_i is the ac resistance[1] of the conductor[2] i in Ω/m, h_i is its average height above ground in m, which can be calculated as tower height $- 2/3$ sag for spans shorter than 500 m (Dommel, 1992), and GMR_i is its geometric mean radius in m.

The quantity $j\omega \frac{\mu_o}{2\pi} \ln \frac{2(h_i + \bar{p})}{\mathrm{GMR}_i}$ has a real part and an imaginary part. The real part will yield the losses in the nonideal ground return. As a result, the self-resistance corrected by ground return R_{ii}, which is the real part of Z_{ii}, will take into account both losses in the conductor i and losses in the nonideal ground return and will be greater than the resistance of the conductor R_i. The imaginary part, on the other hand, will yield the equivalent self-reactance from which the equivalent self-inductance, L_{ii}, in H/m—the self-inductance corrected by ground return—can be derived as $L_{ii} = \mathrm{Im}\{\bar{Z}_{ii}\}/\omega$.

Similarly, the self-impedance, $\bar{Z}_{kk}$, can be obtained by replacing the index i with the index k in the above-mentioned formulae. If the impedances are needed in Ω/mile, the results obtained in Ω/m must be multiplied by 1600.

3. Calculate the Mutual Impedance, $\bar{Z}_{ik}$

The mutual impedance, $\bar{Z}_{ik}$, which is one of the off-diagonal elements of the series impedance matrix, represents the impedance in Ω/m between the loops "conductor i/ground return" and "conductor k/ground return." Such an impedance can be determined from $\bar{Z}_{ik} = j\omega \frac{\mu_o}{2\pi} \ln \frac{D_{ik}}{D_{ik'}}$, where D_{ik} is the distance between the conductor i and the conductor k, and $D_{ik'}$ is the distance between the conductor i and the image of the conductor k.

Given the above-depicted geometry, $D_{ik} = \sqrt{(h - h_k)^2 + d_{ik}^2}$ and $D_{ik'} = \sqrt{(h_i + h_k + \bar{p})^2 + d_{ik}^2}$, where h_i is the average height above ground of the conductor i, h_k is the average height above ground of the conductor k, and d_{ik} is the horizontal distance between the conductors i and k.

The impedance, $\bar{Z}_{ik}$, has a real part and an imaginary part as well. The real part, R_{ik}, represents the phase shift that shows up in the induced voltage for including the nonideal ground return, and the imaginary part yields the equivalent mutual reactance from which the equivalent mutual inductance, L_{ik}—the mutual inductance corrected by ground return—can be derived as $L_{ik} = \mathrm{Im}\{\bar{Z}_{ik}\}/\omega$.

By symmetry, $\bar{Z}_{ki} = \bar{Z}_{ik}$.

4. Eliminate Ground Wires

The order of the series impedance matrix can be reduced since the potentials of the ground wires are equal to zero. To do so, let the set of linear equations be split into subsets of ungrounded conductor equations and subsets of ground wire equations, as indicated in Dommel (1992), and the impedance matrix be divided into the ungrounded conductor submatrices $[Z_{uu}]$ and $[Z_{ug}]$ and the ground wire submatrices $[Z_{gu}]$ and $[Z_{gg}]$. Similarly, let the far-end terminals be short-circuited and voltage drops per unit length

[1]The ac resistance increases as frequency steps up due to the skin effect. For an ACSR conductor, at frequencies higher than power frequency, the ac resistance can be calculated by using an equivalent tubular conductor and Bessel functions (Lewis and Tuttle, 1959). Frequency Dependent Parameters of Stranded Conductors can also be calculated with the formulae given in Galloway et al. (1964).

[2]If i is a bundle of conductors, replace the bundle with one equivalent conductor located at the center of the bundle and use the equivalent resistance of the bundle, the equivalent geometric mean radius of the bundle, and the average height of the bundle instead.

and currents be defined for ungrounded conductors and ground wires by the vectors $[\Delta V_u]$ and $[I_u]$ and $[\Delta V_g]$ and $[I_g]$, respectively:

$$\begin{bmatrix} [\Delta V_u] \\ [\Delta V_g] \end{bmatrix} = \begin{bmatrix} [Z_{uu}] & [Z_{ug}] \\ [Z_{gu}] & [Z_{gg}] \end{bmatrix} \cdot \begin{bmatrix} [I_u] \\ [I_g] \end{bmatrix}$$

Since the vector of voltage drops across the ground wires $[\Delta V_g] = 0$, the system can be written as $[\Delta V_u] = [Z_{\text{red}}] \cdot [I_u]$, where the reduced matrix $[Z_{\text{red}}] = [Z_{uu}] - [Z_{ug}][Z_{gg}]^{-1}$ $[Z_{gu}]$ as a result of a Kron's reduction.

Related Calculations. For a balanced transmission line with only one ground wire, the expression $[Z_{ug}][Z_{gg}]^{-1}[Z_{gu}]$ becomes

$$\frac{\overline{Z}_{ug}^2}{Z_{gg}} \cdot [U],$$

where $\overline{Z}_{ug}$ is the mutual impedance between the loops "ungrounded conductor/ground return" and "ground wire/ground return," $\overline{Z}_{gg}$ is the self-impedance of the loop "ground wire/ground return," and $[U]$ is the unit matrix.

Array manipulations can easily be carried out by using matrix-oriented programs such as MATLAB and MATHCAD. Transmission-line parameters can also be calculated through support routines of EMTP-type programs such as LINE CONSTANTS (Dommel, 1992).

INDUCTIVE SEQUENCE IMPEDANCES OF THREE-PHASE TRANSMISSION LINES

Consider a three-phase transmission line whose series impedance matrix has been calculated and reduced as explained in the subsection "Inductive Impedances of Multiconductor Transmission Lines Including Ground Return Corrections." Determine the series sequence parameters of such a line.

Calculation Procedure

1. Premultiply and Multiply $[\mathbf{Z}_{\text{abc}}]$ *by the Transformation Matrices*

The matrix of sequence impedances $[Z_{012}] = [T]^{-1} \cdot [Z_{abc}] \cdot [T]$, where $[T]$ is the symmetrical component transformation matrix,[3] and $[Z_{abc}]$ is the reduced matrix of ungrounded (phase) conductors that includes the ground return corrections.

$$[T] = \begin{bmatrix} 1 & 1 & 1 \\ 1 & a^2 & a \\ 1 & a & a^2 \end{bmatrix}$$

$$[T]^{-1} = \frac{1}{3}\begin{bmatrix} 1 & 1 & 1 \\ 1 & a & a^2 \\ 1 & a^2 & a \end{bmatrix}$$

[3] $a = e^{+120°}$ and $a^2 = e^{-120°}$ in the transformation matrix.

Related Calculations. When the three-phase transmission line is balanced, the matrix of series sequence impedances is a diagonal matrix, as indicated below:

$$[Z_{012}] = \begin{bmatrix} \overline{Z}_0 & 0 & 0 \\ 0 & \overline{Z}_1 & 0 \\ 0 & 0 & \overline{Z}_2 \end{bmatrix}$$

where the zero sequence impedance $\overline{Z}_0 = \overline{Z}_{self} + 2\,\overline{Z}_{mutual}$, the positive sequence impedance $\overline{Z}_1 = \overline{Z}_{self} - \overline{Z}_{mutual}$, and the negative sequence impedance $\overline{Z}_2 = \overline{Z}_{self} - \overline{Z}_{mutual}$. $\overline{Z}_{self}$ is found by averaging the elements $\overline{Z}_{aa}$, $\overline{Z}_{bb}$, and $\overline{Z}_{cc}$ of $[Z_{abc}]$, and $\overline{Z}_{mutual}$ by averaging the elements $\overline{Z}_{ab}$, $\overline{Z}_{ac}$, and $\overline{Z}_{bc}$ of $[Z_{abc}]$.

INDUCTIVE REACTANCE OF CABLES IN DUCTS OR CONDUIT

Find the reactance per 1000 ft (304.8 m) if three single conductors each having 2-in. (5 cm) outside diameters and 750-cmil cross sections are enclosed in a magnetic conduit.

Calculation Procedure

1. Determine Inductance

Use $L = (2 \times 10^{-7})[{}^{1}\!/_{4} + \ln(D/a)]$ H/m. Common practice dictates that the reactance be given in ohms per thousand feet. Hence, $X_L = [0.0153 + 0.1404 \log(D/a)]$.

2. Use Nomogram for Solution

A nomogram based on the preceding equation is provided in Fig. 9.12. Two factors are used to improve accuracy. The equation for X_L produces a smaller reactance than an open-wire line. If randomly laid in a duct, the value of D is somewhat indeterminate because the outer insulation does not always touch. Therefore, if cables are not clamped on rigid supports, a multiplying factor of 1.2 is used. Further, if confined in a conduit of magnetic material in random lay, a multiplying factor of approximately 1.5 is used. Figure 9.12 includes a correction table for cables bound together rather than randomly laid. Here, sector refers to a cable whose three conductors approximate 120° sectors.

3. Determine X_L Graphically

Draw a line from 750 MCM to 2-in spacing and read the inductive reactance of 0.038 Ω per thousand feet. Then, $X_L = (1.5)(0.038) = 0.057$ Ω per thousand feet $= 0.19 \times 10^{-3}$ Ω/m.

Related Calculations. For a three-phase cable having concentric stranded conductors with a total of 250 MCM each and a diameter of 0.89 in (2.225 cm), the nomogram yields $X_L = 0.0315$ Ω per thousand feet (10^{-4} Ω/m). If the cable is in a magnetic conduit, the tabulated correction factor is used. Thus, $X_L = (1.149)(0.0315) = 0.0362$ Ω per thousand feet, or 1.2×10^{-4} Ω/m.

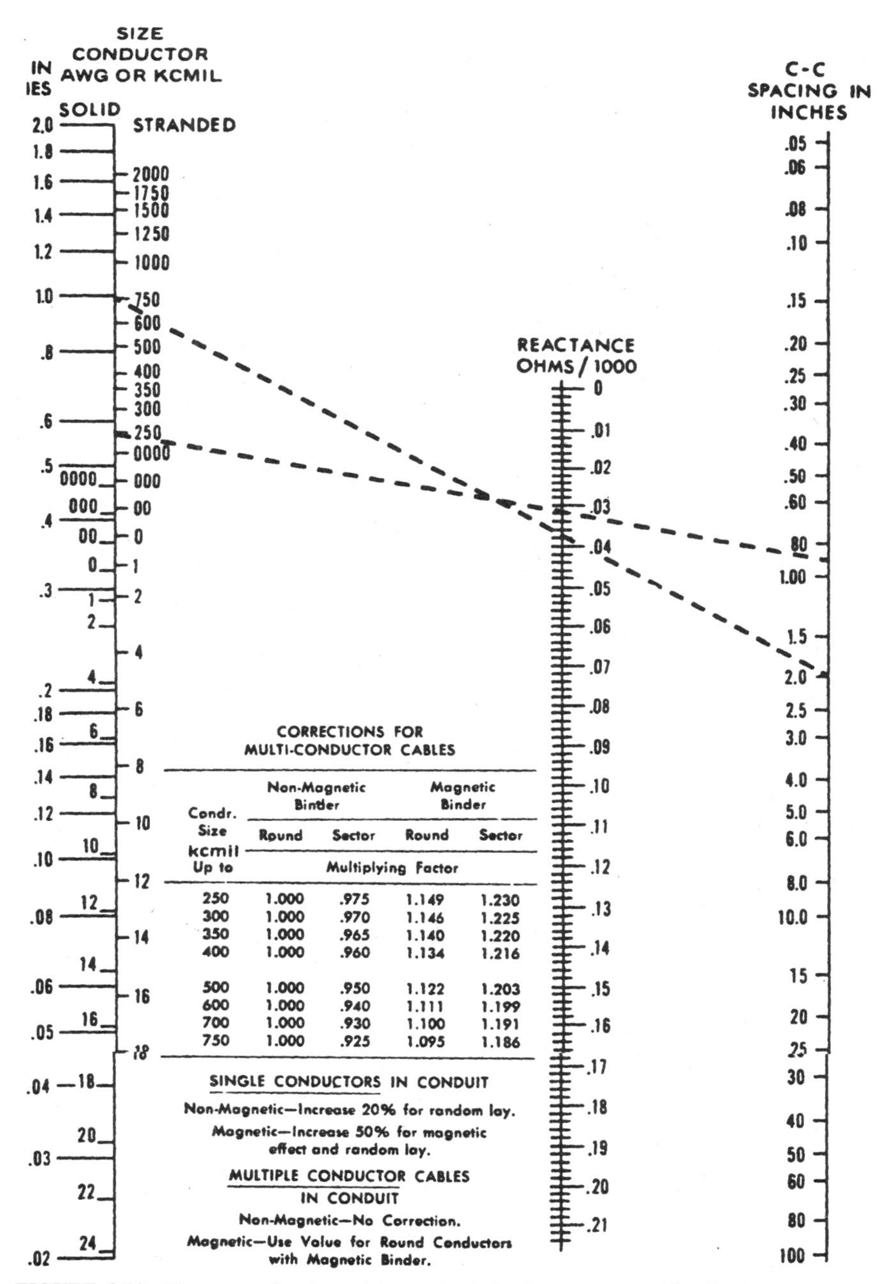

CORRECTIONS FOR MULTI-CONDUCTOR CABLES

Condr. Size kcmil Up to	Non-Magnetic Binder		Magnetic Binder	
	Round	Sector	Round	Sector
	Multiplying Factor			
250	1.000	.975	1.149	1.230
300	1.000	.970	1.146	1.225
350	1.000	.965	1.140	1.220
400	1.000	.960	1.134	1.216
500	1.000	.950	1.122	1.203
600	1.000	.940	1.111	1.199
700	1.000	.930	1.100	1.191
750	1.000	.925	1.095	1.186

SINGLE CONDUCTORS IN CONDUIT

Non-Magnetic—Increase 20% for random lay.

Magnetic—Increase 50% for magnetic effect and random lay.

MULTIPLE CONDUCTOR CABLES IN CONDUIT

Non-Magnetic—No Correction.

Magnetic—Use Value for Round Conductors with Magnetic Binder.

FIGURE 9.12 Nomogram for determining series inductive reactance of insulated conductors to neutral. *(Aluminum Electrical Conductor Handbook)*, (2nd ed.). The Aluminum Association, Inc., 1982, pages 9–11.

INDUCTIVE IMPEDANCES OF MULTICONDUCTOR UNDERGROUND CABLES INCLUDING GROUND-RETURN CORRECTIONS

Consider a transmission system consisting of n buried single-circuit cables, in which each single circuit is made up of a high-voltage conductor and a cable sheath, and in which, for simplicity, only three single circuits, *a, b,* and *c,* have been depicted. Determine the self and mutual series impedances of such a system taking into account the ground return (Fig. 9.13).

Calculation Procedure

1. Calculate the Loop Impedances

The method and the formulae proposed in Wedepohl and Wilcox (1973) and Dommel (1992) are appropriate to describe the electric quantities associated with this cable system. Such a formulation takes into account the skin effect in the conductors, approximates the ground return corrections proposed in Pollaczek (1931), which are the appropriate corrections for underground cables, to closed-form expressions, and allows system studies, if desired, at frequencies higher than power frequency. By assuming $n = 3$ and that the far-end terminals are short-circuited, six coupled equations describe the loop quantities associated with the cable system here under study.

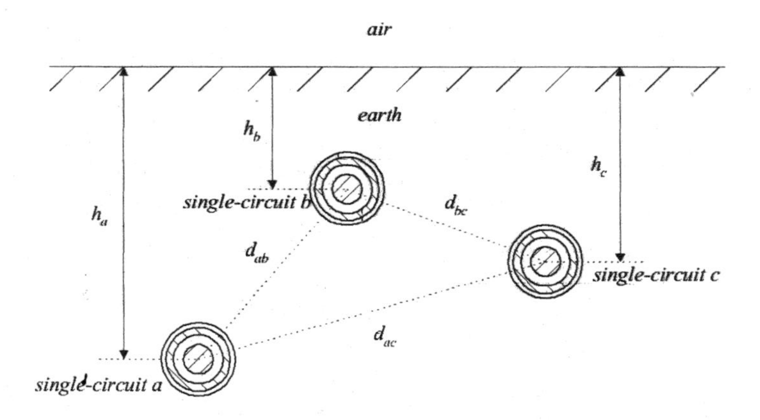

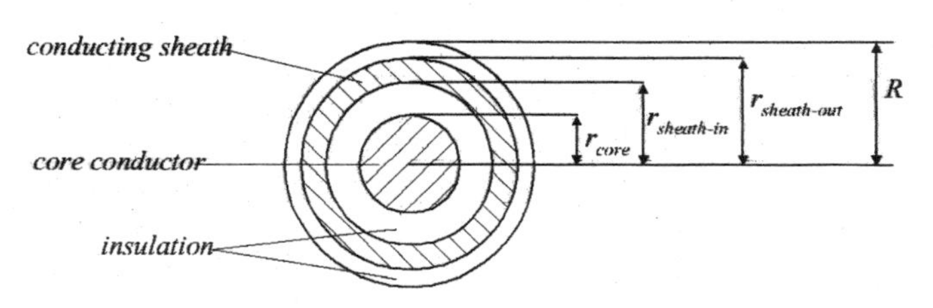

FIGURE 9.13 Cable geometry for parameter calculations.

$$\begin{bmatrix} \Delta\bar{V}_{1a} \\ \Delta\bar{V}_{2a} \\ \Delta\bar{V}_{1b} \\ \Delta\bar{V}_{2b} \\ \Delta\bar{V}_{1c} \\ \Delta\bar{V}_{2c} \end{bmatrix} = \begin{bmatrix} \bar{Z}_{11a} & \bar{Z}_{12a} & 0 & 0 & 0 & 0 \\ \bar{Z}_{12a} & \bar{Z}_{22a} & 0 & \bar{Z}_{ab} & 0 & \bar{Z}_{ac} \\ 0 & 0 & \bar{Z}_{11b} & \bar{Z}_{12b} & 0 & 0 \\ 0 & \bar{Z}_{ab} & \bar{Z}_{12b} & \bar{Z}_{22b} & 0 & \bar{Z}_{bc} \\ 0 & 0 & 0 & 0 & \bar{Z}_{11c} & \bar{Z}_{12c} \\ 0 & \bar{Z}_{ac} & 0 & \bar{Z}_{bc} & \bar{Z}_{12c} & \bar{Z}_{22c} \end{bmatrix} \cdot \begin{bmatrix} \bar{I}_{1a} \\ \bar{I}_{2a} \\ \bar{I}_{1b} \\ \bar{I}_{2b} \\ \bar{I}_{1c} \\ \bar{I}_{2c} \end{bmatrix}$$

where:

$\Delta\bar{V}_1 = \Delta\bar{V}_{\text{core}} - \Delta\bar{V}_{\text{sheath}}$
$\Delta\bar{V}_2 = \Delta\bar{V}_{\text{sheath}}$
$\bar{I}_1 = \bar{I}_{\text{core}}$
$\bar{I}_2 = \bar{I}_{\text{sheath}} + \bar{I}_{\text{core}}$
$\bar{Z}_{11} = \bar{Z}_{\text{core-out}} + \bar{Z}_{\text{core/sheath-insulation}} + \bar{Z}_{\text{sheath-in}}$
$\bar{Z}_{22} = \bar{Z}_{\text{sheath-out}} + \bar{Z}_{\text{sheath/earth-insulation}} + \bar{Z}_{\text{self earth-return}}$
$\bar{Z}_{12} = -\bar{Z}_{\text{sheath-mutual}}$
$\bar{Z}_{ab} = \bar{Z}_{\text{mutual earth-return}_{ab}}$
$\bar{Z}_{ac} = \bar{Z}_{\text{mutual earth-return}_{ac}}$
$\bar{Z}_{bc} = \bar{Z}_{\text{mutual earth-return}_{bc}}$

with:

a, b, c = subscripts denoting quantities associated with the single circuits of phases a, b, and c, respectively.
$\Delta\bar{V}_{\text{core}}$ = voltage drop per unit length across core with respect to ground.
$\Delta\bar{V}_{\text{sheath}}$ = voltage drop per unit length across sheath with respect to ground.
$\bar{I}_{\text{core}}$ = current flowing through core conductor.
$\bar{I}_{\text{sheath}}$ = current flowing through sheath.
$\bar{Z}_{\text{core-out}}$ = internal impedance per unit length of core conductor, calculated from the voltage drop on the outer surface of the core per unit current, when the current returns through the outer conductor. In this case, the outer conductor is the sheath.
$\bar{Z}_{\text{core/sheath-insulation}}$ = impedance per unit length of insulation between core and sheath.
$\bar{Z}_{\text{sheath-in}}$ = internal impedance per unit length of sheath, calculated from the voltage drop on the inner surface of the sheath per unit current, when the current returns through the inner conductor. In this case, the inner conductor is the core.
$\bar{Z}_{\text{sheath-out}}$ = internal impedance per unit length of sheath, calculated from the voltage drop on the outer surface of the sheath per unit current, when the current returns through the outer conductor. In this case, the outer conductor is the earth-return path.
$\bar{Z}_{\text{sheath/earth-insulation}}$ = impedance per unit length of insulation between sheath and earth-return path.
$\bar{Z}_{\text{sheath-mutual}}$ = mutual impedance per unit length of sheath. In this case, it is the mutual impedance between the inside loop "core/sheath" and the outside loop "sheath/earth" of one single circuit.
$\bar{Z}_{\text{self earth-return}}$ = self impedance per unit length of the earth-return path.

$\overline{Z}_{\text{mutual earth-return}}$ = mutual impedance per unit length of the earth-return path. In this case, it is the mutual impedance between the outermost loop "sheath/earth" of one single circuit and the outermost loop "sheath/earth" of another single circuit.

and:

$\overline{Z}_{\text{core-out}} = \dfrac{\rho_{\text{core}}\overline{m}_{\text{core}}}{2\pi r_{\text{core}}}\coth(0.777\overline{m}_{\text{core}}r_{\text{core}}) + \dfrac{0.356\,\rho_{\text{core}}}{\pi r^2_{\text{core}}}$ in Ω/m. This formula yields a maximum error of 4 percent in the resistive part—it occurs when $|\overline{m}r_{\text{core}}| = 5$—and a maximum error of 5 percent in the reactive component—it occurs when $|\overline{m}r_{\text{core}}| = 3.5$ (Wedepohl and Wilcox, 1973). For other values of $|\overline{m}r_{\text{core}}|$, the formula is very accurate and avoids the evaluation of Bessel functions; r_{core} is the radius of the core conductor in m, ρ_{core} is the resistivity of the core conductor in $\Omega \cdot$ m, and $\overline{m}_{\text{core}}$ is the reciprocal of the complex penetration depth of the core; $\overline{m}_{\text{core}} = \sqrt{(j\omega\mu_{\text{core}})/\rho_{\text{core}}}$ in m^{-1}, where μ_{core} is the magnetic permeability of the core in H/m, $\mu_{\text{core}} = \mu_{r_{\text{core}}} \cdot \mu_o$ with $\mu_{r_{\text{core}}} \neq 1$ if the material of the core conductor is magnetic, and ω is the angular frequency in rad/s.

$\overline{Z}_{\text{sheath-in}} = \dfrac{\rho_{\text{sh}}}{2\pi r_{\text{sh-in}}}\left\{\overline{m}_{\text{sh}}\coth(\overline{m}_{\text{sh}}\Delta_{\text{sh}}) - \dfrac{1}{r_{\text{sh-in}} + r_{\text{sh-out}}}\right\}$ in Ω/m. This formula yields a good accuracy if the condition $\dfrac{r_{\text{sh-out}} - r_{\text{sh-in}}}{r_{\text{sh-out}} + r_{\text{sh-in}}} < 1/8$ is satisfied (Wedepohl and Wilcox, 1973); $r_{\text{sh-out}}$ is the outer radius of the sheath in m, $r_{\text{sh-in}}$ is the inner radius of the sheath in m, ρ_{sh} is the resistivity of the sheath in $\Omega \cdot$ m, and $\overline{m}_{\text{sh}}$ is the reciprocal of the complex penetration depth of the sheath; $\overline{m}_{\text{sh}} = \sqrt{(j\omega\mu_{\text{sh}})/\rho_{\text{sh}}}$ in m^{-1}, where μ_{sh} is the magnetic permeability of the sheath in H/m, $\mu_{\text{sh}} = \mu_{r_{\text{sh}}} \cdot \mu_o$ with $\mu_{r_{\text{sh}}} \neq 1$ if the material of the sheath is magnetic, and Δ_{sh} is the thickness of the sheath in m, which can be calculated as $(r_{\text{sh-out}} - r_{\text{sh-in}})$.

$\overline{Z}_{\text{sheath-out}} = \dfrac{\rho_{\text{sh}}}{2\pi r_{\text{sh-out}}}\left\{\overline{m}_{\text{sh}}\coth(\overline{m}_{\text{sh}}\Delta_{\text{sh}}) - \dfrac{1}{r_{\text{sh-in}} + r_{\text{sh-out}}}\right\}$ in Ω/m. This formula yields a good accuracy if the condition between radii mentioned for $\overline{Z}_{\text{sheath-in}}$ is satisfied.

$\overline{Z}_{\text{sheath-mutual}} = \dfrac{\rho_{\text{sh}}\,\overline{m}_{\text{sh}}}{\pi(r_{\text{sh-in}} + r_{\text{sh-out}})}\,\text{cosech}(\overline{m}_{\text{sh}}\Delta_{\text{sh}})$ in Ω/m. This formula yields a good accuracy if the condition between radii mentioned for $\overline{Z}_{\text{sheath-in}}$ and $\overline{Z}_{\text{sheath-out}}$ is satisfied.

$\overline{Z}_{\text{core/sheath-insulation}} = \dfrac{j\omega\mu_1}{2\pi}\ln\left(\dfrac{r_{\text{sh-in}}}{r_{\text{core}}}\right)$ in Ω/m, where μ_1 is the magnetic permeability of the insulation between core and sheath in H/m.

$\overline{Z}_{\text{sheath/earth-insulation}} = \dfrac{j\omega\mu_2}{2\pi}\ln\left(\dfrac{R}{r_{\text{sh-out}}}\right)$ in Ω/m, where μ_2 is the magnetic permeability of the insulation between sheath and earth in H/m, and R is the outside radius of the outermost insulation of the cable in m.

$\overline{Z}_{\text{self earth-return}} = \dfrac{j\omega\mu}{2\pi}\left\{-\ln\left(\dfrac{\gamma\overline{m}R}{2}\right) + 1/2 - 4/3\,\overline{m}h\right\}$ in Ω/m. This formula yields very accurate results at frequencies for which $|\overline{m}R| < 0.25$ (Wedepohl and Wilcox, 1973); μ is the magnetic permeability of the earth-return path, which can be assumed equal to the magnetic permeability in free space (μ_o), $\gamma = 0.577215665$ (Euler's constant), and $\overline{m}$ is the reciprocal of the complex penetration depth of the earth-return path; $\overline{m} = \sqrt{(j\omega\mu)/\rho}$ in m^{-1}, where ρ is the earth return resistivity in $\Omega \cdot$ m, and h is the depth at which the cable is buried in m. The formula is also very accurate if the depth at which the cable is buried is close to 1 m (Wedepohl and Wilcox, 1973).

$\overline{Z}_{\text{mutual earth-return}} = \dfrac{j\omega\mu}{2\pi}\left\{-\ln\left(\dfrac{\gamma\overline{m}d}{2}\right) + \tfrac{1}{2} - \tfrac{2}{3}\overline{m}\ell\right\}$ in Ω/m. This formula yields very accurate results at frequencies for which $|\overline{m}d| < 0.25$ (Wedepohl and Wilcox, 1973); d is the distance between single circuits a and b for $\overline{Z}_{ab}$, the distance between single circuits a and c for $\overline{Z}_{ac}$, and the distance between single circuits b and c for $\overline{Z}_{bc}$; ℓ is the sum of the depths of the single circuits a and b for $\overline{Z}_{ab}$, the sum of the depths of the single circuits a and c for $\overline{Z}_{ac}$, and the sum of the depths of the single circuits b and c for $\overline{Z}_{bc}$. Errors for $\overline{Z}_{\text{self earth-return}}$ and $\overline{Z}_{\text{mutual earth-return}}$ are lower than 1 percent up to frequencies of 100 kHz (Dommel, 1992). The formula is also very accurate if the depths at which the cables are buried are close to 1 m (Wedepohl and Wilcox, 1973).

Related Calculations. If the impedances are needed in Ω per 1000 ft, multiply the impedance in Ω/m by 304.8 m.

If the cable has multiwire concentric neutral conductors, replace the neutral wires with an equivalent concentric sheath and assume that the thickness of such a sheath is equal to the diameter of one of the neutral wires. All neutral wires are assumed to be identical (Smith and Barger, 1972).

If the single circuits have additional conductors, for example, armors, add three more coupled equations and include the corresponding impedances $\overline{Z}_{23}$ and $\overline{Z}_{33}$ into the single circuits a, b, and c, as indicated in Dommel (1992). Move also the impedances $\overline{Z}_{ab}$, $\overline{Z}_{ac}$, and $\overline{Z}_{bc}$ since the outermost loops will be the ones "armor/earth." Derive the formulae for the new impedances by analogy and take care of utilizing the right electric properties as well as the appropriate radii.

2. Transform the Loop Quantities into Conductor Quantities

To transform the loop quantities into conductor quantities, use the procedure recommended in Dommel (1992) as follows. Add row 2 to row 1, add row 4 to row 3, and add row 6 to row 5. By doing so, it is possible to prove that the system is described through the conductor quantities indicated below:

$$\begin{bmatrix} \Delta\overline{V}_{\text{core}a} \\ \Delta\overline{V}_{\text{sheath}a} \\ \Delta\overline{V}_{\text{core}b} \\ \Delta\overline{V}_{\text{sheath}b} \\ \Delta\overline{V}_{\text{core}c} \\ \Delta\overline{V}_{\text{sheath}c} \end{bmatrix} = \begin{bmatrix} \overline{Z}_{cca} & \overline{Z}_{csa} & \overline{Z}_{ab} & \overline{Z}_{ab} & \overline{Z}_{ac} & \overline{Z}_{ac} \\ \overline{Z}_{csa} & \overline{Z}_{ssa} & \overline{Z}_{ab} & \overline{Z}_{ab} & \overline{Z}_{ac} & \overline{Z}_{ac} \\ \overline{Z}_{ab} & \overline{Z}_{ab} & \overline{Z}_{ccb} & \overline{Z}_{csb} & \overline{Z}_{bc} & \overline{Z}_{bc} \\ \overline{Z}_{ab} & \overline{Z}_{ab} & \overline{Z}_{csb} & \overline{Z}_{ssb} & \overline{Z}_{bc} & \overline{Z}_{bc} \\ \overline{Z}_{ac} & \overline{Z}_{ac} & \overline{Z}_{bc} & \overline{Z}_{bc} & \overline{Z}_{ccc} & \overline{Z}_{csc} \\ \overline{Z}_{ac} & \overline{Z}_{ac} & \overline{Z}_{bc} & \overline{Z}_{bc} & \overline{Z}_{csc} & \overline{Z}_{ssc} \end{bmatrix} \cdot \begin{bmatrix} \overline{I}_{\text{core}a} \\ \overline{I}_{\text{sheath}a} \\ \overline{I}_{\text{core}b} \\ \overline{I}_{\text{sheath}b} \\ \overline{I}_{\text{core}c} \\ \overline{I}_{\text{sheath}c} \end{bmatrix}$$

where $\overline{Z}_{cc} = \overline{Z}_{11} + 2\overline{Z}_{12} + \overline{Z}_{22}$, $\overline{Z}_{cs} = \overline{Z}_{12} + \overline{Z}_{22}$, and $\overline{Z}_{ss} = \overline{Z}_{22}$. The diagonal elements $\overline{Z}_{cc}$ and $\overline{Z}_{ss}$ are the self-impedances of the core and sheath with return through earth, respectively. The off-diagonal elements $\overline{Z}_{cs}$, $\overline{Z}_{ab}$, $\overline{Z}_{ac}$, and $\overline{Z}_{bc}$ are the mutual impedances between core and sheath of one cable with return through earth, between sheath a and sheath b with return through earth, between sheath a and sheath c with return through earth, and between sheath b and sheath c with return through earth, respectively. As a result of the above-mentioned arithmetic operations, the system is represented in nodal form, with currents expressed as conductor currents and voltages expressed as voltage drops across the conductors with respect to ground.

Related Calculations. If armors are present, add rows 2 and 3 to row 1 and add row 3 to row 2. Add also rows 5 and 6 to row 4 and add row 6 to row 5. Similarly, add rows 8 and 9 to row 7 and add row 9 to row 8.

Array manipulations can easily be carried out by using matrix-oriented programs such as MATLAB and MATHCAD. Underground cable parameters can also be calculated through support routines of EMTP-type programs such as CABLE CONSTANTS and CABLE PARAMETERS (Ametani, 1980).

3. Eliminate the Sheaths

By interchanging the corresponding rows and columns in the impedance matrix, move voltage drops across sheaths and current flows through sheaths to the bottom of the vectors of voltages and currents, respectively.

Then let the set of linear equations be split into subsets of core conductor equations and subsets of sheath equations, and the impedance matrix be divided into the core conductor submatrices $[Z_{cc}]$ and $[Z_{cs}]$ and the sheath submatrices $[Z_{sc}]$ and $[Z_{ss}]$. Similarly, let the far-end terminals be short-circuited and voltage drops per unit length and currents be defined for core conductors and sheaths by the vectors $[\Delta V_c]$ and $[I_c]$ and $[\Delta V_s]$ and $[I_s]$, respectively:

$$\begin{bmatrix} [\Delta V_c] \\ [\Delta V_s] \end{bmatrix} = \begin{bmatrix} [Z_{cc}] & [Z_{cs}] \\ [Z_{sc}] & [Z_{ss}] \end{bmatrix} \cdot \begin{bmatrix} [I_c] \\ [I_s] \end{bmatrix}$$

Since the vector of voltage drops across the sheaths $[\Delta V_s] = 0$, assuming that both terminals of each sheath are grounded, the system can be written as $[\Delta V_c] = [Z_{\text{red}}] \cdot [I_c]$, where the reduced matrix $[Z_{\text{red}}] = [Z_{cc}] - [Z_{cs}][Z_{ss}]^{-1}[Z_{sc}]$ as a result of a Kron's reduction.

INDUCTIVE SEQUENCE IMPEDANCES OF THREE-PHASE UNDERGROUND CABLES

Consider a three-phase underground transmission system whose series impedance matrix has been calculated and reduced as explained in the subsection "Inductive Impedances of Multiconductor Underground Cables including Ground Return Corrections." Determine the series sequence parameters of such a system.

Calculation Procedure

1. Premultiply and Multiply $[\mathbf{Z}_{\text{abc}}]$ ***by the Transformation Matrices***

The matrix of sequence impedances $[Z_{012}] = [T]^{-1} \cdot [Z_{abc}] \cdot [T]$, where $[T]$ is the symmetrical component transformation matrix and $[Z_{abc}]$ is the reduced matrix of core conductors that includes the sheaths and the ground return corrections.

Related Calculations. Simple formulae for the calculation of sequence impedances of cables, including Carson's ground return corrections, are given in Lewis and Allen (1978). Formulae of sequence impedances of cables are also given in Westinghouse Electric Corporation (1964).

For the calculation of sequence impedances of pipe-type cables, see Neher (1964).

CAPACITANCE ASSOCIATED WITH TRANSMISSION LINES

Determine the balanced charging current fed from one end to a 230-kV, three-phase transmission line having a capacitive reactance of 0.2 MΩ · mi/phase (0.32 MΩ · km phase). The line is 80 mi (128.7 km) long.

Calculation Procedure

1. Determine Capacitive Reactance

The total capacitive reactance per phase, which is assumed to shunt each phase to ground, is $X_C = 0.32/128.7 = 0.0025$ MΩ.

2. Calculate Charging Current

For the voltage-to-neutral value of $230/\sqrt{3} = 133$ kV, the charging current I_c is $(133 \times 10^3)/(0.0025 \times 10^6) = 53.2$ A.

CAPACITANCE OF TWO-WIRE LINE

Determine the capacitance of a long, round conductor carrying a uniform charge density ρ_L on its outer surface (surplus charge always migrates to the outer surface of any conductor). The conductor is surrounded by an outward (for positive charge) vectorial electric field that appears to radiate from the center of the conductor, although it originates with ρ_L on the surface.

Calculation Procedure

1. Determine the Electric Potential

The magnitude of the electric field intensity **E** is given by: $E = \rho_L/(2\pi\epsilon r)$, where ϵ is the permittivity. In free space, $\epsilon = 10^{-9}/36\pi$ F/m. For consistency, the distance r from the center of the conductor is in meters and ρ_L is in coulombs per meter. Integration of E yields the electric potential V between points near the conductor (Fig. 9.14): $V_{ab} = (\rho_L/2\pi\epsilon)\ln(b/a)$ V. The notation V_{ab} indicates that the voltage is the potential at point a with respect to point b.

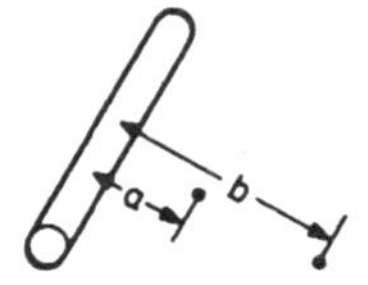

FIGURE 9.14 Long conductor carrying a uniform charge.

2. Consider a Two-Conductor Line

Consider the two conductors as forming a long, parallel conductor system (Fig. 9.15). Each conductor has an equal but opposite charge, typical of two-wire transmission systems. Further, it is assumed that the charge density per unit area is uniform in each conductor, even though a charge attraction exists between conductors, making it nonuniform. This assumption is completely adequate for open-wire lines for which $D \gg a$.

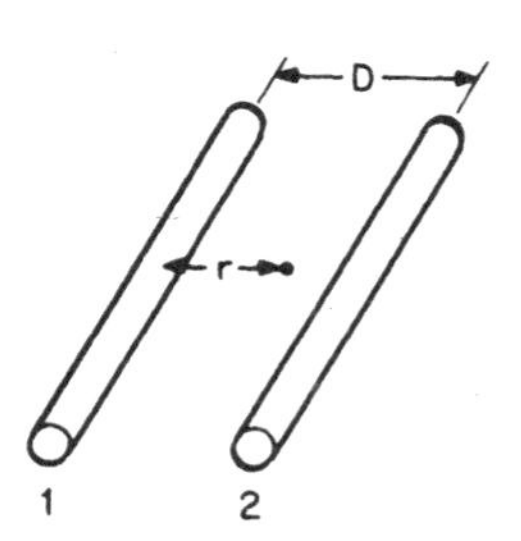

FIGURE 9.15 A charged two-conductor line.

Because the conductors bear charges of opposite polarities, the electric field at point r in the plane of the conductors between them is $E = (\rho_L/2\pi\epsilon)[1/r - 1/(D - r)]$ V/m, where r is the distance from the center of conductor 1 ($r \geq a$), and D is the center-to-center spacing between conductors. By integrating E, potential V_{1r} at conductor 1 with respect to point r is obtained: $V_{1r} = (\rho_L/2\pi\epsilon)\ln[r(D - a)/a(D - r)]$ V.

If r extends to conductor 2 and $D \gg a$, the potential of conductor 1 with respect to conductor 2 is: $(V_{12} = (\rho_L/\pi\epsilon)\ln(D/a)$ V.

3. Calculate Capacitance

The capacitance per unit length is determined from $C' = q/v$ F, where $q = \rho_L l$ and l is the total line length. Dividing by l yields $C = C'/l$ or $C = \pi\epsilon/\ln(D/a)$ F/m, which is the capacitance between conductors per meter length.

4. Determine Capacitance to Vertical Plane between Conductors at **D/2**

The potential for conductor 1 with respect to this plane (or neutral) is $V_{1n} = (\rho_L/2\pi\epsilon)\ln(D/a)$ V for $D \gg a$. It follows that this potential to the so-called neutral plane is one-half the conductor-to-conductor potential. It is easily shown that the potential from conductor 2 is of the same magnitude. If the neutral were grounded, it would not affect the potentials. The capacitance to neutral is then $C = (2\pi\epsilon)/\ln(D/a)$ F/m or $0.0388/\log(D/a)$ μF/mi.

CAPACITIVE REACTANCE OF TWO-WIRE LINE

Find the capacitive reactance to neutral for a two-conductor transmission line if $D = 8$ ft (2.4 m), $a = 0.25$ in. (0.00625 m), and the length of the line is 10 mi (16 km). The frequency is 377 rad/s.

Calculation Procedure

1. Calculate Capacitive Reactance

Recall that $X_C = 1/\omega C$. Substituting for $C = 0.0388 \log(D/a)$, obtain $X_C = 1/[(377)(0.0388)(10)/\log(2.4/0.00625)] = 0.0026$ MΩ to neutral.

Related Calculations. This is a large value of shunt impedance and is usually ignored for a line this short. It also follows that the capacitive reactance between conductors is twice the above value.

CAPACITANCE OF THREE-PHASE LINES

Determine the capacitance for a three-phase line.

Calculation Procedure

1. Consider Capacitance to Neutral

The capacitance to neutral of a three-phase transmission line is best established by considering equilateral spacing of the conductors initially. Other spacing that is unsymmetrical is commonly considered using the geometric mean distance in the equilateral case. The error that results is insignificant, especially when consideration is given to the

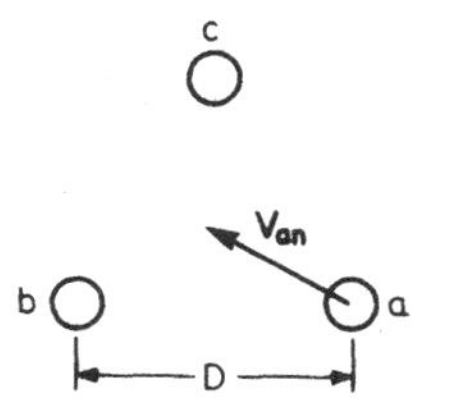

FIGURE 9.16 Equilateral spacing of a three-phase line; a denotes the conductor radii.

uncertainties of an actual line stemming from line towers and terrain irregularities.

2. Determine Phase Voltages

As shown in Fig. 9.16, $\mathbf{V}_{an}$, the potential of phase a with respect to the center of the triangle (the neutral), can be found by superimposing the potentials from all phases along the dimension from phase a to the center. The net charge is zero for any cross section, as it was for the two-wire line. Also, $D \gg a$. Superscripts to identify the three-phase potentials along the dimension for a to b are necessary. Thus, $|\mathbf{V}^a_{an}| = (\rho_{La}/2\pi\epsilon)\ln[(D/\sqrt{3})/D]$ due to phase a, $|\mathbf{V}^b_{an}| = (\rho_{Lb}/2\pi\epsilon)\ln[(D/\sqrt{3})/D]$ due to phase b, and $|\mathbf{V}^c_{an}| = (\rho_{Lc}/2\pi\epsilon)\ln[(D/\sqrt{3})/D]$ due to phase c.

The sum of the above three equations yields $\mathbf{V}_{an}$. Also $\rho_{La} + \rho_{Lb} + \rho_{Lc} = 0$. Thus, $|\mathbf{V}_{an}| = (\rho_{La})/(2\pi\epsilon)\ln(D/a)$ V. This equation has the same form as the equation for the potential to neutral of a two-wire line. The phase-to-neutral potentials of the other phases differ only in phase angle.

3. Determine Capacitance to Neutral

Dividing ρ_{La} by $|\mathbf{V}_{an}|$, we obtain $C = (2\pi\epsilon)/\ln(D/a)$ F/m $= 0.0388/\log(D/a)$ μF/mi.

CAPACITIVE REACTANCE OF THREE-PHASE LINES

Find the capacitive reactance to neutral of a three-phase line at 377 rad/s (Fig. 9.17). The conductors are ACSR Waxwing and the line is 60 mi (96.6 km) long.

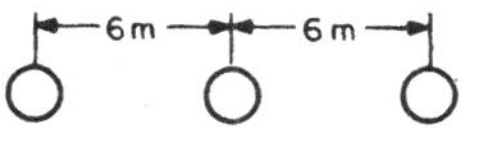

FIGURE 9.17 A three-phase line where conductors are separated by 6 m (20 ft).

Calculation Procedure

1. Calculate Capacitance

From Table 9.2, the external diameter of Waxwing is 0.609 in (0.015 m). Even though the conductors are not in equilateral spacing, use of GMD produces a sufficiently accurate capacitance to neutral. Therefore, GMD $= [(6^2)(12)]^{1/3} = 7.54$ m and $a = 0.015/2 = 0.0075$ m. Hence, $C = 0.0388/\log(7.54/0.0075) = 0.0129$ μF/mi or 0.008 μF/km.

2. Calculate Capacitive Reactance

$X_C = 1/(377)(0.008 \times 10^{-6})(96.6) = 0.0034$ MΩ to neutral.

CAPACITIVE SUSCEPTANCES OF MULTICONDUCTOR TRANSMISSION LINES

Consider a transmission line consisting of n straight conductors (Fig. 9.18) in which only the conductors i and k and their images below earth surface have been shown for simplicity. Determine the self and mutual capacitive susceptances of such a line.

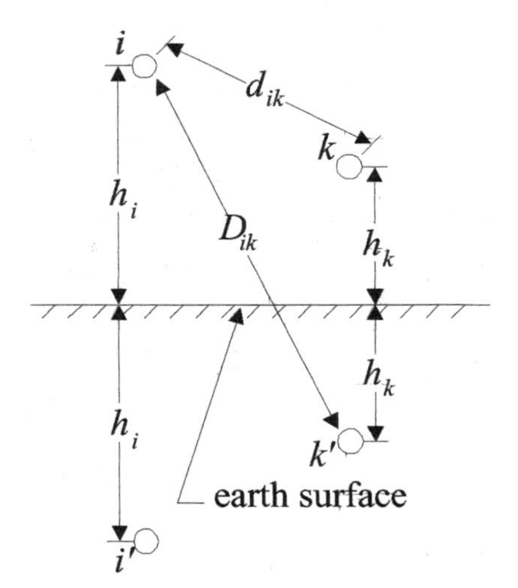

FIGURE 9.18 Transmission line geometry for capacitance calculations.

Calculation Procedure

1. Calculate the Potential Coefficient Matrix

The diagonal elements P_{ii} and the off-diagonal elements P_{ik} in m/F of the potential coefficient matrix $[P]$ can be determined from $P_{ii} = \frac{1}{2\pi\epsilon_o} \ln \frac{2h_i}{r_i}$ and $P_{ik} = \frac{1}{2\pi\epsilon_o} \ln \frac{D_{ik}}{d_{ik}}$, respectively (Dommel, 1985), where h_i is the average height above ground of the conductor i, r_i is the external radius of the conductor i, D_{ik} is the distance between the conductor i and the image below earth surface of the conductor k, d_{ik} is the distance between the conductors i and k, and ϵ_o is the permittivity in free space, which is equal to 8.854×10^{-12} F/m[4].

2. Eliminate the Ground Wires

The potential coefficient matrix $[P]$ can be reduced to a matrix $[P_{\text{red}}]$ by applying a Kron's reduction. To do so, use the same procedure explained in subsection "Impedances of Multiconductor Transmission Lines Including Ground Return Corrections."

3. Calculate the Capacitance Matrix

Find $[C_{\text{red}}]$ in F/m by inverting $[P_{\text{red}}]$ since $[C_{\text{red}}] = [P_{\text{red}}]^{-1}$. However, if the capacitances associated with the ground wires are required, invert $[P]$ to obtain $[C]$.

The capacitance matrices are in nodal form. As a result, the diagonal element C_{ii} stores the sum of the shunt capacitances between conductor i and all other conductors, including ground, and the off-diagonal element C_{ik} stores the negative of the shunt capacitance between conductors i and k, as stated in Dommel (1992).

4. Calculate the Capacitive Susceptance Matrix

Find the reduced matrix of capacitive susceptances $[B_{\text{red}}]$ in Ω^{-1}/m by multiplying $[C_{\text{red}}]$ by ω. If the susceptances are needed in Ω^{-1}/mile, multiply the results obtained in Ω^{-1}/m by 1600.

CAPACITIVE SEQUENCE SUSCEPTANCES OF THREE-PHASE TRANSMISSION LINES

Consider a three-phase transmission line whose shunt susceptance matrix has been calculated and reduced as explained in the subsection "Capacitive Susceptances of Multiconductor Transmission Lines." Determine the shunt sequence parameters of such a line.

[4]If i is a bundle of conductors, replace the bundle with one equivalent conductor located at the center of the bundle and use the equivalent external radius of the bundle and the average height of the bundle instead.

Calculation Procedure

1. Premultiply and Multiply $[\mathbf{B}_{abc}]$ *by the Transformation Matrices*

The matrix of shunt sequence susceptances $[B_{012}] = [T]^{-1} \cdot [B_{abc}] \cdot [T]$, where $[T]$ is the symmetrical component transformation matrix, and $[B_{abc}]$ is the reduced matrix of capacitive susceptances that includes the effect of the ground wires.

Related Calculations. When the three-phase transmission line is balanced, the matrix of shunt sequence susceptances is a diagonal matrix as indicated below:

$$[B_{012}] = \begin{bmatrix} B_0 & 0 & 0 \\ 0 & B_1 & 0 \\ 0 & 0 & B_2 \end{bmatrix}$$

where the zero-sequence susceptance $B_0 = B_{self} + 2B_{mutual}$, the positive-sequence susceptance $B_1 = B_{self} - B_{mutual}$, and the negative-sequence susceptance $B_2 = B_{self} - B_{mutual}$. B_{self} is found by averaging the elements B_{aa}, B_{bb}, and B_{ca} of $[B_{abc}]$, and B_{mutual} by averaging the elements B_{ab}, B_{ac}, and B_{bc} of $[B_{abc}]$.

CAPACITIVE SUSCEPTANCES ASSOCIATED WITH UNDERGROUND CABLES

Consider again the underground cable system described in subsection "Inductive Impedances of Multiconductor Underground Cables Including Ground Return Corrections." Determine the capacitive susceptances of such a system.

1. Calculate the Self and Mutual Susceptances

The method and the formulae proposed in Wedepohl and Wilcox (1973) and Dommel (1992) are also appropriate to calculate the shunt susceptances of this cable system. By assuming that there is no capacitive coupling among the three phases because of shielding effects, the following six no dal equations can be written:

$$\begin{bmatrix} \Delta\bar{I}_{core_a} \\ \Delta\bar{I}_{sheath_a} \\ \Delta\bar{I}_{core_b} \\ \Delta\bar{I}_{sheath_b} \\ \Delta\bar{I}_{core_c} \\ \Delta\bar{I}_{sheath_c} \end{bmatrix} = j \begin{bmatrix} B_{cc_a} & B_{cs_a} & 0 & 0 & 0 & 0 \\ B_{cs_a} & B_{ss_a} & 0 & 0 & 0 & 0 \\ 0 & 0 & B_{cc_b} & B_{cs_b} & 0 & 0 \\ 0 & 0 & B_{cs_b} & B_{ss_b} & 0 & 0 \\ 0 & 0 & 0 & 0 & B_{cc_a} & B_{cs_c} \\ 0 & 0 & 0 & 0 & B_{cs_c} & B_{ss_c} \end{bmatrix} \cdot \begin{bmatrix} \bar{V}_{core_a} \\ \bar{V}_{sheat_a} \\ \bar{V}_{core_b} \\ \bar{V}_{sheath_b} \\ \bar{V}_{core_c} \\ \bar{V}_{sheath_c} \end{bmatrix}$$

where:

a, b, c = subscripts denoting quantities associated with the single circuits of phases a, b, and c, respectively.

$\Delta\bar{I}_{core}$ = charging current per unit length flowing through core conductor.

$\Delta\bar{I}_{\text{sheath}}$ = charging current per unit length flowing through sheath.
$\bar{V}_{\text{core}}$ = voltage of core conductor with respect to ground.
$\bar{V}_{\text{sheath}}$ = voltage of sheath with respect to ground.
B_{cc} = shunt self-susceptance per unit length of core conductor.
B_{cs} = shunt mutual susceptance per unit length between core conductor and sheath.
B_{ss} = shunt self-susceptance per unit length of sheath.

with:

$B_{cc} = B_1$, $B_{cs} = -B_1$, and $B_{ss} = B_1 + B_2$.
B_1 = capacitive susceptance per unit length of insulation layer between core and sheath.
B_2 = capacitive susceptance per unit length of insulation layer between sheath and earth.
$B_i = \omega C_i$ in Ω^{-1}/m and $C_i = 2\pi\epsilon_o\epsilon_{r_i}/\ln(r_i/q_i)$, where C_i is the shunt capacitance of the tubular insulation in F/m, q_i is the inside radius of the insulation, r_i is the outside radius of the insulation, and ϵ_{r_i} is the relative permittivity of the insulation material.

2. Eliminate the Sheaths

The matrix of self and mutual susceptance can be reduced by applying a Kron's reduction. To do so, use the same procedure explained in the subsection "Inductive Impedances of Multiconductor Underground Cables Including Ground-Return Corrections."

3. Calculate the Sequence Susceptances

The matrix of capacitive sequence susceptances can be obtained by applying the symmetrical component transformation matrices to the reduced matrix of self and mutual susceptances. To do so, use the same procedure explained in the subsection "Inductive Sequence Impedances of Three-Phase Underground Cables."

CHARGING CURRENT AND REACTIVE POWER

Determine the input charging current and the charging apparent power in megavars for the example in Fig. 9.17 if the line-to-line voltage is 230 kV.

Calculation Procedure

1. Calculate Charge Current, $\mathbf{I_c}$

$I_c = (V/\sqrt{3})X_C$, where V is the line-to-line voltage. Then, $I_c = (230 \times 10^3)/(\sqrt{3})(0.0034 \times 10^6) = 39.06$ A.

2. Calculate Reactive Power, Q

$Q = \sqrt{3}VI = (\sqrt{3})(230 \times 10^3)(39.06) = 15.56$ MVAR for the three-phase line.

TRANSMISSION-LINE MODELS FOR POWER-FREQUENCY STUDIES[5]

Short transmission lines [up to 80 km (50 mi)] are represented by their series impedance consisting of the line resistance R_L and inductive reactance X_L. In cases where R_L is less than 10 percent of X_L, R_L is sometimes ignored.

The model for medium lines [up to 320 km (200 mi)] is represented in Fig. 9.19, where the line capacitance C_L is considered. Expressions for admittance $\mathbf{Y_L}$ and impedance $\mathbf{Z_L}$ are: $\mathbf{Y_L} = j\omega C_L$ and $\mathbf{Z_L} = R_L + jX_L$. The voltage $\mathbf{V_S}$ and current $\mathbf{I_S}$ for the sending end of the line are: $\mathbf{V_S} = (\mathbf{V_R}\mathbf{Y_L}/2 + \mathbf{I_R})\mathbf{Z_L} + \mathbf{V_R}$ and $\mathbf{I_S} = \mathbf{V_S}\mathbf{Y_L}/2 + \mathbf{V_R}\mathbf{Y_L}/2 + \mathbf{I_R}$, where $\mathbf{V_R}$ is the receiving-end voltage and $\mathbf{I_R}$ is the receiving-end current. The above equations may be written as: $\mathbf{V_S} = \mathbf{A}\mathbf{V_R} + \mathbf{B}\mathbf{I_R}$ and $\mathbf{I_S} = \mathbf{C}\mathbf{V_R} + \mathbf{D}\mathbf{I_R}$, where $\mathbf{A} = \mathbf{D} = \mathbf{Z_L}\mathbf{Y_L}/2 + 1$, $\mathbf{B} = \mathbf{Z_L}$, and $\mathbf{C} = \mathbf{Y_L} + \mathbf{Z_L}\mathbf{Y_L^2}/4$.

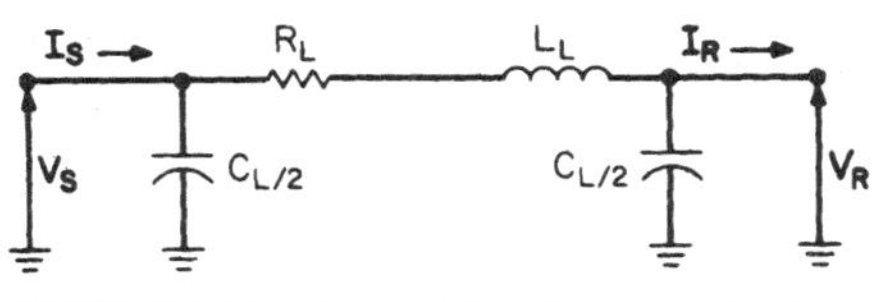

FIGURE 9.19 A model of a single-phase, medium-length line.

For long transmission lines, $\mathbf{V_S} = (\mathbf{V_R} + \mathbf{I_R}\mathbf{Z_c})e^{\gamma x}/2 + (\mathbf{V_R} - \mathbf{I_R}\mathbf{Z_c})e^{-\gamma x}/2$ and $\mathbf{I_S} = (\mathbf{V_R}/\mathbf{Z_c} + \mathbf{I_R})e^{\gamma x}/2 + (\mathbf{V_R}/\mathbf{Z_c} - \mathbf{I_R})e^{-\gamma x}/2$, where the characteristic impedance $\mathbf{Z_c} = \sqrt{\mathbf{Z}/\mathbf{Y}}\ \Omega$ and the propagation constant $\gamma = \sqrt{\mathbf{ZY}}$ per kilometer (or per mile).

MEDIUM TRANSMISSION-LINE MODELS FOR POWER-FREQUENCY STUDIES

Calculate the sending-end voltage and current for a 320-km (200 mi) transmission line. The receiving-end line-to-line voltage is 230 kV and the current is 200 A, at a power factor of 0.8 lagging. The line parameters per kilometer are: $R = 0.2\ \Omega$, $L = 2$ mH, $C = 0.01\ \mu$F, and $f = 60$ Hz.

Calculation Procedure

1. Determine Y_L and Z_L

$$\mathbf{Y_L} = j\omega C_L \frac{1}{\text{km}} \cdot l = \frac{j377(0.01)(10^{-6})}{\text{km}} \cdot 320 \text{ km} = j1206 \text{ microsiemens } (\mu\text{S})$$

$$\mathbf{Z_L} = \frac{R_L}{\text{km}} \cdot l + \frac{X_L}{\text{km}} \cdot l = \frac{0.2\,\Omega}{\text{km}}(320 \text{ km}) + \frac{j377(2)(10^{-3})\,\Omega}{\text{km}} \cdot 320 \text{ km}$$

$$= -64 + j241.3\ \Omega$$

[5]Transmission line models for electromagnetic transient studies can be found in Dommel (1992).

2. Calculate A, B, C, and D

$$\mathbf{A} = \mathbf{D} = \mathbf{Y}_L\mathbf{Z}_L/2 + 1 = j1206(64 + j241.3)/2 + 1 = 0.8553\underline{/2.5869^\circ}$$

$$\mathbf{B} = \mathbf{Z}_L = 64 + j241.3 = 249.64\ \underline{/75.1455^\circ}$$

$$\mathbf{C} = \mathbf{Y}_L + \mathbf{Z}_L\mathbf{Y}_L^2/4 = j1206 + (64 + j241.3)(j1206)^2/4 = 0.0011\underline{/91.1926^\circ}$$

3. Calculate V_S and I_S

The receiving-end phase voltage is $230/\sqrt{3} = 132.8$ kV.

The sending-end voltage $(\mathbf{V}_S) = (0.8553\underline{/2.5864^\circ})(132.8 \times 10^3\underline{/0^\circ}) + (249.64\underline{/75.1455^\circ})(200\underline{/-36.9^\circ}) = 156.86\underline{/13.2873}$ kV. The magnitude of the sending-end line voltage is 271.69 kV.

The calculation shows $\mathbf{I}_S$ is 147.77 A, which is less than the receiving-end current. (This effect is due to the reactive compensation introduced by the line capacitances).

Related Calculations. In case the sending-end values $\mathbf{V}_S$ and $\mathbf{I}_S$ are known, the receiving-end voltage and current may be calculated by: $\mathbf{V}_S = \dfrac{D\mathbf{V}_s - B\mathbf{I}_s}{\mathbf{AD} - \mathbf{BC}}$ and $\mathbf{I}_R = \dfrac{-C\mathbf{V}_s + A\mathbf{I}_s}{\mathbf{AD} - \mathbf{BC}}$

LONG TRANSMISSION-LINE MODELS FOR POWER-FREQUENCY STUDIES

Recompute the medium-line model example (320 km long) with the same values of potential, current, and line parameters using the long-line model.

Calculation Procedure

1. Calculate Z and Y

Because $R = 0.2$ Ω/km and $L = 2$ mH/km, the series impedance per unit length is $\mathbf{Z} = 0.2 + j0.754 = 0.78\underline{/75.1^\circ}$ Ω/km. Since $C = 0.01$ μF/km, the shunt admittance per kilometer is $\mathbf{Y} = j3.77$ S/km.

2. Calculate Z_c

The characteristic (surge) impedance is $\mathbf{Z}_c = [0.78\underline{/75.1^\circ}/(3.77 \times 10^{-6}\underline{/90^\circ})]^{1/2} = 455\underline{/-7.45^\circ}$ Ω. If the resistance is less than one-tenth the inductive reactance per unit length, the characteristic impedance approaches a real number.

3. Calculate the Propagation Constant, γ

$\gamma = [(0.78\underline{/75.1^\circ})(3.77 \times 10^{-6}\underline{/90^\circ})]^{1/2} = 1.72 \times 10^{-3}\underline{/82.55^\circ}$. A small resistance results in a value approaching an imaginary number. To be useful, γ must be in rectangular form. Thus, $\gamma = 0.223 \times 10^{-3} + j1.71 \times 10^{-3}$.

The real part of γ is the *attenuation factor* α. $\alpha = 0.223 \times 10^{-3}$ nepers/km. The imaginary part is the *phase-shifting constant* β. $\beta = 1.71 \times 10^{-3}$ rad/km.

4. Calculate V_S and I_S

The per-phase receiving voltage to neutral is 132.8 kV. Substituting the above values in the equations for $\mathbf{V}_S$ and $\mathbf{I}_S$ yields: $\mathbf{V}_S = [(132.8 \times 10^3\angle 0°)/2]\,[\exp(0.223 \times 10^{-3})(200)]\,[\exp(j1.71 \times 10^{-3})(200)] + [(200\angle -36.9°)(455\angle -7.45°)/2]\,[\exp\,(0.223 \times 10^{-3})(200)]\,[\exp(j1.71 \times 10^{-3})(200)] + [(132.8 \times 10^3\angle 0°)/2]\,[\exp\,(-0.223 \times 10^{-3})(200)]\,[\exp(-j1.71 \times 10^{-3})(200)] - [(200\angle -36.9°)(455\angle -7.45°/2]\,[\exp(-0.223 \times 10^{-3})(200)]\,[\exp(-j1.71 \times 10^{-3}\,(200)]$ and $\mathbf{I}_S = \{[(132.8 \times 10^3\angle 0°\,/(455\angle -7.45°)]/2\}\,[\exp(0.223 \times 10^{-3})(200)]\,[\exp(j1.71 \times 10^{-3})(200)] + (200\angle -36.9°/2)\,[\exp(0.223 \times 10^{-3})(200)]\,[\exp(j1.71 \times 10^{-3})(200)] - [(132.8 \times 10^3\angle 0°)/455\angle -7.45°]/2\}\,[\exp(-0.223 \times 10^{-3})(200)]\,[\exp(j1.71 \times 10^{-3})(200)] + (200\angle -36.9°/2\,[\exp(-0.223 \times 10^{-3})(200)]\,[\exp(-j1.71 \times 10^{-3})(200)]$. When terms are combined, $\mathbf{V}_S = 150.8\angle 8.06°$ kV (phase to neutral) and $\mathbf{I}_S = 152.6\angle -4.52°$ A (line). The magnitude of the line voltage at the input is 261.2 kV. These results correspond to the results of the medium-line model of the previous example with little difference for a 320-km line.

Related Calculations. Equations for $\mathbf{V}_S$ and $\mathbf{I}_S$ are most easily solved by suitable FORTRAN, C, or C++ algorithms, by EMTP–type programs, or by matrix-oriented programs such as MATLAB and MATHCAD. With such programs, the value of x can be incremented outward from the receiving end to display the behavior of the potential and current throughout the line. Such programs are suitable for lines of any length.

The first term in each equation for $\mathbf{V}_S$ and $\mathbf{I}_S$ may be viewed as representing a traveling wave from the source to the load end of the line. If x is made zero, the wave is incident at the receiving end. The second term in each equation represents a wave reflected from the load back toward the source. If x is made zero, the value of this wave is found at the receiving end. The sum of the two terms at the receiving end should be 132.8 kV and 200 A in magnitude.

If the impedance at the load end is equal to the characteristic (surge) impedance $\mathbf{Z}_c$, the reflected terms (second terms in equations for $\mathbf{V}_S$ and $\mathbf{I}_S$) are zero. The line is said to be matched to the load. This is hardly possible in transmission line, but is achieved at much higher (e.g., radio) frequencies. This eliminates the so-called standing waves stemming from the summation of terms in the equations for $\mathbf{V}_S$ and $\mathbf{I}_S$. Such quantities at standing-wave ratio SWR and reflection coefficient σ are easily calculated but are beyond the needs of power studies.

COMPLEX POWER

Determine the complex power **S** at both ends of the 320-km (200-mi) transmission line using the results of the preceding example.

Calculation Procedure

1. Calculate S at Receiving End

Use $\mathbf{S} = 3\mathbf{VI}^*$, where $\mathbf{V}$ is the phase-to-neutral voltage and $\mathbf{I}^*$ is the complex conjugate of the line current under balanced conditions. Then, at the receiving end, $\mathbf{S} = (3)(132.8 \times 10^3\angle 0°)\,(200)\angle 36.9°) = 63{,}719\text{ kW} + j47{,}841\text{ kVAR}$.

2. Calculate S at Sending End

At the sending end, $\mathbf{S} = (3)(150.8 \times 10^3\angle 8.06°)(152.6\angle 4.52°) = 67{,}379$ kW + $j15{,}036$ kVAR.

Related Calculations. The transmission line must, in this case, be furnishing some of the receiving end's requirements for reactive power from the supply of stored charge, because the apparent power input in kVAR is less than the output. Power loss of the line may be determined by $Q = 47{,}841 - 15{,}036 = 32{,}905$ kVAR made up by stored line charge and $P = 67{,}379 - 63{,}719 = 3660$ kW line-resistance losses.

SURGE IMPEDANCE LOADING

A convenient method of comparing the capability of transmission lines to support energy flow (but not accounting for resistance-loss restrictions) is through the use of *surge impedance loading,* SIL. If the line is assumed terminated in its own surge impedance value as a load (preferably a real number), then a hypothetical power capability is obtained that can be compared with other lines.

Compare two 230-kV lines for their power capability if $Z_{c1} = 500\ \Omega$ for line 1 and $Z_{c2} = 400\ \Omega$ for line 2.

Calculation Procedure

1. Determine Expression for SIL

If Z_c is considered as a load, the load current I_L may be expressed by $I_L = V_L/\sqrt{3}Z_c$ kA, where V_L is the magnitude of the line-to-line voltage in kilovolts. Then, SIL $= \sqrt{3}V_L I_L = V_L^2/Z_c$.

2. Calculate SIL Values

$\text{SIL}_1 = 230^2/500 = 106$ MW and $\text{SIL}_2 = 230^2/400 = 118$ MW. Line 2 has greater power capability than line 1.

BIBLIOGRAPHY

The Aluminum Association. 1982. *Aluminum Electrical Conductor Handbook*, 2nd ed. Washington, D.C.: The Aluminum Association.

Ametani, A. 1980. "A General Formulation of Impedance and Admittance of Cables," *IEEE Transactions on Power Apparatus and Systems*, Vol. PAS-99, No. 3, pp. 902–910, May/June.

Carson, J. R. 1926. "Wave Propagation in Overhead Wires with Ground Return," *Bell Syst. Tech. J.*, Vol. 5, pp. 539–554, October.

Deri, A., G. Tavan, A. Semlyen, and A. Castanheira. 1981. "The Complex Ground Return Plane, a Simplified Model for Homogeneous and Multi-layer Earth Return," *IEEE Transactions on Power Apparatus and Systems*, Vol. PAS-100, pp. 3686–3693, August.

Dommel, H. W. 1992. *EMTP Theory Book*, 2nd ed. Vancouver, B.C., Canada: Microtran Power Systems Analysis Corporation.

Dommel, H. W. 1985. "Overhead Line Parameters from Handbook Formulas and Computer Programs," *IEEE Transactions on Power Apparatus and Systems*, Vol. PAS-104, No. 2, pp. 366–372, February.

Dubanton, C. 1969. "Calcul Approché des Paramètres Primaires et Secondaires d'Une Ligne de Transport, Valeurs Homopolaires, ("Approximate Calculation of Primary and Secondary Transmission Line Parameters, Zero Sequence Values", in French)," *EDF Bulletin de la Direction des Études et Reserches*, pp. 53–62, Serie B—Réseaux Électriques. Matériels Électriques No. 1.

Elgerd, O. I. 1982. *Electric Energy Systems Theory: An Introduction*, 2nd ed. New York: McGraw-Hill.

Galloway, R. H., W. B. Shorrocks, and L. M. Wedepohl. 1964. "Calculation of Electrical Parameters for Short and Long Polyphase Transmission Lines," *Proceedings of the Institution of Electrical Engineers*, Vol. 111, No. 12, pp. 2051–2059, December.

Gary, C. 1976. "Approache Complète de la Propagation Multifilaire en Haute Fréquence par Utilisation des Matrices Complexes, ("Complete Approach to Multiconductor Propagation at High Frequency with Complex Matrices," in french)," *EDF Bulletin de la Direction des Études et Reserches*, pp. 5–20, Serie B—Réseaux Électriques. Matériels Électriques No. 3/4.

Grainger, J. J. and W. D. Stevenson, Jr. 1994. *Power System Analysis*. New York: McGraw-Hill.

Gross, C. A. 1986. *Power System Analysis*, 2nd ed. New York: Wiley.

Lewis, W. A. and G. D. Allen, "Symmetrical-Component Circuit Constants and Neutral Circulating Currents for Concentric-Neutral Underground Distribution Cables," *IEEE Transactions on Power Apparatus and Systems*, Vol. PAS-97, No. 1, pp. 191–199, January/February 1978.

Lewis, W. A. and P. D. Tuttle. 1959. "The Resistance and Reactance of Aluminum Conductors, Steel Reinforced," *AIEE Transactions*, Vol. 78, Pt. III, pp. 1189–1215, February.

Neher, J. H. 1964. "The Phase Sequence Impedance of Pipe-Type Cables," *IEEE Transactions on Power Apparatus and Systems*, Vol. PAS-83, pp. 795–804, August.

Neuenswander, J. R. 1971. *Modern Power Systems*. Scranton: International Textbook Co.

Pollaczek, F. 1931. "Sur le Champ Produit par un Conducteur Simple Infiniment Long Parcouru par un Courant Alternatif ("On the Field Produced by an Infinitely Long Wire Carrying Alternating Current," French translation by J. B. Pomey)," *Revue Générale de l'Electricité*, Vol. 29, No. 2, pp. 851–867.

Smith, D. R. and J. V. Barger. 1972. "Impedance and Circulating Current Calculations for the UD Multi-Wire Concentric Neutral Circuits," *IEEE Transactions on Power Apparatus and Systems*, Vol. PAS-91, No. 3, pp. 992–1006, May–June.

Wedepohl, L. M. and D. J. Wilcox. 1973. "Transient Analysis of Underground Power-Transmission Systems—System-Model and Wave-Propagation Characteristics," *Proceedings of the Institution of Electrical Engineers*, Vol. 120, No. 2, pp. 253–260, February.

Westinghouse Electric Corporation. 1964. *Electrical Transmission and Distribution Reference Book*. East Pittsburgh: Westinghouse Electric Corporation.

SECTION 10

ELECTRIC-POWER NETWORKS

Lawrence J. Hollander, P.E.
Dean of Engineering Emeritus
Union College

POWER SYSTEM REPRESENTATION: GENERATORS, MOTORS, TRANSFORMERS, AND LINES

The following components comprise a simplified version of a power system, listed in sequential physical order from the generator location to the load: (1) two steam-electric generators, each at 13.2 kV; (2) two step-up transformers, 13.2/66 kV; (3) sending-end, high-voltage bus at 66 kV; (4) one long transmission line at 66 kV; (5) receiving-end bus at 66 kV; (6) a second 66-kV transmission line with a center-tap bus; (7) step-down transformer at receiving-end bus, 66/12 kV, supplying four 12-kV motors in parallel; and (8) a step-down transformer, 66/7.2 kV, off the center-tap bus, supplying a 7.2-kV motor. Draw a one-line diagram for the three-phase, 60-Hz system, including appropriate oil circuit breakers (OCBs).

Calculation Procedure

1. Identify the Appropriate Symbols

For electric-power networks an appropriate selection of graphic symbols is shown in Fig. 10.1.

2. Draw the Required System

The system described in the problem is shown in Fig. 10.2. The oil circuit breakers are added at the appropriate points for proper isolation of equipment.

Related Calculations. It is the general procedure to use one-line diagrams for representing three-phase systems. When analysis is done using symmetrical components, different

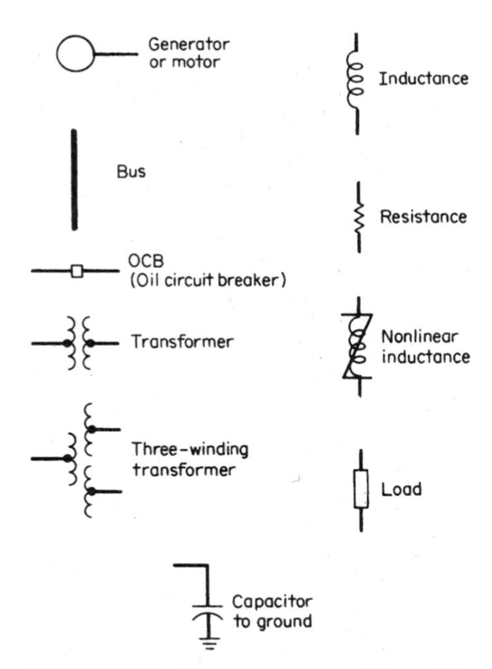

FIGURE 10.1 Common power symbols used in one-line diagrams.

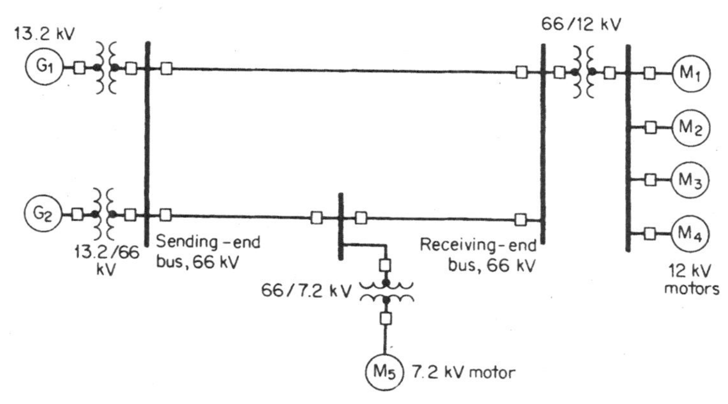

FIGURE 10.2 Three-phase power system represented by one-line diagram.

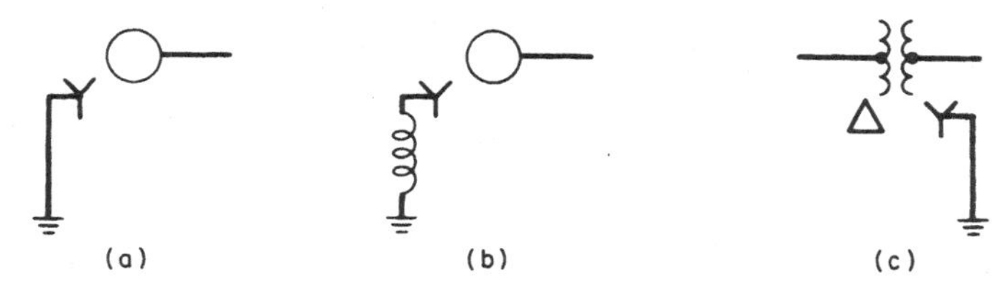

FIGURE 10.3 Identification for wye-connected generator or motor. (*a*) Solidly grounded. (*b*) Grounded through an inductance. (*c*) The transformer is identified as being delta-wye, with the wye side solidly grounded.

diagrams may be drawn that will represent the electric circuitry for positive, negative, and zero-sequence components. Additionally, it is often necessary to identify the grounding connection, or whether the device is wye- or delta-connected. This type of notation is shown in Fig. 10.3.

PER-UNIT METHOD OF SOLVING THREE-PHASE PROBLEMS

For the system shown in Fig. 10.4, draw the electric circuit or reactance diagram, with all reactances marked in per-unit (p.u.) values, and find the generator terminal voltage assuming both motors operating at 12 kV, three-quarters load, and unity power factor.

Generator	Transformers (each)	Motor A	Motor B	Transmission line
13.8 kV	25,000 kVA	15,000 kVA	10,000 kVA	$X = 65\ \Omega$
25,000 kVA	13.2/69 kV	13.0 kV	13.0 kV	
Three-phase	X_L = 11 percent	X'' = 15 percent	X'' = 15 percent	
X'' = 15 percent				

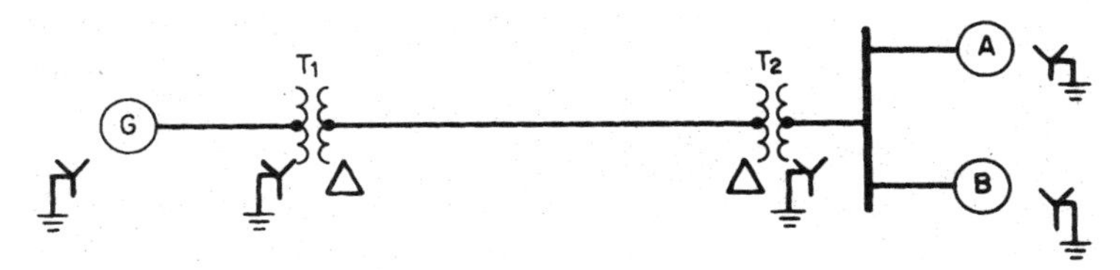

FIGURE 10.4 One-line diagram of electric-power system supplying motor loads. The table on previous page gives specifications.

Calculation Procedure

1. Establish Base Voltage through the System

By observation of the magnitude of the components in the system, a base value of apparent power S is chosen; it should be of the general magnitude of the components, and the choice is arbitrary. In this problem, 25,000 kVA is chosen as the base S, and simultaneously, at the generator end 13.8 kV is selected as a base voltage V_{base}.

The base voltage of the transmission line is then determined by the turns ratio of the connecting transformer: (13.8 kV)(69 kV/13.2 kV) = 72.136 kV. The base voltage of the motors is determined likewise but with the 72.136-kV value: thus, (72.136 kV)(13.2 kV/69 kV) = 13.8 kV. The selected base S value remains constant throughout the system, but the base voltage is 13.8 kV at the generator and at the motors, and 72.136 kV on the transmission line.

2. Calculate the Generator Reactance

No calculation is necessary for correcting the value of the generator reactance because it is given as 0.15 p.u. (15 percent), based on 25,000 kVA and 13.8 kV. If a different S base were used in this problem, then a correction would be necessary as shown for the transmission line, motors, and transformers.

3. Calculate the Transformer Reactance

It is necessary to make a correction when the transformer nameplate reactance is used because the calculated operation is at a different voltage, 13.8 kV/72.136 kV instead of 13.2 kV/69 kV. Use the equation for correction: per-unit reactance = (nameplate per-unit reactance)(base kVA/nameplate kVA)(nameplate kV/base kV)^2 = (0.11)(25,000/25,000)(13.2/13.8)^2 = 0.101 p.u. This applies to each transformer.

4. Calculate the Transmission-Line Reactance

Use the equation: $X_{\text{per unit}}$ = (ohms reactance)(base kVA)/(1000)(base kV)^2 = (65)(25,000)/(1000)(72.1)^2 = 0.313 p.u.

5. Calculate the Reactance of the Motors

Corrections need to be made in the nameplate ratings of both motors because of differences of ratings in kVA and kV as compared with those selected for calculations in this problem. Use the correcting equation from Step 3, above. For motor A, X''_A = (0.15 p.u.)(25,000 kVA/15,000 kVA)(13.0 kV/13.8 kV)^2 = 0.222 p.u. For motor B, similarly, X''_B = (0.15 p.u.)(25,000 kVA/10,000 kVA)(13.0 kV/13.8 kV)^2 = 0.333 p.u.

6. Draw the Reactance Diagram

The completed reactance diagram is shown in Fig. 10.5.

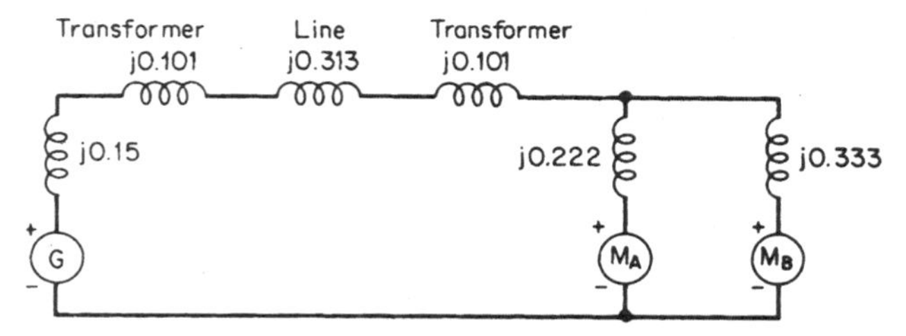

FIGURE 10.5 One-line reactance circuit diagram (reactances shown on a per-unit basis).

7. Calculate Operating Conditions of the Motors

If the motors are operating at 12 kV, this represents 12 kV/13.8 kV = 0.87 per-unit voltage. At unity power factor, the load is given as three-quarters or 0.75 p.u. Thus, expressed in per unit, the combined motor current is obtained by using the equation $I_{\text{per unit}}$ = per-unit power/per-unit voltage = 0.75/0.87 = $0.862\angle 0°$ p.u.

8. Calculate the Generator Terminal Voltage

The voltage at the generator terminals, $\mathbf{V}_G = \mathbf{V}_{\text{motor}}$ + the voltage drop through transformers and transmission line = $0.87\angle 0° + 0.862\angle 0°(j0.101 + j0.313 + j0.101) = 0.87 + j0.444 = 0.977\angle 27.03°$ p.u. In order to obtain the actual voltage, multiply the per-unit voltage by the base voltage at the generator. Thus, $\mathbf{V}_G = (0.977\angle 27.03°)(13.8 \text{ kV}) = 13.48\angle 27.03°$ kV.

Related Calculations. In the solution of these problems, the selection of base voltage and apparent power are arbitrary. However, the base voltage in each section of the circuit must be related in accordance with transformer turns ratios. The base impedance may be calculated from the equation Z_{base} = (base kV)^2(1000)/(base kVA). For the transmission-line section in this problem, $Z_{\text{base}} = (72.136)^2(1000)/(25{,}000) = 208.1\ \Omega$; thus the per-unit reactance of the transmission line equals (actual ohms)/(base ohms) = 65/208.1 = 0.313 p.u.

PER-UNIT BASES FOR SHORT-CIRCUIT CALCULATIONS

For the system shown in Fig. 10.6, assuming S bases of 30,000 and 75,000 kVA, respectively, calculate the through impedance in ohms between the generator and the output terminals of the transformer.

Calculation Procedure

1. Correct Generator 1 Impedance (Reactance)

Use the equation for changing S base: per-unit reactance = (nameplate per-unit reactance)(base kVA/nameplate kVA) = $X'' = (0.20)(30{,}000/40{,}000) = 0.15$ p.u. on a 30,000-kVA base. Similarly, $X'' = (0.20)(75{,}000/40{,}000) = 0.375$ p.u. on a 75,000-kVA base.

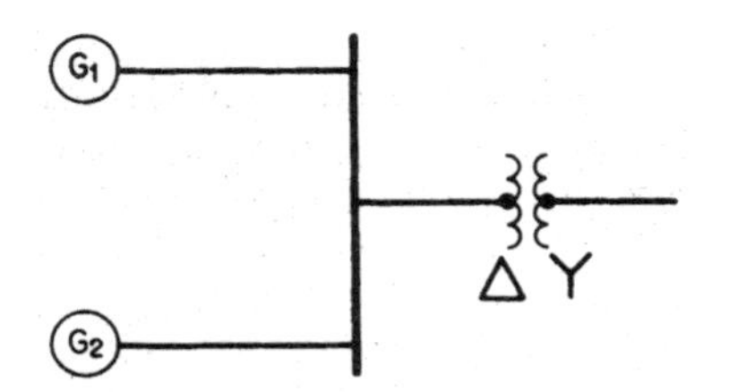

FIGURE 10.6 System of paralleled generators and transformer with calculations demonstrating different apparent-power bases.

Generator 1	Generator 2	Transformer
40,000 kVA	30,000 kVA	75,000 kVA
13.2 kV	13.2 kV	13.2 kV (delta)
$X'' = 0.20$ p.u.	$X'' = 0.25$ p.u.	66 kV (wye)
		$X = 0.10$ p.u.

2. Correct Generator 2 Impedance (Reactance)

Use the equation as in Step 1, above: $X'' = (0.25)(30{,}000/30{,}000) = 0.25$ p.u. on a 30,000-kVA base. $X'' = (0.25)(75{,}000/30{,}000) = 0.625$ p.u. on a 75,000-kVA base.

3. Correct the Transformer Impedance (Reactance)

Use the same equation as for the generator corrections: $X = (0.1)(30{,}000/75{,}000) = 0.04$ p.u. on a 30,000-kVA base; it is 0.10 on a 75,000-kVA base (Fig. 10.7).

4. Calculate the Through Impedance per Unit

In either case, the through impedance is equal to the parallel combination of generators 1 and 2, plus the series impedance of the transformer. For the base of 30,000 kVA, $jX_{\text{total}} = (j0.15)(j0.25)/(j0.15 + j0.25) + j0.04 = j0.134$ p.u. For the base of 75,000 kVA, $jX_{\text{total}} = (j0.375)(j0.625)/(j0.375 + j0.625) + j0.10 = j0.334$ p.u.

5. Convert Impedances (Reactances) to Ohms

The base impedance (reactance) in ohms is (1000)(base kV)^2/(base kVA). Thus, for the case of the 30,000-kVA base, the base impedance is $(1000)(13.2)^2/30{,}000 = 5.808\ \Omega$.

The actual impedance in ohms of the given circuit is equal to (per-unit impedance)(base impedance) $= (j0.134)(5.808) = 0.777\ \Omega$, referred to the low-voltage side of the transformer.

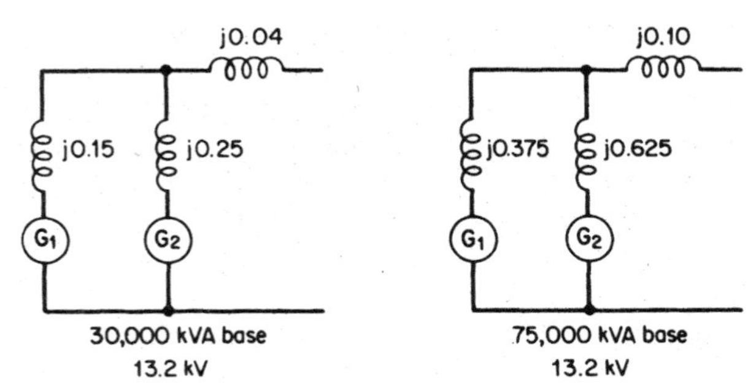

FIGURE 10.7 Equivalent electric network diagrams for solution on a 30,000-kVA base and on a 75,000-kVA base.

The base impedance (reactance) in ohms for the case of the 75,000-kVA base is $(1000)(13.2)^2/75{,}000 = 2.32\ \Omega$. The actual impedance in ohms of the given circuit is equal to $(j0.334)(2.32) = 0.777\ \Omega$, referred to the low-voltage side of the transformer.

6. Compare Results of Different* S *Bases

It is seen that either S base, selected arbitrarily, yields the same answer in ohms. When working on a per-unit basis, the same per-unit values hold true for either side of the transformers; in actual ohms, amperes, or volts, turns ratio adjustments are necessary depending on the reference side for which calculations are being made. In this problem, the total reactance is $0.777\ \Omega$, referred to the low-voltage side of the transformers; it is $(0.777)(66\ \text{kV}/13.2\ \text{kV})^2 = 19.425\ \Omega$, referred to the high-voltage side. Expressed in per unit it is 0.134 (on a 30,000-kVA base) or 2.32 (on a 75,000-kVA base) on either side of the transformer.

Related Calculations. This problem illustrates the arbitrary selection of the S base, provided the value selected is used consistently throughout the circuit. Likewise, the kV base may be selected in one part of the circuit, but in all other parts of the circuit the kV base values must be related in accordance with transformer turns ratios. In both cases, a little experience and general observation of the given information will suggest appropriate selections of per-unit bases that will yield comfortable numbers with which to work.

CHANGING THE BASE OF PER-UNIT QUANTITIES

The reactance of a forced-oil-cooled 345-kV/69-kV transformer is given as 22 percent; it has a nameplate capacity of 450 kVA. Calculations for a short-circuit study are being made using bases of 765 kV and 1 MVA. Determine (1) the transformer reactance on the study bases and (2) the S base for which the reactance will be 10 percent on a 345-kV base.

Calculation Procedure

1. Convert the Nameplate Reactance to Study Bases

The study bases are 765 kV and 1000 kVA. Let subscript 1 denote the given nameplate conditions and subscript 2 denote the new or revised conditions. Use the equation: X_2 p.u. $= (X_1$ p.u.$)[(\text{base kVA}_2)/(\text{base kVA}_1)] \times [(\text{base kV}_1)/(\text{base kV}_2)]^2$. Thus, $X_2 = (0.22)[(1000)/(450)][(345)/(765)]^2 = 0.099$ p.u.

2. Calculation of Alternative* S *Base

Assume it is desired that the per-unit reactance of the transformer be 0.10 on a 345-kV base; use the same equation as in the first step and solve for base kVA_2. Thus, base $\text{kVA}_2 = (X_2/X_1)(\text{base kVA}_1)[(\text{base kV}_2)/(\text{base kV}_1)]^2 = [(0.10)/(0.22)](450)[(345)/(345)]^2 = 204.5$ kVA.

Related Calculations. The equation used in this problem may be used to convert any impedance, resistance, or reactance expressed in per unit (or percent) from one set of kV, kVA bases to any other set of kV, or kVA bases.

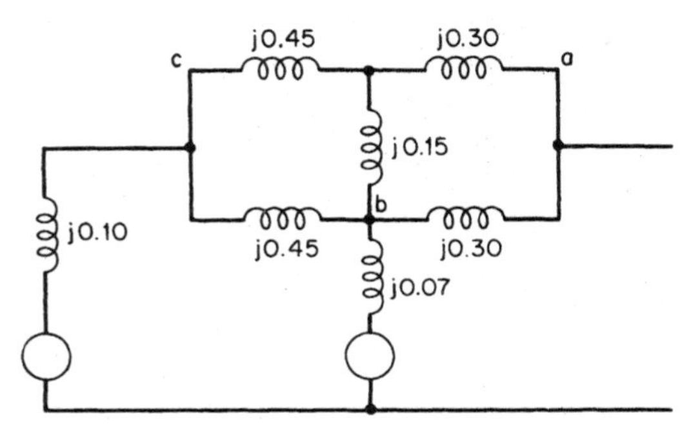

FIGURE 10.8 Portion of electrical network diagram (reactances shown on a per-unit basis).

WYE-DELTA AND DELTA-WYE CONVERSIONS

A one-line portion of an electrical network diagram is shown in Fig. 10.8. Using wye-delta and/or delta-wye conversions, reduce the network to a single reactance.

Calculation Procedure

1. Convert Wye to Delta

The network reduction may be started at almost any point. One starting point is to take the wye-formation of reactances *a*, *b*, and *c* and convert to a delta. Refer to Fig. 10.9 for the appropriate equation. Thus, substituting *X*'s for *Z*'s in the equations yields $X_{ab} =$

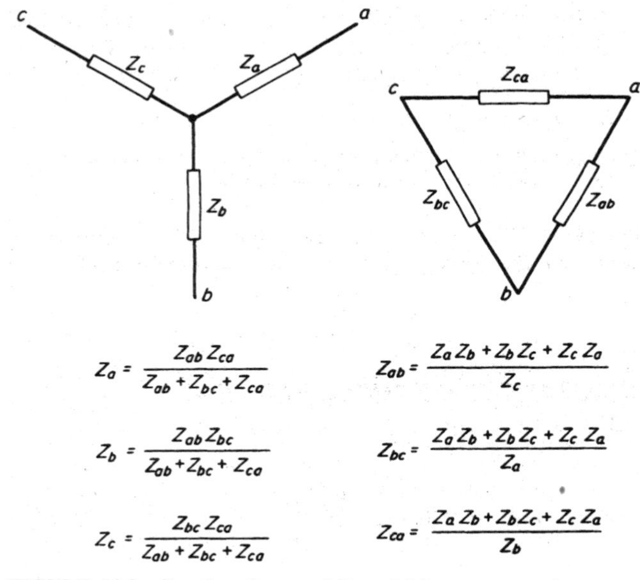

FIGURE 10.9 Equations for wye-delta and delta-wye conversions.

$(X_aX_b + X_bX_c + X_cX_a)/X_c = [(j0.30)(j0.15) + (j0.15)(j0.45) + (j0.45)(j0.30)]/(j0.45) = [j^2 0.045 + j^2 0.0675 + j^2 0.135]/j0.45 = (j^2 0.2475)/j0.45 = j0.55$ p.u. (Fig. 10.10).

$X_{bc} = (X_aX_b + X_bX_c + X_cX_a)/X_a = j^2 0.2475/j0.30 = j0.825$ p.u. and $X_{ca} = (X_aX_b + X_bX_c + X_cX_a)/X_b = j^2 0.2475/j0.15 = j1.65$ p.u.

2. Combine Parallel Reactances

It is noted that after the first wye-delta conversion, points a and b are connected by two reactances in parallel, yielding a combined reactance of $(j0.55)(j0.30)/(j0.55 + j0.30) = j0.194$ p.u. Points b and c are connected by two reactances in parallel, yielding a combined reactance of $(j0.825)(j0.45)/(j0.825 + j0.45) = j0.290$ p.u.

3. Convert Delta to Wye

Points a, b, and c now form a new delta connection that will be converted to a wye. $X_a = X_{ab}X_{ca}/(X_{ab} + X_{bc} + X_{ca}) = (j0.194)(j1.65)/(j0.194 + j0.290 + j1.65) = j0.150$ p.u. $X_b = X_{ab}X_{bc}/(X_{ab} + X_{bc} + X_{ca}) = (j0.194)(j0.290)/(j0.194 + j0.290 + j1.65) = j0.026$ p.u. And $X_c = X_{bc}X_{ca}/(X_{ab} + X_{bc} + X_{ca}) = (j0.290)(j1.65)/(j0.194 + j0.290 + j1.65) = j0.224$ p.u.

4. Combine Reactances in Generator Leads

In each generator lead there are now two reactances in series. In the c branch the two reactances total $j0.10 + j0.224 = j0.324$ p.u. In the b branch the total series reactance is $j0.07 + j0.026 = j0.096$ p.u. These may be paralleled, leaving one equivalent generator and one reactance, $(j0.324)(j0.096)/(j0.324 + j0.096) = j0.074$ p.u.

5. Combine Remaining Reactances

The equivalent generator reactance is now added to the reactance in branch a, yielding $j0.074 + j0.150 = j0.224$ p.u., as illustrated in Fig. 10.10.

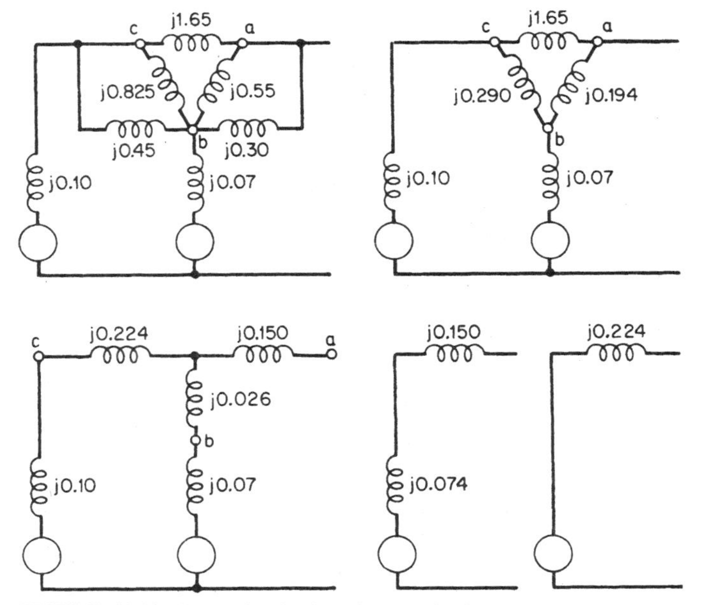

FIGURE 10.10 Network reduction of power circuit.

Related Calculations. This problem illustrates the delta-wye and wye-delta conversion used for network reductions. Such conversions find many applications in network problems.

PER-UNIT REACTANCES OF THREE-WINDING TRANSFORMERS

For the three-phase, 60-Hz system shown in Fig. 10.11, the leakage reactances of the three-winding transformer are: $X_{ps} = 0.08$ p.u. at 50 MVA and 13.2 kV, $X_{pt} = 0.07$ p.u. at 50 MVA and 13.2 kV, and $X_{st} = 0.20$ p.u. at 20 MVA and 2.2 kV, where subscripts p, s, and t refer to primary, secondary, and tertiary windings. Using a base of 50,000 kVA and 13.2 kV, calculate the various reactances in the circuit and draw the simplified-circuit network.

Calculation Procedure

1. Correct the Generator Reactance

The generator reactance is given as 0.10 p.u. on the base of 50,000 kVA and 13.2 kV. No correction is needed.

2. Correct the Three-Winding Transformer Reactances

X_{ps} and X_{pt} are given as 0.08 and 0.07 p.u., respectively, on the base of 50,000 kVA and 13.2 kV. No correction is needed. However, X_{st} is given as 0.20 p.u. at 20,000 kVA and 2.2 kV; it must be corrected to a base of 50,000 kVA and 13.2 kV. Use the equation $X_2 = X_1(\text{base} \quad S_2/\text{base} \quad S_1)(V_{\text{base }1}/V_{\text{base }2})^2 = (0.20)[(50{,}000)/(20{,}000)][(2.2)/(13.2)]^2 = 0.014$ p.u.; this is the corrected value of X_{st}.

3. Calculate the Three-Phase, Three-Winding Transformer Reactances as an Equivalent Wye

Use the equation $X_p = \frac{1}{2}(X_{ps} + X_{pt} - X_{st}) = \frac{1}{2}(0.08 + 0.07 - 0.014) = 0.068$ p.u., $X_s = \frac{1}{2}(X_{ps} + X_{st} - X_{pt}) = \frac{1}{2}(0.08 + 0.014 - 0.07) = 0.012$ p.u., and $X_t = \frac{1}{2}(X_{pt} + X_{st} - X_{ps}) = \frac{1}{2}(0.07 + 0.014 - 0.08) = 0.002$ p.u.

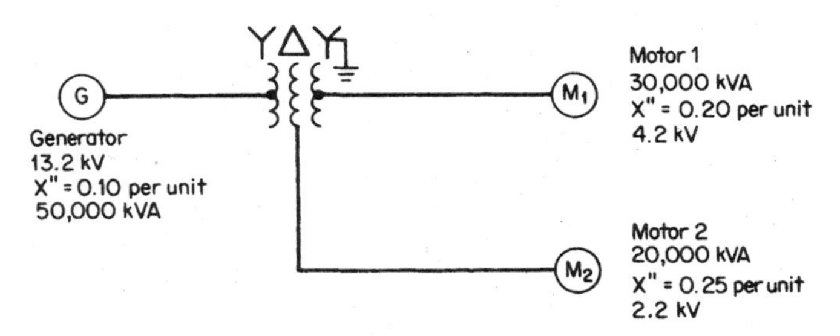

Three-winding transformer

Primary	Secondary	Tertiary
13.2 kV	4.2 kV	2.2 kV
50 MVA	30 MVA	20 MVA

FIGURE 10.11 Three-winding transformer interconnecting motor loads to generator.

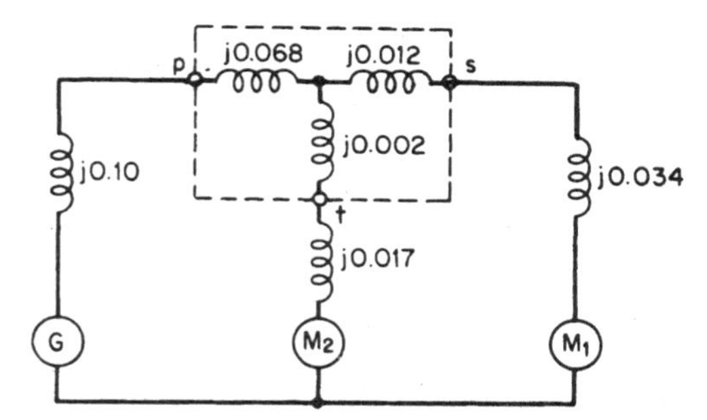

FIGURE 10.12 Three-winding transformer equivalent circuit.

4. Correct the Motor Reactances

X'' of motor load 1 is given as 0.20 on a base of 30,000 kVA and 4.2 kV. The corrected value is $(0.20)(50{,}000/30{,}000)(4.2/13.2)^2 = 0.034$ p.u.

X'' of motor load 2 is given as 0.25 on a base of 20,000 kVA and 2.2 kV. The corrected value is determined similarly: $(0.25)(50{,}000/20{,}000)(2.2/13.2)^2 = 0.017$ p.u.

5. Draw the Simplified-Circuit Network

See Fig. 10.12.

Related Calculations. Three-winding transformers are helpful in suppressing third-harmonic currents; the tertiary winding is connected in closed delta for that purpose. The third-harmonic currents develop from the exciting current. This problem illustrates the handling of the reactances of three-winding transformers.

CALCULATION OF COMPLEX POWER P + jQ

At a certain point in the solution of a balanced three-phase problem, the current per phase is equal to $5.0\underline{/-37^\circ}$ A and the line voltage is $69\underline{/0^\circ}$ kV. With a per-unit base of 1000 kVA and 72 kV, determine the complex power.

Calculation Procedure

1. Change Voltage to a Per-Unit Basis

The voltage is given as $69\underline{/0^\circ}$ kV. Since the base voltage is 72 kV, the per-unit voltage $= 69/72 = 0.96\underline{/0^\circ}$ p.u.

2. Change Current to a Per-Unit Basis

Use the equation: $S = \sqrt{3}V_L I_L$; thus, $I_L = S/\sqrt{3}V_L$. For the base conditions specified, $I_L = 1{,}000{,}000$ VA/$(\sqrt{3})(72{,}000$ V$) = 8.02$ A. The given current is $5\underline{/-37^\circ}$ A/8.02 $= 0.623\underline{/-37^\circ}$ p.u.

3. Calculate Complex Power

Use the equation for complex power: $\mathbf{S} = P + jQ = \mathbf{VI^*}$, where the asterisk indicates the conjugate value of the current. $\mathbf{S} = (0.96\underline{/0^\circ})(0.623\underline{/37^\circ}) = 0.598\underline{/37^\circ} = 0.478 +$

$j0.360$ p.u. The real power $P = 0.478$ p.u. or (0.478)(1000 kVA) = 478 kW. The reactive power $Q = 0.360$ p.u. or (0.360)(1000 kVA) = 360 kVAR. $P + jQ = 478$ kW + $j360$ kVAR, or $0.478 + j0.360$ p.u.

Related Calculations. This problem illustrates the finding of complex power using the conjugate of the current and per-unit notation. Alternatively, the real power may be obtained from the equation: $P = \sqrt{3}V_L I_L \cos\theta = (\sqrt{3})(69{,}000\text{ V})(5.0\text{ A})\cos 37° = 478$ kW. Reactive power is $Q = \sqrt{3}V_L I_L \sin\theta = (\sqrt{3})(69{,}000\text{ V})(5.0\text{ A})\sin 37° = 360$ kVAR.

CHECKING VOLTAGE PHASE SEQUENCE WITH LAMPS

Given the three leads of a 120-V, three-phase, 60-Hz system, two identical lamps and an inductance are connected arbitrarily, as shown in Fig. 10.13. Determine the phase sequence of the unknown phases for the condition that one lamp glows brighter than the other.

Calculation Procedure

1. Assume Per-Unit Values of Lamps and Inductance

For whatever kVA, kV bases may be assumed, let each lamp be a pure resistance of 1.0 p.u., and let the inductance be without resistance, or $1.0\underline{/90°}$ p.u. That is, $R_A = R_B = 1.0\underline{/0°}$ p.u., and $X = 1.0\underline{/90°}$ p.u. Practically, the inductive angle will not be 90°, but slightly less because of resistance; this solution depends upon the inductance having a high ratio of X_L/R.

2. Assume a Phase Rotation

Assume that the phase rotation is *xy, zx, yz,* or that $\mathbf{V}_{xy} = 1.0\underline{/0°}$, $\mathbf{V}_{zx} = 1.0\underline{/-120°}$, and $\mathbf{V}_{yz} = 1.0\underline{/120°}$ p.u.

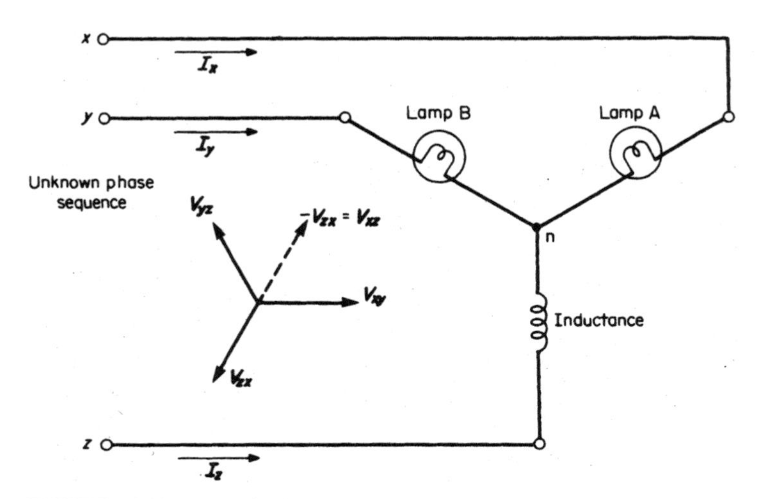

FIGURE 10.13 Determination of phase sequence with two resistance lamps and an inductance.

3. Develop Voltage Equations for Lamps

Three equations may be written for the circuit: (1) $\mathbf{I}_x + \mathbf{I}_y + \mathbf{I}_z = 0$; (2) $\mathbf{V}_{zx} - R_A\mathbf{I}_x + X\mathbf{I}_z = 0$; and (3) $\mathbf{V}_{zy} - R_B\mathbf{I}_y + X\mathbf{I}_z = 0$. Substitute the assumed values for $\mathbf{V}_{zx}$, $\mathbf{V}_{zy}$ $(= -\mathbf{V}_{yz})$, R_A, and R_B. Equation (1) remains unchanged; $\mathbf{I}_x + \mathbf{I}_y + \mathbf{I}_z = 0$. Equation (2) becomes $1\angle -120° - \mathbf{I}_x + 1\angle 90°\ \mathbf{I}_z = 0$, or $-\mathbf{I}_x + 1\angle 90°\ \mathbf{I}_z = -(1\angle -120°) = 1\angle 60°$. Equation (3) becomes $-(1\angle 120°) - \mathbf{I}_y + 1\angle 90°\ \mathbf{I}_z = 0$, or $-\mathbf{I}_y + 1\angle 90°\ \mathbf{I}_z = 1\angle 120°$.

4. Determine the Voltage across Lamp A

The voltage across lamp A is proportional to the current through it, namely $\mathbf{I}_x$. The solution of the simultaneous equations for $\mathbf{I}_x$ yields

$$\mathbf{I}_x = \frac{\begin{vmatrix} 0 & 1 & 1 \\ 1\angle 60° & 0 & j \\ 1\angle 120° & -1 & j \end{vmatrix}}{\begin{vmatrix} 1 & 1 & 1 \\ -1 & 0 & j \\ 0 & -1 & j \end{vmatrix}} = \frac{-0.50 - j1.866}{1 + j2} = 0.863\ \angle 191.6°\ \text{p.u.}$$

5. Determine the Voltage across Lamp B

The voltage across lamp B is proportional to the current through it, namely $\mathbf{I}_y$. The solution of the simultaneous equations for $\mathbf{I}_y$ yields

$$\mathbf{I}_y = \frac{\begin{vmatrix} 1 & 0 & 1 \\ -1 & 1\angle 60° & j \\ 0 & 1\angle 120° & j \end{vmatrix}}{1 + j2} = \frac{0.5 + j0.134}{1 + j2} = 0.232\ \angle -48.4°\ \text{p.u.}$$

6. Confirm the Voltage Sequence

From the previous two steps it is determined that the current through lamp A is greater than through lamp B. Therefore, lamp A is brighter than lamp B, indicating that the phase sequence is as was assumed, *xy, zx, yz*.

7. Let the Phase Sequence Be Reversed

The reversed phase sequence is *xy, yz, zx*. In this case, the assumed voltages become: $\mathbf{V}_{xy} = 1.0\angle 0°$, $\mathbf{V}_{zx} = 1.0\angle 120°$, and $\mathbf{V}_{yz} = 1.0\angle -120°$.

8. Determine the Voltage across Lamp A for Reversed Sequence

The voltage across lamp A, by solution of simultaneous equations, is proportional to

$$\mathbf{I}_x = \frac{\begin{vmatrix} 0 & 1 & 1 \\ 1\angle -60° & 0 & j \\ 1\angle -120° & -1 & j \end{vmatrix}}{1 + j2} = \frac{-0.5 - j0.134}{1 + j2} = 0.232\ \angle 131.5°\ \text{p.u.}$$

9. Determine the Voltage across Lamp B for Reversed Sequence

The voltage across lamp B, by solution of simultaneous equations, is proportional to

$$\mathbf{I}_y = \frac{\begin{vmatrix} 1 & 0 & 1 \\ -1 & 1\angle -60^\circ & j \\ 0 & 1\angle -120^\circ & j \end{vmatrix}}{1 + j2} = \frac{0.5 + j1.866}{1 + j2} = 0.863\angle 11.6^\circ \text{ p.u.}$$

10. Confirm the Voltage Sequence

Thus, for the connection of Fig. 10.13, lamp B will glow brighter than lamp A for the voltage sequence pattern *xy, yz, zx*.

Related Calculations. This problem indicates one method of checking phase sequence of voltages. Another method is based on a similar analysis and uses a combination of a voltmeter, a capacitor, and an inductance. These two methods rely on the use of an unbalanced load impedance.

TOTAL POWER IN BALANCED THREE-PHASE SYSTEM

A three-phase balanced 440-V, 60-Hz system has a wye-connected per-phase load impedance of $22\angle 37^\circ\ \Omega$. Determine the total power absorbed by the load.

Calculation Procedure

1. Determine the Line-to-Neutral Voltage

The given voltage class, 440, is presumed to be the line-to-line voltage. The line-to-neutral voltage is $440\ \text{V}/\sqrt{3} = 254\ \text{V}$.

2. Determine the Current per Phase

The current per phase for a wye connection is the line current $= V_{\text{phase}}/Z_{\text{phase}} = 254\angle 0^\circ\ \text{V}/22\angle 37^\circ\ \Omega = 11.547\angle -37^\circ\ \text{A}$.

3. Calculate the Power per Phase

The power per phase $P_{\text{phase}} = V_{\text{phase}} I_{\text{phase}} \cos\theta = (254\angle 0^\circ\ \text{V})(11.547\ \text{A}) \cos 37^\circ = 2342.3\ \text{W}$.

4. Calculate the Total Power

The total power is $P_{\text{total}} = 3P_{\text{phase}} = (3)(2342.3\ \text{W}) = 7027\ \text{W}$. Alternatively, $P_{\text{total}} = \sqrt{3}V_L I_L \cos\theta = (\sqrt{3})(440\ \text{V})(11.547\ \text{A}) \cos 37^\circ = 7027\ \text{W}$.

Related Calculations. This problem illustrates two approaches to finding the total power in a three-phase circuit. It is assumed herein that the system is balanced. For unbalanced systems, symmetrical components are used and the analysis is done separately for the zero-sequence, positive-sequence, and negative-sequence networks.

DIVISION OF LOAD BETWEEN TRANSFORMERS IN PARALLEL

Two single-phase transformers are connected in parallel on both the high- and low-voltage sides and have the following characteristics: transformer A, 100 kVA, 2300/120 V, 0.006-Ω resistance, and 0.025-Ω leakage reactance referred to the low-voltage side; transformer B, 150 kVA, 2300/115 V, 0.004-Ω resistance, and 0.015-Ω leakage reactance referred to the low-voltage side. A 125-kW load at 0.85 pf (power factor), lagging, is connected to the low-voltage side, with 125 V at the terminal of the transformers. Determine the primary voltage and the current supplied by each transformer.

Calculation Procedure

1. Determine the Admittance Y_A of Transformer A

The impedance referred to the low-voltage side is given as $0.006 + j0.025\ \Omega$. Convert the impedance from rectangular form to polar form ($0.0257\underline{/76.5°}$) and take the reciprocal to change the impedance into an admittance ($38.90\underline{/-76.5°}$); in rectangular form this is equal to $9.08 - j37.82$ S.

2. Determine the Admittance Y_B of Transformer B

The impedance referred to the low-voltage side is given as $0.004 + j0.015\ \Omega$. Convert the impedance from rectangular form to polar form ($0.0155\underline{/75.07°}$) and take the reciprocal to change the impedance into an admittance ($64.42\underline{/-75.07°}$); in rectangular form this is equal to $16.60 - j62.24$ S.

3. Determine the Total Admittance of the Paralleled Transformers

The total admittance, $\mathbf{Y}_{\text{total}}$, is the sum of the admittance of transformer A and that of transformer B, namely $25.68 - j100.06$ S, as referred to the low-voltage side of the transformers. In polar form this is $103.3\underline{/-75.6°}$.

4. Determine the Total Current

For the total current, use the equation $P/(V \cos \theta) = 125{,}000\text{ W}/(125\text{ V})(0.85) =$ 1176 A.

5. Calculate the Primary Voltage

Assume that the secondary voltage is the reference phasor, $125 + j0$ V. The load current lags the secondary voltage by an angle whose cosine is 0.85, the power-factor angle. The load current in polar form is $1176\underline{/-31.79°}$ and in rectangular form is $999.6 - j619.5$ A. The primary voltage is obtained from the equation $\mathbf{V}_1 = a_{\text{total}}\mathbf{V}_2 + (a_{\text{total}}\mathbf{I}_1/\mathbf{Y}_{\text{total}})$, where a_{total} = turns ratio for the two transformers in parallel. This same equation may be developed into another form: $\mathbf{V}_1 = (\mathbf{V}_2\mathbf{Y}_{\text{total}} + \mathbf{I}_1)/(\mathbf{Y}_{\text{total}}/a_{\text{total}})$, where $\mathbf{Y}_{\text{total}}/a_{\text{total}} = Y_A/a_A + Y_B/a_B$, a_A is the turns ratio of transformer A (i.e., $a_A = 2300/120 = 19.17$), and a_B is the turns ratio of transformer B (i.e., $a_B = 2300/115 = 20.0$). $\mathbf{Y}_{\text{total}}/a_{\text{total}} = (9.08 - j37.82)/19.17 + (16.60 - j62.24)/20 = 0.47 - j1.97 + 0.83 - j3.11 = 1.30 - j5.08 = 5.24\underline{/-75.65°}$. Thus, $\mathbf{V}_1 = [(125)(25.68 - j100.06) + 999.6 - j619.5]/5.24\underline{/-75.65°} = 13{,}787\underline{/-72.2°}/5.24\underline{/-75.65°} = 2631\underline{/3.45°}$ V (primary-side voltage of two transformers in parallel).

6. Calculate the Division of Load between Transformers

First, calculate the current through transformer A using the equation:

$$\mathbf{I}_A = \frac{\mathbf{I}_{\text{total}} + \mathbf{V}_1\,[(\mathbf{Y}_{\text{total}}/a_A) - (\mathbf{Y}_{\text{total}}/a_{\text{total}})]}{\mathbf{Z}_A\mathbf{Y}_{\text{total}}}$$

$$= \frac{(999.6 - j619.5) + 2631\angle 3.45^\circ\{[(25.68 - j100.06)/19.17] - (1.30 - j5.08)\}}{(0.0257\angle 76.5^\circ)(103.3\angle -75.6^\circ)}$$

$$= 1494\angle -41.02^\circ/(0.0257\angle 76.5^\circ)(103.3\angle 75.6^\circ)$$

$$= 562.6\angle 41.92^\circ = 418.7 - j375.9 \text{ A}$$

Similarly, the current through transformer B is calculated as

$$\mathbf{I}_B = \frac{\mathbf{I}_{\text{total}} + \mathbf{V}_1\,[(\mathbf{Y}_{\text{total}}/a_B) - (\mathbf{Y}_{\text{total}}/a_{\text{total}})]}{\mathbf{Z}_B\mathbf{Y}_{\text{total}}}$$

$$= 643.9\angle -22.97^\circ = 593 - j251 \text{ A}$$

The portion of load carried by transformer A = (562.6 A)(0.120 kV) = 67.5 kVA = 67.5 percent of its rating of 100 kVA. The portion of load carried by transformer B = (643.9 A)(0.115 kV) = 74.0 kVA, or (74.0 kVA)(100 percent)/150 kVA = 49.3 percent of its rating of 150 kVA.

Related Calculations. This problem illustrates the approach to calculating the division of load between two single-phase transformers. The method is applicable to three-phase transformers under balanced conditions and may be extended to three-phase transformers under unbalanced conditions by applying the theory of symmetrical components.

PHASE SHIFT IN WYE-DELTA TRANSFORMER BANKS

A three-phase 300-kVA, 2300/23,900-V, 60-Hz transformer is connected wye-delta as shown in Fig. 10.14. The transformer supplies a load of 280 kVA at a power factor of 0.9

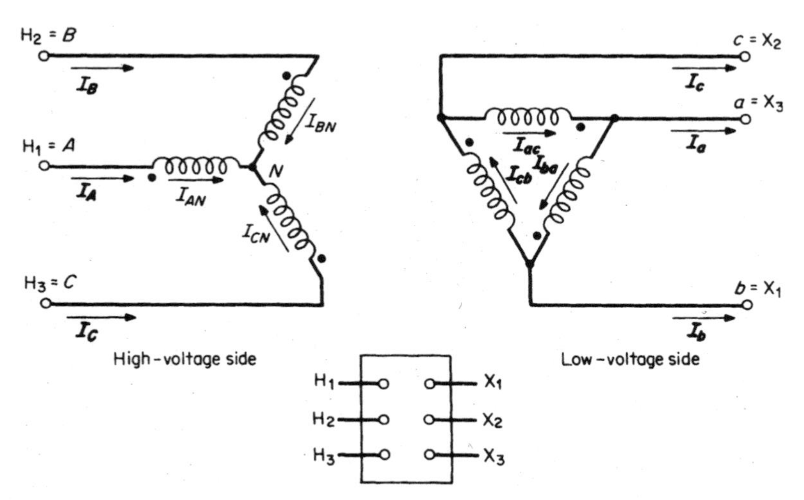

FIGURE 10.14 Wye-delta transformer connections for phase-shift analysis.

lagging. The supply voltages (line-to-neutral) on the high-voltage side are $\mathbf{V}_{AN} = 13{,}800\angle 0^\circ$, $\mathbf{V}_{BN} = 13{,}800\angle -120^\circ$, and $\mathbf{V}_{CN} = 13{,}800\angle 120^\circ$. Find the phasor voltages and currents for the transformer.

Calculation Procedure

1. Determine the Turns Ratio

The turns ratio of windings that are drawn in parallel (linked magnetically) is the same for each pair of windings in a three-phase transformer. In this case, the turns ratio $a = 13{,}800\ \text{V}/2300\ \text{V} = 6$.

2. Determine the Low-Voltage-Side Phasor Voltages

On the delta side, the voltage $\mathbf{V}_{ab} = \mathbf{V}_{AN}/a = 13{,}800\angle 0^\circ/6 = 2300\angle 0^\circ$ V, $\mathbf{V}_{bc} = \mathbf{V}_{BN}/a = 13{,}800\angle -120^\circ/6 = 2300\angle -120^\circ$ V, and $\mathbf{V}_{ca} = \mathbf{V}_{CN}/a = 13{,}800\angle 120^\circ/6 = 2300\angle 120^\circ$ V.

3. Determine the Line-to-Line Voltages on the High-Voltage Side

The supply voltages were given on the basis of the line-to-neutral voltages in each phase. The line-to-line voltages are obtained from phasor additions: $\mathbf{V}_{AB} = \mathbf{V}_{AN} - \mathbf{V}_{BN} = 13{,}800\angle 0^\circ - 13{,}800\angle -120^\circ = 13{,}800 + 6900 + j11{,}951 = 20{,}700 + j11{,}951 = 23{,}900\angle 30^\circ$ V, $\mathbf{V}_{BC} = \mathbf{V}_{BN} - \mathbf{V}_{CN} = 13{,}800\angle -120^\circ - 13{,}800\angle 120^\circ = -6900 - j11{,}951 + 6900 - j11{,}951 = -j23{,}900 = 23{,}900\angle -90^\circ$ V, and $\mathbf{V}_{CA} = \mathbf{V}_{CN} - \mathbf{V}_{AN} = 13{,}800\angle 120^\circ - 13{,}800\angle 0^\circ = -6900 + j11{,}951 - 13{,}800 = -20{,}700 + j11{,}951 = 23{,}900\angle 150^\circ$ V. The high-voltage line-to-line voltages lead the low-voltage line-to-line voltages by 30°; that is the convention for both wye-delta and delta-wye connections.

4. Determine the Load Current on the High-Voltage Side

Use the equation: $I_L = \text{VA}/3V_{\text{phase}}$. The magnitude of the current on the high-voltage side is $\mathbf{I}_{AN} = \mathbf{I}_{BN} = \mathbf{I}_{CN} = 300{,}000\ \text{VA}/(3)(13{,}800\ \text{V}) = 7.25$ A. The power-factor angle $= \cos^{-1} 0.9 = 25.84^\circ$. With a lagging power-factor angle, the current lags the respective voltage by that angle. $\mathbf{I}_{AN} = 7.25\angle -25.84^\circ$ A, $\mathbf{I}_{BN} = 7.25\angle -20^\circ - 25.84^\circ = 7.25\angle -145.84^\circ$ A, and $\mathbf{I}_{CN} = 7.25\angle 120^\circ - 25.84^\circ = 7.25\angle 94.16^\circ$ A.

5. Determine the Load Current on the Low-Voltage Side

$\mathbf{I}_{ab} = a\mathbf{I}_{AN}$, $\mathbf{I}_{bc} = a\mathbf{I}_{BN}$, and $\mathbf{I}_{ca} = a\mathbf{I}_{CN}$. Thus, $\mathbf{I}_{ab} = (6)(7.25\angle -25.84^\circ) = 43.5\angle -25.84^\circ$ A, $\mathbf{I}_{bc} = (6)(7.25\angle -145.84^\circ) = 43.5\angle -145.84^\circ$ A, and $\mathbf{I}_{ca} = (6)(7.25\angle 94.16^\circ) = 43.5\angle 94.16^\circ$ A. The line currents may be obtained by phasor addition. Thus, $\mathbf{I}_a = \mathbf{I}_{ac} - \mathbf{I}_{ba} = \mathbf{I}_{ab} - \mathbf{I}_{ca} = 43.5\angle -25.84^\circ - 43.5\angle 94.16^\circ = 39.15 - j18.96 + 3.16 - j43.4 = 42.31 - j62.36 = 75.4\angle -55.84^\circ$ A. $\mathbf{I}_b = \mathbf{I}_{ba} - \mathbf{I}_{cb} = \mathbf{I}_{bc} - \mathbf{I}_{ab} = 43.5\angle -145.84^\circ - 43.5\angle -25.84^\circ = -35.99 - j24.43 - 39.15 + j18.96 = -75.14 - j5.47 = 75.4\angle -175.84^\circ$ A. $\mathbf{I}_c = \mathbf{I}_{cb} - \mathbf{I}_{ac} = \mathbf{I}_{ca} - \mathbf{I}_{bc} = 43.5\angle 94.16^\circ - 43.5\angle -145.84^\circ = -3.16 + j43.4 + 35.99 + j24.43 = 32.83 + j67.83 = 75.4\angle 64.16^\circ$ A.

6. Compare Line Currents on High- and Low-Voltage Sides

Note that the line currents on the low-voltage side (delta) lag the line currents on the high-voltage side (wye) by 30° in each phase, respectively.

High-voltage side	Low-voltage side
$\mathbf{I}_A = \mathbf{I}_{AN} = 7.25\angle -25.84^\circ$ A	$\mathbf{I}_a = 75.4\angle -55.84^\circ$ A
$\mathbf{I}_B = \mathbf{I}_{BN} = 7.25\angle -145.84^\circ$ A	$\mathbf{I}_b = 75.4\angle -175.84^\circ$ A
$\mathbf{I}_C = \mathbf{I}_{CN} = 7.25\angle 94.16^\circ$ A	$\mathbf{I}_c = 75.4\angle 64.16^\circ$ A

Related Calculations. This problem illustrates the method of calculating the phase shift through transformer banks that are connected wye-delta or delta-wye. For the standard connection shown, the voltage and current on the high-voltage side will lead the respective quantities on the low-voltage side by 30°; it makes no difference whether the connection is wye-delta or delta-wye.

CALCULATION OF POWER, APPARENT POWER, REACTIVE POWER, AND POWER FACTOR

For the equivalent circuit shown in Fig. 10.15, generator A is supplying 4000 W at 440 V and 0.90 pf, lagging. The motor load is drawing 9500 W at 0.85 pf, lagging. Determine the apparent power in VA, the reactive power in VARS, the power in W, and power factor of each generator and of the motor load.

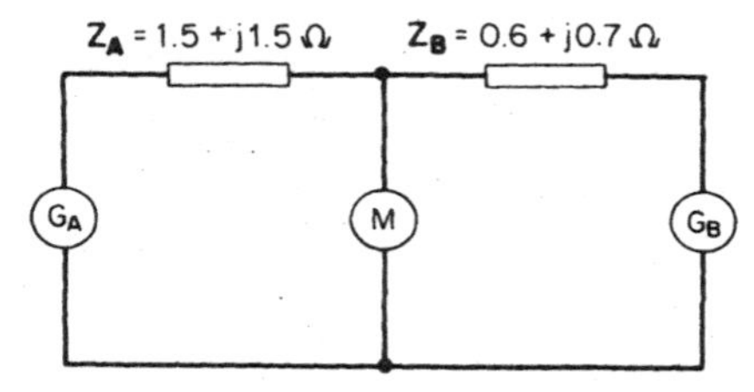

FIGURE 10.15 Equivalent circuit of motor load fed by two generators.

Calculation Procedure

1. Determine the Current from Generator A

Use the equation: $P = EI \cos \theta$ and solve for the current. $I_A = P_A/E_A \cos \theta_A$ = 4000 W/(440 V)(0.90) = 10.1 A.

2. Determine the Reactive Power and Apparent Power Values for Generator A

To find the apparent power, divide the power in watts by the power factor: VA = W/pf = 4000 W/0.90 = 4444.4 VA. The reactive power is W tan ($\cos^{-1}$ pf) = (4000 W) tan ($\cos^{-1}$ 0.90) = 1937.3 VARS.

3. Calculate the Losses in Z_A

The power loss in watts in Z_A is $I_A^2 R_A = (10.1)^2(1.5\ \Omega) = 153$ W. The loss in VARS in Z_A is $I_A^2 X_A = (10.1)^2(1.5\ \Omega) = 153$ VARS. The quantities of power and reactive power delivered to the motor from generator A are 4000 − 153 = 3847 W, and 1937.3 − 153 = 1784.3 VARS, respectively.

4. Calculate the Requirements of the Motor Load

At the motor, the apparent power = W/pf = 9500 W/0.85 = 11,176.5 VA. The reactive power = W tan ($\cos^{-1}$ pf) = (9500 W) tan ($\cos^{-1}$ 0.85) = 5887.6 VARS.

5. Calculate the Motor Requirements to Be Supplied by Generator B

The power (in watts) needed from generator B = 9500 − 3847 = 5653 W. The reactive power (in VARS) needed from generator B = 5887.6 VARS − 1784.3 VARS = 4103.3 VARS. The apparent power = $\sqrt{5653^2 + 4103.3^2}$ = 6985.2 VA.

6. Calculate the Current from Generator B

First calculate the voltage at the load as represented by the voltage drop from generator A. $V_{\text{load}} = \text{VA}_{\text{genA}}/I_A = \sqrt{3847^2 + 1784.3^2}/10.1 = 419.9$ V. Therefore, I_B = 6985.2 VA/419.9 V = 16.64 A.

7. Calculate the Losses in Z_B

The loss in watts in $Z_B = I_B^2 R_B = (16.64)^2(0.6) = 166.1$ W. The loss in VARS in $Z_B = I_B^2 X_B = (16.64)^2(0.7) = 193.8$ VARS. Thus, generator B must supply $5653 + 166.1 = 5819.1$ W, $4103.3 + 193.8 = 4297.1$ VARS, and $\sqrt{(5819.1)^2 + (4297.1)^2} = 7233.7$ VA.

8. Summarize the Results

	Apparent power, VA	Reactive power, VAR	Power, W
Generator A	4,444	1,937	4,000
Generator B	7,234	4,297	5,819
Motor load	11,177	5,888	9,500
Losses:			
Z_A	0.216	0.153	0.153
Z_B	0.255	0.193	0.166

It is noted that the sum of the KW and KVARS values generated equals the sum of the KW and KVAR values of the motor load and losses in Z_A and Z_B, respectively.

Related Calculations. This problem illustrates the relationship of KVA, KVAR, and KW values and power factor. For the three-phase case, the factor $\sqrt{3}$ must be included. For example, $P = \sqrt{3}EI \cos \theta$ and VARS $= \sqrt{3}\, EI \sin \theta$.

THE POWER DIAGRAM: REAL AND APPARENT POWER

A series circuit, as shown in Fig. 10.16, has a resistance of 5 Ω and a reactance $X_L = 6.5$ Ω. The power source is 120 V, 60 Hz. Determine the real power, the reactive power, the apparent power, and the power factor; construct the power diagram.

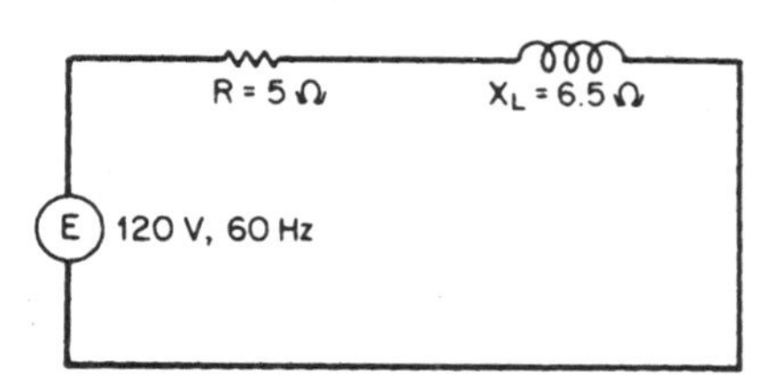

FIGURE 10.16 Series power circuit used to illustrate real, reactive, and apparent power.

Calculation Procedure

1. Determine the Impedance of the Circuit

The total impedance of the circuit is $\mathbf{Z_T} = R + jX_L = 5 + j6.5 = 8.2\angle 52.43^\circ$ Ω.

2. Determine the Phasor Current

The phasor current is determined from the equation: $\mathbf{I} = \mathbf{E}/\mathbf{Z_T} = 120\angle 0^\circ$ V/8.2 $\angle 52.43^\circ$ Ω $= 14.63\angle -52.43^\circ$ A $= 8.92 - j11.60$ A.

3. Calculate the Real Power

Use the equation: $P = EI \cos \theta = (120)(14.63) \cos 52.43^\circ = 1070.0$ W, where $\cos \theta =$ power factor $= 0.61$.

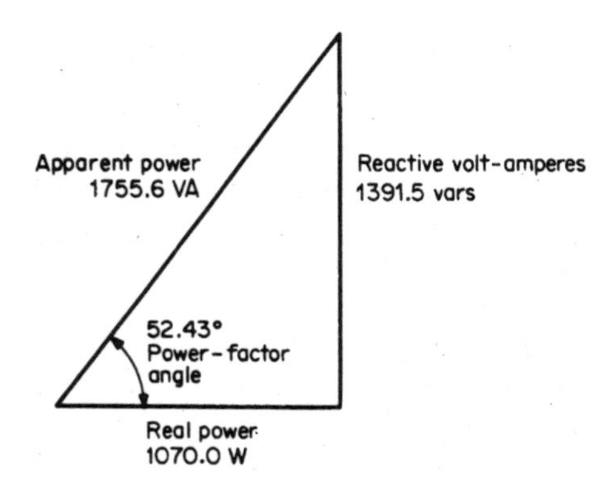

FIGURE 10.17 The power diagram (or power triangle).

4. Calculate the Reactive Power

Use the equation: VARS = $EI \sin \theta = (120)(14.63) \sin 52.43° = 1391.5$ VARS. The $\sin \theta$ is sometimes known as the reactive factor; in this case, $\sin \theta = \sin 52.43° = 0.793$.

5. Calculate the Apparent Power

The apparent power is $EI = (120)(14.63) = 1755.6$ VA. Also, it may be calculated from the equation: real power + j(reactive power) = 1070.0 W + j1391.5 VARS = $1755.6\underline{/52.43°}$ VA.

6. Construct the Power Diagram

The power diagram is shown in Fig. 10.17. Also, it is evident that power factor = real power/apparent power = 1,070.0 W/1755.6 VA = 0.61.

Related Calculations. The procedure is similar for a three-phase circuit where the appropriate equation for real power is $P = \sqrt{3} E_{line} I_{line} \cos \theta$. The apparent power is VA = $\sqrt{3} E_{line} I_{line}$, and the reactive power is VARS = $\sqrt{3} E_{line} I_{line} \sin \theta$.

STATIC CAPACITORS USED TO IMPROVE POWER FACTOR

A manufacturing company has several three-phase motors that draw a combined load of 12 kVA, 0.60 power factor, lagging, from 220-V mains. It is desired to improve the power factor to 0.85 lagging. Calculate the line current before and after adding the capacitors and determine the reactive power rating of the capacitors that are added.

Calculation Procedure

1. Calculate the Line Current before Adding the Capacitors

Use the equation: $P = \sqrt{3} E_{line} I_{line} \cos \theta$, solving for I_{line}. $I_{line} = P/\sqrt{3} E_{line} \cos \theta$ = 7.2 kW/($\sqrt{3}$)(220 V)(0.6) = 31.5 A.

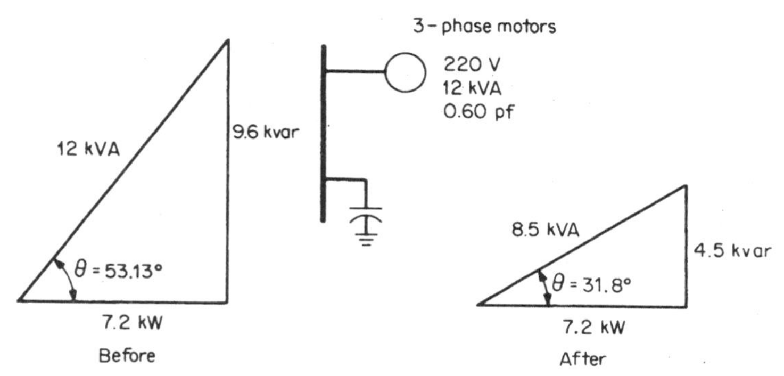

FIGURE 10.18 The effect of adding capacitors to improve a lagging power factor.

2. *Calculate the Line Current after Adding the Capacitors*

Use the same equation as in the first step. $I_{\text{line}} = P/\sqrt{3}E_{\text{line}} \cos \theta = 7.2 \text{ kW}/(\sqrt{3})(220 \text{ V})(0.85) = 22.2$ A. Note that the improved power factor has resulted in less current; 31.5 A is required at 0.60 pf, and 22.2 A is required at 0.85 pf.

3. *Calculate the Rating in kVAR of Capacitors*

The reactive power in kVAR needed in the circuit before the capacitors are added is determined by multiplying the apparent power in kVA by the sine of the power-factor angle: (12 kVA) sin $(\cos^{-1} 0.60) = 9.6$ kVAR. The real power = kVA × power factor = (12 kVA)(0.60) = 7.2 kW; this remains constant before and after the capacitor bank is added (see Fig. 10.18). The rating of the capacitor bank is the difference in kVAR between the before and after cases, namely, 9.6 kVAR − 4.5 kVAR = 5.1 kVAR at 220 V (three-phase).

Related Calculations. Lagging power factors may be improved by the addition of sources of leading reactive power; these sources are usually capacitor banks and sometimes synchronous capacitors (i.e., synchronous motors on the line that are operated in the overexcited state). In either case, leading reactive power is supplied at the installation so that the current and losses are reduced over long transmission or distribution lines.

THREE-PHASE SYNCHRONOUS MOTOR USED TO CORRECT POWER FACTOR

A three-phase synchronous motor is rated 2200 V, 100 kVA, 60 Hz. It is being operated to improve the power factor in a factory; overexcitation results in its operating at a leading power factor of 0.75 at 75 kW full load. (1) Determine the reactive power in kVAR delivered by the motor, (2) draw the power triangle, and (3) determine the power factor needed for the motor to deliver 50 kVAR with a 75-kW load.

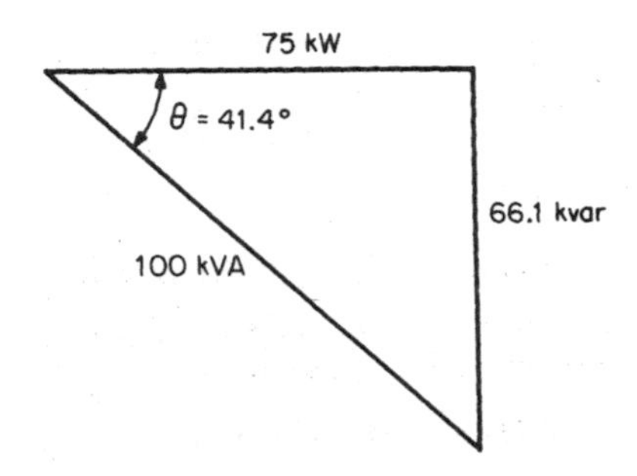

FIGURE 10.19 Power triangle for synchronous motor operating at a leading power factor.

Calculation Procedure

1. Calculate the Power-Factor Angle

The power-factor angle is determined from $\theta = \cos^{-1}$ (pf) $= \cos^{-1}$ 0.75 = 41.4°, leading.

2. Calculate the Number of kVAR Delivered

Reactive power = kVA × sin θ = (kW/cos θ) sin θ = kW tan θ = 75 tan 41.4° = 66.1 kVAR. The power triangle is shown in Fig. 10.19.

3. Calculate the Power Factor for Delivering 50 kVAR

If the load is 75 kW and the reactive power is 50 kVAR, $\theta = \tan^{-1}$ (reactive power/load kW) $= \tan^{-1}$ (50 kVAR/75 kW) = 33.7°. The power factor = cos 33.7° = 0.832 leading.

Related Calculations. Synchronous motors operated in the overexcited mode deliver reactive power instead of absorbing it. In this manner they act as synchronous capacitors and are useful in improving power factor. In a factory having many induction motors (a lagging power-factor load), it is often desirable to have a few machines act as synchronous capacitors to improve the overall power factor.

POWER CALCULATION OF TWO-WINDING TRANSFORMER CONNECTED AS AN AUTOTRANSFORMER

A single-phase transformer is rated 440/220 V, 5 kVA, at 60 Hz. Calculate the capacity in kVA of the transformer if it is connected as an autotransformer to deliver 660 V on the load side with a supply voltage of 440 V.

Calculation Procedure

1. Draw the Connection Diagram

See Fig. 10.20 for the connection diagram.

2. Calculate Rated Current in the Windings

The rated current in the H_1-H_2 winding is I_H = 5000 VA/440 V = 11.36 A. The rated current in the X_1-X_2 winding is I_X = 5000 VA/220 V = 22.73 A.

3. Calculate the Apparent Power Delivered to the Load

The voltage across the load is 660 V and the current through it is 22.73 A. Thus, the load receives (660 V)(22.73 A) = 15,000 VA. For a 2:1-ratio two-winding transformer connected as an autotransformer, the delivered apparent power is tripled.

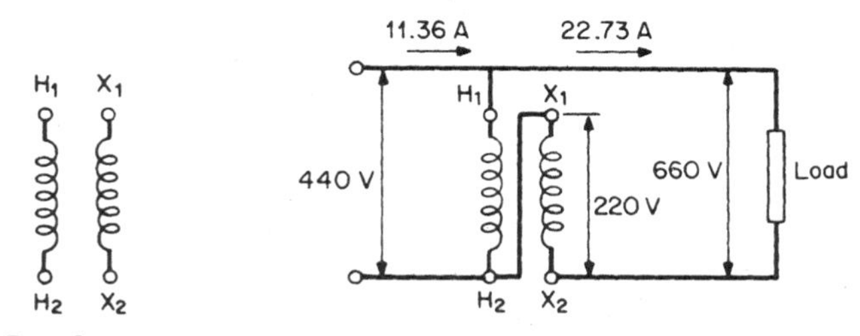

FIGURE 10.20 Single-phase transformer connected as autotransformer.

Related Calculations. It can be shown that for autotransformer connection of single-phase transformers the capacity increases with a certain relationship to the turns ratio. For example, for a 1:1-ratio two-winding transformer the autotransformer connection produces a 1:2 ratio and the capacity is doubled; for a 2:1-ratio two-winding transformer the autotransformer connection becomes a 2:3-ratio transformer and the capacity is tripled; for a 3:1-ratio two-winding transformer the autotransformer connection becomes a 3:4-ratio transformer and the capacity increases 4 times. A serious consideration in using the autotransformer connection between two transformer windings is the solid electrical connection between them that may necessitate special insulation requirements.

TWO-WATTMETER METHOD FOR DETERMINING THE POWER OF A THREE-PHASE LOAD

In the circuit shown in Fig. 10.21, $\mathbf{E}_{ab} = 220\angle 0°$ V, $\mathbf{E}_{bc} = 220\angle -120°$ V, and $\mathbf{E}_{ca} = 220\angle 120°$ V. Determine the reading of each wattmeter and show that the sum of W_1 and W_2 = the total power in the load.

Calculation Procedure

1. Calculate I_a

Use the relation: $\mathbf{I}_a = \mathbf{I}_{ac} - \mathbf{I}_{ba}$, where the delta load currents are each equal to the voltage divided by the load impedance. $\mathbf{E}_{ac} = -\mathbf{E}_{ca} = 220\angle -60°$, and $\mathbf{E}_{ba} = -\mathbf{E}_{ab} = 220\angle 180°$. $\mathbf{I}_a = (220\angle -60°/(12 + j14) - (220\angle 180°)/(16 + j10) = 11.96\angle -109.4° - 11.64\angle 148° = -3.97 - j11.28 + 9.87 - j6.17 = 5.9 - j17.45 = 18.42\angle -71.32°$ A.

2. Calculate I_c

Use the relation $\mathbf{I}_c = \mathbf{I}_{cb} - \mathbf{I}_{ac} = 220\angle 60°/(8 - j8) - 220\angle -60°/(12 + j14) = 19.47\angle 105° - 11.96\angle -109.4° = -5.04 + j18.81 + 3.97 + j11.28 = -1.07 + j30.09 = 30.11\angle 92.04°$ A.

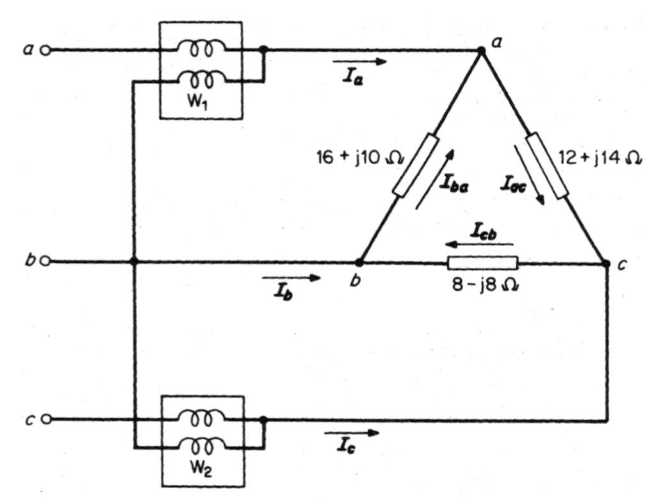

FIGURE 10.21 Two-wattmeter connection for measuring three-phase power.

3. Calculate Power Reading of Wattmeter 1

Use the equation: $W_1 = \mathbf{E_{ab}I_a} \cos\theta = (220\text{ V})(18.42\text{ A})\cos(0° + 71.32°) = 1297.9$ W.

4. Calculate Power Reading of Wattmeter 2

Use the equation $W_2 = \mathbf{E_{cb}I_c} \cos\theta = (220\text{ V})(30.11\text{ A})\cos(60° - 92.04°) = 5615.2$ W.

5. Calculate the Total Power of the Load

The total power $= W_1 + W_2 = 1297.9 + 5615.2 = 6913.1$ W.

6. Calculate the Power Loss in Each Load Resistance

Use the equation $P = I^2R$. For resistance in delta-side *ac*, $R = 12\ \Omega$; $P = (11.96)^2(12) = 1716.5$ W. For resistance in delta-side *ab*, $R = 16\ \Omega$; $P = (11.64)^2(16) = 2167.8$ W. The total power absorbed by the load $= 1716.5 + 3032.6 + 2167.8 = 6916.9$ W, which compares with the sum of W_1 and W_2 (the slight difference results from rounding off the numbers throughout the problem solution).

Related Calculations. When the two-wattmeter method is used for measuring power in a three-phase circuit, care must be given to the proper connections of the voltage and current coils. The common potential lead must be connected to the phase not being used for the current-coil connections.

OPEN-DELTA TRANSFORMER OPERATION

A new electric facility is being installed wherein the loading is light at this time. The economics indicate that two single-phase transformers be connected in open-delta and that the future load be handled by installing the third transformer to close the delta whenever the load growth reaches the necessary level. Each single-phase transformer is rated 2200/220 V, 200 kVA, 60 Hz.

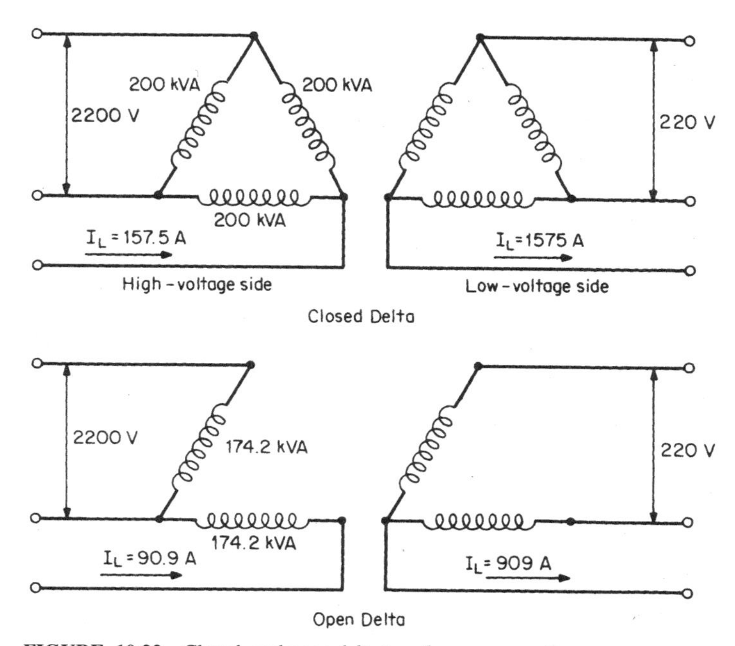

FIGURE 10.22 Closed- and open-delta transformer connections.

If for the open-delta connection (using two transformers) each transformer is fully loaded and for the closed-delta connection (using three transformers) each transformer is fully loaded, what is the ratio of three-phase loading of the open delta as compared with the closed delta? Refer to Fig. 10.22.

Calculation Procedure

1. Calculate the Line Currents for Closed Delta

Use the three-phase equation kVA $= \sqrt{3}E_{\text{line}}I_{\text{line}}$ solving for I_{line}. On the high-voltage side, $I_{\text{line}} = 600{,}000\ \text{VA}/(\sqrt{3})(2200\ \text{V}) = 157.5\ \text{A}$. Similarly, on the low-voltage side, $I_{\text{line}} = 600{,}000\ \text{VA}/(\sqrt{3})(220\ \text{V}) = 1575\ \text{A}$.

2. Calculate the Transformer Currents for the Closed Delta

The transformer currents $= I_{\text{line}}/\sqrt{3} = 157.5/\sqrt{3} = 90.9$ A on the high-voltage side or 909 A on the low-voltage side. Another method of determining these currents is to divide the transformer kVA value by the voltage; thus [200 kVA (per transformer)]/2.2 kV = 90.9 A on the high-voltage side, and [200 kVA (per transformer)]/0.22 kV = 909 A on the low-voltage side.

3. Calculate the Transformer Current for the Open-Delta

In the open-delta case, the magnitude of the line currents must be the same as that of the transformer currents. Again, if each transformer is carrying 200 kVA, the line currents on the high-voltage side must be 90.9 A and on the low-voltage side, 909 A. That is, the line currents are reduced from 157.5 to 90.9 A, and 1575 A is reduced to 909 A, respectively.

4. Calculate the Apparent Power of the Open Delta

The three-phase apparent power of the open delta = $\sqrt{3}E_L I_L$ = ($\sqrt{3}$)(2200 V)(90.9 A) = 346.4 kVA. A comparison may be made with the 600 kVA loading of the closed delta; thus, (346.4 kVA)(100 percent)/600 kVA = 57.7 percent. The 346.4 kVA is carried by only two transformers (174.2 kVA each). Thus, each transformer in the open-delta case carries only (174.2 kVA)(100 percent)/200 kVA = 87.1 percent of the loading for the closed-delta case.

Related Calculations. Calculations involving the open-delta connection are often required, sometimes for load growth planning as in this problem, and sometimes for contingency planning. For the open delta, the flow-through three-phase kVA value is reduced to 57.7 percent of the closed-delta value. The individual transformer loading is reduced to 87.1 percent on the two transformers that form the open delta as compared with 100 percent for each of three transformers in a closed delta. The basic assumptions in these calculations are a balanced symmetrical system of voltages and currents and negligible transformer impedances.

REAL AND REACTIVE POWER OF A THREE PHASE MOTOR IN PARALLEL WITH A BALANCED-DELTA LOAD

In the system shown in Fig. 10.23, the balanced-delta load has an impedance on each side of 12 − j10 Ω. The three-phase induction motor is rated 230 V, 60 Hz, 8 kVA at 0.72 power factor, lagging. It is wye-connected. Determine (1) the line current, (2) the power factor, and (3) the power requirements of the combined load.

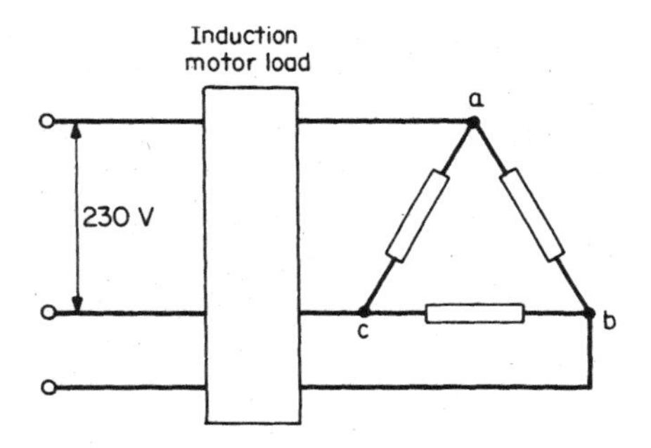

FIGURE 10.23 Three-phase motor load paralleled with a balanced-delta load.

Calculation Procedure

1. Calculate the Motor Current

The motor is wye-connected; thus, the line current is equal to the phase current. Use the equation: $I_{\text{line}} = \text{kVA}/\sqrt{3}\text{kV}_{\text{line}}$ = 8 kVA($\sqrt{3}$)(0.230 kV) = 20.08 A.

2. Calculate the Equivalent Impedance of the Motor

Use the equation: $(V_{\text{line}}/\sqrt{3})/I_{\text{phase}} = Z_{\text{phase}} = (230/\sqrt{3})/20.08 = 6.61$ Ω. The angle $\theta = \cos^{-1}(\text{pf}) = \cos^{-1} 0.72 = 43.95°$. $R_{\text{motor}} = Z \cos\theta = 6.61 \cos 43.95° = 4.76$ Ω. $X_{\text{motor}} = Z \sin\theta = 6.61 \sin 43.95° = 4.59$ Ω. Thus, $\mathbf{Z}_{\text{motor}} = 4.76 + j4.59$ Ω per phase $= 6.61\underline{/43.95°}$ Ω per phase.

3. Convert Balanced-Delta Load to Equivalent Wye

Use the equations $\mathbf{Z}_a = (\mathbf{Z}_{ab}\mathbf{Z}_{ca})/(\mathbf{Z}_{ab} + \mathbf{Z}_{bc} + \mathbf{Z}_{ca}) = (12 - j10)^2/(3)(12 - j10) = \frac{1}{3}(12 - j10) = 4.00 - j3.33 = 5.21\underline{/-39.8°}$ Ω.

4. Combine the Two Balanced-Wye Loads

The motor load of $4.76 + j4.59\ \Omega$ per phase is to be paralleled with the load of $4.00 - j3.33\ \Omega$ per phase. Use the equation $\mathbf{Z}_{total} = \mathbf{Z}_{motor}\mathbf{Z}_{load}/(\mathbf{Z}_{motor} + \mathbf{Z}_{load}) = (6.61\underline{/43.95°})(5.21\underline{/-39.8°})/(4.76 + j4.59 + 4.00 - j3.33) = 3.88\underline{/-4.0°}\ \Omega$ per phase (equivalent-wye connection).

5. Calculate the Total Line Current and Power Factor

Use the equation $I_{total} = (V_{line}/\sqrt{3})/Z_{total} = (230\text{ V}/\sqrt{3})/3.88\ \Omega = 34.2$ A. The pf = $\cos\theta = \cos -4.0° = 0.999$ (essentially, unity power factor).

6. Calculate the Power Requirements

Use the equation $W_{total} = 3I^2_{total}R_{total} = (3)(34.2\text{ A})^2(3.88\ \Omega) = 13{,}615$ W. The apparent power $= \sqrt{3}V_{line}I_{line} = (\sqrt{3})(230\text{ V})(34.2\text{ A}) = 13{,}624$ VA. The apparent power is approximately equal to the real power; the power factor is approximately unity.

Related Calculations. This problem illustrates one technique in handling paralleled loads. There are other approaches to this problem. For example, the current could be obtained for each load separately and then added to obtain a combined, or total, current.

BIBLIOGRAPHY

El-Hawary, Mohamed E. 1995. *Electrical Power Systems.* New York: IEEE.

Fitzgerald, Arthur E., Charles Kingsley, Jr., and Stephen D. Umans. 1990. *Electric Machinery*. New York: McGraw-Hill.

Gönen, Turan. 1988. *Electric Power Transmission Engineering*. New York: Wiley.

Grainger, John J., and William D. Stevenson, Jr. 1994. *Power System Analysis*. New York: McGraw-Hill.

Institute of Electrical and Electronics Engineers. 1993. *Electrical Machinery.* New York: IEEE.

Knable, Alvin H. 1982. *Electrical Power Systems Engineering*. Malabar, Fla.: Kreiger.

Matsch, Leander W. 1986. *Electromagnetic and Electromechanical Machines,* 3rd ed. New York: Harper & Row.

Nasar, Syed A. 1995. *Electric Machines and Power Systems.* New York: McGraw-Hill.

Nasar, Syed A. 1998. *Schaum's Outline of Theory and Problems of Electric Machines and Electromechanics.* New York: McGraw-Hill.

Nasar, Syed A. and L. E. Unnewehr. *Electromechanics and Electric Machines,* 2nd ed. New York: Wiley.

Pansini, Anthony J. 1975. *Basic Electrical Power Transmission*. Rochelle Park, N.J.: Hayden Book Co.

Slemon, Gordon R. and A. Straughen. 1992. *Electric Machines and Drives*. Reading, Mass.: Addison-Wesley Publishing Co.

Stein, Robert. 1979. *Electric Power System Components*. New York: Van Nostrand Reinhold.

Sullivan, Robert L. 1977. *Power System Planning*. New York: McGraw-Hill.

Wildi, Théodore. 2000. *Electrical Machines, Drives, and Power Systems*. Saddle River, N.J.: Prentice Hall.

LOAD-FLOW ANALYSIS IN POWER SYSTEMS

Badrul H. Chowdhury
Professor
Electrical & Computer Engineering Department
University of Missouri-Rolla

INTRODUCTION

The load-flow problem models the nonlinear relationships among bus power injections, power demands, and bus voltages and angles, with the network constants providing the circuit parameters. It is the heart of most system-planning studies and also the starting point for transient and dynamic stability studies. This section provides a formulation of the load-flow problem and its associated solution strategies. An understanding of the fundamentals of three-phase systems is assumed, including per-unit calculations, complex power relationships, and circuit-analysis techniques.

There are two popular numerical methods for solving the power-flow equations. These are the Gauss-Seidel (G-S) and the Newton-Raphson (N-R) Methods (Grainger and Stevenson, 1994; Elgerd, 1982; Glover and Sharma, 1994). The N-R method is superior to the G-S method because it exhibits a faster convergence characteristic. However, the N-R method suffers from the disadvantage that a "flat start" is not always possible since the solution at the beginning can oscillate without converging toward the solution. In order to avoid this problem, the load-flow solution is often started with a G-S algorithm followed by the N-R algorithm after a few iterations.

There is also an approximate but faster method for the load-flow solution. It is a variation of the N-R method, called the fast-decoupled method, which was introduced by Stott and Alsac (1974). We will not be covering this method in this section.

NOMENCLATURE

S_D = complex power demand

S_G = complex power generation

S = complex bus power

P_D = real power demand, MW

P_G = real power generation

P = real bus power

Q_D = reactive power demand in MVAR

Q_G = reactive power generation

Q = reactive bus power

$|V|$ = bus voltage magnitude

δ = bus voltage angle

$\hat{V}$ = complex voltage

B = shunt susceptance

y_p = shunt admittance

y_s = series admittance

R = series resistance

X = series reactance

Z_s = series impedance

X_G = synchronous reactance

Y_{ii} = driving point admittance at bus i

Y_{ij} = transfer admittance between busses i and j

$|Y_{ij}|$ = magnitude of Y_{ij}

γ_{ij} = angle of Y_{ij}

E = synchronous machine-generated voltage

$\hat{I}$ = complex current

Y_{bus} = bus admittance matrix

DEVELOPING POWER-FLOW EQUATIONS

A two-bus example, shown in Fig. 11.1, is used to simplify the development of the power-flow equations. The system consists of two busses connected by a transmission line. One can observe that there are six electrical quantities associated with each bus: P_D, P_G, Q_D, Q_G, $|V|$, and δ. This is the most general case, in which each bus is shown to have both generation and demand. In reality, not all busses will have power generation. The impedance diagram of the two-bus system is shown in Fig. 11.2. The transmission line is represented by a π-model and the synchronous generator is represented by a source behind a synchronous reactance. The loads are assumed to be constant impedance for the sake of representing them on the impedance diagram. Typically, the load is represented by a constant power device, as shown in subsequent figures.

Figure 11.3 is the same as Fig. 11.2 but with the generation and demand bundled together to represent "bus power," which represents bus power injections. Bus power is defined as

$$S_1 = S_{G1} - S_{D1} = (P_{G1} - P_{D1}) + j(Q_{G1} - Q_{D1}) \tag{1}$$

and

$$S_2 = S_{G2} - S_{D2} = (P_{G2} - P_{D2}) + j(Q_{G2} - Q_{D2}) \tag{2}$$

Also, injected current at bus 1 is

$$\hat{I}_1 = \hat{I}_{G1} - \hat{I}_{D1} \tag{3}$$

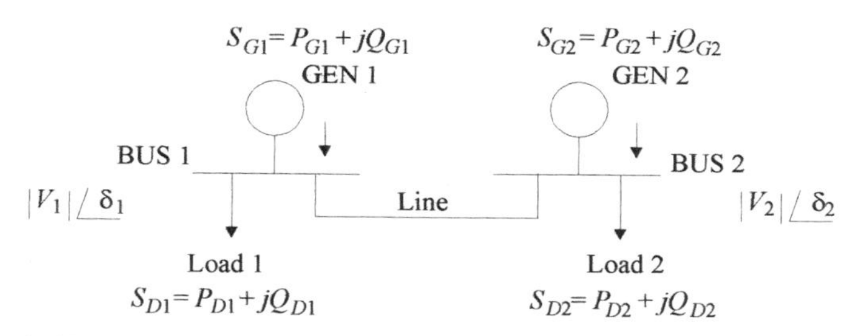

FIGURE 11.1 A two-bus power system.

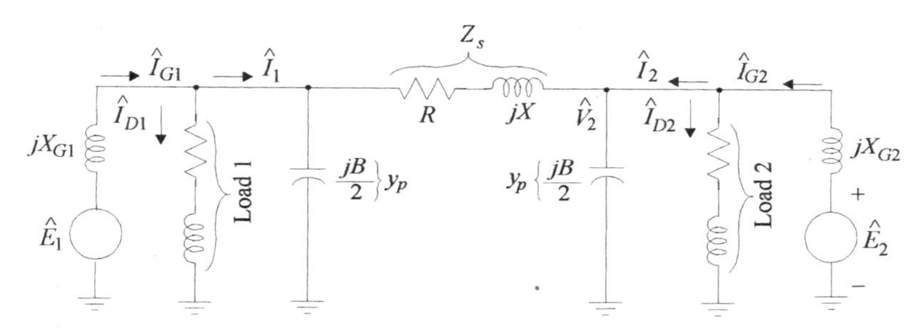

FIGURE 11.2 Impedance diagram for the two-bus power system.

and injected current at bus 2 is

$$\hat{I}_2 = \hat{I}_{G2} - \hat{I}_{D2} \tag{4}$$

All quantities are assumed to be per unit. Then, since

$$S_1 = \hat{V}_1\hat{I}_1^* \Rightarrow P_1 + jQ_1 = \hat{V}_1\hat{I}_1^* \Rightarrow (P_1 - jQ_1) = \hat{V}_1^*\hat{I}_1 \tag{5}$$

and, since

$$S_2 = \hat{V}_2\hat{I}_2^* \Rightarrow P_2 + jQ_2 = \hat{V}_2\hat{I}_2^* \Rightarrow (P_2 - jQ_2) = \hat{V}_2^*\,\hat{I}_2 \tag{6}$$

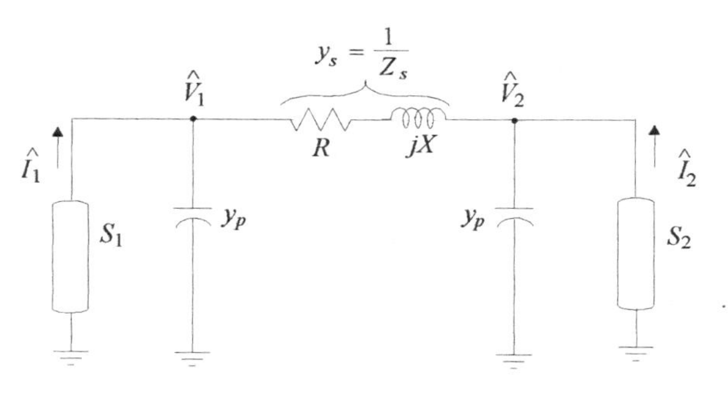

FIGURE 11.3 Bus powers with transmission line π-model for the two-bus system.

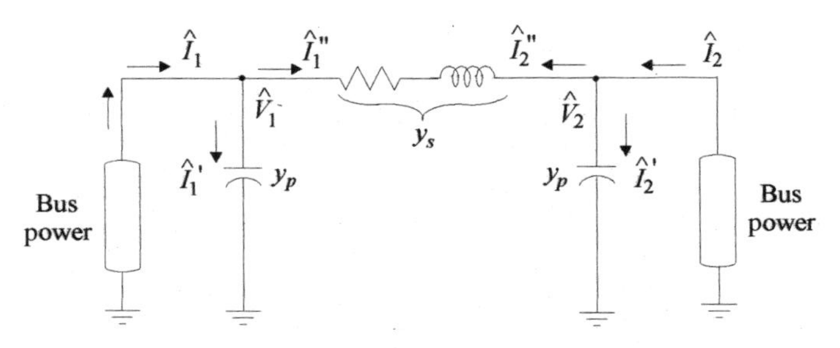

FIGURE 11.4 Current flows in the network model.

Let us define current flows in the circuit as shown in Fig. 11.4. Therefore, at bus 1

$$\hat{I}_1 = \hat{I}_1' + \hat{I}_1''$$

$$= \hat{V}_1 y_p + (\hat{V}_1 - \hat{V}_2) y_s$$

$$\hat{I}_1 = (y_p + y_s)\hat{V}_1 + (-y_s)\hat{V}_2 \tag{7}$$

$$\therefore \hat{I}_1 = Y_{11}\hat{V}_1 + Y_{12}\hat{V}_2 \tag{8}$$

where $Y_{11} \triangleq$ sum of admittances connected at bus 1 $= y_p + y_s$ (9)

$Y_{12} \triangleq$ negative of the admittance between busses 1 and 2 $= -y_s$ (10)

Similarly, at bus 2

$$\hat{I}_2 = \hat{I}_2' + \hat{I}_2''$$

$$= \hat{V}_2 y_p + (\hat{V}_2 - \hat{V}_1) y_s$$

$$\hat{I}_2 = (-y_s)\hat{V}_1 + (y_p + y_s)\hat{V}_2 \tag{11}$$

$$\hat{I}_2 = Y_{21}\hat{V}_1 + Y_{22}\hat{V}_2 \tag{12}$$

$Y_{22} \triangleq$ sum of all admittances connected at bus 2 $= y_p + y_s$ (13)

$Y_{21} \triangleq$ negative of the admittance between busses 2 and 1 $= -y_s = Y_{12}$ (14)

Hence, for the two-bus power system, the current injections are

$$\begin{bmatrix} I_1 \\ I_2 \end{bmatrix} = \begin{bmatrix} Y_{11} & Y_{12} \\ Y_{21} & Y_{22} \end{bmatrix} \begin{bmatrix} \hat{V}_1 \\ \hat{V}_2 \end{bmatrix} \tag{15}$$

In matrix notation,

$$I_{\text{bus}} = Y_{\text{bus}} V_{\text{bus}} \tag{16}$$

The two-bus system can easily be extended to a larger system. Consider an n-bus system. Figure 11.5*a* shows the connections from bus 1 of this system to all the other busses. Figure 11.5*b* shows the transmission line models. Equations (5) through (16) that were

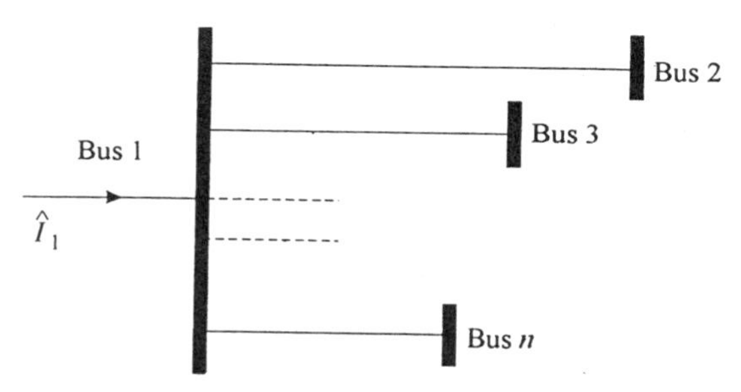

FIGURE 11.5a Extending the analysis to an n-bus system.

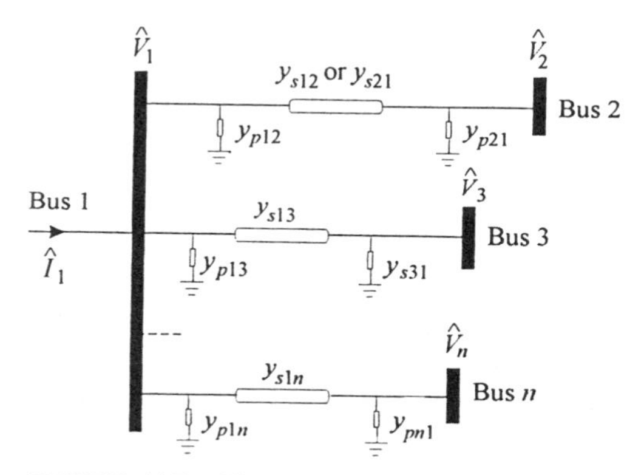

FIGURE 11.5b The π-model for the n-bus system.

derived for the two-bus system can now be extended to represent the n-bus system. This is shown next.

$$\hat{I}_1 = \hat{V}_1 y_{p12} + \hat{V}_1 y_{p13} + \cdots + \hat{V}_1 y_{p1n} + (\hat{V}_1 - \hat{V}_2) y_{s12} + (\hat{V}_1 - \hat{V}_3) y_{s13} + \cdots + (\hat{V}_1 - \hat{V}_2) y_{s1n}$$

$$= (y_{p12} + y_{p13} + \cdots + y_{p1n} + y_{s12} + y_{s13} + \cdots + y_{s1n})\hat{V}_n - y_{s12}\hat{V}_2 - y_{s13}\hat{V}_3 + \cdots - y_{s1n}\hat{V}_n \quad (17)$$

$$\hat{I}_1 = Y_{11}\hat{V}_1 + Y_{12}\hat{V}_2 + Y_{13}\hat{V}_3 + \cdots + Y_{1n}\hat{V}_n \quad (18)$$

where $$Y_{11} = (y_{p12} + y_{p13} + \cdots + y_{p1n} + y_{s12} + y_{s13} + \cdots + y_{s1n}) \quad (19)$$

$= $ sum of all admittances connected to bus 1

$$Y_{12} = -y_{S12};\ Y_{13} = -y_{S13};\ Y_{1n} = -y_{s1n} \quad (20)$$

$$\therefore \hat{I}_1 = \sum_{j=1}^{n} Y_{ij}\hat{V}_j \quad (21)$$

Also, extending the power Eq. (5) to an n-bus system,

$$P_1 - jQ_1 = \hat{V}_1^* I_1 = \hat{V}_1^* \sum_{j=1}^{n} Y_{1j} \hat{V}_j \tag{22}$$

Equation (22) can be written for any generic bus i:

$$P_i - jQ_i = \hat{V}_i^* \sum_{j=1}^{n} Y_{ij} \hat{V}_j \qquad i = 1, 2, \ldots, n \tag{23}$$

Equation (23) represents the nonlinear power-flow equations. Equation (15) can also be rewritten for an n-bus system:

$$\begin{bmatrix} \hat{I}_1 \\ \hat{I}_2 \\ \vdots \\ \hat{I}_n \end{bmatrix} = \begin{bmatrix} Y_{11} & Y_{12} & \cdots & Y_{1n} \\ Y_{21} & Y_{22} & \cdots & Y_{2n} \\ \vdots & \vdots & & \vdots \\ Y_{n1} & Y_{n2} & \cdots & Y_{nn} \end{bmatrix} \begin{bmatrix} \hat{V}_1 \\ \hat{V}_2 \\ \vdots \\ \hat{V}_n \end{bmatrix} \cdots \tag{24}$$

or

$$I_{\text{bus}} = Y_{\text{bus}} V_{\text{bus}} \tag{25}$$

where

$$Y_{\text{bus}} = \begin{bmatrix} Y_{11} & Y_{12} & \cdots & Y_{1n} \\ Y_{21} & Y_{22} & \cdots & Y_{2n} \\ \vdots & \vdots & & \vdots \\ Y_{n1} & Y_{n2} & \cdots & Y_{nn} \end{bmatrix} = \text{bus admittance matrix} \tag{26}$$

POWER-FLOW SOLUTION

Let us take a generic bus as shown in Fig. 11.6. As mentioned earlier, each bus has six quantities or variables associated with it. They are $|V|$, δ, P_G, Q_G, P_D, and Q_D. Assuming that there are n busses in the system, there would be a total of $6n$ variables.

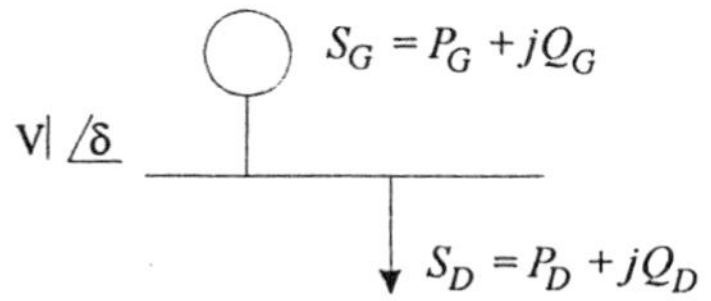

$S_G = P_G + jQ_G$

$V| \angle \delta$

$S_D = P_D + jQ_D$

FIGURE 11.6 A generic bus.

TABLE 11.1 Bus Classifications

Bus classification	Prespecified variables	Unknown variables
Slack or swing	$\lvert V\rvert$, δ, P_D, Q_D	P_G, Q_G
Voltage-controlled	$\lvert V\rvert$, P_G, P_D, Q_D	δ, Q_G
Load	P_G, Q_D, P_D, Q_D	$\lvert V\rvert$, δ

The power-flow Eq. (23) can be resolved into the real and reactive parts as follows:

$$\therefore P_i = \text{Real}\left[\hat{V}_i^* \sum_{j=1}^{n} Y_{ij}\hat{V}_j\right] \qquad i = 1, 2, \ldots, n \tag{27}$$

$$Q_i = -\text{Imag}\left[\hat{V}_i^* \sum_{j=1}^{n} Y_{ij}\hat{V}_j\right] \qquad i = 1, 2, \ldots, n \tag{28}$$

Thus, there are $2n$ equations and $6n$ variables for the n-bus system. Since there cannot be a solution in such case, $4n$ variables have to be prespecified. Based on parameter specifications, we can now classify the busses as shown in Table 11.1.

We will now describe the methods used in solving the power-flow equations.

Gauss-Seidel (G-S) Algorithm for Power-Flow Solution

Note that the power-flow equations are

$$P_i - jQ_i = \hat{V}_i^* \sum_{j=1}^{n} Y_{ij}\hat{V}_j \qquad i = 1, 2, \ldots, n \tag{23}$$

$$= \hat{V}_i^* Y_{ii}\hat{V}_i + \sum_{j=1, j\neq i}^{n} \hat{V}_i^* Y_{ij}\hat{V}_j \tag{29}$$

$$\Rightarrow \hat{V}_i^* Y_{ii}\hat{V}_i = (P_i - jQ_i) - \sum_{j=1, j\neq i}^{n} \hat{V}_i^* Y_{ij}\hat{V}_j \tag{30}$$

$$\Rightarrow Y_{ii}\hat{V}_i = \frac{P_i - jQ_i}{\hat{V}_i^*} - \sum_{j=1, j\neq i}^{n} Y_{ij}\hat{V}_j \tag{31}$$

$$\Rightarrow \hat{V}_i = \frac{\dfrac{P_i - jQ_i}{\hat{V}_i^*} - \displaystyle\sum_{j=1, j\neq i}^{n} Y_{ij}\hat{V}_j}{Y_{ii}} \tag{32}$$

Also, from (29),

$$P_i = \text{Re}\left[\hat{V}_i^* Y_{ii}\hat{V}_i + \sum_{j=1, j\neq i}^{n} \hat{V}_i^* Y_{ij}\hat{V}_j\right] \tag{33}$$

and

$$Q_i = -\text{Imag}\left[\hat{V}_i^* Y_{ii} \hat{V}_i + \sum_{j=1, j\neq i}^{n} \hat{V}_i^* Y_{ij} \hat{V}_j \right] \tag{34}$$

where Y_{ij} are elements of the Y_{bus}.

The G-S Algorithm

Step 0. Formulate and Assemble $\mathbf{Y}_{bus}$ in Per Unit

Step 1. Assign Initial Guesses to Unknown Voltage Magnitudes and Angles

$$|V| = 1.0, \quad \delta = 0$$

Step 2a. For Load Buses, Find $\hat{\mathbf{V}}_i$ form Eq. (32)

$$V_i^{(k+1)} = \left[(P_i - jQ_i) / V_i^{*(k)} - \sum_{j=1, j\neq i}^{n} Y_{ij} \hat{V}_j^{(k)} \right] / Y_{ii}$$

where k = iteration no. For voltage-controlled busses, find $\hat{V}_i$ using (34) and (32) together. That is, find Q_i first.

$$Q_i^{(k+1)} = -\text{Imag}\left[\hat{V}_i^{*(k)} \left\{ V_i^{(k)} Y_{ii} + \sum_{j=1, j\neq i}^{n} Y_{ij} V_j^{(k)} \right\} \right]$$

Then

$$V_i^{(k+1)} = \left[(P_i - jQ_i) / \hat{V}_i^{*(k)} - \sum_{j=1, j\neq i}^{n} Y_{ij} \hat{V}_j^* \right] / Y_{ii}$$

However, $|V_i|$ is specified for voltage-controlled busses. So, $V_i^{(k+1)} = |V_i, \text{spec}| \angle \delta_{i,\text{calc}}^{(k+1)}$

In using Eqs. (32) and (34), one must remember to use the most recently calculated values of bus voltages in each iteration. So, for example, if there are five busses in the system being studied, and one has determined new values of bus voltages at busses 1–3, then during the determination of bus voltage at bus 4, one should use these newly calculated values of bus voltages at 1, 2, and 3; busses 4 and 5 will have the values from the previous iteration.

Step 2b. For Faster Convergence, Apply Acceleration Factor to Load Buses

$$V_{i,\text{acc}}^{(k+1)} = V_{i,\text{acc}}^{(k)} + \alpha (V_i^{(k)} - V_{i,\text{acc}}^{(k)}) \tag{35}$$

where α = acceleration factor.

Step 3. Check Convergence

$$|\text{Re}|\, [\hat{V}_i^{(k+1)}] - \text{Re}\, [\hat{V}_i^{(k)}] \leq \varepsilon \tag{36}$$

That is, the absolute value of the difference of the real part of the voltage between successive iterations should be less than a tolerance value ε. Typically, $\varepsilon \leq 10^{-4}$, and also,

$$|\text{Imag}\, [\hat{V}_i^{(k+1)}] - \text{Imag}[\hat{V}_i^{(k)}]| \leq \varepsilon \tag{37}$$

That is, the absolute value of the difference of the imaginary value of the voltage should be less than a tolerance value ε.

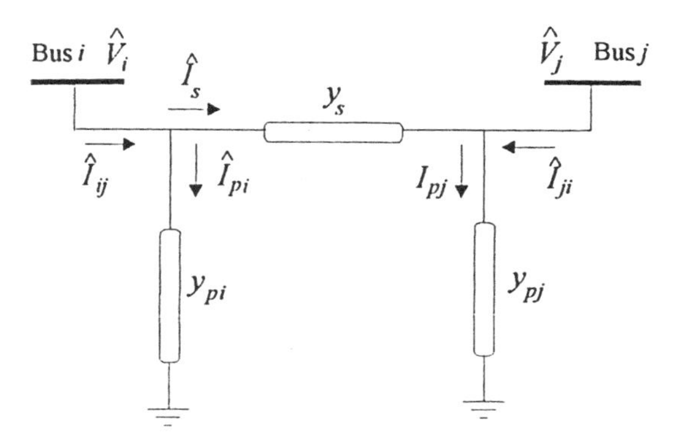

FIGURE 11.7 A two-bus system illustrating line-flow computation.

If the difference is greater than tolerance, return to Step 3. If the difference is less than tolerance, the solution has converged; go to Step 4.

Step 4. Find Slack Bus Power P_G ***and*** Q_G ***from Eqs. (27) and (28)***

Step 5. Find All Line Flows as Described in the Next Section

Computing Line Flows. As the last step in any power-flow solution, one has to find the line flows. This is illustrated by the two-bus system shown in Fig. 11.7. Line current, $\hat{I}_{ij}$, at bus i is defined positive in the direction $i \rightarrow j$.

$$\hat{I}_{ij} = \hat{I}_s + \hat{I}_{pi} (\hat{V}_i - \hat{V}_j)_{ys} + \hat{V}_i y_{pi} \tag{38}$$

Let S_{ij}, S_{ji} be line powers defined positive into the line at bus i and j, respectively.

$$S_{ij} = P_{ij} + jQ_{ij} = \hat{V}_i \hat{I}_{ij}^* = \hat{V}_i \left(\hat{V}_i^* - \hat{V}_j^* \right) y_s^* + |V_i|^2 y_{pi}^* \tag{39}$$

$$S_{ji} = P_{ji} + jQ_{ji} = \hat{V}_j \hat{I}_{ji}^* = \hat{V}_j \left(\hat{V}_j^* - \hat{V}_i^* \right) y_s^* + |V_j|^2 y_{pi}^* \tag{40}$$

The power loss in line $(i - j)$ is the algebraic sum of the power flows determined from (39) and (40).

$$\therefore S_{Lij} = S_{ij} + S_{ji} \tag{41}$$

Newton-Raphson (N-R) Method for Power-Flow Solution

The Newton-Raphson method enables us to replace the nonlinear set of power-flow equations of (23) with a linear set. We will show this after the basis for the method is explained.

The Taylor series expansion of a function $f(x)$ of a single variable, x, around the point $(x - a)$ is given by

$$f(x) = f(a) + (x - a) \left. \frac{\partial f}{\partial x} \right|_a + \frac{(x - a)^n}{2!} \left. \frac{\partial^2 f}{\partial x^2} \right|_a + \cdots + \frac{(x - a)^n}{n!} \frac{\partial^n f}{\partial x^n} + \Re_n \tag{42}$$

where $\left. \frac{\partial f}{\partial x} \right|_a$ = value of the derivative evaluated at $x = a$.

The series converges if $\lim_{n \rightarrow \infty} \Re_n = 0$.

If $(x - a) \ll 1$ then we can neglect the higher-order terms and write (42) as

$$f(x) \approx f(a) + (x - a)\left.\frac{\partial f}{\partial x}\right|_a \tag{43}$$

For a function of n variables, one can expand around the point: $(x_1 - a_1)$, $(x_2 - a_2)$, $(x_n - a_n)$ with $(x_k - a_k) \ll 1$ and $k = 1, 2, \ldots, n$. Then, Eq. (42) becomes

$$f(x_1, x_2, \ldots, x_n) \approx f(a_1, a_2, \ldots, a_n) + (x_1 - a_1)\left.\frac{\partial f}{\partial x_1}\right|_{a_1} + (x_2 - a_2)\left.\frac{\partial f}{\partial x_2}\right|_{a_2} + \cdots + (x_n - a_n)\left.\frac{\partial f}{\partial x_n}\right|_{a_n} \tag{44}$$

Let us consider a set of nonlinear equations, each a function of n variables:

$$\begin{aligned} f_1(x_1, x_2, \ldots, x_n) &= y_1 \\ f_2(x_1, x_2, \ldots, x_n) &= y_2 \\ &\vdots \\ f_n(x_1, x_2, \ldots, x_n) &= y_n \end{aligned} \tag{45}$$

or

$$f_k(x_1, x_2, \ldots, x_n) = y_k \qquad k = 1, 2, \ldots, n$$

Assume initial values $x_k^{(0)}$ and some correction, Δx_k, which when added to $x_k^{(0)}$ yield $x_k^{(1)}$. When $x_k^{(0)}$ are close to the solution, x_k, the Δx_k^s are small.

Using the approximate Taylor's series, we have

$$f_k(x_1, x_2, \ldots, x_n) = f_k(x_1^{(0)}, x_2^{(0)}, \ldots, x_n^{(0)}) + \Delta x_1\left.\frac{\partial f_k}{\partial x_1}\right|_{x_1^{(0)}} + \Delta x_2\left.\frac{\partial f_k}{\partial x_2}\right|_{x_2^{(0)}} + \cdots + \Delta x_n\left.\frac{\partial f_k}{\partial x_n}\right|_{x_n^{(0)}} = y_k \qquad k = 1, 2, \ldots, n \tag{46}$$

or, in matrix form,

$$\begin{bmatrix} y_1 - f_1(x_1^{(0)}, x_2^{(0)}, \ldots, x_n^{(0)}) \\ y_2 - f_2(x_1^{(0)}, x_2^{(0)}, \ldots, x_n^{(0)}) \\ \vdots \\ y_n - f_n(x_1^{(0)}, x_2^{(0)}, \ldots, x_n^{(0)}) \end{bmatrix} = \begin{bmatrix} \left.\frac{\partial f_1}{\partial x_1}\right|_{x_1^{(0)}} & \left.\frac{\partial f_1}{\partial x_2}\right|_{x_2^{(0)}} & \cdots & \left.\frac{\partial f_1}{\partial x_n}\right|_{x_n^{(0)}} \\ \left.\frac{\partial f_2}{\partial x_1}\right|_{x_1^{(0)}} & \left.\frac{\partial f_2}{\partial x_1}\right|_{x_2^{(0)}} & \cdots & \left.\frac{\partial f_2}{\partial x_n}\right|_{x_n^{(0)}} \\ \vdots & \vdots & \vdots & \\ \left.\frac{\partial f_n}{\partial x_1}\right|_{x_1^{(0)}} & \left.\frac{\partial f_n}{\partial x_2}\right|_{x_2^{(0)}} & \cdots & \left.\frac{\partial f_n}{\partial x_n}\right|_{x_n^{(0)}} \end{bmatrix} \begin{bmatrix} \Delta x_1 \\ \Delta x_2 \\ \vdots \\ \Delta x_n \end{bmatrix} \tag{47}$$

or

$$[\Delta U]^{(0)} = [J]^{(0)}[\Delta X]^{(0)} \tag{48}$$

where $[J]$ is the Jacobian matrix.

$$\therefore [\Delta X] = ([J]^{(0)})^{-1}[\Delta U]^{(0)} \tag{49}$$

To continue iteration, find $[X]^{(1)}$ from

$$[X]^{(1)} = [X]^{(0)} + [\Delta X]^{(0)} \tag{50}$$

Generally,

$$[X]^{(k+1)} = [X]^{(k)} + [\Delta X]^{(k)} \tag{51}$$

where k = iteration number.

The Newton-Raphson Method Applied to Power-Flow Equations

The N-R method is typically applied on the real form of the power-flow equations:

$$\left.\begin{aligned} P_i &= \sum_{k=1}^{n} |V_i||V_k||y_{ik}|\cos(\delta_k - \delta_i + \gamma_{ik}) = f_{ip} & (52)\\ Q_i &= -\sum_{k=1}^{n} |V_i||V_k||y_{ik}|\sin(\delta_k - \delta_i + \gamma_{ik}) = f_{iq} & (53) \end{aligned}\right\} \quad i = 1, \ldots, n$$

Assume, temporarily, that all busses, except bus 1, are of the "load" type. Thus, the unknown parameters consist of the $(n - 1)$ voltage phasors, $\hat{V}_2, \ldots, \hat{V}_n$. In terms of real variables, these are:

Angles	$\delta_2, \delta_3, \ldots, \delta_n$	$(n - 1)$ variables
Magnitudes	$\lvert V_2\rvert, \lvert V_3\rvert, \ldots, \lvert V_n\rvert$	$(n - 1)$ variables

Rewriting (47) for the power-flow equations,

$$\begin{bmatrix} \Delta P_2^{(0)} \\ \Delta P_3^{(0)} \\ \vdots \\ \Delta P_n^{(0)} \\ \Delta Q_2^{(0)} \\ \Delta Q_3^{(0)} \\ \vdots \\ \Delta Q_n^{(0)} \end{bmatrix} = \begin{bmatrix} \left.\frac{\partial f_{2p}}{\partial \delta_2}\right|^{(0)} & \left.\frac{\partial f_{2p}}{\partial \delta_3}\right|^{(0)} & \cdots & \left.\frac{\partial f_{2p}}{\partial \delta_n}\right|^{(0)} & \left.\frac{\partial f_{2p}}{\partial |V_2|}\right|^{(0)} & \cdots & \left.\frac{\partial f_{2p}}{\partial |V_n|}\right|^{(0)} \\ \left.\frac{\partial f_{3p}}{\partial \delta_2}\right|^{(0)} & \left.\frac{\partial f_{3p}}{\partial \delta_3}\right|^{(0)} & \cdots & \left.\frac{\partial f_{3p}}{\partial \delta_n}\right|^{(0)} & \left.\frac{\partial f_{3p}}{\partial |V_2|}\right|^{(0)} & \cdots & \left.\frac{\partial f_{3p}}{\partial |V_n|}\right|^{(0)} \\ \vdots & \vdots & \vdots & \vdots & \vdots & \vdots & \vdots \\ \left.\frac{\partial f_{np}}{\partial \delta_2}\right|^{(0)} & \left.\frac{\partial f_{np}}{\partial \delta_3}\right|^{(0)} & \cdots & \left.\frac{\partial f_{np}}{\partial \delta_n}\right|^{(0)} & \left.\frac{\partial f_{np}}{\partial |V_2|}\right|^{(0)} & \cdots & \left.\frac{\partial f_{np}}{\partial |V_n|}\right|^{(0)} \\ \left.\frac{\partial f_{2q}}{\partial \delta_2}\right|^{(0)} & \left.\frac{\partial f_{2q}}{\partial \delta_3}\right|^{(0)} & \cdots & \left.\frac{\partial f_{2q}}{\partial \delta_n}\right|^{(0)} & \left.\frac{\partial f_{2q}}{\partial |V_2|}\right|^{(0)} & \cdots & \left.\frac{\partial f_{2q}}{\partial |V_n|}\right|^{(0)} \\ \left.\frac{\partial f_{3q}}{\partial \delta_2}\right|^{(0)} & \left.\frac{\partial f_{3q}}{\partial \delta_3}\right|^{(0)} & \cdots & \left.\frac{\partial f_{3q}}{\partial \delta_n}\right|^{(0)} & \left.\frac{\partial f_{3q}}{\partial |V_2|}\right|^{(0)} & \cdots & \left.\frac{\partial f_{3q}}{\partial |V_n|}\right|^{(0)} \\ \vdots & \vdots & \vdots & \vdots & \vdots & \vdots & \vdots \\ \left.\frac{\partial f_{nq}}{\partial \delta_2}\right|^{(0)} & \left.\frac{\partial f_{nq}}{\partial \delta_3}\right|^{(0)} & \cdots & \left.\frac{\partial f_{nq}}{\partial \delta_n}\right|^{(0)} & \left.\frac{\partial f_{nq}}{\partial |V_2|}\right|^{(0)} & \cdots & \left.\frac{\partial f_{nq}}{\partial |V_n|}\right|^{(0)} \end{bmatrix} \begin{bmatrix} \Delta \delta_2^{(0)} \\ \Delta \delta_3^{(0)} \\ \vdots \\ \Delta \delta_n^{(0)} \\ \Delta |V_2|^{(0)} \\ \Delta |V_3|^{(0)} \\ \vdots \\ \Delta |V_n|^{(0)} \end{bmatrix} \tag{54}$$

Before proceeding any further, we need to account for voltage-controlled busses. For every voltage-controlled bus in the system, delete the corresponding row and column from the Jacobian matrix. This is done because the mismatch element for a voltage-controlled bus is unknown.

Writing Eq. (54) in matrix form,

$$\underline{\Delta U}^{(0)} = \underline{J}^{(0)} \cdot \underline{\Delta X}^{(0)} \tag{55}$$

where $\Delta \underline{U}^{(0)}$ = vector of power mismatches at initial guesses

$\underline{J}^{(0)}$ = the Jacobian matrix evalutated at the initial guesses

$\Delta \underline{X}^{(0)}$ = the error vector at the zeroth iteration

The N-R Algorithm

Step 0. Formulate and Assemble $\mathbf{Y}_{bus}$ in Per Unit

Step 1. Assign Initial Guesses to Unknown Voltage Magnitudes and Angles for a Flat Start

$$|V| = 1.0, \quad \delta = 0$$

***Step 2. Determine the Mismatch Vector ΔU for Iteration* k**

***Step 3. Determine the Jacobian Matrix J for Iteration* k**

Step 4. Determine Error Vector ΔX from Eq. (55)

Set $\mathbf{X}$ at iteration $(k + 1)$: $\mathbf{X}^{(k+1)} = \mathbf{X}^{(k)} + \Delta\mathbf{X}^{(k)}$. Check if the power mismatches are within tolerance. If so, go to Step 5. Otherwise, go back to Step 2.

Step 5. Find Slack Bus Power $\mathbf{P_G}$ and $\mathbf{Q_G}$ from Eqs. (27) and (28)

Step 6. Compute Line Flows Using Eqs. (39) and (40) and the Total Line Losses from Eq. (41)

EXAMPLE OF N-R SOLUTION

Consider the three-bus system shown in Fig. 11.8. Known quantities are also shown. Given: $\hat{V}_1 1.0\angle 0°$ p.u., $|V_2| = 1.0$ p.u., $P_2 = 0.6$ p.u., $P_3 = -0.8$ p.u., $Q_3 = -0.6$.

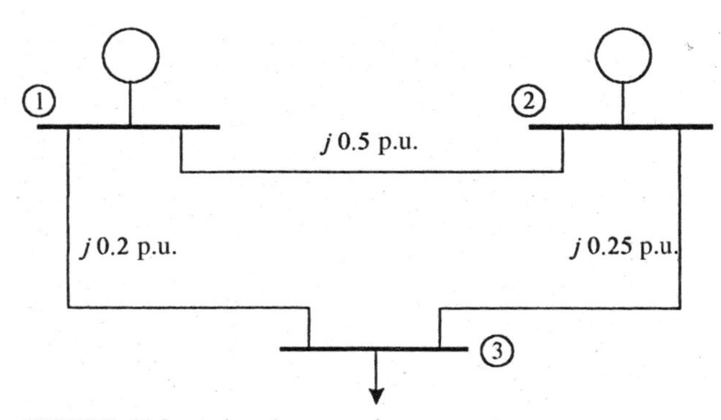

FIGURE 11.8 A three-bus example power system.

Step 0.

$$Y_{\text{bus}} = \begin{bmatrix} -j7 & j2 & j5 \\ j2 & -j6 & j4 \\ j5 & j4 & -j9 \end{bmatrix}$$

Step 1.

$$\underline{X}^{(0)} = \begin{bmatrix} \delta_2^{(0)} \\ \delta_3^{(0)} \\ |V_2|^0 \\ |V_3|^0 \end{bmatrix} = \begin{bmatrix} 0 \\ 0 \\ 1.0 \\ 1.0 \end{bmatrix}$$

Step 2.

$$P_2 = f_{2p} = |y_{21}|\,|V_2|\,|V_1| \cos(\delta_1 - \delta_2 + \gamma_{21}) + |y_{22}|\,|V_2|\,|V_2| \cos(\delta_2 - \delta_2 + \gamma_{22}) + |y_{23}|\,|V_2|\,|V_3| \cos(\delta_3 - \delta_2 + \gamma_{23}) \quad (56)$$

$$Q_2 = f_{2q} = |y_{21}|\,|V_2|\,|V_1| \sin(\delta_1 - \delta_2 + \gamma_{21}) + |y_{22}|\,|V_2|\,|V_2| \sin(\delta_2 - \delta_2 + \gamma_{22}) + |y_{23}|\,|V_2|\,|V_3| \sin(\delta_3 - \delta_2 + \gamma_{23}) \quad (57)$$

$$P_3 = f_{3p} = |y_{31}|\,|V_3|\,|V_1| \cos(\delta_1 - \delta_3 + \gamma_{31}) + |y_{32}|\,|V_3|\,|V_2| \cos(\delta_2 - \delta_3 + \gamma_{32}) + |y_{33}|\,|V_3|^2 \cos(\delta_3 - \delta_3 + \gamma_{33}) \quad (58)$$

$$Q_3 = f_{3q} = -\left[|y_{31}|\,|V_3|\,|V_1| \sin(\delta_1 - \delta_3 + \gamma_{31}) + |y_{32}|\,|V_3|\,|V_2| \sin(\delta_2 - \delta_3 + \gamma_{32}) + |y_{33}|\,|V_3|^2 \sin\gamma_{23}\right] \quad (59)$$

The specified bus powers are: $P_{2S} = 0.6$, $P_{3S} = -0.8$, $Q_{3S} = -0.6$. The calculated bus powers at this iteration are: $P_2^0 = 2.1.1 \cos(0 - 0 + 90) + 6.1.1 \cos(-90°) + 4.1.1 \cos(0 - 0 + 90) = 0$, $P_3^0 = 5.1.1 \cos 90 + 4.1.1 \cos 90 + 9.1^2 \cos(-90) = 0$, $Q_3^0 = -(5.1.1 \sin 90 + 4.1.1 \sin 90° + 9.1^2 \sin(-90°)) = 0$. Therefore, the mismatches are $\Delta P_2 = 0.6$; $\Delta P_3 = -0.8$; $\Delta Q_3 = -0.6$.

Step 3.

$$\frac{\partial f_{2p}}{\partial \delta_2} = |y_{21}|\,|V_1|\,|V_2| \sin(\delta_2 - \gamma_{21}) - |y_{23}|\,|V_2|\,|V_3| \sin(\delta_2 - \delta_3 - \gamma_{23}) \quad (60)$$

$$\frac{\partial f_{2q}}{\partial \delta_3} = |y_{23}|\,|V_2|\,|V_3| \sin(\delta_3 - \delta_2 + \gamma_{23}) \quad (61)$$

$$\left.\boxed{\begin{matrix} \dfrac{\partial f_{2q}}{\partial \delta_2}, & \dfrac{\partial f_{2q}}{\partial |V_2|}, & \dfrac{\partial f_{2q}}{\partial |V_3|}, & \dfrac{\partial f_{2q}}{\partial \delta_3} \\ \dfrac{\partial f_{2p}}{\partial |V_2|}, & \dfrac{\partial f_{3p}}{\partial |V_2|}, & \dfrac{\partial f_{3q}}{\partial |V_2|} & \end{matrix}}\right\}$$

No need to evaluate these since bus 2 is a voltage-controlled bus.

$$\frac{\partial f_{2p}}{\partial |V_3|} = |y_{23}|\,|V_2| \cos(\delta_3 - \delta_2 + \gamma_{23}) \tag{62}$$

Therefore, at the initial guesses,

$$\left.\frac{\partial f_{2p}}{\partial \delta_2}\right|^0 = 6 \qquad \left.\frac{\partial f_{2p}}{\partial \delta_3}\right|^0 = -4 \qquad \frac{\partial f_{2p}}{\partial |V_3|} = 0$$

$$\left.\frac{\partial f_{3p}}{\partial \delta_2}\right|^0 = -|y_{32}|\,|V_3|\,|V_2| \sin(\delta_2 - \delta_3 + \gamma_{32}) = -4 \tag{63}$$

$$\left.\frac{\partial f_{3p}}{\partial \delta_3}\right|^0 = -\,[|y_{31}|\,|V_3|\,|V_1| \sin(\delta_3 - \gamma_{31}) - |y_{32}|\,|V_3|\,|V_2| \sin(\delta_3 - \delta_2 + \gamma_{32})] = 9 \tag{64}$$

$$\left.\frac{\partial f_{3p}}{\partial |V_3|}\right|^0 = -\,[|y_{31}|\,|V_1| \cos(\delta_1 - \delta_3 + \gamma_{31}) + |y_{32}|\,|V_2| \cos(\delta_2 - \delta_3 + \gamma_{32}) + 2|y_{33}|\,|V_3| \cos\gamma_{33}] = 0 \tag{65}$$

$$\left.\frac{\partial f_{3q}}{\partial \delta_2}\right|^0 = -[|y_{32}|\,|V_3|\,|V_2| \cos(\delta_2 - \delta_3 + \gamma_{32})] = 0 \tag{66}$$

$$\left.\frac{\partial f_{3q}}{\partial d_2}\right|^0 = -[|y_{31}|\,|V_3|\,|V_1| \cos(\delta_3 - \delta_1 - \gamma_{31}) - |y_{32}|\,|V_3|\,|V_2| \cos(\delta_3 - \delta_2 - \gamma_{32})] = 0 \tag{67}$$

$$\left.\frac{\partial f_{3q}}{\partial |V_3|}\right|^0 = -\,[|y_{31}|\,|V_1| \sin(\gamma_1 - \gamma_3 + \gamma_{31}) + |y_{32}|\,|V_2| \sin(\gamma_2 - \gamma_3 + \gamma_{32}) + 2|y_{33}|\,|V_3| \sin\gamma_{33}] = 9 \tag{68}$$

$$\begin{bmatrix} \Delta P_2^{(0)} \\ \Delta P_3^{(0)} \\ \Delta Q_2^{(0)} \\ \Delta Q_3^{(0)} \end{bmatrix} = \begin{bmatrix} 6 & -4 & 0 & 0 \\ -4 & 9 & 0 & 0 \\ 0 & 0 & 10 & -4 \\ 0 & 0 & -4 & 9 \end{bmatrix} \begin{bmatrix} \Delta\delta_2 \\ \Delta\delta_3 \\ \Delta|V_2| \\ \Delta|V_3| \end{bmatrix}$$

After the row and column corresponding to bus 2 are eliminated:

$$\therefore \begin{bmatrix} 0.6 \\ -0.8 \\ -0.6 \end{bmatrix} = \begin{bmatrix} 6 & -4 & 0 \\ -4 & 9 & 0 \\ 0 & 0 & 0 \end{bmatrix} \begin{bmatrix} \Delta\delta_2 \\ \Delta\delta_3 \\ \Delta|V_3| \end{bmatrix}$$

Step 4.

To solve the preceding equations, one can resort to inversion of the Jacobian matrix. However, computationally, it is more efficient to apply a numerical technique such as the

Gaussian elimination technique. The latter can be found in any textbook dealing with numerical analysis. This technique is applied next.

$$\begin{bmatrix} 0.6 \\ -0.8 \\ -0.6 \end{bmatrix} = \begin{bmatrix} 6 & -4 & 0 \\ -4 & 9 & 0 \\ 0 & 0 & 0 \end{bmatrix} \begin{bmatrix} \Delta\delta_2 \\ \Delta\delta_3 \\ \Delta|V_3| \end{bmatrix} \begin{matrix} \text{Divide by 6} \\ \text{Divide by 4} \\ \, \end{matrix}$$

$$\begin{bmatrix} 0.1 \\ -0.2 \\ -0.6 \end{bmatrix} = \begin{bmatrix} 1 & -0.667 & 0 \\ -1 & 2.25 & 0 \\ 0 & 0 & 9 \end{bmatrix} \begin{bmatrix} \Delta\delta_2 \\ \Delta\delta_3 \\ \Delta|V_3| \end{bmatrix} \begin{matrix} \, \\ \text{Add this row to row 1} \\ \, \end{matrix}$$

$$\begin{bmatrix} 0.1 \\ -0.1 \\ -0.6 \end{bmatrix} = \begin{bmatrix} 1 & -0.667 & 0 \\ 0 & 1.583 & 0 \\ 0 & 0 & 9 \end{bmatrix} \begin{bmatrix} \Delta\delta_2 \\ \Delta\delta_3 \\ \Delta|V_3| \end{bmatrix} \begin{matrix} \, \\ \text{Divide by 1.583} \\ \, \end{matrix}$$

$$\begin{bmatrix} 0.1 \\ -0.063 \\ -0.6 \end{bmatrix} = \begin{bmatrix} 1 & 0.667 & 0 \\ 0 & 1 & 0 \\ 0 & 0 & 9 \end{bmatrix} \begin{bmatrix} \Delta\delta_2 \\ \Delta\delta_3 \\ \Delta|V_3| \end{bmatrix}$$

By back substitution,

$$\begin{aligned} \Delta|V_3| &= -\frac{0.6}{9} = -0.067 \leftarrow \\ \Delta\delta_3 &= -0.063 \leftarrow \\ \Delta\delta_2 &= -0.1 - 0.667 \times \Delta\gamma_3 \\ &= 0.058 \leftarrow \\ \delta_2 &= 0 + 0.058 = 0.058 \end{aligned}$$

$$\begin{aligned} |V_3|^{(1)} &= |V_2|^{(0)} + \Delta|V_3| \\ &= 1.0 - 0.067 \\ &= 0.933 \\ \delta_2 &= 0 - 0.063 = -0.063 \end{aligned}$$

Continue further iterations until convergence is achieved.

CONCLUDING REMARKS

The two solution strategies described here comprise the basic steps in a load-flow solution. The reader should be reminded that, occasionally, an off-nominal transformer, a capacitor, or other network devices also have to be modeled. Most of these models can be represented in the bus admittance matrix. Another practical consideration that one needs to bear in mind is that all generators have upper and lower limits of reactive power generation. Hence, if during a load flow iteration it is found that any one of the generators is violating its limits, then that particular bus where the generator is located is said to have lost voltage control and, thus, should be treated as a load bus in subsequent iterations.

As is obvious from the two methods, computer-based analysis is essential for obtaining accurate load-flow solutions of any realistically sized power system. A computer-based analysis typically utilizes many numerical techniques, such as optimal ordering and sparsity techniques, in order to reduce memory and storage requirements. There are several excellent load-flow programs available that are widely used by engineers in utility companies for frequent system studies. While industry-grade load-flow software tends to be very expensive, there are now many educational versions of load-flow software available that are inexpensive and quite adequate for classroom use or for studying small-scale systems.

BIBLIOGRAPHY

Elgerd, O. I. 1982. *Electric Energy Systems Theory—An Introduction*, 2nd ed. New York: McGraw-Hill.

Glover, J. D., and M. Sharma. 1994. *Power System Analysis, and Design*, 2nd ed. Boston: PWS Publishing.

Grainger, J. J., and W. D. Stevenson. 1994. *Power System Analysis*. New York: McGraw-Hill.

Stott, B., and O. Alsac. 1974. "Fast Decoupled Load Flow," *IEEE Transactions on Power Apparatus & Systems*, Vol. PAS-93, pp. 859–869.

SECTION 12

POWER-SYSTEMS CONTROL

Marija Ilić
Senior Research Scientist
Electrical Engineering and Computer Science
Massachusetts Institute of Technology

INTRODUCTION

The National Electric Reliability Council's Operating Committee (NERC-OC) is responsible for the promulgation of rules for its member companies so that power flow to consumers is reliable, especially on a regional basis. The *control area* is the basic unit recognized by NERC-OC. It may consist of a single large private company, a government-operated system such as the Tennessee Valley Authority, or several investor-owned companies banded together in a power pool.

The distinguishing feature of control areas is a single control center entrusted with the authority to operate the system within its area. The primary responsibility of a control area is to match its load with its own power generation. However, through interties with neighboring control areas, it is also prepared for mutual aid and scheduled sales of energy. Such mutual aid in emergencies may take the form of the flow of energy to a control area not directly adjacent to it. Even normally scheduled sales of energy may find intervening control areas between buyer and seller.

Frequency, measured in hertz (Hz), is the direct indication of the status of interconnected control areas. In North America, the normal frequency is 60 Hz. However, fluctuations in demand against generation cause all the interconnected control areas to witness the same variation in frequency (normally no more than ± 0.1 Hz). The frequency falls below 60 Hz when the demand exceeds the generation. Conversely, when the generation is more than necessary, the frequency rises.

In this dynamic system, constant regulation is needed to maintain the frequency within narrow limits. A single control area may be responsible for excess demand. In that case, the interties will show a net flow of energy above scheduled sales. This net inflow is called the area control error (ACE). The dispatcher will take steps within 1 minute to increase generation and return the ACE to zero (an NERC-OC rule). The dispatcher has 10 min to accomplish this. An outflow requiring reduced generation in some other case may also be observed.

POWER-SYSTEMS CONTROL

The electric power industry is currently undergoing major technological and organizational changes. The increasing presence of cost-effective, small, flexible power plants, customer automation, electronic control of wire equipment and the revolutionary changes in communications and computing all offer possibilities for active reliance on closed-loop control of electric power systems. At the same time, the industry has moved from a top-down, systemwide decision making into active, decentralized decision making by the system users (power suppliers and users) with increased interest in minimizing systemwide coordination.

These changes present a power-systems-control designer with the need to revisit the methods by which the system is operated today and to possibly establish new control paradigms more amenable to a decentralized operation.

In this brief chapter, we provide a concise summary of the control design presently practiced in the regulated industry first. We next describe the ongoing conceptual changes in the power-systems-control design for the competitive industry. We emphasize conceptual issues and point the reader to other references related to specific designs and calculations necessary for implementing controllers for meeting the established performance criteria.

OBJECTIVES OF POWER-SYSTEMS CONTROL

The ultimate objective of power-systems control is to balance time-varying load demand so that the power at the customer side is of sufficiently high quality and independent of any disturbances present on the system. The quality of the ac power is measured with respect to the variations in voltage magnitude and frequency of its sinusoidal waveform. More recently, an increased presence of higher-order harmonics in the basic sinusoidal waveforms to voltage, current, and instantaneous power has also become a concern related to power quality.

Under normal conditions, variations in load demand are small and random based on the forecasted load patterns. The system is also exposed to the small deviations in parameters and state from the nominal, resulting in dynamic transitions from the perturbed conditions to the stationary operation characterized by the sinusoidal stationary waveforms.

Under large, unexpected equipment outages (emergency mode), the system dynamics is potentially unstable, implying that if the fault is not eliminated within a certain "critical" clearing time, the system may lose synchronism and/or may experience voltage-collapse conditions. In the case of asymmetric faults, an additional concern is keeping a three-phase system balanced.

The role of control is to react to the deviations from the desired, baseline (nominal) voltage and frequency patterns as power is delivered to the users. The system control is well developed and often automated for regulating these relevant system outputs in response to the relatively small deviations from the nominal conditions. Power-systems control is much less developed and automated for emergency conditions; it is often system-specific, based on offline studies and human decisions.

Control Performance Criteria

The basic objectives of power-systems control can be approached in at least two different ways, that is, either by specifying the control-design criteria top-down or by specifying the criteria at the component level (both users and producers) and providing for their minimal coordination so that the system integrity is ensured at the same time. We begin by defining control performance criteria for a typical power system to differentiate these two qualitatively different cases.

Given the basic objectives of power-system operation as described, one needs to differentiate between the control performance criteria set in a top-down, systemwide fashion, typical of today's regulated industry, and the performance criteria set by the system users themselves (primary level). Moreover, in a horizontally structured interconnection one differentiates between the control performance objectives at the highest (interconnection, tertiary) level and the performance objectives at the specific subsystem (control area, secondary) level.

As the power system operations and planning become more decentralized and competitive, it is important to specify the "product," its sellers and buyers, its "value" to the buyers, and price by the sellers of control technologies, much in the same way as basic power purchase is carried out. This cannot be done without a clear definition of control performance criteria.

Given this need, differentiate among the following:

- Control criteria at the individual (or group of) customers level
- Control performance criteria at the power-suppliers level
- Control performance criteria at a subsystem (control area or some other aggregate entity) level
- Control performance criteria at the interconnection (system) level

Generally, there could be more than one provider of control services to meet the specified performance criteria. For example, a load-voltage control criterion could be met by means of generation-, transmission/distribution-, and/or load-control design. However, the cost of meeting the same performance criteria could vary drastically depending on who is providing this control and to which groups of users.

The most general power-systems-control design should allow for these choices and for clear incentives to implement the most economic technology among the ones capable of meeting the same technical criteria. The methods by which these technologies would be valued are currently under development. Two things are clear, however: (1) Many of these control services will be provided through markets separable from the basic power market. Such experiments, are already in place, particularly in Norway (Ilić et al., 1998) and California (Gaebe, 1997). In these experiments, many of the control-related services are generally referred to as ancillary service markets. (2) Over time, much more power-systems control will be provided in a distributed way by small power plants, controllable wires, and online load control. In order for this to become a reality, one needs a clear understanding of the control-design problems and the assumptions under which the design may become ineffective. This is particularly critical as the power-system operations begin to rely on control instead of on robust design. The tradeoffs between overdesigning the system and the value of control for avoiding some of this must be studied.

Finally, we observe here that the control performance criteria questions are closely related to the question of technical standards for the newly evolving power industry (Interconnected Operations Services, 1996).

Control Performance Criteria at a Customer Level. As different customers purchase electricity some of them may be willing to have slight deviations in power quality if the price of electricity reflects this willingness to be adaptive to the overall system condition. Some customers, in the not-so-distant future, will be interested in providing their own power and/or exercising partial control over power quality (e.g., power-factor control). Nevertheless, many of these customers will remain connected to the power grid for various reasons. This situation indicates a definite need for specifying the control performance criteria at the user's level unbundled from the system-level control performance criteria. Related to this unbundling are rules, responsibilities, and regulations for using the system to which the users are connected.

Examples of this would be specification of acceptable voltage and frequency thresholds by different groups of system users in normal operation and/or specifications of the rate of interruptions when the system is under stress. The important feature of this type of criteria is that for a given power system with, say, *nd* customers, one could have a different nominal (baseline) voltage at each node, that is, a *vector criteria:*

$$\underline{E}_L^{\text{nom}}(t) = [E_{L1}^{\text{nom}}(t)\ E_{L2}^{\text{nom}}(t) \cdots E_{Lnd}^{\text{nom}}(t)] \tag{1}$$

with the acceptable thresholds in voltage deviations

$$\underline{E}_L^{\text{thr}}(t) = [E_{L1}^{\text{thr}}(t)\ E_{L2}^{\text{thr}}(t) \cdots E_{Lnd}^{\text{thr}}(t)] \tag{2}$$

Similarly, the users could specify acceptable rates of interruption[1] when the system experiences difficult conditions, for example,

$$\underline{R}^{\text{int}}(t) = [R_{L1}^{\text{int}}(t)\ R_{L2}^{\text{int}}(t) \cdots R_{Lnd}^{\text{int}}(t)] \tag{3}$$

The frequency deviations are harder to differentiate in stationary operation; however, it is possible to have them characterized in principle as $f^{\text{nom}} = \dfrac{\omega^{\text{nom}}}{2\pi} = 60$ Hz and have different acceptable thresholds of deviation as a vector criteria (Ilić and Liu, 1996):

$$\underline{\omega}_L^{\text{thr}}(t) = [\omega_{L1}^{\text{thr}}(t)\ \omega_{L2}^{\text{thr}}(t) \cdots \omega_{Lnd}^{\text{thr}}(t)] \tag{4}$$

Control Performance Criteria at a Power-Producer Level. It is important to recognize here that with the influx of distributed generation there will be an increased presence of small power suppliers in large numbers, often close to the electricity users. Moreover, not all of these sources of electricity are easy to control; for example, solar, wind, and many other future sources of electricity are of this type.

[1]The interruption specifications could take different forms (Tan and Varaiya, 1993).

This situation raises a basic question concerning the technical standards and control-performance-criteria specifications for power suppliers. These criteria are slightly more complicated than the customer-level criteria. Usually, the customers are already connected to the grid. Instead, one may need to distinguish between the connection (access) standards for new power supply entries and their operating (control) performance criteria. For purposes of systems-control design, we recognize a potential nonuniformity of specifications, including the fact that some power suppliers cannot control their power output at all (assuming no storage facitlities). To take into consideration this situation when defining a general power-systems control problem it is necessary to specify the ranges of acceptable voltage variations and rates of interruption for uncontrollable power suppliers (much in the same way as in Eqs. (1) and (2)), treating them as negative loads.

Power suppliers capable of controlling power output within certain capacity and at a certain rate of response (MW/minute) are basically control inputs subject to constraints:

$$\underline{P}_G^{\min} = [P_{G1}^{\min}\ P_{G2}^{\min} \cdots P_{Gn}^{\min}] \tag{5}$$

$$\underline{P}_G^{\max} = [P_{G1}^{\max}\ P_{G2}^{\max} \cdots P_{Gn}^{\max}] \tag{6}$$

Moreover, if the power suppliers are willing to provide voltage/reactive power control, these specifications need to be defined as a vector criteria, defining the ranges within which voltage control is provided (control limits):

$$\underline{E}_G^{\min} = [E_{G1}^{\min}\ E_{G2}^{\min} \cdots E_{Gn}^{\min}] \tag{7}$$

and

$$\underline{E}_G^{\max} = [E_{G1}^{\max}\ E_{G2}^{\max} \cdots E_{Gn}^{\max}] \tag{8}$$

It is assumed that the excitation systems are very fast, and it is therefore not necessary to specify the rate of response.

Control Performance Criteria at a (Sub)System Level. A qualitatively different approach is to define control criteria at a system level and design control to guarantee this performance. Examples of control specifications at a system level would be the requirements that total power scheduled to be generated, $\sum_{i=1}^{i=n} P_{Gi}[kT_H]$, equals total anticipated load demand, $\hat{P}_L^{\text{sys}}[kT_H]$ at, say, each hour, kT_H, where $k = 1, 2, \ldots$, or at any other time sample when generation scheduling is computed in the control centers. This systemwide criterion for each power system is expressed mathematically as

$$\sum_{i=1}^{i=nd} P_{Gi}[kT_H] = \hat{P}_L^{\text{sys}}[kT_H] \tag{9}$$

Note: To understand the dependence of industry organization on the operating and control objectives observe here that the total anticipated system load, $\hat{P}_L^{\text{sys}}[kT_H]$, is not necessarily the sum of the anticipated loads by individual customers, $\sum_{i=1}^{i=nd} P_{Li}[kT_H]$. Typically, individual power-load patterns are not observable in the regulated industry online; only the energy consumed is tracked. This is drastically changing under competition, where much emphasis is on active pricing of electricity and the demand elasticity.

The second typical criteria at a system level concerns acceptable frequency deviations, $f^{\text{sys,max}}[kT_s] = \dfrac{\omega^{\text{sys,max}}[kT_s]}{2\pi}$, created by the deviations of the actual system load pattern, $P_L^{\text{actual}}[kT_s]$, from the anticipated system load, $\hat{P}_L^{\text{sys}}[kT_H]$, and the generation inertia in actually producing the scheduled generation.[2]

Specifying acceptable deviations in load voltage is much less rigorous at a system level. The industry has developed notions of "optimal" voltage profiles, characterized as voltages that lead, for example, to minimal transmission losses. However, this notion of optimal voltage profile defined at a system level is not unique and it is hard to quantify, since it is, among other factors, system-specific (Carpasso et al., 1980). The net result here is that utilities attempt to maintain voltages as close to 1 p.u. as possible, with the allowable uniform deviations of ±2 percent. Under stress, this threshold is relaxed uniformly to ±5 percent.

Finally, the utilities have a conservative system-level criterion that requires sufficient power-supply reserve and strong network design to supply users in an uninterrupted way when any single equipment (generator, transmission line) outage takes place. This is a so-called $(n - 1)$ reliability criterion at a transmission-system level. At a distribution-system level, similar criteria exist that require the average length or frequency of power interruption to remain uniformly within a certain prespecified tolerance.

The Relations Between Control-Criteria Specifications. Theoretically, the two types of criteria should be related; however, quantifying these relations on a very large power system may be difficult. The typical approach in the regulated power industry has been a so-called top-down approach in which the control criteria are defined at a system level.

The distinction between the two types of performance-criteria specifications may appear insignificant at first sight, since it is well known that under stationary conditions frequency is the same everywhere on the system and the objective is to provide sufficient control (generation) so as not to interrupt service to any customer for some prespecified time following a major equipment outage. In this case, the only difference in service could be noted with respect to the voltage variations at different system locations (typically, customers at the end of the feeders are likely to have worse voltage support than those closer to the substation).

The distinction in quality of service in the evolving industry may require much more differentiation in specifications at different system locations; some groups of customers may be willing to be interrupted more frequently when system reserve is low and pay less in return. Others may require very high-quality, uninterrupted power delivery. Similarly, different users may require a different quality of voltage support, again at different prices for electricity.[3] Moreover, as the generation, transmission/distribution, and load services begin to form as separate corporate businesses, it becomes necessary to clearly identify objectives for each entity separately. In this situation, it becomes necessary to design control capable of meeting these distributed performance specifications. Consequently, each control function ends up having a different value to different participants in the competitive power industry.

---

[2]This criterion is more transparent in the multiarea systems, where each control area has a specification for meeting so-called area control error (ACE), defined below in the context of the automatic generation control (AGC) scheme.

[3]It is conceptually possible to differentiate among users for quality of frequency variations as well; however, this would require much more refined measurement techniques in order to differentiate stationary deviations according to location (Ilić and Liu, 1996).

Therefore, the control objectives should be defined for generation, transmission/distribution, and load-serving entities separately. The overall performance of the physically interconnected system could be assessed by the system operators asking how these distributed performance criteria impact the systemwide performance in terms of the traditional measures, such as system reserve necessary to meet the $(n - 1)$ reliability criterion, that is, the system ability to serve any customer without interruption when any single equipment outage occurs, or the area control-error quality, that is, frequency deviations in each control area caused by imbalances in its total generation and demand, or, furthermore, in the time-error correction reflecting a cumulative frequency shift at the interconnection level as a result of the overall system generation/demand imbalance.

It turns out that in an environment driven by profit/benefit maximization and financial risk management, the ability to deliver the product and the quality of products (frequency, voltage, rate of interruption, and harmonic contents) need to be specified in order to be met. Moreover, a different quality of service is provided at different prices. Currently, much debate is taking place in terms of defining the value and means of providing frequency and voltage regulation in the competitive power industry.

An interesting new control problem concerns grouping (aggregation) of different users of control equipment to maximize benefits at the least cost. This question is closely related to the creation of markets for control services.

CLASSIFICATION OF CONTROL TOOLS

Historically, the majority of power-system control has been generation-based. In addition, there are controllers that directly regulate transmission-line power flows as well as some controllers at the system users' end. This classification is important to keep in mind as generation, transmission, and user services are becoming separate functional and corporate entities.

Generation-Based Control

Generation scheduling of real (active) power, P_G, for supplying anticipated real-power demand, $\hat{P}_L$, is the main means of balancing supply and demand and, consequently, maintaining system frequency close to its nominal value. The remaining generation-demand imbalance created by the deviations, $P_L(t)$, of the actual load demand, $P_L^{\text{actual}}(t) - \hat{P}_L(t)$ from the anticipated, $\hat{P}_L(t)$, results in frequency deviations, $\omega(t)$, and these are corrected for by means of closed-loop governor control as described in some detail next.

Generation control for load voltage control $(E_{L1} \ . \ . \ . \ E_{Lnd})$ is done by controlling the excitation system so that terminal voltage of the generators $(E_{G1} \ . \ . \ . \ E_{Gn})$ remains close to its set point values. The closed-loop excitation control is automated; an excitation system is basically a constant gain, fully decentralized, proportional-differential (PD) controller.

The generation-based voltage control is necessary to compensate for reactive power losses associated with the real-power delivery and also with the reactive power consumed at the customers' end, so that despite this reactive power consumption, load voltage remains within the limits acceptable to the system users, as specified in Eqs. (1) and (12) above. The reactive power scheduling and/or voltage regulation is not as systematic as the real-power scheduling and control. Some presently practiced approaches are described in Ilić and Liu(1996); Ilić and Zaborszky (2000).

Transmission/Distribution Control

A typical electric-power system is currently equipped with a variety of switched-type reactive devices (capacitors, inductors, transformers) intended for direct voltage control at the user's side and for direct regulation of power flows on the selected transmission lines. These devices, when connected in series with the transmission lines, change their natural transfer characteristics; an inductive transmission line with a controllable series capacitor is a direct line-flow controller, since the flow is directly proportional to the line conductance. The switching of the reactive devices has typically been mechanically controlled and therefore slow. Recently so-called flexible ac transmission systems (FACTS) technologies have made it possible to change the characteristics of selected transmission lines by power electronic switching, once each cycle, the number of controllable reactive power devices connected to them. These new technologies have opened opportunities for systematic changes of transfer characteristics along particular transmission paths.

Load-Demand Control

It is important to keep in mind the existence of a variety of control devices capable of directly controlling voltage at the power user's side. These are shunt capacitors and/or inductors, which are connected in parallel with the load. These controls, like the transmission/distribution-type controllers, have historically been based on slow mechanical switching. Most typical devices of this type are on-load-changing transformers (OLTCs) and switched capacitor banks. The more recent FACTS-type technologies are very fast, power electronically switched controllers. Depending on the voltage level for which they are intended, they range from powerful static VAR controllers (SVCs) through much smaller, HV-DC lite technologies.

ASSUMPTIONS UNDERLYING CONTROL DESIGN IN NORMAL OPERATION

A close look into typical control design for large power systems reveals an unexpected simplicity. The output variables of direct interest (frequency and voltage) are controlled in a somewhat independent way; each is in turn, controlled using a spatial and temporal hierarchical approach. Most of the disturbance is suppressed in an open-loop mode, statically. The rest is controlled at each subsystem (utility, control area) level as well as at each major component (generator) level in an entirely decentralized way.

It should be obvious that such a simplified approach does not work when the system is subjected to very large contingencies, mainly-because real power cannot be decoupled from reactive power/voltage and a spatial/temporal separation cannot hold under such conditions. More generally, any constant-gain control design (such as in presently implemented excitation systems and governors) is effective only when model linearization is valid around a given operating point.

Moreover, systematic, generation-based control under normal conditions is developed without much consideration for the effects of other types of controllers (transmission and/or load); the latter are primarily structural, in the sense that they alter input/output transfer functions of the transmission system. As such, they are difficult to design systematically. To avoid possible operating problems created by the interplay of generation-based control with other types of control, current design has been such that the rates at which different controllers activate are sufficiently different so that one can assume that their effects are separable. This is again true in principle only when the system is subjected to relatively small disturbances, and it is generally not true under contingencies.

This leaves us with the conclusion that presently practiced automated control design in electric power systems could be expected to perform well only for operating conditions close to nominal, for which the system and control were initially designed. Consequently, when the system is subject to large equipment outages, it becomes necessary to rely on human expert knowledge, which is often system-specific.

HIERARCHICAL CONTROL IN THE REGULATED ELECTRIC POWER INDUSTRY

In a regulated industry, generation scheduling and control to meet system load demand are done in a hierarchical manner. The hierarchies are based on both temporal and spatial separation. To briefly review the decision and control approaches in the regulated industry, consider two types of power systems architectures: (1) isolated systems comprising a single utility (control area) and (2) an interconnection consisting of several horizontally structured subsystems (utilities, control areas) electrically interconnected via the tie-lines. These two designs are conceptually different because a single control area case is characterized by a two-level decision and control hierarchy, one being at the component level and the other at the entire system level. In the case of an interconnection consisting of several control areas, the general structure has three levels, component, control area (subsystem), and the interconnection.

Single Control Area Case

Consider an electric power system with n nodes. Its net generation/demand $[P_{G1}(t)\ .\ .\ .\ P_{Gn}(t)]$ is controllable; the power injections $[P_{L1}(t)\ .\ .\ .\ P_{Lnd}(t)]$ at the remaining n_d nodes represent uncertain load demand. Assume that network topology and generation capacity $[K_{G1}\ .\ .\ .\ K_{Gn}]$ are known over the entire time period of interest, T, and that the generation/demand imbalance is created by slow variations in demand, $P_Lj[kT_H]$, $k = 1, . . .$ (say, hourly, where T_H stands for one hour of sampling time). The balancing of this estimated hourly demand is done using an open-loop scheduling control scheme. Its formulation is briefly given here.

Short-Term Generation Scheduling (System-Level Control). At the control-area (system) level, the short-term generation scheduling problem is the problem of optimizing total operating cost:

$$\min_{P_{Gi}[kT_H]} E\left\{\sum_{k=t0}^{T/T_H}\sum_{i=1}^{n} c_i(P_{Gi}[kT_H])\right\} \tag{10}$$

so that the anticipated demand is balanced, that is,

$$\sum_{i=1}^{n} P_{Gi}[kT_H] = \sum_{j=1}^{nd} P_{Lj}[kT_H] \tag{11}$$

within the power-generation-capacity limits,

$$P_{Gi}[kT_H] \le K_{Gi} : \sigma_i(t) \tag{12}$$

and so that the transmission-line flow in each line l, $\forall l \in 1, \ldots, L$,

$$F_l[kT_H] = \sum_{i=1}^{(n+nd)} H_{li}(P_{Gi}[kT_H] - P_{Li}[kT_H]) \tag{13}$$

remains within the transmission-line-capacity limits, K_l, that is,

$$F_l = K_l : \mu_l\,[kT_H] \tag{14}$$

If the load is assumed uncertain, this short-term generation-scheduling problem requires dynamic programming-type tools for adequate turning on and off of the power plants. This problem is known as the unit-commitment problem (Allen and Ilić,1998). If the load is assumed to be known (estimated), the generation scheduling requires solving a deterministic static optimization problem, known as the optimal-power-flow problem (Lugtu, 1978). Most control centers have software for solving this constrained cost-optimization problem as frequently as each half hour. At the optimum, assuming the load is known, the locational cost (spot price (Schweppe et al., 1988)), $p_i(t) = \dfrac{dc_i}{dP_i(t)}$, is different at each node i and it can be expressed as

$$p_i[kT_H] = p[kT_H] - \sum_{l=1}^{l=L} H_{li}\mu_l\,[kT_H] \tag{15}$$

The term, $\Sigma_{l=1}^{l=L}\ \mathrm{H}_{li}\mu_l\,[kT_H]$, where L is the total number of transmission lines in the system, reflects locational differences in optimal costs of electricity production caused by the active transmission constraints ("congestion") relative to the unconstrained electricity price $p[kT_H]$ (see Eq. (14)).

Primary Control for Stabilization (Equipment-Level Control). The net (systemwide) real-time mismatch between generation produced, according to this short-term scheduling method and the actual demand consumed at each instant t, results in small, systemwide stationary deviations. Different systems regulate these stationary deviations in frequency different ways. Single control areas with much flexible generation (such as hydro) are capable of correcting for cumulative frequency deviations by manually changing the set point values of the generator-turbine-governor (G-T-G) units when so-called time-error (proportional to the integral of frequency deviations) exceeds a certain preagreed-upon threshold, $\Delta f^{\max}$, without necessarily automating the process. In multiarea systems, such as the one in the United States, the regulation is done by an automatic generation control (AGC) scheme, which is described in the section devoted to multiarea systems.

Finally, the fastest frequency variations around the values resulting after AGC (or an equivalent corrective scheme at a control-area level) has acted each τ_s seconds are stabilized by the local controllers, governor systems in particular. At present, these controllers are entirely decentralized, constant-gain output controllers. It can be shown that the dynamics of any power plant, i, no matter how complex, amount to being modeled as

$$x_i(t) = \tilde{f}_i(x_i(t), u_i(t), y_i(t)) \tag{16}$$

(Ilić and Zaborszky, 2000), where $x_i(t)$, $u_i(t)$, and $y_i(t)$ are local state, primary control, local output vectors, respectively. A local continuous primary control, $u_i(t)$, such as the governor, is typically designed to stabilize a local error signal:

$$e_i(t) = y_i(t) - y_i[k\tau_s] \tag{17}$$

If a local controller is of the switching type, a closed-loop model defined as Eq. (16) with a local control law is obtained instead:

$$u_i[(k + 1)\tau] = u_i[k\tau] - d_i r_i(e_i[k\tau]) \tag{18}$$

which acts only at discrete times $k\tau$, $k = 1, \ldots,$ where $r_i()$ is a relay-type function. Capacitor/inductor switching on controllable loads and onload tab-changing transformers are typically used for load voltage control according to this control law. (Generally, there is no explicit relation between τ, the timing at which primary controllers is switched, and the rate at which the set point values of these controllers are changed at each control-area level $[k\tau_s]$.)

In summary, in the single control-area system, one can identify two levels of control, online systemwide generation scheduling for meeting total anticipated demand and a very fast stabilization level at each individual generator (component). These two levels are implemented by physically changing the set points of the governors each $[kT_H]$ and the real-time t stabilization in response to the fastest deviations so that the frequency of each generator is kept close to its set point value (Ilić and Liu, 1996). Observe that the separation of the two control levels has been driven by the temporal separation of load-demand deviations evolving at significantly different rates, one anticipated each hour $[kT_H]$ and the other being much faster, real-time t load dynamics. The system level is open loop in anticipation of a disturbance, and the component-level control is automated, closed-loop, responding to the system output deviations (generation frequency) caused by the actual generation/demand imbalances.

Multiarea Control Case

The same reasoning behind separating a single control area into two levels applies to spatial separation of an interconnected power system (consisting of several control areas—utilities—that are electrically interconnected) into three control levels. This leads to the basis for hierarchical operation by decision and control within an interconnection. Clearly, these hierarchies work only by assuming not much interaction between the subsystems (Ilić and Liu, 1996). The main decision concerns whether the system is in normal operating mode, characterized by relatively small changes and weak interactions between the control areas or in an operating mode that requires special procedures.

The three-level control concept in the regulated U.S. industry of operating the system under normal conditions has been based on giving autonomy to each subsystem to plan and schedule generation to meet its own (connected) load demand for some assumed tie-line flow exchange with its neighbors. Each subsystem, having its own control center, then employs the short-term scheduling method just described, with the exception of attempting to meet the preagreed-upon tie-line flow schedules of its neighbors. These agreements have historically been bilateral but also based on cooperating in observng the $(n - 1)$ reliability criterion of the entire interconnection.

Open-Loop Generation Scheduling (System-Level Control)

The overall (interconnection, tertiary) level is not coordinated online at present (Ilić and Liu, 1996); instead, each subsystem has a preassigned participation in time-error correction resulting from the cumulative systemwide frequency deviations. To avoid excessive time-error, each subsystem (control area) is equipped with its own decentralized AGC (secondary-level control). The principles of this scheme are ingenious and are briefly summarized next. Important for purposes of this paper is the fact that the fully decentralized AGC works perfectly only when its tuning is done very carefully. Finally, each generator is equipped with primary stabilization (governor), which is designed to work in the same way as the control system of the single control area described earlier.

Automatic Generation Control (AGC) (Subsystem-Level Control). The remaining generation-demand imbalance created by unpredictable, typically small and fast, load-demand variations as well as the inertia of the G-T-G units not capable of producing the scheduled electrical output instantaneously has been controlled in an automated way by means of AGC. This is a very simple, powerful concept which is based on the fact that in stationary operation power-system frequency is observable at each node, i, and it reflects the total system generation-demand imbalance. The accepted industry standard has been until recently to rely on AGC, which is effectively a decentralized, output control-area scheme, with the Area Control Error (ACE) of each area, I, being the output variable of interest defined as

$$ACE^I[k\tau_3] = F^I[k\tau_3] - 10B^I \frac{\omega[k\tau_3]}{2\pi} \tag{19}$$

where $F^I[k\tau_s]$ is the deviation of a net power flow from the area I, and B^I is known as the area bias, which is chosen as close as possible to the so-called natural response of the area β^I. Assuming the system is at equilibrium, the basic power-to-frequency static characteristic is

$$\omega^I = \frac{P_G^I}{\beta^I} \tag{20}$$

The principle of the entirely decentralized AGC, in which each subsystem (control area) regulates its own ACE^I, relies on the fact that if the frequency bias B^I is chosen to be close to the natural response of the area β^I, then each area will effectively balance its own generation demand and the entire interconnection will be balanced. However, when the system is presented by a load-demand change, $P_d[k\tau_s]$, it can be shown that the stationary frequency changes driven by this disturbance could be modeled as

$$\omega[(k+1)\tau_s] = \frac{(B^I - B^K)}{(\beta^I + \beta^K)}\,\omega[k\tau_s] - \frac{P_d[k\tau_s]}{10(\beta^I + \beta^K)} \tag{21}$$

(Ilić et al., 1998). The main observation here is that the system frequency depends on the sum of the subsystem biases $(B^I + B^K)$. This is the foundation for so-called dynamic scheduling in today's industry, in which power plants forming one area could participate in frequency regulation of the other area. This, however, no longer implies that each area balances its own supply and demand.

Moreover, when the $ACE[k\tau_s]$ is tuned as described, entirely decentralized AGC regulates deviations of tie-line flows back to their scheduled values and brings system frequency very close to its nominal value under the stationary conditions. Consequently, at present there is no even minimal close-loop tertiary level coordination; instead, a cumulative frequency error is corrected for by each area participating in eliminating so-called time-error.

POWER-SYSTEMS CONTROL IN THE CHANGING INDUSTRY

An important premise in the hierarchical control of the regulated electric power systems is that each subsystem within an otherwise electrically interconnected system plans and operates its generation without relying much on the neighboring subsystems. Recall from a brief summary of AGC that as long as each control area tunes its frequency bias term, B, (Eq. (19)) to compensate for load deviations in its own area, the entire interconnection will be balanced.

The changing power industry is characterized by its striking open-access requirements; namely, power producers could supply customers who are not necessarily in the same control area as long as the economic reasons are favorable to do this. An immediate implication of this is that the role of the control area (secondary level) within an interconnected system comprising several control areas must be revisited.

As the notion of a control area is analyzed under the new industry rules, two different questions arise:

- Congestion management, that is, maintaining the power flows within the capacity limits defined in Eq. (14)
- Frequency control

In the regulated industry, the inter-area congestion management is done in a decentralized way. Each control area schedules its generation to meet its own demand for the pre-agreed-upon tie-line flow exchange with the neighboring control areas, and each control area observes the transmission-line flow limits in its own area. Although the tie-line flow exchange is not directly controllable, it is regulated to preagreed values with the aid of AGC, which responds to the ACE signal (and not only to the frequency deviation signal). The problem of inadvertent energy exchange is closely related to the conceptual impossibility of directly controlling the tie-line flows by means of generation control only. In an industry based on cooperation, much effort has been made to maintain the tie-line flow exchanges close to their scheduled values. This way a congestion problem related to the violation of line power-flow limits of tie-lines (interarea flows) has been avoided most of the time. As we have, explained the system frequency is regulated at the same time by each control area regulating its ACE signal.

This hierarchical separation of tie-lines from the other line flows becomes questionable under the open-access operating rules. Individual subsystems do not have any direct control over tie-line flows; these are often determined by wheeling power across the control areas. The congestion-control problem under open access remains an open research problem at this time. For a survey of techniques recently proposed, see Ilić and Zaborszky (2000, Chapter 13).

Similarly, the system frequency control may need qualitatively different solutions under open access. In particular, if the tie-line flows are not going to be tightly controlled, the concept of ACE may lose its fundamental role of regulating power generation so that the tie-line flows remain as scheduled. All indications are that market implementations of what amounts to the flat frequency control will be necessary in the near future, see Ilić and Zaborszky (2000, Chapter 12). It is important to notice that such a solution is already being implemented in Norway.

CONCLUSIONS

We conclude by referring the reader to a much more elaborate treatment of the very broad topic of power-systems control. The recently published reference Ilić and Zaborszky (2000, Chapters 12, 13, and 14) provides many details of specific design approaches as well as of the fundamental changes brought about by industry restructuring. The most relevant parts in proceeding with systematic control design concern clear posing of control performance criteria and the establishment of the mathematical models that naturally lend themselves to a control design of interest. The overall power-systems control system is a mix of fast and slower controllers located at the sites of power producers, users, and the power network. In order for their effects to be useful, one must understand their interdependence. This makes the overall power-systems control design very complex. The real challenge is posing the practical problem in terms of well-known general control-design problems applicable to many other dynamic systems. Much more work is needed for establishing a systematic control design so that the desired performance criteria are met at various levels of a complex power system.

BIBLIOGRAPHY

Allen, E., and M. Ilić, *Price-based Commitment Decisions in the Electricity Market.* 1998. London: Springer.

Carpasso, A., E. Mariani, and C. Sabelli. 1980. "On the Objective Functions for Reactive Power Optimization," IEEE Winter Power Meeting, Paper No. A 80WM 090-1.

G. B. Gaebe. 1997. California's Electric Power Restructuring," Proc. EPRI Workshop on Future of Power Delivery in the 21st Century, La Jolla, C.A.

Ilić, M., and S. X. Liu. 1996. *Hierarchical Power Systems Control: Its Value in a Changing Industry.* London: Springer.

Ilić, M., P. Skantze, L. H. Fink, and J. Cardell, 1998. "Power Exchange for Frequency Control (PXFC)," Proc. Bulk Power Syst, Dynamics Control—Restructuring, Santorini, Greece.

Ilić, M., and J. Zaborszky. 2000. *Dynamics and Control of Large Electric Power Systems.* NewYork: Wiley & Sons, Interscience.

Interconnected Operations Services, NERC Report 1996.

Lugtu, R. "Security Constrained Dispatch," IEEE paper No. F 78 725-4.

Schweppe, F. C., M. C. Caramanis, and R. D. Tabors. 1988. *Spot Pricing of Electricity*. Boston, M.A.: Kluwer Academic Publishers.

Tan, C-W., and P. Varaiya. 1993. "Interruptible Electric Power Service Contracts," *Journal of Economic Dynamics and Control,* Vol. 17, pp. 495–517.

SECTION 13

SHORT-CIRCUIT COMPUTATIONS

Lawrence J. Hollander, P.E.
Dean of Engineering Emeritus
Union College

TRANSFORMER REGULATION DETERMINED FROM SHORT-CIRCUIT TEST

A single-phase transformer has the following nameplate data: 2300/220 V, 60 Hz, 5 kVA. A short-circuit test (low-voltage winding short-circuited) requires 66 V on the high-voltage winding to produce rated full-load current; 90 W is measured on the input. Determine the transformer's percent regulation for a load of rated current and a power factor of 0.80, lagging.

Calculation Procedure

1. Compute the Rated Full-Load Current (High-Voltage Side)

For a single-phase ac circuit use the relation kVA $= VI/1000$, where kVA = apparent power in kilovolt-amperes, V = potential in volts, and I = current in amperes. Rearranging the equation yields $I = 1000 \text{ kVA}/V = (1000)(5)/2300 = 2.17$ A.

2. Compute the Circuit Power Factor for the Short-Circuit Test

Use the relation for power factor: pf = W/VA, where W is real power in watts and VA is apparent power in volt-amperes (volts × amperes); pf $= 90/(66)(2.17) = 0.628$. The power factor is the cosine of the angle between voltage and current; $\cos^{-1} 0.628 = \theta = 51.1°$ lagging. This is the pf angle of the transformer's internal impedance.

3. Compute the Circuit Power-Factor Angle for the Operating Condition

As in Step 2, use the relation $\cos^{-1} 0.80 = \theta = 36.9°$ lagging. This is the pf angle of the load served by the transformer.

4. Compute the Transformer Output Voltage for Serving an 0.80-pf Load at Rated Full-Load Current

Use the relation for IR drop in the transformer: $V_{IR} = V_{sc} \cos(\theta_{ii} - \theta_{load}) = 66 \cos(51.1° - 36.9°) = 66 \cos 14.2° = 64.0$ V, where V_{sc} = short-circuit test voltage, θ_{ii} = internal-impedance phase angle, and θ_{load} = load phase angle. Refer to Fig. 13.1. Use the relation for IX drop in the transformer: $V_{IX} = V_{sc} \sin(\theta_{ii} - \theta_{load}) = 66 \sin(51.1° - 36.9°) = 66 \sin 14.2° = 16.2$ V. Thus, $V^2_{input} = (V_{output} + V_{IR})^2 + V^2_{IX} = 2300^2 = (V_{output} + 64)^2 + 16.2^2$. Solution of the equation for V_{output} yields 2236 V.

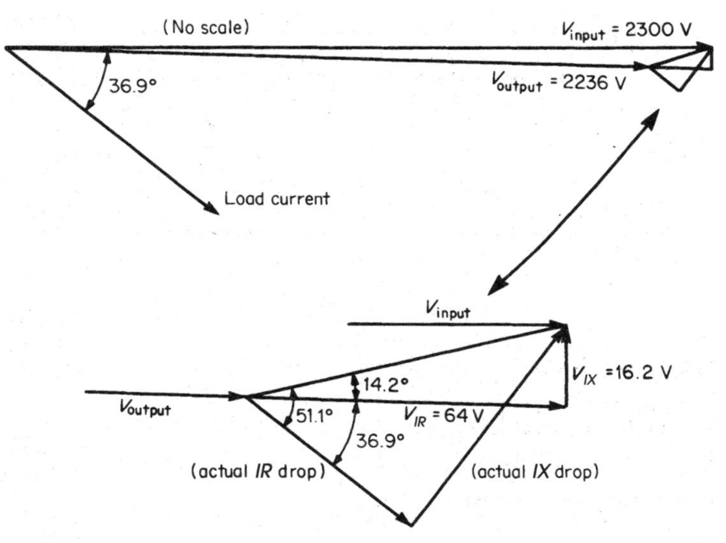

FIGURE 13.1 Transformer internal impedance-drop triangle. In accordance with trigonometric relations, the actual IR and IX drops are changed to values more easily handled mathematically.

5. Compute the Transformer Regulation

Use the relation, percent regulation = $[(V_{\text{input}} - V_{\text{output}})/V_{\text{input}}](100 \text{ percent}) = [(2300 - 2236)/2300](100 \text{ percent}) = 2.78$ percent.

Related Calculations. The general method presented here is valid for calculating transformer regulation with both leading and lagging loads.

TERMINAL VOLTAGE OF SINGLE-PHASE TRANSFORMER AT FULL LOAD

A single-phase transformer has the following nameplate data: 2300/440 V, 60 Hz, 10 kVA. The ohmic constants of the equivalent circuit are primary resistance referred to the primary, $r_1 = 6.1$; secondary resistance referred to the secondary, $r_2 = 0.18$; primary leakage reactance referred to the primary, $x_1 = 13.1$; and secondary leakage reactance referred to the secondary, $x_2 = 0.52$. With the primary supply voltage set at nameplate rating, 2300 V, calculate the secondary terminal voltage at full load, 0.80 pf, lagging.

Calculation Procedure

1. Compute the Turns Ratio

The turns ratio, a, is determined from the equation, a = primary voltage/secondary voltage = 2300/440 = 5.23.

2. Compute the Total Resistance of the Transformer Referred to the Primary

The equation for the total series resistance of the simplified equivalent circuit (see Fig. 13.2) referred to the primary is $R = r_1 + a^2 r_2 = 6.1 + (5.23)^2 (0.18) = 11.02\ \Omega$.

3. Compute the Total Leakage Reactance of the Transformer Referred to the Primary

The equation for the total leakage reactance of the simplified equivalent circuit (see Fig. 13.2) referred to the primary is $X = x_1 + a^2 x_2 = 13.1 + (5.23)^2(0.52) = 27.32\ \Omega$.

4. Compute the Load Current

The approximate or simplified equivalent circuit is referred to the primary side of the transformer. Use the equation, $I_1 = I_2$ = kVA rating/kV rating = 10/2.3 = 4.35 A, and assume this to be the reference phasor. Thus, in phasor notation, $\mathbf{I}_1 = \mathbf{I}_2 = 4.35 + j0$. It is given that the load current lags the output voltage, $\mathbf{V}_2$, by a pf angle of $\cos^{-1} 0.80 = \theta = 36.87°$.

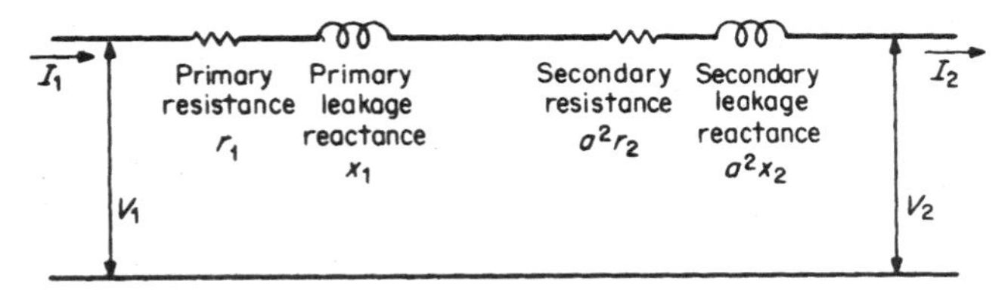

FIGURE 13.2 Transformer approximate equivalent circuit referred to the primary side.

5. Compute the Output Voltage

Use the equation, $\mathbf{V}_1 = \mathbf{V}_2 + \mathbf{I}_2(R + jX)$, where all quantities are referred to the primary side of the transformer. With $\mathbf{I}_2$ as the reference phasor, $\mathbf{V}_2$ is written as $(0.8 + j0.6)V_2$. Thus, $|\mathbf{V}_1| = (0.8 + j0.6)V_2 + 4.35(11.02 + j27.32) = 2300$. This is rewritten in the form, $2300^2 = (0.8V_2 + 47.94)^2 + (0.6V_2 + 118.84)^2$; by further rearrangement of this equation, V_2 may be determined from the solution for a quadratic equation, $x = [-b \pm (b^2 - 4ac)^{1/2}]/2a$. The quadratic equation is $0 = V_2^2 + 219.3V_2 - 5{,}273{,}578.8$; the solution in primary terms yields $V_2 = 2189.4$ V.

The actual value of the secondary voltage (load voltage) is computed by dividing V_2 (referred to the primary side) by the turns ratio, a. Thus, the actual value of $V_2 = 2189.4/5.23 = 418.62$ V.

Related Calculations. The general method presented here is valid for calculating the terminal voltage for both leading and lagging loads and may be used for three-phase transformers having balanced loads. For unbalanced loads, the same analysis may be based on the concept of symmetrical components.

VOLTAGE AND CURRENT IN BALANCED THREE-PHASE CIRCUITS

The line-to-line voltage of a balanced three-phase circuit is 346.5 V. Assuming the reference phasor, $\mathbf{V}_{ab}$, to be at 0°, determine all the voltages and the currents in a load that is wye-connected and has an impedance $\mathbf{Z}_L = 12\angle 25°\ \Omega$ in each leg of the wye. If the same loads were connected in delta, what would be the currents in the lines and in the legs of the delta?

Calculation Procedure

1. Draw the Phasor Diagram of Voltages

Assuming a phase sequence of a, b, c, the phasor diagram of voltages is as shown in Fig 13.3.

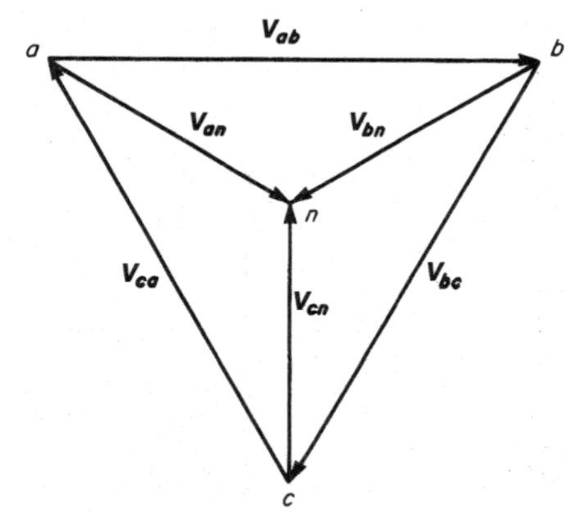

FIGURE 13.3 Phasor diagram of voltages based on a phase sequence of *a*, *b*, *c*.

2. Identify All Voltages in Polar Form

From the phasor diagram for a balanced system, the polar form of the voltages may be written: $\mathbf{V}_{an} = 200\angle 330°$ V, $\mathbf{V}_{bn} = 200\angle 210°$ V, $\mathbf{V}_{cn} = 200\angle 90°$ V, $\mathbf{V}_{ab}$ (reference) $= 346.5\angle 0°$ V, $\mathbf{V}_{bc} = 346.5\angle 240°$ V, $\mathbf{V}_{ca} = 346.5\angle 120°$ V.

3. Calculate the Currents for the Wye Connection

Calculate the current in the load for each branch of the wye connection; these currents must lag the respective voltages by the power-factor angle of the load. Thus, $\mathbf{I}_{an}$ must lag $\mathbf{V}_{an}$ by 25°; $\mathbf{I}_{an} = \mathbf{V}_{an}/\mathbf{Z}_L = 200\angle 330°\text{ V}/12\angle 25°\ \Omega = 16.7\angle 305°$ A. In a

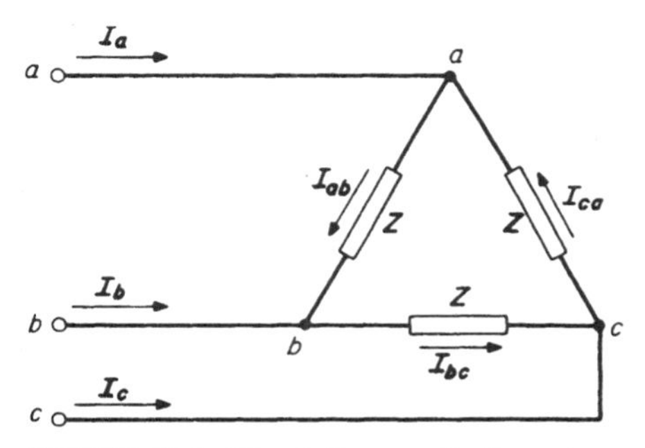

FIGURE 13.4 Circuit diagram of a three-phase delta-connected load.

similar manner, $\mathbf{I_{bn}} = 16.7\angle 185°$ A and $\mathbf{I_{cn}} = 16.7\angle 65°$ A.

4. Calculate Internal Delta Currents for Delta Connection

For the delta-connected load, the voltage across each leg (side) of the delta is the line-to-line voltage rather than the line-to-neutral voltage (see Fig. 13.4). $\mathbf{I_{ab}} = \mathbf{V_{ab}}/\mathbf{Z_L} = 346.5\angle 0°$ V$/12\angle 25°\ \Omega = 28.9\angle -25°$ A. In a similar manner, $\mathbf{I_{bc}} = 28.9\angle 215°$ A and $\mathbf{I_{ca}} = 28.9\angle 95°$ A.

5. Calculate the Line Currents for the Delta Connection

Calculate the line currents for the delta-connected load using the relation: the sum of the currents into a node must equal the sum of the currents leaving a node. At node a, $\mathbf{I_a} + \mathbf{I_{ca}} = \mathbf{I_{ab}}$ or $\mathbf{I_a} = \mathbf{I_{ab}} - \mathbf{I_{ca}} = 28.9\angle -25° - 28.9\angle 95° = 26.2 - j12.2 + 2.5 - j28.8 = 28.7 - j41.0 = 50.0\angle -55°$ A. Similarly, $\mathbf{I_b} = 50.0\angle 185°$ A and $\mathbf{I_c} = 50.0\angle 65°$ A. In this case, for the delta-connected load, the line currents are $\sqrt{3}$ greater than the phase currents within the delta and lag them by 30°.

Related Calculations. This procedure may be used for all balanced loads having any combinations of resistance, inductance, and capacitance. Further, it is applicable to both wye- and delta-connected loads. Through the use of symmetrical components, unbalanced loads may be handled in a similar manner. Symmetrical components are most helpful for making the calculations of balanced loads with unbalanced voltages applied to them.

THREE-PHASE SHORT-CIRCUIT CALCULATIONS

Three alternators are connected in parallel on the low-voltage side of a wye-wye three-phase transformer as shown in Fig. 13.5. Assume that the voltage on the high-voltage side of the transformer is adjusted to 132 kV, the transformer is unloaded, and no currents are flowing among the alternators. If a three-phase short circuit occurs on the high-voltage side of the transformer, compute the subtransient current in each alternator.

Calculation Procedure

1. Select the Base Quantities for Per-Unit Calculations

The selection of base quantities is arbitrary, but usually these quantities are selected to minimize the number of conversions from one set of base quantities to another. In this example, the following base quantities are chosen: 50,000 kVA or 50 MVA, 13.2 kV (low-voltage side), and 138 kV (high-voltage side).

2. Convert Per-Unit Reactances to Selected Base Quantities

Alternators 1 and 2 are equal in all respects; the per-unit (p.u.) reactance of each is corrected in accordance with the equation, $X''_{\text{new base}} = X''_{\text{old base}}\text{MVA}_{\text{new base}}\text{kV}^2_{\text{old base}}/\text{MVA}_{\text{old base}}\text{kV}^2_{\text{new base}} = (0.14)(50)(13.2)^2/(25)(13.2)^2 = 0.28$ p.u. Because both the old

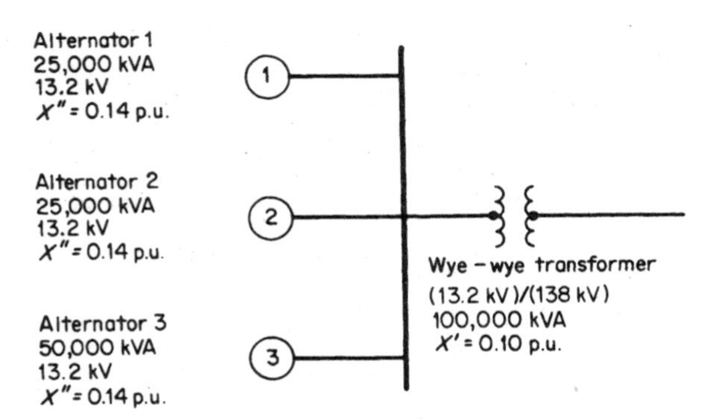

FIGURE 13.5 Paralleled alternators connected to a wye-wye transformer.

and new base voltages in this equation are 13.2 kV, the base voltage has no effect on the calculation in this example; this is not always the case.

The correction for the per-unit reactance of alternator 3 is done in the same manner except that in this case the per-unit reactance that is given is already on the selected base of 50,000 kVA and 13.2 kV. The correction of the per-unit reactance of the transformer is done similarly, $X_{\text{new base}} = (0.10)(50)/100 = 0.05$ p.u.

3. *Calculate the Internal Voltage of the Alternators*

The internal voltage of the alternators must be calculated according to the actual voltage conditions on the system, just prior to the fault. Because the alternators are unloaded (in this problem), there is no voltage drop; the internal voltage is the same as the voltage on the high-voltage side of the transformer when the transformer ratio is taken into account. $E_{\text{alt 1}} = E_{\text{alt 2}} =$ (actual high-side voltage)/(base high-side voltage) $= 132$ kV/138 kV $= 0.957$ p.u. For alternator 3, the calculation is the same, yielding 0.957 p.u. for the internal generated voltage.

4. *Calculate the Subtransient Current in the Short Circuit*

The subtransient current in the short circuit is determined from the equation $\mathbf{I}'' = \mathbf{E}_{\text{alt}}/$reactance to fault $= 0.957/(j0.07 + j0.05) = 0.957/j0.12 = -j7.98$ p.u. (see Fig. 13.6).

5. *Calculate the Voltage on the Low-Voltage Side of the Transformer*

The voltage on the low-voltage side of the transformer is equal to the voltage rise from the short-circuit location (zero voltage) through the transformer reactance, $j0.05$ per unit. $\mathbf{V}_{\text{low side}} = \mathbf{I}''\mathbf{X}_{\text{trans}} = (-j7.98)(j0.05) = 0.399$ p.u.

6. *Calculate Individual Alternator Currents*

The general equation for individual alternator currents is $\mathbf{I}''_{\text{alt}} =$ (voltage drop across alternator reactance)/(alternator reactance). Thus, $\mathbf{I}''_{\text{alt}}\ 1 = (0.957 - 0.399)/j0.28 = -j1.99$ p.u. $\mathbf{I}''_{\text{alt}}\ 2$ is the same as that of alternator 1. $\mathbf{I}''_{\text{alt}}\ 3 = (0.957 - 0.399)/j0.14 = -j3.99$ p.u.

7. *Convert Per-Unit Currents to Amperes*

The base current = base kVA/($\sqrt{3}$)(base kV) = 50,000/($\sqrt{3}$) (13.2) = 2187 A. Thus, $\mathbf{I}''_{\text{alt}}\ 1 = \mathbf{I}''_{\text{alt}}\ 2 = (2187)(-j1.99) = 4352\angle{-90^\circ}$ A. $\mathbf{I}''_{\text{alt}}\ 3 = (2187)(-j3.99) = 8726\angle{-90^\circ}$ A.

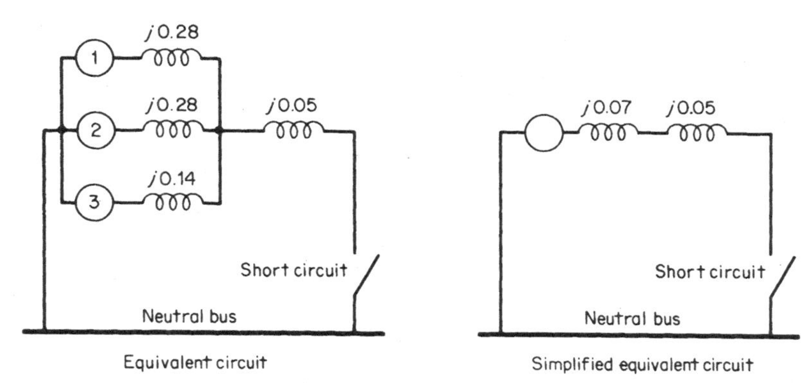

FIGURE 13.6 Development of equivalent circuit where all reactances are on a per-unit basis.

Related Calculations. These calculations have yielded subtransient currents because subtransient reactances were used for the alternators. For calculations of transient currents or synchronous currents, the respective reactances must be transient or synchronous, as the case may be.

SUBTRANSIENT, TRANSIENT, AND SYNCHRONOUS SHORT-CIRCUIT CURRENTS

A pumped-storage facility using hydroelectric generators is connected into a much larger 60-Hz system through a single 138-kV transmission line, as shown in Fig. 13.7. The transformer leakage reactances are each 0.08 p.u., the larger system is assumed to be an infinite bus, and the inductive reactance of the transmission line is 0.55 p.u. The 50,000-kVA, 13.8-kV pumped-storage generating station is taken as the base of the given per-unit values. A solid three-phase short circuit occurs on the transmission line adjacent to the sending-end circuit breaker. Before the short circuit the receiving-end bus was at 100 percent value, unity power factor, and the hydroelectric generators were 75 percent loaded, on the basis of kVA rating. Determine the subtransient, transient, and synchronous short-circuit currents.

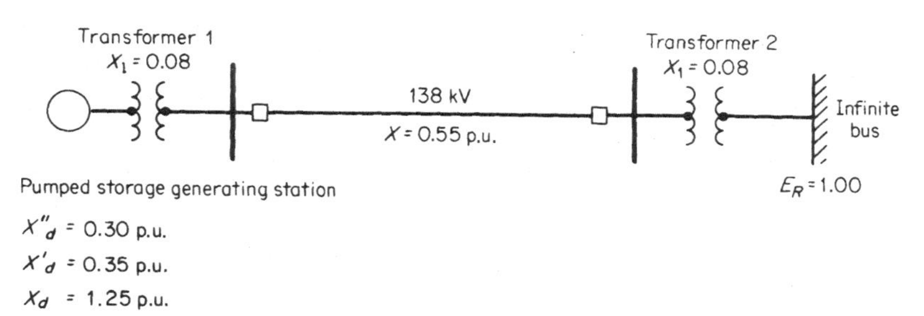

FIGURE 13.7 One-line diagram of a generating and transmission system.

Calculation Procedure

1. Compute the Prefault Voltage behind the Subtransient Reactance

In this first calculation, the subtransient short-circuit current is determined; the whole series of calculations will be repeated for determining the transient short-circuit current and repeated again for determining the synchronous short-circuit current. The total impedance from the internal voltage of the hydroelectric generators to the infinite bus for the subtransient calculation is $X''_d + X_l$ of transformer 1 + X of the transmission line + X_l of transformer 2, or 0.30 + 0.08 + 0.55 + 0.08 = 1.01 p.u. The receiving-end voltage $\mathbf{E_R} = 1.00 + j0$, and the current from the hydroelectric generators is 0.75 p.u (i.e., 75 percent loaded on the basis of kVA rating). Thus, $\mathbf{E''_{int}} = 1.00 + j0 + (0.75)(j1.01) = 1.00 + j0.76 = 1.26\underline{/37.2°}$ p.u. voltage, where $\mathbf{E''_{int}}$ is the subtransient internal voltage.

2. Compute the Subtransient Current to the Fault

The impedance from the internal voltage of the hydroelectric generators to the fault location = $X_{gf} = X''_d + X_{l1} = 0.30 + 0.08 = 0.38$ p.u., where X_{gf} is the generator-to-fault impedance. The subtransient current from the hydroelectric generators to the fault = $E''_{int}/X_{gf} = 1.26/0.38 = 3.32$ p.u. The subtransient current from the infinite bus to the fault = $E_R/X_{bf} = 1.00/(0.08 + 0.55) = 1.00/0.63 = 1.59$ p.u., where X_{bf} is the impedance from the infinite bus to the fault. The total subtransient short-circuit current at the fault = 3.32 from the hydroelectric generators + 1.59 from the infinite bus = 4.91 p.u.

3. Compute the Effect of the Maximum dc Component Offset

The maximum possible dc offset is taken to be $\sqrt{2}$ times the symmetrical wave, and the value of the total offset wave is the short-circuit current, $I_{sc} = \sqrt{I_n^2 + I_w^2}$, where I_n is the current with dc offset neglected and I_w is the current with dc offset. From the hydroelectric generators, current (with maximum dc component) is $3.32\sqrt{2}$, and from the infinite bus, current is $1.59\sqrt{2}$, for a total of $4.91\sqrt{2} = 6.94$ p.u. The greatest rms value of $I_{sc} = \sqrt{4.91^2 + 6.94^2} = 8.5$ p.u.

4. Convert the Per-Unit Current to Amperes

The base current = $(50{,}000)/\sqrt{3}(138) = 209.2$ A. Therefore, I_{sc} = (8.5 p.u.)(209.2) = 1778 A, the subtransient current at the fault.

5. Compute the Prefault Voltage behind the Transient Reactance

Use the procedure as for the subtransient case, the total impedance from the internal voltage of the hydroelectric generators to the infinite bus for the transient calculation is $X'_d + X_l$ of transformer 1 + X of the transmission + X_l of transformer 2, or 0.35 + 0.08 + 0.55 + 0.08 = 1.06 p.u. The transient internal voltage of the hydroelectric generators $\mathbf{E'_{int}} = 1.00 + j0 + 0.75(j1.06) = 1.00 + j0.80 = 1.28\underline{/38.7°}$ p.u. voltage.

6. Calculate the Transient Current to the Fault

The total transient reactance from the internal voltage of the hydroelectric generators to the fault location is $X'_{gf} = X'_d + X_l$ of transformer 1 = 0.35 + 0.08 = 0.43 p.u. The transient current from the hydroelectric generators to the fault is $I'_g = E'_{int}/X'_{gf} = 1.28/0.43 = 2.98$ p.u. (only the magnitude is considered, not the angle).

The transient current from the infinite bus to the fault, $I'_b = E_R/X'_{bf} = 1.00/(0.08 + 0.55) = 1.00/0.63 = 1.59$ p.u. Thus, the total transient short-circuit current at the fault location I'_t= 2.98 (from the hydroelectric generators) + 1.59 (from the infinite bus) = 4.57 p.u.

7. Convert the Per-Unit Current in Amperes

As before, the base current is 209.2 A; the transient current at the fault location is I_{sc} = (4.57 p.u.)(209.2 A) = 956 A.

8. Calculate the Synchronous Current

Use the same procedure as for subtransient and transient current but substitute the synchronous reactance of the generator rather than the subtransient or transient reactance. $\mathbf{E}_{int} = 1.00 + j0 + 0.75(j1.96) = 1.00 + j1.47 = 1.78\angle 55.8°$ p.u. voltage. $X_{gf} = X_d + X_{l1} = 1.25 + 0.08 = 1.33$ p.u. $I_g = E_{int}/X_{gf} = 1.78/1.33 = 1.34$ p.u. (only the magnitude is considered, not the angle). $I_b = E_R/X_{bf} = 1.00(0.08 + 0.55) = 1.00/0.63 = 1.59$ p.u. The total synchronous short-circuit current at the fault location, I_t = 1.33 (from the hydroelectric generators) + 1.59 (from the infinite bus) = 2.92 p.u. Thus, as before, I_t = (2.92 p.u.)(209.2 A) = 610.9 A.

Related Calculations. For all situations, the calculations for subtransient, transient, and synchronous currents are done with the respective reactances of the generator. The subtransient reactance is the smallest of the three and yields the largest short-circuit current. In addition, the dc offset is used for the subtransient condition, usually considered to be within the first three cycles. Synchronous conditions may be considered to prevail after the first 60 cycles (1 s).

POWER IN UNBALANCED THREE-PHASE CIRCUITS

A balanced three-phase distribution system has 240 V between phases; a 20-Ω resistive load is connected from phase *b* to *c*; phase *a* is open. Using symmetrical components, calculate the power delivered to the resistor. Actually, the resistor represents an unbalanced load on a balanced three-phase system. See Fig. 13.8.

Calculation Procedure

1. Determine the Voltages at the Load

In order to determine the voltages of the system at the point of load, let the voltage between phases *b* and *c* be the reference and assume a sequence of rotation of *a*, *b*, *c*. By inspection, write $\mathbf{E}_{cb} = 240\angle 0°$, $\mathbf{E}_{ba} = 240\angle 120°$, $\mathbf{E}_{ac} = 240\angle 240°$, or $\mathbf{E}_a = (240/\sqrt{3})\angle 90°$, $\mathbf{E}_b = (240/\sqrt{3})\angle -30°$, and $\mathbf{E}_c = (240/\sqrt{3})\angle 210°$, where the phase voltage $240/\sqrt{3} = 138.6$ V.

2. Determine the Corresponding Branch Currents

$\mathbf{I}_{cb} = \mathbf{E}_{cb}/\mathbf{Z} = (240\angle 0°)/20\angle 0°) = 12\angle 0°$ A, and both $\mathbf{I}_{ba}$ and $\mathbf{I}_{ac} = 0$, because of the open circuit from phase *a* to *b* and from phase *a* to *c*.

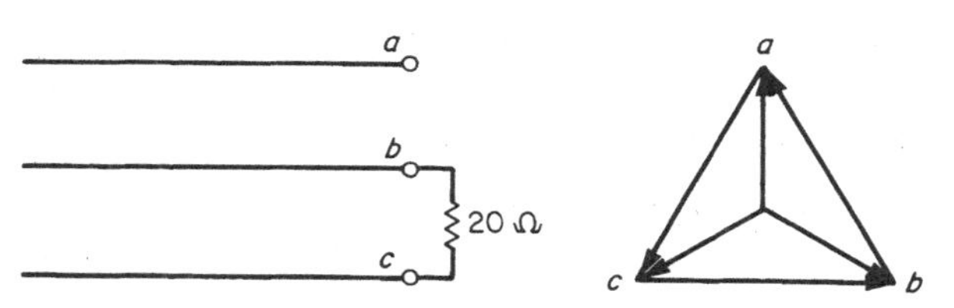

FIGURE 13.8 Unbalanced resistor load on a balanced voltage system.

3. Prepare to Calculate the Power Delivered

Assuming that in an unbalanced three-phase circuit the total power is the sum of the powers represented by the separate phase-sequence-component products, use the relation for total power, $P_t = 3E_{a1}I_{a1}\cos\theta_1 + 3E_{a2}I_{a2}\cos\theta_2 + 3E_{a0}I_{a0}\cos\theta_0$. In this problem, the positive-sequence voltage is the only voltage to be considered because the input voltages are balanced; negative- and zero-sequence voltages are not present. Consequently, to solve the power equation it is not necessary to calculate I_{a2} nor I_{a0} because those terms will go to zero; only I_{a1} need be calculated. The positive-sequence component of the voltage, thus, is the same as the phase voltage, $E_a = E_{a1} = 240/\sqrt{3} = 138.6\underline{/90^\circ}$. Before proceeding further, calculate the positive sequence component of the current.

4. Calculate the Positive Sequence Component of Current

Use the equation: $\mathbf{I_{a1}} = (\mathbf{I_a} + a\mathbf{I_b} + a^2\mathbf{I_c})/3$, where in this problem $\mathbf{I_a} = 0$ and $\mathbf{I_b} = -\mathbf{I_c} = 12\underline{/0^\circ}$. $\mathbf{I_{a1}} = (0 + 12\underline{/120^\circ} + 12\underline{/180^\circ + 240^\circ})/3 = (12\underline{/120^\circ} + 12\underline{/60^\circ})/3 = (-6 + j10.4 + 6 + j10.4)/3 = j20.8/3 = j6.93 = 6.93\underline{/90^\circ}$ A.

5. Calculate the Power Delivered

Use the equation $P_t = 3E_{a1}I_{a1}\cos\theta = (3)(138.6)(6.93)\cos(90^\circ - 90^\circ) = 2880$ W.

Related Calculations. It should be recognized that the answer to this problem is obtained immediately from the expression for power in a resistance, $P = E^2/R = 240^2/20 = 2880$ W. However, the more powerful concept of symmetrical components is used in this case to demonstrate the procedure for more complex situations. The general solution shown is most appropriate for cases where there are not only positive-sequence components but also negative- and/or zero-sequence components. It is important to recognize that the total power is the sum of the powers represented by the separate phase-sequence-component products.

DETERMINATION OF PHASE-SEQUENCE COMPONENTS

A set of unbalanced line currents in a three-phase, four-wire system is as follows: $\mathbf{I_a} = -j12$, $\mathbf{I_b} = -16 + j10$, and $\mathbf{I_c} = 14$. Find the positive-, negative-, and zero-sequence components of the current.

Calculation Procedure

1. Change Cartesian Form of Current to Polar Form

Use the standard trigonometric functions (sin, cos, and tan) to convert the given line currents into polar form: $\mathbf{I_a} = -j12 = 12\underline{/-90^\circ}$, $\mathbf{I_b} = -16 + j10 = 18.9\underline{/148^\circ}$, $\mathbf{I_c} = 14\underline{/0^\circ}$.

2. Calculate Positive-Sequence Components

Use the equation for positive-sequence components of current: $\mathbf{I_{a1}} = (\mathbf{I_a} + a\mathbf{I_b} + a^2\mathbf{I_c})/3 = (12\underline{/-90^\circ} + 18.9\underline{/148^\circ + 120^\circ} + 14\underline{/240^\circ})/3 = 0 - j12 - 0.66 - j18.89 - 7.0 - j12.12)/3 = (-7.66 - j43.01)/3 = 14.56\underline{/259.9^\circ}$. Thus, the a-phase positive-sequence component $\mathbf{I_{a1}} = 14.56\underline{/259.9^\circ}$, the b-phase positive-sequence component $\mathbf{I}_{b1} = 14.56\underline{/259.9^\circ - 120^\circ} = 14.56\underline{/139.9^\circ}$, and the c-phase positive-sequence component $\mathbf{I}_{c1} = 14.56\underline{/259.9^\circ + 120^\circ} = 14.56\underline{/19.9^\circ}$. See Fig. 13.9.

3. Calculate Negative-Sequence Components

Use the equation for negative-sequence components of currents: $\mathbf{I}_{a2} = (\mathbf{I}_a + a^2\mathbf{I}_b + a\mathbf{I}_c)/3 = (12\underline{/-90^\circ} + 18.9\underline{/148^\circ + 240^\circ} + 14\underline{/120^\circ})/3 = 4.41\underline{/42.9^\circ}$. Thus, the a-phase negative sequence component, $\mathbf{I}_{a2} = 4.41\underline{/42.9^\circ}$, the b-phase negative sequence component, $\mathbf{I}_{b2} = 4.41\underline{/42.9^\circ + 120^\circ} = 4.41\underline{/162.9^\circ}$, and the c-phase negative sequence component, $\mathbf{I}_{c2} = 4.41\underline{/42.9^\circ + 240^\circ} = 4.41\underline{/-77.1^\circ}$. See Fig. 13.9.

4. Calculate the Zero-Sequence Components

Use the equation for zero-sequence components of current, $\mathbf{I}_{a0} = (\mathbf{I}_a + \mathbf{I}_b + \mathbf{I}_c)/3 = (-j12 - 16 + j10 + 14)/3 = 0.94\underline{/225^\circ}$. The zero-sequence component for each phase is the same, $\mathbf{I}_{a0} = \mathbf{I}_{b0} = \mathbf{I}_{c0} = 0.94\underline{/225^\circ}$. See Fig. 13.9.

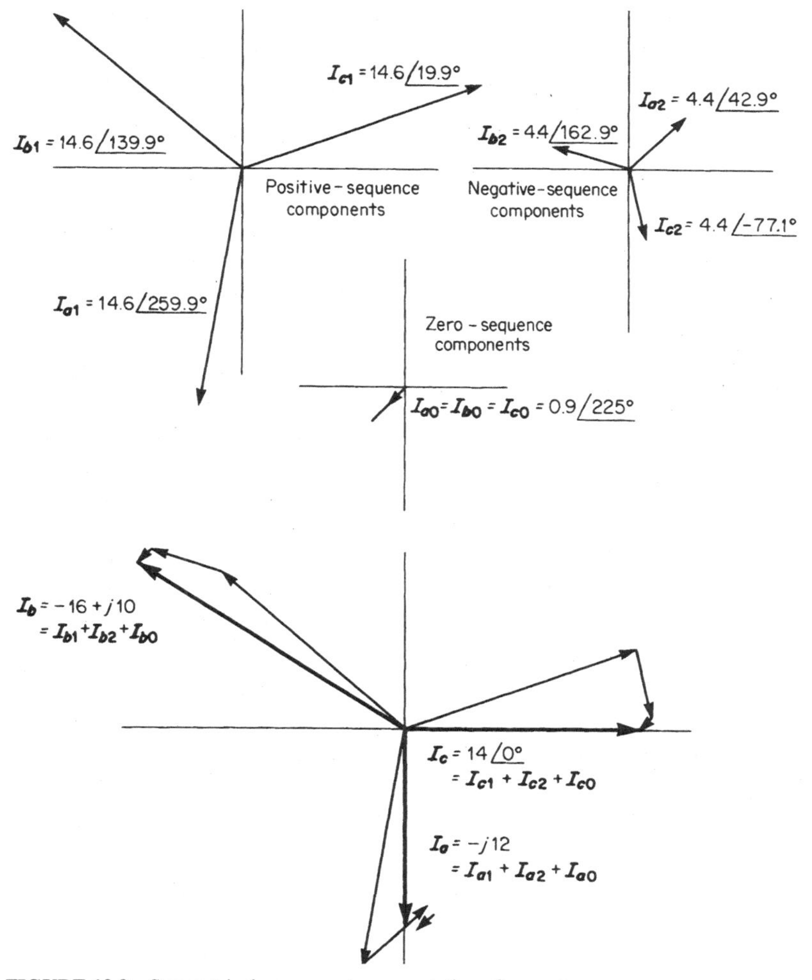

FIGURE 13.9 Symmetrical component representation of currents.

5. Calculate the Phase Currents

This is merely a final check and demonstrates the procedure for determining the phase currents if the sequence components are known; the calculation may be done graphically as shown in Fig. 13.9 or mathematically. The mathematical solution follows. The a-phase current $\mathbf{I}_a = \mathbf{I}_{a0} + \mathbf{I}_{a1} + \mathbf{I}_{a2} = 0.94\angle 225^\circ + 14.56\angle 259.9^\circ + 4.41\angle 42.9^\circ = -j12$. The b-phase current $\mathbf{I}_b = \mathbf{I}_{a0} + a^2\mathbf{I}_{a1} + a\mathbf{I}_{a2} = \mathbf{I}_{b0} + \mathbf{I}_{b1} + \mathbf{I}_{b2} = 0.94\angle 225^\circ + 14.56\angle 139.9^\circ + 4.41\angle 162.9^\circ = -16 + j10$. The c-phase current $\mathbf{I}_c = \mathbf{I}_{a0} + \mathbf{a}\mathbf{I}_{a1} + a^2\mathbf{I}_{a2} = \mathbf{I}_{c0} + \mathbf{I}_{c1} + \mathbf{I}_{c2} = 0.94\angle 225^\circ + 14.56\angle 19.9^\circ + 4.41\angle -77.1^\circ = 14 + j0$.

Related Calculations. The procedure shown here is applicable for determining the nine symmetrical components, $\mathbf{I}_{a0}$, $\mathbf{I}_{b0}$, $\mathbf{I}_{c0}$, $\mathbf{I}_{a1}$, $\mathbf{I}_{b1}$, $\mathbf{I}_{c1}$, $\mathbf{I}_{a2}$, $\mathbf{I}_{b2}$, and $\mathbf{I}_{c2}$, when one is given unbalanced phase currents. By substitution of voltages for currents, the same equation forms may be used for determining the nine symmetrical components of unbalanced phase voltages. On the other hand, if the symmetrical components are known, the equations used in Step 5 give the unbalanced phase voltages. In many instances, one or more of the negative- or zero-sequence components may not exist (i.e., may equal zero). For perfectly balanced three-phase systems with balanced voltages, currents, impedances, and loads, there will be no negative- and zero-sequence components; only positive-sequence components will exist.

PROPERTIES OF PHASOR OPERATORS, j AND a

Determine the value of each of the following relations and express the answer in polar form: ja, $1 + a + a^2$, $a + a^2$, $a^2 + ja + ja^2 3$.

Calculation Procedure

1. Determine the Value of the Operators

A phasor on which a operates is rotated 120° counterclockwise (positive or forward direction), and there is no change in magnitude. See Fig. 13.10. A phasor on which j operates is rotated by 90° counterclockwise. Although $-j$ signifies rotation of $-90°$ because that position is exactly 180° out of phase with $+j$, the similar situation is not true for $-a$ as compared with $+a$. If $+a = 1\angle 120^\circ$, the position of 180° reversal is at $-60°$; therefore, $-a = 1\angle -60^\circ$.

2. Determine the Value of **ja**

Use the relations $j = 1\angle 90^\circ$ and $a = 1\angle 120^\circ$. Thus, $ja = 1\angle 210^\circ$.

3. Determine the Value of **1 + a + a²**

This is a very common expression occurring with symmetrical components. It represents three balanced phasors of magnitude 1, each displaced 120° from the other. The sum is equal to $1\angle 0^\circ + 1\angle 120^\circ + 1\angle 240^\circ = 0$.

4. Determine the Value of **a + a²**

Use the relations $a = 1\angle 120^\circ$ and $a^2 = 1\angle 240^\circ$. Change each to Cartesian form: $a = 1\angle 120^\circ = -0.5 + j0.866$, $a^2 = 1\angle 240^\circ = -0.5 - j0.866$. Therefore, $a + a^2 = -0.5 + j0.866 - 0.5 - j0.866 = -1.0 + j0 = 1\angle 180^\circ = (1\angle 90^\circ)(1\angle 90^\circ) = j^2$.

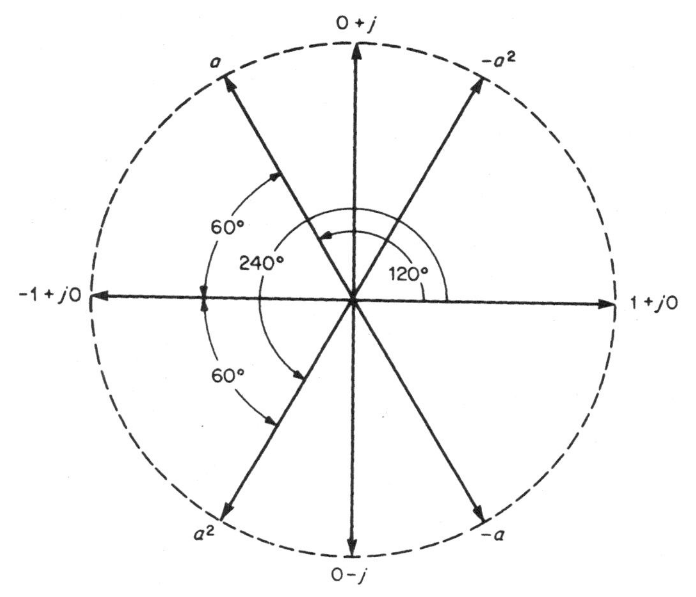

FIGURE 13.10 Properties of phasor operator a.

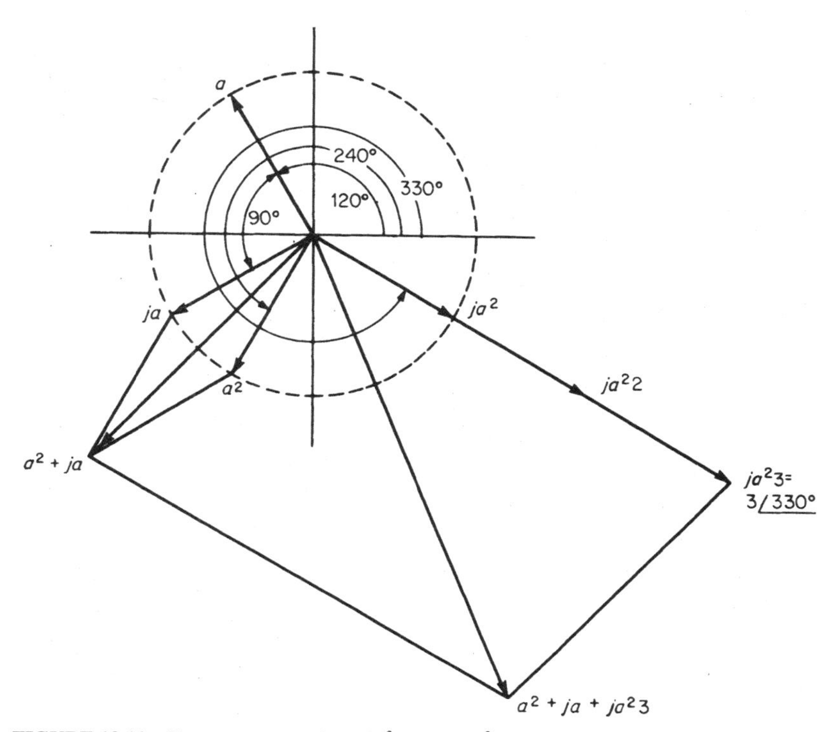

FIGURE 13.11 Phasor representation of $a^2 + ja + ja^23$.

5. Determine the Value of $a^2 + ja + ja^2 3$

Use the relation, $a^2 = 1\angle 240° = -0.5 - j0.866$, $ja = (1\angle 90°)(1\angle 120°) = 1\angle 210° = -0.866 - j0.5$, and $ja^2 3 = (1\angle 90°)(1\angle 240°)(3) = 3\angle 330° = 2.6 - j1.5$. The sum is $-0.5 - j0.866 - 0.866 - j0.5 + 2.6 - j1.5 = 1.234 - j2.866 = 3.12\angle -66.7°$. See Fig. 13.11.

Related Calculations. The j operator is very common in all power calculations; the a operator is used in all calculations involving symmetrical components.

COMPLEX POWER CALCULATED WITH SYMMETRICAL COMPONENTS

The resolution of a set of three-phase unbalanced voltages into symmetrical components yields the following: $\mathbf{V}_{a1} = 150\angle 0°$ V, $\mathbf{V}_{a2} = 75\angle 30°$ V, and $\mathbf{V}_{a0} = 10\angle -20°$ V. The component currents are $\mathbf{I}_{a1} = 12\angle 18°$ A, $\mathbf{I}_{a2} = 6\angle 30°$ A, and $\mathbf{I}_{a0} = 12\angle 200°$ A. Determine the complex power represented by these voltages and currents.

Calculation Procedure

1. Calculate Zero-Sequence Complex Power

Use the relation zero-sequence complex power $\mathbf{S}_0 = P_0 + jQ_0 = 3\mathbf{V}_{a0}\mathbf{I}^*_{a0}$, where the asterisk signifies the conjugate of the quantity (i.e., if $\mathbf{I}_{a0} = 12\angle 200°$, the conjugate of the quantity $\mathbf{I}^*_{a0} = 12\angle -200°$). The zero-sequence complex power is $3\mathbf{V}_{a0}\mathbf{I}^*_{a0} = (3)(10\angle -20°)(12\angle -200°) = 360\angle -220° = (-275.8 + j231.4)$ VA $= -275.8$ W $+ j231.4$ VARS.

2. Calculate Positive-Sequence Complex Power

Use the relation positive-sequence complex power $\mathbf{S}_1 = P_1 + jQ_1 = 3\mathbf{V}_{a1}\mathbf{I}^*_{a1} = (3)(150\angle 0°)(12\angle -18°) = 5400\angle -18° = (5135.7 - j1668.7)$ VA $= 5135.7$ W $- j1688.7$ VARS.

3. Calculate Negative-Sequence Complex Power

Use the relation negative-sequence complex power $\mathbf{S}_2 = P_2 + jQ_2 = 3\mathbf{V}_{a2}\mathbf{I}^*_{a2} = (3)(75\angle 30°)(6\angle -30°) = 1350\angle 0°$ VA $= 1350$ W $+ j0$ VARS.

4. Calculate Total Complex Power

Use the relation, $\mathbf{S}_t = P_t + jQ_t = 3\mathbf{V}_{a0}\mathbf{I}^*_{a0} + 3\mathbf{V}_{a1}\mathbf{I}^*_{a1} + 3\mathbf{V}_{a2}\mathbf{I}^*_{a2} = (-275.8 + 5135.7 + 1350)$ W $+ j(231.4 - 1668.7)$ VARS $= 6209.9$ W $- j1437.3$ VARS.

5. Alternative Solution: Compute the Phase Voltage

As an alternative solution and as a check, compute the phase voltages and currents: $\mathbf{V}_a = \mathbf{V}_{a0} + \mathbf{V}_{a1} + \mathbf{V}_{a2} = 10\angle -20° + 150\angle 0° + 75\angle 30° = 224.4 + j34.1 = 226.9\angle 8.6°$. In this calculation, the intervening mathematical steps, wherein the polar form of the phasors are converted to Cartesian form, are not shown. Similarly, $\mathbf{V}_b = \mathbf{V}_{b0} + \mathbf{V}_{b1} + \mathbf{V}_{b2} = \mathbf{V}_{a0} + a^2\mathbf{V}_{a1} + a\mathbf{V}_{a2} = 10\angle -20° + 150\angle 240° + 75\angle 150° = 161.9\angle 216.3°$. $\mathbf{V}_c = \mathbf{V}_{c0} + \mathbf{V}_{c1} + \mathbf{V}_{c2} = \mathbf{V}_{a0} + a\mathbf{V}_{a1} + a^2\mathbf{V}_{a2} = 10\angle -20° + 150\angle 120° + 75\angle 270° = 83.4\angle 141.9°$.

6. Compute the Phase Currents

Use the same relations as for the phase voltages: $\mathbf{I}_a = \mathbf{I}_{a0} + \mathbf{I}_{a1} + \mathbf{I}_{a2} = 12\angle 200° + 12\angle 18° + 6\angle 30° = 5.9\angle 26.0°$ and $\mathbf{I}^*_a = 5.9\angle -26.0°$. $\mathbf{I}_b = \mathbf{I}_{b0} + \mathbf{I}_{b1} + \mathbf{I}_{b2} =$

$\mathbf{I}_{a0} + a^2\mathbf{I}_{a1} + a\mathbf{I}_{a2} = 12\angle 200° + 12\angle 258° + 6\angle 150° = 22.9\angle 214.1°$ and $\mathbf{I}_b^* = 22.9\angle -214.1°$. $\mathbf{I}_c = \mathbf{I}_{c0} + \mathbf{I}_{c1} + \mathbf{I}_{c2} = \mathbf{I}_{a0} + a\mathbf{I}_{a1} + a^2\mathbf{I}_{a2} = 12\angle 200° + 12\angle 138° + 6\angle 270° = 20.3\angle 185.9°$ and $\mathbf{I}_c^* = 20.3\angle -185.9°$.

7. Compute the Complex Power

Use the relation, $\mathbf{S}_t = \mathbf{V}_a\mathbf{I}_a^* + \mathbf{V}_b\mathbf{I}_b^* + \mathbf{V}_c\mathbf{I}_c^* = (226.9\angle 8.6°)(5.9\angle -26.0°) + (161.9\angle 216.3°)(22.9\angle -214.1°) + (83.4\angle 141.9°)(20.3\angle -185.9°)$. Completing the mathematical solution of this equation will yield the same result (6210 W $- j$1436 VARS) as did the solution using the symmetrical components.

Related Calculations. This problem illustrates two methods of finding the complex power $P + jQ$; namely, (1) by symmetrical components and (2) by unbalanced phase components. In either case, the complex power is obtained by summation of the products of the respective phasor voltages by the conjugate phasor currents.

IMPEDANCES AND REACTANCES TO DIFFERENT SEQUENCES

A salient-pole generator is connected to a system having a reactance of 9.0 p.u., as shown in Fig. 13.12; the system base values are 15,000 kVA and 13.2 kV. Draw the positive-, negative-, and zero-sequence diagrams for a three-phase short circuit at the load.

Calculation Procedure

1. Assign Impedance and/or Reactance Values to the Generator

When impedance and/or reactance values are not given for parts of a system, it is necessary to make estimates. The literature of the power industry contains extensive listings of typical values of reactance; in most cases, the resistance is neglected and only the reactance is used. Typical values for generator reactance are subtransient, $X''_d = 0.10$ p.u., transient, $X'_d = 0.20$ p.u., and synchronous, X_d 1.20 p.u. (each of these being positive-sequence values).

The negative-sequence reactance can vary between 0.10 p.u. for large two-pole turbine generators to 0.50 for salient-pole machines; the zero-sequence reactance can vary similarly from 0.03 to 0.20 p.u.

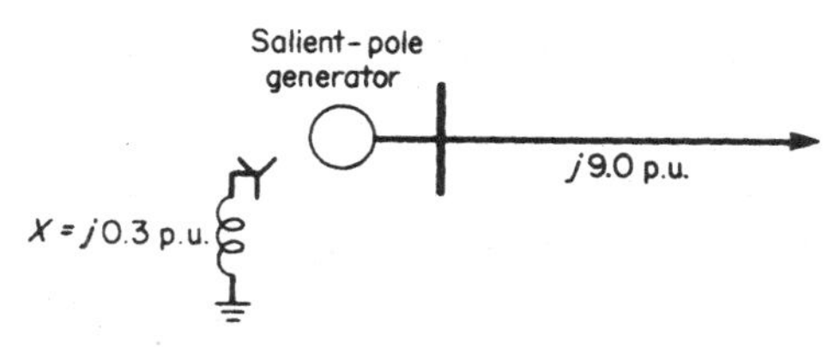

FIGURE 13.12 Representation of salient-pole generator connected to a system.

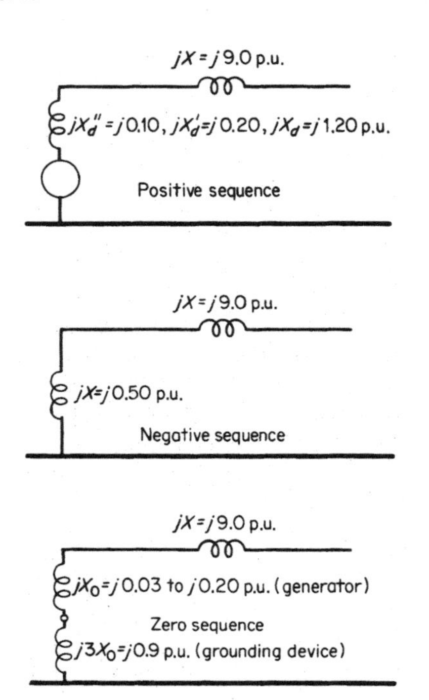

FIGURE 13.13 Posititve- , negative- , and zero-sequence diagrams; approximate and typical values of generator reactances are included.

2. Draw the Positive-Sequence Diagram

The generated voltage, being of positive sequence *a*, *b*, *c*, is shown in the positive-sequence diagram only. The grounding reactance of the generator does not appear in the positive-sequence diagram. See Fig. 13.13.

3. Draw the Negative-Sequence Diagram

No generated voltage appears in the negative-sequence diagram; neither is there shown the reactance of the grounding device. See Fig. 13.13.

4. Draw the Zero-Sequence Diagram

No generated voltage appears in the zero-sequence diagram, but it should be noted that the reactance of the grounding device is multiplied by 3.

Related Calculations. If component reactances are not known, it is possible to make estimates of these values in the per-unit system on the basis of typical values found in handbook tables and manufacturers' literature. For short-circuit calculations, estimated values of reactance will give satisfactory results.

LINE-TO-LINE SHORT-CIRCUIT CALCULATIONS

An ungrounded wye-connected generator having a subtransient reactance, $X''_d = 0.12$ p.u., a negative-sequence reactance, $X_2 = 0.15$ p.u., and a zero-sequence reactance $X_0 = 0.05$ p.u. is faulted at its terminals with a line-to-line short circuit. Determine (1) the line-to-line subtransient short-circuit current and (2) the ratio of that current with respect to three-phase short-circuit current. The generator is rated 10 MW, 13.8 kV, and operates at 60 Hz.

Calculation Procedure

1. Draw the Sequence Networks

It is necessary to draw only the positive- and negative-sequence diagrams because zero-sequence currents are not involved with line-to-line short circuits. See Fig. 13.14.

2. Connect the Sequence Networks for Line-to-Line Fault

For a line-to-line fault, the positive and negative sequence networks are connected in parallel, as shown in Fig. 13.15.

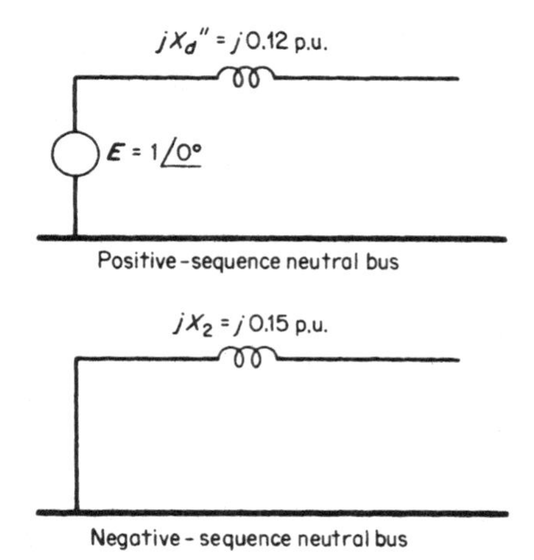

FIGURE 13.14 Positive- and negative-sequence network diagram.

3. Calculate Line-to-Line Fault Current

Use the equation for phase-a positive-sequence current, $\mathbf{I_{a1}} = \mathbf{E/Z} = 1\underline{/0°}/(j0.12 + j0.15) = 1\underline{/0°}/j0.27 = -j3.70$ p.u. The phase-a negative-sequence current $\mathbf{I_{a2}} = -\mathbf{I_{a1}} = +j3.70$ p.u. The phase-a fault current $\mathbf{I_a} = \mathbf{I_{a0}} + \mathbf{I_{a1}} + \mathbf{I_{a2}} = 0 - j3.70 + j3.70 = 0$. The phase-$b$ fault current $\mathbf{I_b} = \mathbf{I_{a0}} + a^2\mathbf{I_{a1}} + a\mathbf{I_{a2}} = 0 + 3.70\underline{/-90° + 240°} + 3.70\underline{/90° + 120°} = -3.20 + j1.85 - 3.20 - j1.85 = -6.40$ p.u. Similarly, the phase-c fault current $\mathbf{I_c} = \mathbf{I_{a0}} + a\mathbf{I_{a1}} + a^2\mathbf{I_{a2}} = 0 + 3.70\underline{/-90° + 120°} + 3.7\underline{/90° + 240°} = 6.40$ p.u. Thus, it is calculated that $\mathbf{I_b} = -\mathbf{I_c} = -6.40$ p.u.

4. Convert the Per-Unit Current to Amperes

First determine the base current using the equation power $= \sqrt{3}V_{\text{line}}I_{\text{line}}\cos\theta$ or $I_{\text{line}} = $ power$/\sqrt{3}V_{\text{line}}\cos\theta$. Power$/\cos\theta$ = volt-amperes. Thus, volt-amperes$/\sqrt{3}V_{\text{line}}$ = 10,000 kVA/($\sqrt{3}$)(13.8 kV) = 418.4 A (base current). Therefore, the fault-current magnitude in phase a and phase b = (418.4 A)(6.40 p.u.) = 2678 A.

5. Calculate the Three-Phase Short-Circuit Current

For the three-phase case, only the positive-sequence network diagram is used. $\mathbf{I_a} = \mathbf{E/Z} = 1\underline{/0°}/j0.12 = -j8.33$ p.u. Converting this to amperes yields (magnitude only) I_a = (8.33 p.u.)(418.4 A) = 3485 A.

6. Calculate Ratio of Short-Circuit Currents

The ratio of line-to-line short-circuit current with respect to three-phase short-circuit current (magnitudes only) is 2678 A/3485 A = 0.768. The same calculation may be done with the per-unit values, namely, 6.40 p.u./8.33 p.u. = 0.768.

Related Calculations. In order to calculate the line-to-line short-circuit currents, it is necessary to establish the positive- and negative-sequence network diagrams; these two networks are connected in parallel and the calculation of sequence components of current proceeds from that point. It matters not how extensive the networks, as long as each net-

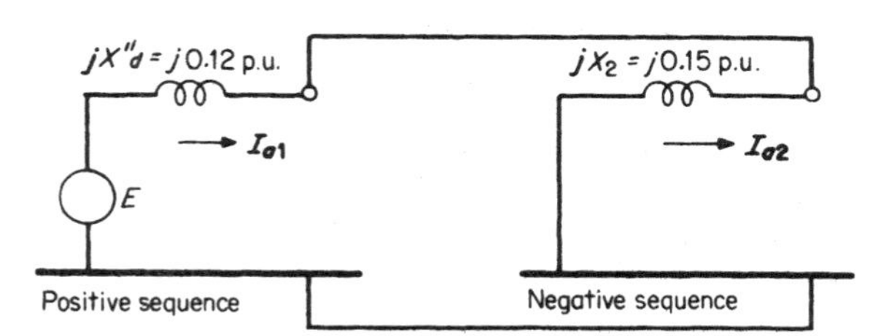

FIGURE 13.15 Positive- and negative-sequence diagrams connected for a line-to-line fault.

work may be reduced to its simplest form. In the usual case with a number of generators, the generated voltages are paralleled in the positive-sequence network.

IMPEDANCE TO ZERO SEQUENCE FOR GENERATORS, TRANSFORMERS, AND TRANSMISSION LINES

A balanced three-phase system is shown in Fig. 13.16. Draw the zero-sequence network.

Calculation Procedure

1. Determine the Zero-Sequence Treatment of Transformers

The zero-sequence equivalent circuit for transformers is shown in Fig. 13.17. It should be noted that zero-sequence current cannot flow in the secondary of the transformer if it does not flow in the primary (provided the transformer itself is not at fault).

2. Determine Zero-Sequence Treatment of Grounding Devices

Use the relation that the reactance of grounding devices appears in the zero-sequence network diagrams at three times the actual value. See Fig. 13.18.

3. Draw the Complete Zero-Sequence Network Diagram

Note that since each end of the transmission line is connected to delta-connected transformers, the line is isolated from zero-sequence currents in the network diagram. See Fig. 13.19.

Related Calculations. This problem illustrates a number of different possibilities that exist in deriving paths for zero-sequence currents; most of the situations that occur are demonstrated. Here again, once the sequence network is drawn, it may be simplified by combining elements.

LINE-TO-GROUND SHORT-CIRCUIT CALCULATIONS

A 13.2-kV, 30,000-kVA generator has positive- , negative- , and zero-sequence reactances of 0.12, 0.12, and 0.08 p.u., respectively. The generator neutral is grounded through a reactance of 0.03 p.u. For the given reactance, determine the line-to-line voltages and short-circuit currents when a single line-to-ground fault occurs at the generator terminals. It may be assumed that the generator was unloaded before the fault.

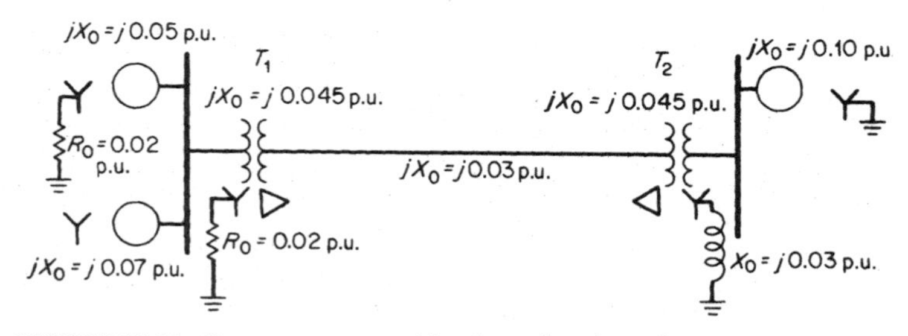

FIGURE 13.16 Zero-sequence quantities shown for a three-phase system.

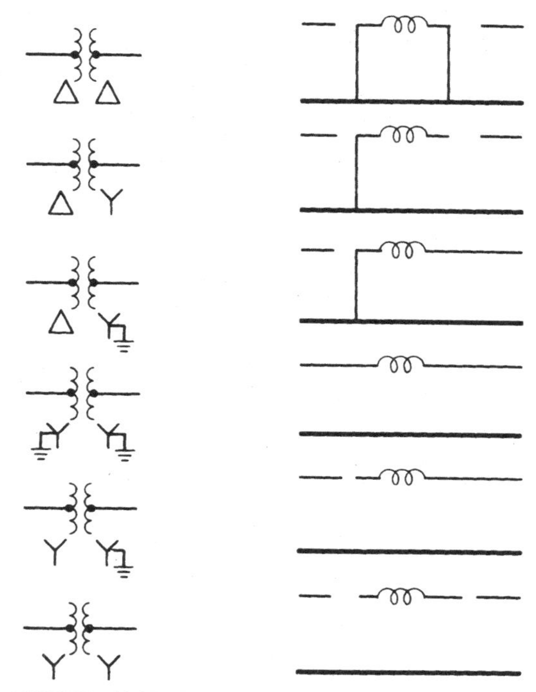

FIGURE 13.17 Zero-sequence equivalent circuits for transformers.

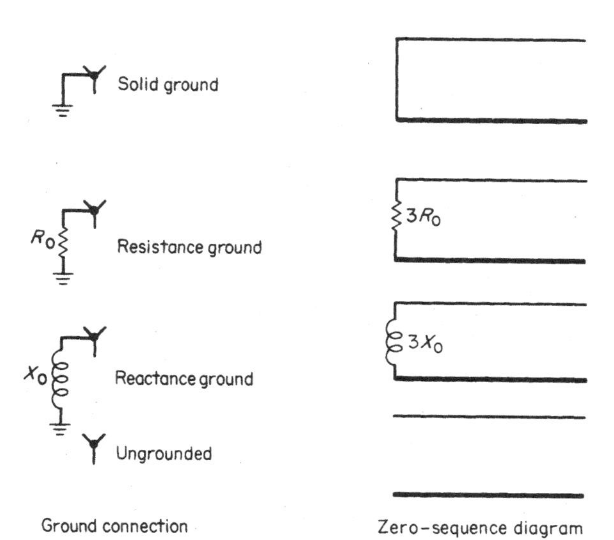

FIGURE 13.18 Zero-sequence diagrams for ground connections.

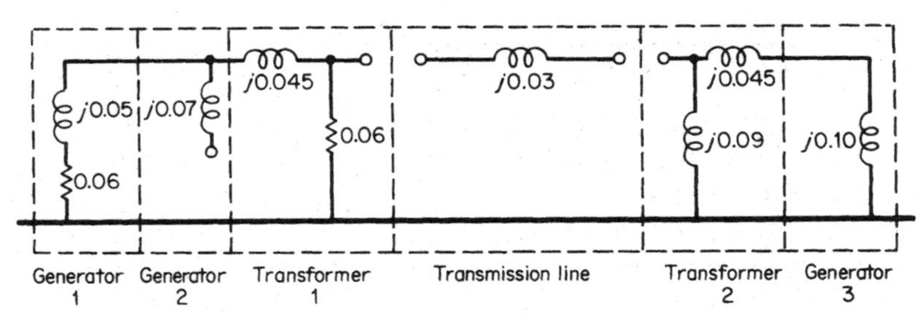

FIGURE 13.19 Complete zero-sequence network diagram where all reactances are on a per-unit basis.

Calculation Procedure

1. Draw the Sequence Network Diagram

The internal generator voltage before the fault is equal to the terminal voltage, because the generator at that time is unloaded; it is equal to $\mathbf{E_g} = 1.0 + j0$ p.u. The sequence network diagram for a single line-to-ground fault is shown in Fig. 13.20.

2. Calculate the Total Series Impedance

The total series impedance $= \mathbf{Z_1} + \mathbf{Z_2} + \mathbf{Z_0}$. In the neutral or ground connection, if it exists, the impedance in the zero-sequence network will be three times the actual value. Thus, $\mathbf{Z_1} + \mathbf{Z_2} + \mathbf{Z_0} = j0.12 + j0.12 + j0.08 + (3)(j0.03) = j0.41$ p.u.

3. Calculate the Positive-, Negative-, and Zero-Sequence Components

The positive-sequence current component $\mathbf{I_{a1}} = \mathbf{E_g}/(\mathbf{Z_1} + \mathbf{Z_2} + \mathbf{Z_0}) = (1.0 + j0)/j0.41 = -j2.44$ p.u. For the single line-to-ground fault, $\mathbf{I_{a1}} = \mathbf{I_{a2}} = \mathbf{I_{a0}}$; thus, $\mathbf{I_{a2}} = -j2.44$ and $\mathbf{I_{a0}} = -j2.44$ p.u.

4. Calculate the Base Current

The base current $= \text{kVA}_{\text{basic}}/(\sqrt{3}\ \text{kV}_{\text{base}}) = 30{,}000/(\sqrt{3})(13.2) = 1312$ A.

5. Calculate the Phase Currents

The phase a current $= \mathbf{I_{a1}} + \mathbf{I_{a2}} + \mathbf{I_{a0}} = 3\mathbf{I_{a1}} = (3)(-j2.44) = -j7.32$ p.u. In SI units, the current in phase a is (7.32 p.u.)(1312 A) = 9604 A. when only the magnitude is of interest, the $-j$ may be neglected in the final step.

The phase b current $= a^2\mathbf{I_{a1}} + a\mathbf{I_{a2}} + \mathbf{I_{a0}} = a^2\mathbf{I_{a1}} + a\mathbf{I_{a1}} + \mathbf{I_{a1}} = \mathbf{I_{a1}}(a^2 + a + 1) = 0$. Similarly, the phase c current $= a\mathbf{I_{a1}} + a^2\mathbf{I_{a2}} + \mathbf{I_{a0}} = a\mathbf{I_{a1}} + a^2\mathbf{I_{a1}} + \mathbf{I_{a1}} = \mathbf{I_{a1}}(a + a^2 + 1) = 0$. Because only phase a is shorted to ground at the fault, phases b and c are open-circuited and carry no current.

6. Calculate the Sequence Voltage Components

With phase a as the reference point at the fault, $\mathbf{V_{a1}} = \mathbf{E_g} - \mathbf{I_{a1}}\mathbf{Z_1} = 1.0 - (-j2.44)(j0.12) = 1.0 - 0.293 = 0.707$ p.u. $\mathbf{V_{a2}} = -\mathbf{I_{a2}}\mathbf{Z_2} = -(-j2.44)(j0.12) = -0.293$ p.u. $\mathbf{V_{a0}} = -\mathbf{I_{a0}}\mathbf{Z_0} = -(-j2.44)(j0.08 + j0.09) = -0.415$ p.u.

7. Convert the Sequence Voltage Components to Phase Voltages

$\mathbf{V_a} = \mathbf{V_{a1}} + \mathbf{V_{a2}} + \mathbf{V_{a0}} = 0.707 - 0.293 - 0.415 = 0$. $\mathbf{V_b} = a^2\mathbf{V_{a1}} = a\mathbf{V_{a2}} + \mathbf{V_{a0}} = 0.707\underline{/240^\circ} - 0.293\underline{/120^\circ} - 0.415 = -0.622 - j0.866$ p.u. $\mathbf{V_c} = a\mathbf{V_{a1}} + a^2\mathbf{V_{a2}} + \mathbf{V_{a0}} = 0.707\underline{/120^\circ} - 0.293\underline{/240^\circ} - 0.415 = -0.622 + j0.866$ p.u.

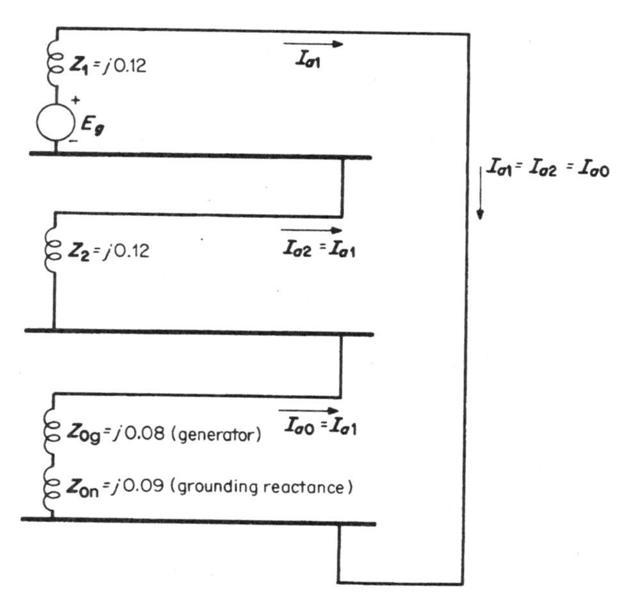

FIGURE 13.20 Sequence network representation of a single line-to-ground fault of unloaded generator; phase a is grounded.

8. Convert the Phase Voltages to Line Voltages

$\mathbf{V}_{ab} = \mathbf{V}_a - \mathbf{V}_b = 0 - (-0.622 - j0.866) = 0.622 + j0.866 = 1.07\underline{/54.3^\circ}$ p.u. $\mathbf{V}_{bc} = \mathbf{V}_b - \mathbf{V}_c = -0.622 - j0.866 - (-0.622 + j0.866) = -j1.732 = 1.732\underline{/270^\circ}$ p.u. $\mathbf{V}_{ca} = \mathbf{V}_c - \mathbf{V}_a = -0.622 + j0.866 = 1.07\underline{/125.7^\circ}$ p.u.

9. Convert the Line Voltages to SI Units

In this problem, the generator voltage *per phase*, $\mathbf{E}_g$, was assumed to be 1.0 p.u. Therefore, 1.0 p.u. voltage = 13.2 kV/$\sqrt{3}$ = 7.62 kV. The line voltage is SI units become

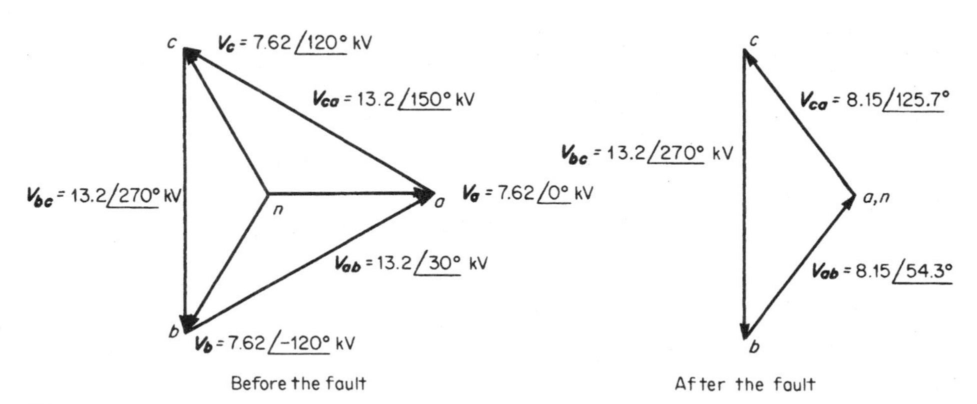

FIGURE 13.21 Voltage phasor diagrams for a single line-to-ground fault on phase a.

$\mathbf{V}_{ab} = (1.07\underline{/54.3^\circ})(7.62\ \text{kV}) = 8.15\underline{/54.3^\circ}$ kV, $\mathbf{V}_{bc} = (1.732\underline{/270^\circ})(7.62\ \text{kV}) = 13.2\underline{/270^\circ}$ kV, and $\mathbf{V}_{ca} = (1.07\underline{/125.7^\circ})(7.62\ \text{kV}) = 8.15\underline{/125.7^\circ}$ kV.

10. Draw the Voltage Phasor Diagram

See Fig. 13.21.

Related Calculations. The procedure shown for a single line-to-ground fault is applicable to any other type of fault provided the sequence networks are connected in the proper manner. For example, for a line-to-line fault, the positive-sequence network and the negative-sequence network are connected in parallel without the zero-sequence network.

SUBTRANSIENT-CURRENT CONTRIBUTION FROM MOTORS; CIRCUIT-BREAKER SELECTION

Consider the system shown in Fig. 13.22, wherein a generator (20,000 kVA, 13.2 kV, $X''_d = 0.14$ p.u. (is supplying two large induction motors (each being 7500 kVA, 6.9 kV, $X''_d = 0.16$ p.u.). The three-phase step-down transformer is rated 20,000 kVA, 13.2/6.9 kV, and its leakage reactance is 0.08 p.u. A three-phase short circuit occurs on the bus. Find the subtransient fault current and the symmetrical short-circuit interrupting current.

Calculation Procedure

1. Convert Subtransient Motor Reactances to Generator-Base kVA

To convert the motor reactance from a 7500-kVA to a 20,000-kVA base, use the relation: $X''_d = (0.16)(20{,}000/7500) = 0.427$ p.u. Thus, the reactance of each motor is $j0.427$ p.u., and the combined reactance of the two motors connected in parallel is $j0.427/2 = j0.214$ p.u.

2. Draw the Network Diagram

The network diagram is shown in Fig. 13.23. The total reactance from the generator to the bus is $j0.14 + j0.08 = j0.22$ p.u., and the total reactance from the motors to the bus is $j0.214$ p.u.

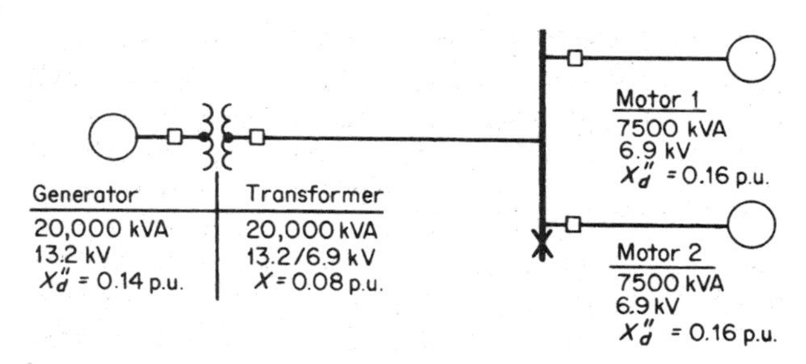

FIGURE 13.22 One-line diagram of a generator supplying two motors.

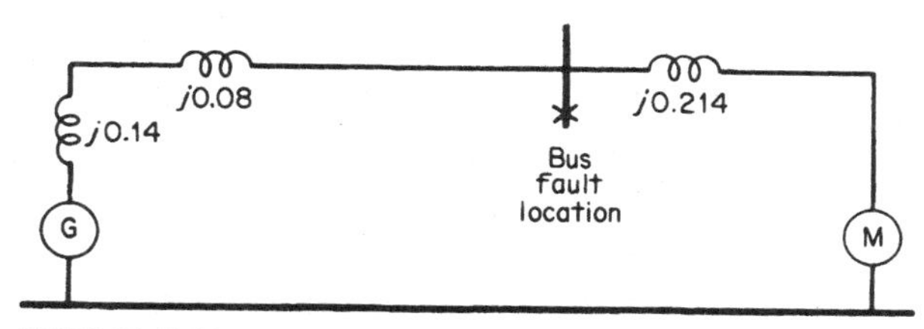

FIGURE 13.23 Equivalent network diagram of a generator supplying a motor load.

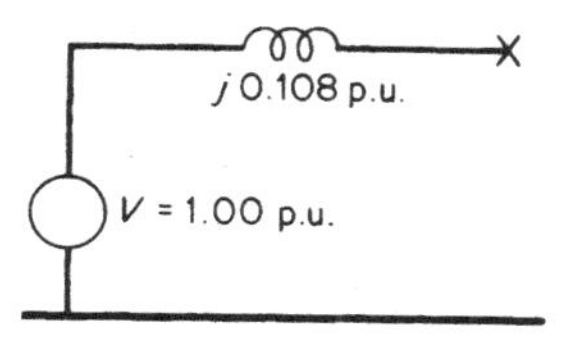

FIGURE 13.24 Reduced network diagram.

3. Reduce the Network Diagram

The network diagram may be reduced by paralleling the generator and motor voltages (1.00 p.u. each) and paralleling the reactance path of (1) generator to faulted bus, and (2) motor to faulted bus, yielding a combined reactance of $(j0.22)\,(j0.214)/(j0.22 + j0.214) = j0.108$ p.u. The reduced network diagram is shown in Fig. 13.24.

4. Calculate Subtransient Fault Current

The subtransient symmetrical short-circuit current $= V/jX = 1.00/j0.108 = -j9.22$ p.u. In terms of amperes on the 6.9-kV bus, the base current $= 20{,}000$ kVA/($\sqrt{3}$)(6.9 kV) $= 1674$ A. Thus, the subtransient fault current $= (-j9.22)(1674) = 15{,}429$ A (the $-j$ is usually ignored at this point because only the magnitude is of interest).

5. Calculate Symmetrical Short-Circuit Interrupting Current

The interrupting current is related to the speed of operation of the circuit breaker and includes the contribution of current from the motors at the time of current interruption. Were the motors of the synchronous type, the reactance that would be used in the network diagram would be 1.5 times subtransient reactance; this, in effect, represents an approximate transient reactance.

For induction motors, the symmetrical short-circuit current interrupted is the same as the subtransient symmetrical short-circuit current, namely 15,429 A.

Related Calculations. The breaker must be selected to handle the *interrupting* current indicated by the calculation of the symmetrical short-circuit current. It is necessary to consult manufacturers' literature for selecting an appropriate breaker on the basis of the symmetrical current that can be interrupted.

INDUCTION-MOTOR INRUSH CURRENT

A three-phase, 240-V, wye-connected, 15-hp, six-pole, 60-Hz, wound-rotor induction motor has the following equivalent-circuit constants referred to the stator: $r_1 = 0.30$, $r_2 = 0.15$, $x_1 = 0.45$, $x_2 = 0.25$, and $x_\phi = 15.5\ \Omega$ per phase. At all loads, the core losses, friction, and windage are 500 W. Compare the inrush (starting) current with the load current at 3 percent slip, assuming that the rotor windings are short-circuited.

Calculation Procedure

1. Draw the Equivalent Circuit

The equivalent circuit is shown in Fig. 13.25.

2. Calculate the Total Impedance of the Equivalent Circuit for a Slip of 3 Percent

The secondary or rotor impedance (referred to the stator) is $r_2/s + jx_2 = 0.15/0.03 + j0.25 = 5 + j0.25\ \Omega$ per phase. This impedance in parallel with jx_ϕ is $(5 + j0.25)(j15.5)/(5 + j0.25 + j15.5) = 4.4 + j1.6\ \Omega$ per phase. The total impedance of the equivalent circuit is $4.4 + j1.6 + r_1 + jx_1 = 4.4 + j1.6 + 0.3 + j0.45 = 4.7 + j2.05 = 5.12\angle 23.6°\ \Omega$ per phase.

3. Calculate the Stator (Input) Current for a Slip of 3 Percent

The input current at the running condition of 3 percent slip is $I_1 = V/Z = 240\ V/(\sqrt{3})(5.12\ \Omega) = 27.06$ A per phase at a power factor of cos 23.6° = 0.917.

4. Calculate the Total Impedance of the Equivalent Circuit for a Slip of 100 Percent

The slip at starting condition is 100 percent. This value is used to calculate the total impedance of the equivalent circuit. Thus, the secondary or rotor impedance referred to the stator is $r_2/s + jx_2 = 0.15/1.0 + j0.25 = 0.15 + j0.25\ \Omega$ per phase. This impedance in parallel with jx_ϕ is $(0.15 + j0.25)(j15.5)/(0.15 + j0.25 + j15.5) = 0.145 + j0.248\ \Omega$ per phase. The total impedance of the equivalent circuit is $0.145 + j0.248 + r_1 + jx_1 = 0.145 + j0.248 + 0.30 + j0.45 = 0.445 + j0.698 = 0.83\angle 57.48°\ \Omega$ per phase.

5. Calculate the Starting Current

The input current at the condition of start (slip = 100 percent) is $I_1 = V/Z_{start} = 240\ V/(\sqrt{3})(0.83\ \Omega) = 166.9$ A per phase at a power factor of cos 57.48° = 0.538.

6. Compare Starting (Inrush Current) to Running Current at 3 Percent Slip

The ratio of starting current to running current (at 3 percent slip) is 166.9 A/27.06 A = 6.2.

Related Calculations. The procedure shown is applicable for all values of slip and for both squirrel-cage and wound-rotor induction motors. Although the assumption in this problem was that the rotor winding was short-circuited, other conditions may be calculated when the rotor circuit external resistance or reactance is known.

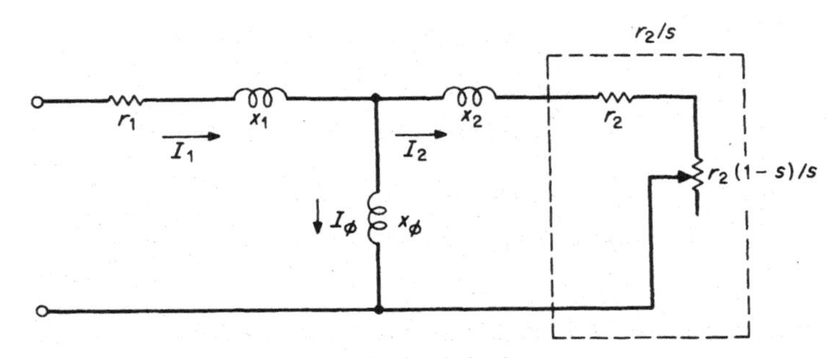

FIGURE 13.25 Equivalent circuit of an induction motor.

INDUCTION-MOTOR SHORT-CIRCUIT CURRENT

A 1000-hp, 2200-V, 25-Hz, wye-connected, 12-pole wound-rotor induction motor has a full-load efficiency of 94.5 percent and a power factor of 92 percent. Referred to the stator, the constants of the machine in ohms per phase are: $r_1 = 0.102$, $r_2 = 0.104$, $x_1 = 0.32$, $x_2 = 0.32$, $x_\phi = 16.9$. Determine the motor short-circuit current at the time of occurrence, assuming that immediately before the fault, full-load rated conditions existed.

Calculation Procedure

1. Draw the Equivalent Circuit Diagram

For the running condition with the rotor winding short-circuited, the circuit is as shown in Fig. 13.25.

2. Calculate the Prefault Stator Current

$I_{stator} = I_1 = (1000 \text{ hp})(746 \text{ W/hp})/(0.92)(0.945)(\sqrt{3})(2200 \text{ V}) = 225.2$ A. The angle equals $\cos^{-1} 0.92$, or 23.1°.

3. Calculate the Motor Transient Reactance

The motor transient reactance is determined by neglecting the rotor resistance and using the relation $x_1' = x_1 + x_\phi x_2/(x_\phi + x_2)$. Thus, $x_1' = 0.32 + (16.9)(0.32)/(16.9 + 0.32) = 0.634\ \Omega$ per phase.

4. Calculate the Voltage behind the Transient Reactance

Refer to Fig. 13.26. $\mathbf{E}_1' = \mathbf{V}_1 - (r_1 + jx_1')\mathbf{I}_1 = 2200/\sqrt{3} - (0.102 + j0.634)(225.2\underline{/-23.1°}) = 1270.2 - (0.642\underline{/80.86°})(225.2\underline{/-23.1°}) = 1270.2 - 144.6\underline{/57.76°} = 1270.2 - 77.2 - j122.3 = 1193 - j122.3 = 1199.3\underline{/-5.85°}$ V.

5. Calculate the Short-Circuit Current at the Instant of Occurrence

The initial short-circuit current is considered to be equal to the voltage behind the transient reactance divided by the transient reactance, or 1199.3 V/0.634 Ω = 1891.6 A. This is the rms initial short-circuit current per phase from the motor.

Related Calculations. This procedure is used to calculate the initial short-circuit current as dependent upon the voltage behind a calculated transient reactance. This initial short-circuit current decays very rapidly. The calculation is based on the machine running at rated conditions with very small slip.

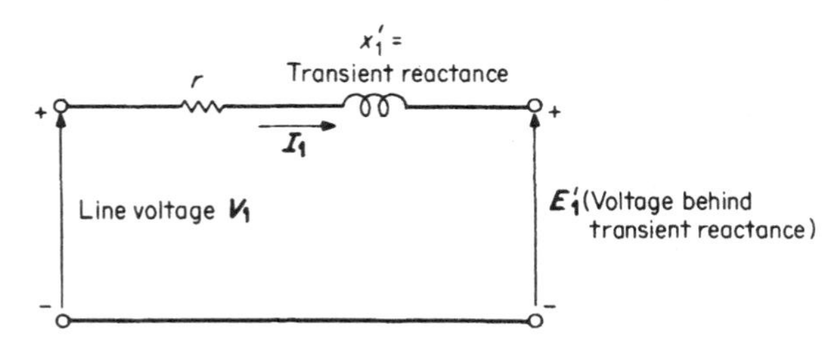

FIGURE 13.26 Transient-reactance diagram of an induction motor.

BUS VOLTAGES CALCULATED BY MATRIX EQUATION AND INVERSION

The three sequence components of bus voltage for a given bus are given as $V_0 = -0.105$, $V_1 = 0.953$, and $V_2 = -0.230$ p.u. Determine the three phase voltages and the three sequence components of the phase voltages.

Calculation Procedure

1. Write the Phase-Voltage Equations

The three separate equations will be written first in order to show the relation to the matrix equation, $\mathbf{V_a} = \mathbf{V_0} + \mathbf{V_1} + \mathbf{V_2} = -0.105 + 0.953 - 0.230$. $\mathbf{V_b} = \mathbf{V_0} + a^2\mathbf{V_1} + a\mathbf{V_2} = -0.105 + 0.953\underline{/240^\circ} - 0.230\underline{/120^\circ}$, and $\mathbf{V_c} = \mathbf{V_0} + a\mathbf{V_1} + a^2\mathbf{V_2} = -0.105 + 0.953\underline{/120^\circ} - 0.230\underline{/240^\circ}$.

2. Write the Matrix Equation

In matrix form, the individual equations become

$$\begin{bmatrix} \mathbf{V_a} \\ \mathbf{V_b} \\ \mathbf{V_c} \end{bmatrix} = \begin{bmatrix} 1 & 1 & 1 \\ 1 & a^2 & a \\ 1 & a & a^2 \end{bmatrix} \begin{bmatrix} -0.105 \\ +0.953 \\ -0.230 \end{bmatrix}$$

3. Solve for V_a

$\mathbf{V_a} = -0.105 + 0.953 - 0.230 = 0.618$.

4. Solve for V_b

$\mathbf{V_b} = -0.105 + 0.953\underline{/240^\circ} - 0.230\underline{/120^\circ} = -0.4665 - j1.0243 = 1.1255\underline{/245.5^\circ}$.

5. Solve for V_c

$\mathbf{V_c} = -0.105 + 0.953\underline{/120^\circ} - 0.230\underline{/240^\circ} = -0.4665 + j1.0245 = 1.1255\underline{/114.5^\circ}$.

6. Write the Full Matrix Equation

In matrix form, the full equation becomes

$$\begin{bmatrix} \mathbf{V_a} \\ \mathbf{V_b} \\ \mathbf{V_c} \end{bmatrix} = \begin{bmatrix} 1 & 1 & 1 \\ 1 & a^2 & a \\ 1 & a & a^2 \end{bmatrix} \begin{bmatrix} -0.105 \\ +0.953 \\ -0.230 \end{bmatrix} = \begin{bmatrix} 0.618\underline{/0^\circ} \\ 1.1255\underline{/245.5^\circ} \\ 1.1255\underline{/114.5^\circ} \end{bmatrix}$$

$\mathbf{V_{abc}} = [\mathbf{T}]\mathbf{V_{012}}$, where

$$[\mathbf{T}] = \begin{bmatrix} 1 & 1 & 1 \\ 1 & a^2 & a \\ 1 & a & a^2 \end{bmatrix}$$

7. Write the Matrix Equation to Determine Sequence Components

$\mathbf{V}_{012} = \frac{1}{3}[T]^{-1}\mathbf{V}_{abc}$, where

$$[\mathbf{T}]^{-1} = \begin{bmatrix} 1 & 1 & 1 \\ 1 & a & a^2 \\ 1 & a^2 & a \end{bmatrix}$$

$$\begin{bmatrix} \mathbf{V}_0 \\ \mathbf{V}_1 \\ \mathbf{V}_2 \end{bmatrix} = \frac{1}{3}\begin{bmatrix} 1 & 1 & 1 \\ 1 & a & a^2 \\ 1 & a^2 & a \end{bmatrix} \begin{bmatrix} 0.618\angle 0° \\ 1.126\angle 245.5° \\ 1.126\angle 114.5° \end{bmatrix}$$

The solution yields: $V_0 = -0.105$, $V_1 = 0.953$, and $V_2 = -0.230$ p.u.

Related Calculations. The matrix style of writing and solving equations is most useful in all types of three-phase problems. It is applicable particularly for problems wherein symmetrical components are used.

POWER FLOW THROUGH A TRANSMISSION LINE; ABCD CONSTANTS

The **ABCD** constants of a three-phase transmission line (nominal pi circuit) are $\mathbf{A} = 0.950 + j0.021 = 0.950\angle 1.27°$, $\mathbf{B} = 21.0 + j90.0 = 92.4\angle 76.87°\ \Omega$, $\mathbf{C} = 0.0006\angle 90°$ S, and $\mathbf{D} = \mathbf{A}$. Find the steady-state stability limit of the line if both the sending and receiving voltages are held to 138 kV: (1) with the **ABCD** constants as given, (2) with the shunt admittances neglected, and (3) with both the series resistance and the shunt admittances neglected. Refer to Fig. 13.27.

Calculation Procedure

1. Calculate the Steady-State Stability Limit for Nominal pi Circuit

The equation for steady-state stability limit is $P_{max} = |\mathbf{V}_s||\mathbf{V}_r|/|\mathbf{B}| - (|\mathbf{A}||\mathbf{V}_r|^2 \neq |\mathbf{B}|)$ $\cos(\beta - \alpha)$, where $|\mathbf{V}_s|$ = magnitude of sending-end voltage = 138 kV and $|\mathbf{V}_r|$ = magnitude of receiving-end voltage = 138 kV. $P_{max} = (138)(138)(10^6)/92.4 - [(0.950)(138)^2(10^6)/92.4]\cos(76.87° - 1.27°) = 206.1 \times 10^6 - 48.7 \times 10^6 = 157.4$ MW.

2. Calculate the Steady-State Stability Limit with Series Impedance Only

The shunt admittances at the sending end ($\mathbf{Y}_s$) and at the receiving end ($\mathbf{Y}_r$) are both equal to zero; therefore, in the equation $\mathbf{A} = 1 + \mathbf{Y}_r\mathbf{Z}$, the second term on the right is equal to zero and $\mathbf{A} = 1$. Similarly, $\mathbf{D} = 1$. $\mathbf{B}$ remains unchanged at $92.4\angle 76.87°$, and $\mathbf{C} = 0$. The steady-state stability limit is determined from the same equation used in Step 1. $P_{max} = |\mathbf{V}_s||\mathbf{V}_r|/|\mathbf{B}| = (|\mathbf{A}||\mathbf{V}_r|^2/|\mathbf{B}|)\cos(\beta - \alpha) = (138)(138)(10^6)/92.4 - [(1)(138)^2(10^6)/92.4]\cos 76.87° = 206.1 \times 10^6 - 46.8 \times 10^6 = 159.3$ MW.

3. Calculate the Steady-State Stability Limit with Series Reactance Only

Both the shunt admittances and the series resistance are neglected by letting them equal zero. Therefore, $\mathbf{A} = 1$, $\mathbf{B} = j90 = 90\angle 90°$, $\mathbf{C} = 0$, and $\mathbf{D} = 1$. Again, the same

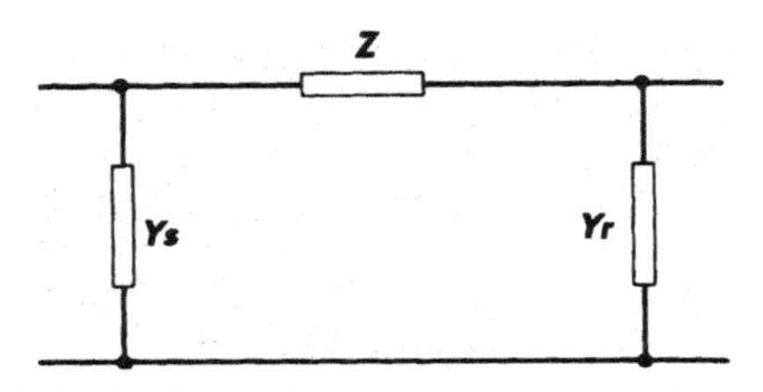

FIGURE 13.27 **ABCD** constants for a nominal pi circuit. $\mathbf{Y}_s$ = sending-end shunt admittance in siemens; $\mathbf{Y}_r$ = receiving-end shunt admittance in siemens; **Z** = series impedance in ohms; $\mathbf{A} = 1 + \mathbf{Y}_r\mathbf{Z}$ (unitless), $\mathbf{A}\angle\alpha$; $\mathbf{B} = \mathbf{Z}$ in Ω, $\mathbf{B}\angle\beta$; $\mathbf{C} = \mathbf{Y}_s + \mathbf{Y}_r + \mathbf{Z}\mathbf{Y}_s\mathbf{Y}_r$ in siemens; $\mathbf{D} = 1 = \mathbf{Y}_s\mathbf{Z}$ (unitless).

equation is used to determine the steady-state stability limit: $P_{max} = (138)(138)(10^6)/90.0 - [(1)(138)^2(10^6)/90.0] \cos (90° - 0°) = 211.6 - 0 = 211.6$ MW.

Related Calculations. The equation for maximum power presents the steady-state stability limit and may be used for all power transmission problems where **ABCD** constants are known or can be determined from the circuit parameters.

BIBLIOGRAPHY

Anderson, Paul M. 1995. *Analysis of Faulted Power Systems.* New York: IEEE Press.

Beeman, Donald (ed.). 1955. *Industrial Power Systems Handbook.* New York: McGraw-Hill.

Freeman, Peter John. 1968. *Electric Power Transmission and Distribution.* London, Toronto: Harrap.

Grainger, John J. 1994. *Power System Analysis.* New York: McGraw-Hill.

Greenwood, Allan. 1991. *Electrical Transients in Power Systems.* New York: Wiley.

Guile, Alan Elliott and W. Paterson. 1977. *Electrical Power Systems.* New York: Pergamon Press.

Hubert, Charles I. 1969. *Preventive Maintenance of Electrical Equipment.* New York: McGraw-Hill.

Knable, Alvin H. 1967. *Electrical Power Systems Engineering: Problems and Solutions.* Malabar, Fla.: Krieger.

Knight, Upton, George. 1972. *Power Systems Engineering and Mathematics.* Oxford, New York: Pergamon Press.

Ragaller, Klaus (ed.). 1978. *Proceedings of the Brown-Boveri Symposium on Current Interruptions in High Voltage Networks.* New York: Plenum Press.

Seiver, J. R., and John Paschal. 1999. *Short-Circuit Calculations: The Easy Way.* Overland Park, Kan.: EC&M Books.

Sullivan, Robert Lee. 1977. *Power System Planning.* New York: McGraw-Hill.

SECTION 14

SYSTEM GROUNDING

David R. Stocking
Lyncole XIT Grounding

Elizabeth Robertson
Lyncole XIT Grounding

SELECTION OF GROUNDING SYSTEM

Determine what factors are significant in the selection of a grounding system.

Calculation Procedure

1. Consider Grounding Impedance

The different levels of grounding impedance are:

a. Solidly grounded: No intentional grounding impedance.

b. Effectively grounded: $R_0 \leq X_1$, $X_0 \leq 3X_1$, where R is the system fault resistance and X is the system fault reactance. (Subscripts 1, 2, and 0 refer to positive-, negative-, and zero-sequence symmetrical components, respectively.)

c. Reactance grounded: $X_0 \leq 10X_1$.

d. Resistance grounded: Intentional insertion of resistance into the system grounding connection; $R_0 \geq 2X_0$.

e. High-resistance grounded: The insertion of nearly the highest permissible resistance into the grounding connection; $R_0 \leq X_{0c}/3$, where X_{0c} is the capacitive zero-sequence reactance.

f. Grounded for serving line-to-neutral loads: $Z \leq Z_1$, where Z is the system fault impedance.

2. Evaluate Disadvantages and Advantages

a. Solidly grounded: Provides for the highest level of fault current to permit maximum ability for overcurrent protection for isolation of faulted circuit. Fault current may need to be limited if equipment ratings are to be met. Will trip on first fault (it is this factor that occasionally leads to use of ungrounded circuits). Provides greatest ability for protection against arcing faults. Provides maximum protection against system overvoltages because of lightning, switching surges, static, contact with another (high) voltage system, line-to-ground faults, resonant conditions, and restriking ground faults. Limits the difference of electric potential between all uninsulated conducting objects in a local area.

b. Effectively grounded: Permits the use of lower-rated (80 percent) surge arresters. Reduces fault current in comparison with solidly grounded circuits. The reactance limitations provide a fault-relaying current of at least 60 percent of the three-phase short-circuit value.

c. Reactance grounded: In order to limit the transient overvoltage, $X_0 \leq 10X_1$. This usually results in higher fault currents than resistance-grounded systems. Used to reduce zero-sequence fault current to generator fault-current rating (normally line-to-line rating).

d,e. Resistance grounded: At high resistance, extreme transient overvoltages are limited to 250 percent of normal. This system is intended to reduce fault damage, mechanical stresses, stray currents, and flash hazards. It requires sophisticated relaying. At the low-resistance end, this system allows fairly large fault current so it minimizes high-resistance grounding advantages. It is easier to relay.

f. Grounded line-to-neutral load: Used for single-phase loads.

Related Calculations. Some proponents of not grounding circuits suggest that the floating circuit provides a degree of safety because the first accidental contact with a live line would not present a shock hazard. However, the capacitance of lines may allow a dangerous level of current to be conducted through a person.

RECOMMENDED GROUND RESISTANCE FOR SOLIDLY GROUNDED SYSTEM

Determine the appropriate maximum ground resistance permitted.

Calculation Procedure

1. Decide If System Is Residential

Compare the projected system with Fig. 14.1. If loads are 120/240-V single phase, the projected system is classified as residential-commercial. A maximum ground resistance of 25 Ω is recommended.

2. Decide If System Is Light Industrial

Compare the projected system with Figs. 14.2 and 14.3. If loads are primarily single phase on a three-phase system, the system is classified as light industrial. The maximum ground resistance recommended is 5 Ω.

3. Decide If System Is for a Substation

Compare the projected system with Fig. 14.3. If the loads are primarily three-phase as in the figure, then the maximum ground resistance recommended is 1 Ω.

4. Consider Lightning Protection

When a significant lightning threat is expected, use lightning arresters coupled with a maximum ground resistance of 10 Ω or less.

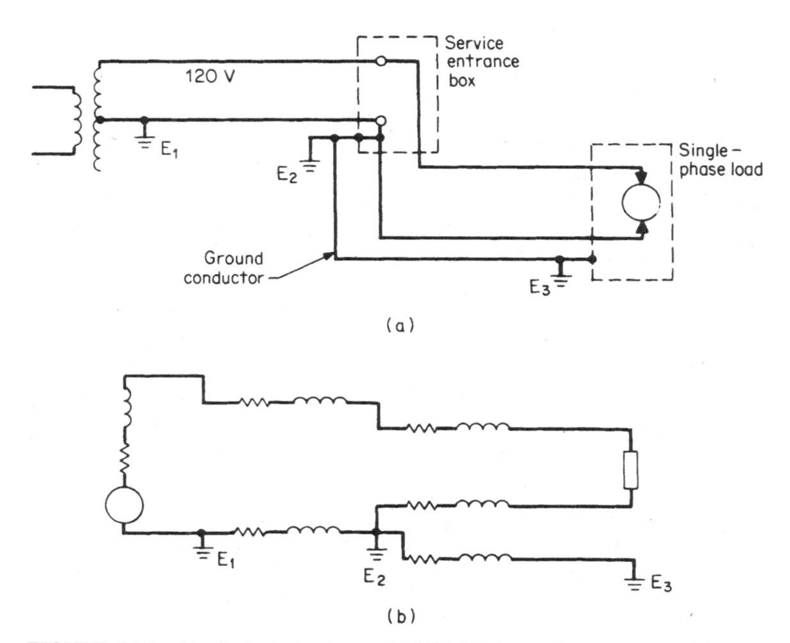

FIGURE 14.1 Standard single-phase, 240/120 V, three-wire system used in the United States. *(a)* Circuit. *(b)* Equivalent circuit.

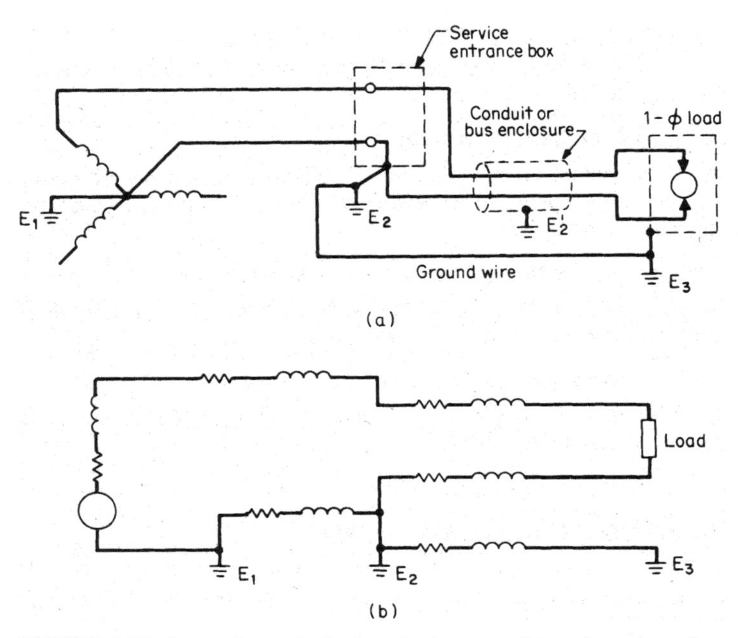

FIGURE 14.2 Low-voltage, single-phase load connected to a three-phase, four-wire circuit. *(a)* Circuit. *(b)* Equivalent circuit.

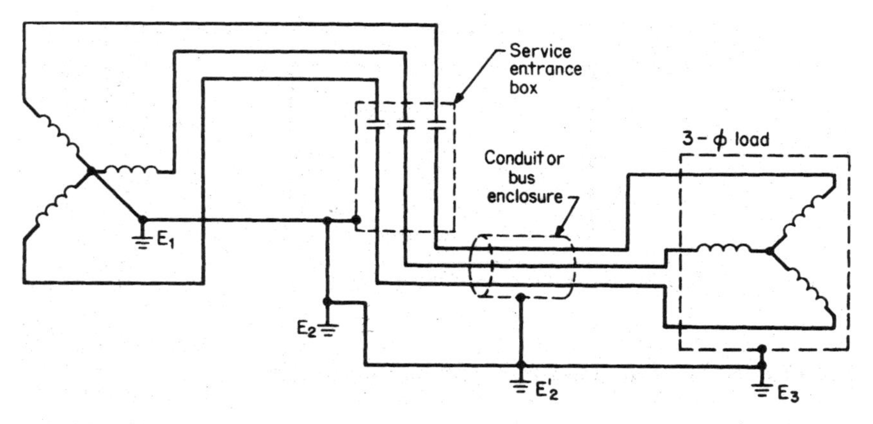

FIGURE 14.3 480/277 V, four-wire, three-phase to three-phase load. E_1, E_2, E_2', and E_3 are ground connections to earth.

FORTUITOUS CONSTRUCTION GROUNDS

Underground metallic water pipe, well casings, metallic building frames, and concrete piers provide a resistance connection to ground. Decide if these features provide the minimum required resistance.

Calculation Procedure

1. Consider Underground Piping

Although some data indicate metallic underground water systems or metallic underground sewers which may have low earth ground resistances, we suggest first testing them due to the implementation of insulative PVC fittings and seal tape.

2. Consider Well Casings and Metallic Building Frames

These two types of systems have resistances to ground of less than 25 Ω; however, they typically cannot be accurately tested.

3. Consider Nonmetallic Underground Construction

Wooden water pipes, plastic pipes, and nonconductive gaskets provide a circuit interruption. It is recommended that in all grounding construction, a ground-resistance measurement be performed.

CONCRETE PIER GROUNDS

Determine the ground resistance of one or multiple concrete piers each formed of four rebars with spacer rings. Each rebar is 3 m (10 ft) long. The soil resistivity is 15,800 Ω · cm.

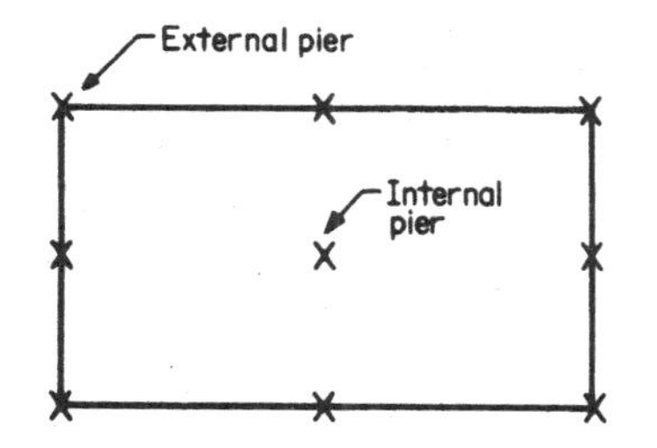

FIGURE 14.4 Multiple concrete pier arrangement.

Calculation Procedure

1. Calculate the Ground Resistance of One Pier

A reinforced-concrete pier of four rebars will have approximately one-half the resistance of a simple driven rod 1.59 cm ($^5/_8$ in.) in diameter of the same length.

Use the earth resistance of a driven 3-m rod (from calculation of driven ground in the next section), which is 64.8 Ω. One reinforced-concrete pier is one-half this value, or 32.4 Ω earth resistance.

2. Calculate the Ground Resistance for a Multiple Pier

In the case of multiple piers (arrangement of Fig. 14.4) or footings, divide the resistance of a single pier by half the number of outside piers. *Do not* include interior piers. For Fig. 14.4 we have eight exterior piers. Thus, the resistance is $32.4/(\frac{1}{2})(8) = 8.1\ \Omega$.

GROUND RESISTIVITY AND DESIGN OF DRIVEN GROUNDS

Select and design a ground rod for a commercial building in sandy loam of moderate sandiness in a climate with a 61-cm (2-ft) frost line and 1.27 m (50 in.) of rain a year.

Calculation Procedure

1. Choose First Calculation-Attempt Configuration

Table 14.1 shows some of the various shapes of manufactured ground rods. The most common is a 3-m (10-ft) cylindrical rod. National Electrical Code (NEC®) requires a minimum of 2.4-m (8-ft) driven length; a 3-m rod, therefore, is long enough to meet the code. As to diameter, NEC® requires 1.59-cm ($^5/_8$-in.) minimum diameter for steel rods and 1.27-cm ($^1/_2$-in.) minimum diameter for copper or copper-clad steel rods. Minimum practical diameters for driving limitations for 3-m rods are:

1.27 cm ($^1/_2$ in.)—average soil

1.59 cm ($^5/_8$ in.)—most soils

1.91 cm ($^3/_4$ in.)—very hard soils or more than 3-m driving depth

A practical selection, suitable for most soils and NEC® specifications, is a rod 1.59 cm ($^5/_8$ in.) in diameter by 3 m (10 ft) long.

2. Determine Resistivity

It is strongly recommended to use the soil resistivity test method described later in this chapter to perform grounding calculations. The following discussion is an example of one type of soil.

Table 14.2 indicates a variation of resistivity for sandy loam of 1020 to 135,000 Ωcm with an average of 15,800 Ωcm. To decide which to choose, examine the effect of moisture in Table 14.3. Because 15,800 Ωcm lies in the 10 to 15 percent moisture range, our 1.27 m (50 in.) per year indicates this will be a conservative estimate. If there is an indication the climate is seasonal with long drought periods, implying that the soil moisture may drop very low, one would want experimental data, or to use a resistivity closer to the 135,000 Ωcm end of the resistivity range.

The data indicate our design will be subject to possibly significant variations in soil temperature. Table 14.4 shows, as the temperature rises from 10°C (50°F) to 20°C (68°F), a decrease of 27 percent in resistivity. Within the accuracy of our selection of the appropriate resistivity, this variation may be ignored. However, the variation from 20°C (68°F) to a temperature below freezing shows an increase in resistivity. To be conservative, we will assume the frozen 61 cm (2 ft) of soil is completely insulating.

3. Compute R *for Single Rod*

From Table 14.1 for one ground rod of length L and radius a, we have $R = (\rho/2\pi L)[\ln(4L/a) - 1]$, where L is 297 cm (9.75 ft) [since we have a 3-m (10-ft) rod driven with 3 cm of the rod exposed] minus 61 cm (2 ft), equal to 236 cm (7.75 ft); a is 1.59 cm ($^5/_8$ in.); and $\rho = 15{,}800$ Ωcm. Substituting values, we calculate R as 64.8 Ω. This is significantly greater than the value indicated under the "Recommended Ground Resistance for Solidly Grounded Commercial System" of 25 Ω maximum. Our first design is unsatisfactory; we therefore examine an alternative design.

4. Compute R *for Two Rods, Close Spacing* (s < L)

From Table 14.1, we have $R = (\rho/4\pi L)[\ln(4L/a) + \ln(4L/s) - 2 + s/2L - s^2/16L^2 + s^4/512L^4 \cdots]$, where $L = 236$ cm, $a = 1.59$ cm, $s = 100$ cm ($< L = 239$ cm), an arbitrary selection, and $\rho = 15{,}800$ Ωcm.

We calculate $R = 40.1$ Ω. Because this resistance still exceeds the recommended 25 Ω, we can consider a further variation.

TABLE 14.1 Formulas for Calculation of Resistances to Ground*

	Configuration	Formula
	Hemisphere, radius a	$R = \frac{\rho}{2\pi a}$
•	One ground rod, length, L, radius, a	$R = \frac{\rho}{2\pi L}\left(\ln\frac{4L}{a} - 1\right)$
• •	Two ground rods, $s > L$; spacing, s	$R = \frac{\rho}{4\pi L}\left(\ln\frac{4L}{a} - 1\right) + \frac{\rho}{4\pi s}\left(1 - \frac{L^2}{3s^8} + \frac{2L^4}{5s^4}\cdots\right)$
••	Two ground rods, $s < L$; spacing, s	$R = \frac{\rho}{4\pi L}\left(\ln\frac{4L}{a} + \ln\frac{4L}{s} - 2 + \frac{s^2}{2L} - \frac{s^2}{16L^2} + \frac{s^4}{512L^4}\cdots\right)$
—	Buried horizontal wire, length, $2L$, depth, $s/2$	$R = \frac{\rho}{4\pi L}\left(\ln\frac{4L}{a} + \ln\frac{4L}{s} - 2 + \frac{s^2}{2L} - \frac{s^2}{16L^2} + \frac{s^4}{512L^4}\cdots\right)$
	Right-angle turn of wire, length of arm, L, depth, $s/2$	$R = \frac{\rho}{4\pi L}\left(\ln\frac{2L}{a} + \ln\frac{2L}{s} - 0.2373 + 0.2146\frac{s}{L} + 0.1035\frac{s^4}{L^2} - 0.424\frac{s^4}{L}\cdots\right)$
	Three-point star, length of arm, L, depth, $s/2$	$R = \frac{\rho}{6\pi L}\left(\ln\frac{2L}{a} + \ln\frac{2L}{s} + 1.071 - 0.209\frac{s}{L} + 0.238\frac{s^3}{L^8} - 0.054\frac{s^4}{L^4}\cdots\right)$
+	Four-point star, length of arm, L, depth, $s/2$	$R = \frac{\rho}{8\pi L}\left(\ln\frac{2L}{a} + \ln\frac{2L}{s} + 2.912 - 1.071\frac{s}{L} + 0.645\frac{s^3}{L^8} - 0.145\frac{s^4}{L^4}\cdots\right)$
	Six-point star, length of arm, L, depth, $s/2$	$R = \frac{\rho}{12\pi L}\left(\ln\frac{2L}{a} + \ln\frac{2L}{s} + 6.851 - 3.128\frac{s}{L} + 1.758\frac{s^2}{L^3} - 0.409\frac{s^4}{L^4}\cdots\right)$
	Eight-point star, length of arm, L, depth, $s/2$	$R = \frac{\rho}{16\pi L}\left(\ln\frac{2L}{a} + \ln\frac{2L}{s} + 10.98 - 5.51\frac{s}{L} + 3.26\frac{s^2}{L^3} - 1.17\frac{s^4}{L^4}\cdots\right)$

continued

TABLE 14.1 Formulas for Calculation of Resistances to Ground* *(Continued)*

○	Ring of wire, diameter of ring, D, diameter of wire, d, depth, $s/2$	$R = \frac{\rho}{2\pi^2 D}\left(\ln\frac{8D}{d} + \ln\frac{4D}{s}\right)$
—	Buried horizontal strip, length, $2L$, section a by b, depth, $s/2$, $b < a/8$	$R = \frac{\rho}{4\pi L}\left[\ln\frac{4L}{a} + \frac{a^2 - \pi ab}{2(a+b)^2} + \ln\frac{4L}{s} - 1 + \frac{s}{2L} - \frac{s^2}{16L^2} + \frac{s^4}{512L^4} \cdots\right]$
⊘	Buried horizontal round plate radius, a, depth, $s/2$	$R = \frac{\rho}{8a} + \frac{\rho}{4\pi s}\left(1 - \frac{7}{12}\frac{a^2}{s^2} + \frac{33}{40}\frac{a^4}{s^4} \cdots\right)$
	Buried vertical round plate radius, a, depth, $s/2$	$R = \frac{\rho}{8a} + \frac{\rho}{4\pi s}\left(1 + \frac{7}{24}\frac{a^2}{s^2} + \frac{99}{320} + \frac{a^4}{s^4} + \cdots\right)$

*Approximate formulas. Dimensions must be in centimeters to give resistance in Ω. ρ = resistivity of earth in Ω cm.
Source: H. B. Wright, *AIEE*, vol. 55, 1936, pp. 1319–1328.

TABLE 14.2 Resistivity of Different Soils

Soil	Resistivity, Ωcm Minimum	Average	Maximum
Ashes, cinders, brine, and waste	590	2370	7000
Clay, shale, gumbo, and loam	340	4060	16,300
Same, with varying proportions of sand and gravel	1020	15,800	135,000
Gravel, sand, and stones with little clay or loam	59,000	94,000	458,000

TABLE 14.3 Effect of Moisture Content on Resistivity of Soil

Moisture content, percent by weight	Resistivity, Ωcm Top soil	Sandy loam
0	$>10^9$	$<10^9$
2.5	250,000	150,000
5	165,000	43,000
10	53,000	18,500
15	19,000	10,500
20	12,000	6300
30	6400	4200

TABLE 14.4 Effect of Temperature on Resistivity of Sandy Loam, 15.2 Percent Moisture

Temperature °C	°F	Resistivity, Ωcm
20	68	7200
10	50	9900
0 (water)	32	13,800
0 (ice)	32	30,000
−5	23	79,000
−15	14	330,000

5. Compute **R** *for Two Rods, Wide Spacing* (s > L)

From Table 14.1, we have $R = (\rho/4\pi L)[\ln(4L/a) - 1] + (\rho/4\pi s)(1 - L^2/3s^8 + 2L^4/5s^4 \cdots)$.

We select an arbitrary spacing of $s = 400$ cm ($13\frac{1}{8}$ ft), obtaining $R = 35.5\ \Omega$.

6. Decide If Two Rods at 400 cm Are Satisfactory

Recognizing the wide variability of sandy loam resistivity, one uses judgment in selecting the value of ρ. If the power system is a small commercial operation with a low-capacity transformer (and fault current will be moderate) or if the soil tends to be more loam than sand (4060 Ωcm average resistivity as compared with 15,800 Ωcm for sandy loam), then R for two widely spaced rods would be 9.12 Ω, which would be satisfactory. On the other hand, if the soil tends to be of high resistivity or if the substation transformer is of large capacity, there would be a need to examine one of the lower-resistance systems, such as the "star" counterpoise shown in Table 14.1 or a ground grid (see next problem).

Related Calculations. Tables 14.2, 14.3, and 14.4 indicate the variability of the resistivity of earth as a function of earth type, temperature, moisture, and chemical content. Resistivity is also sensitive to backfill compaction, earth pressure against the grounding metal, and the magnitude of fault current. It is common practice to assume for most ground calculations an earth resistivity of 10,000 Ωcm, as compared with the 15,800 Ωcm used in the preceding calculation.

Because R is directly proportional to resistivity, significant design variations from actual tested results can occur if a realistic resistivity is not selected. This can have severe consequences. For instance, a common lightning-stroke current of 1000 A through the 35.5 Ω ground resistance (see Step 5 above), would generate a 35,000-V transient on the electrical system. If the resistivity in the actual installation is four times our assumed value, a voltage transient of 142,000 V would result. For these reasons, it is prudent to test the finished installation as described later in this section.

GROUNDING WITH GROUND GRIDS

If a grounding location consists of a thin overburden of soil over a rock substrate, the deep constructions from Table 14.1 are impractical. Design a horizontal-grid (four-mesh) grounding system for a light industrial application.

Calculation Procedure

1. Define an Initial Design

Use a conductor radius r of 0.0064 m ($\frac{1}{4}$ in.), a conductor depth below earth surface, s, of 0.30 m (1 ft), and an overall grid width, w, of 6.1 m (20 ft). Use a four-mesh grid and a resistivity of 10,000 Ωcm. (By way of illustration, Fig. 14.5 indicates a sketch of a nine-mesh grounding grid.)

2. Calculate Parameter for Abscissa of Fig. 14.6

a. $A = w^2 = 37.2\ \text{m}^2$ (400 ft^2).
b. $2rs/A = 1.03 \times 10^{-4}$.

3. Read Ordinate from Fig. 14.6

Corresponding to an abscissa of 1.03×10^{-4}, the ordinate is 0.097 for a four-mesh grid.

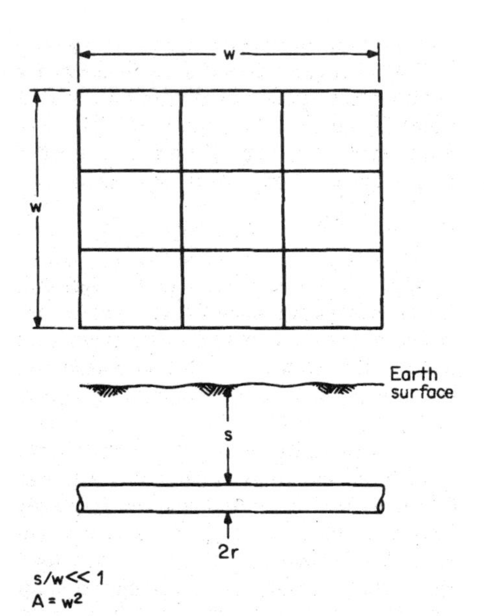

FIGURE 14.5 Nine-mesh shallow grid; w = width of grid in meters; A = area of grid in square meters; r = radius of conductor in meters, ρ = resistivity of soil in Ωcm; and s = depth of grid below earth surface in meters.

4. Calculate **R**

Solving the equation $6.56\pi wR/\rho = 0.097$, find $R = 7.7\ \Omega$. This is close enough to the recommended 5 Ω for industrial plant design.

GROUNDING WITH A SIX-POINT STAR

The star horizontal-ground system of Table 14.1 is a possible alternative ground mat to the four-mesh grid designed in the previous section. Using the same length of ground conductors and other dimensions as the four-mesh ground grid in the previous design, calculate the ground resistance of the star.

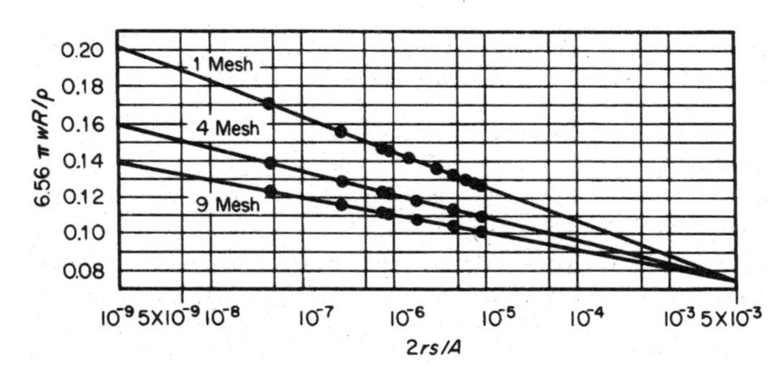

FIGURE 14.6 Grounding grid parameters.

Calculation Procedure

1. Calculate the Equivalent **L** *of the Star Arm*

The total ground conductor in the four-element ground mesh is 36.6 m (120 ft). Each of the six star arms is 36.6/6 = 6.1 m long.

2. Calculate **R** *for the Star*

Use $R = (\rho/12\pi L)[\ln(2L/a) + \ln(2L/s) + 6.851 - 3.128s/L + 1.758s^2/L^3 - 0.49s^4/L^4 \ldots]$, where ρ = 10,000 Ωcm, L = 610-cm (20-ft) arm length, s = 30-cm (2-ft) depth, and a = 1.59-cm (⅝-in.) diameter. Substituting values, find $R = 7.7\ \Omega$, the same value as in the mesh design in the preceding problem.

EFFECT OF GROUND-FAULT DISTANCE FROM GROUND POINT

The formulas in Table 14.1 are based on the resistance from a ground electrode of the given geometry to a hemisphere at an infinite radius. Calculate the variation in resistance for a hemispherical ground electrode of radius a as a function of distance to the outer hemispherical electrode, the distance being less than infinity.

Calculation Procedure

1. Express Resistance between Two Hemispheres Using Ohm's Law

Ohm's law indicates $R = \rho L/A$. For the two hemispheres, the incremental resistance dR between two spheres of area $A = 2\pi r^2$ separated by a distance dr is $dR = \rho(dr/2\pi r^2)$. Integration yields $R = (\rho/2\pi)(1/a - 1/r_2)$, where a is the radius of the inner sphere and r_2 is the radius of the outer sphere.

2. Calculate R as a Function of r₂

Define r_2 as na, where $n = 1, 2, 3, \ldots, \infty$; hence, $R = (\rho/2\pi a)(1 - 1/n)$. Substituting for various values of $n = 1, 2, \ldots, \infty$, we obtain:

n	R
1	$0 \times \dfrac{\rho}{2\pi a}$
2	$\dfrac{1}{2} \times \dfrac{\rho}{2\pi a}$
5	$\dfrac{4}{5} \times \dfrac{\rho}{2\pi a}$
10	$\dfrac{9}{10} \times \dfrac{\rho}{2\pi a}$
20	$\dfrac{19}{20} \times \dfrac{\rho}{2\pi a}$
50	$\dfrac{49}{50} \times \dfrac{\rho}{2\pi a}$
100	$\dfrac{99}{100} \times \dfrac{\rho}{2\pi a}$

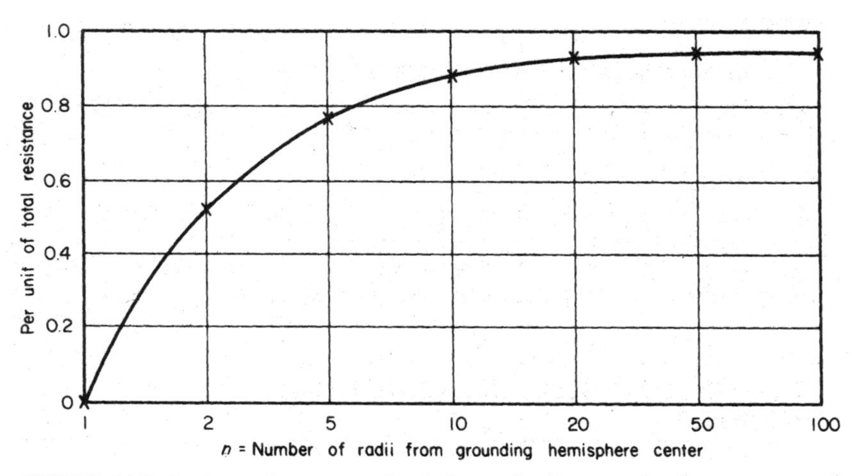

FIGURE 14.7 Resistance between two hemispheres of radius, *a,* and radius, *na,* compared with hemisphere of radius, *a,* and radius infinity.

These results are plotted in Fig. 14.7. Note that 90 percent of the resistance is developed in the ground voltage drop over a distance of the first 10 radii.

EFFECT OF RESISTIVITY OF EARTH AND INDUCTANCE ON DEPTH OF GROUND CURRENT

Figure 14.8 indicates the distributed path of ground current between *A* and *B*. Accepted engineering practice represents the distributed current path as a single conductor of 30-cm (1-ft) radius with the same inductive and resistive voltage drop as the distributed current case. Calculate the distance *P* for resistivities of 340, 10,000, and 458,000 Ωcm from Table 14.2.

Calculation Procedure

1. Calculate* P *for ρ of 340 Ωcm

$P = 8.5\sqrt{\rho}$ m = 157 m.

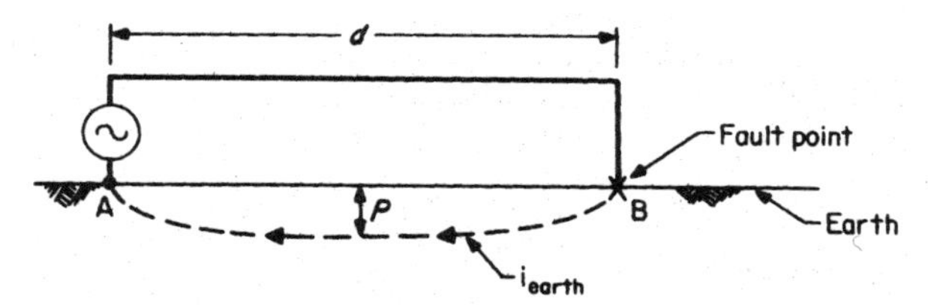

FIGURE 14.8 Path of fault current: d = distance between two ground points; P = depth of equivalent conductor when $d = \infty$.

2. Calculate **P** *for* ρ *of 10,000 Ωcm*

$P = 8.5\sqrt{\rho}$ m = 850 m.

3. Calculate **P** *for* ρ *of 458,000 Ωcm*

$P = 8.5\sqrt{\rho}$ m = 5750 m.

IMPEDANCE OF LONG-DISTANCE FAULTS COMPARED WITH GROUNDING RESISTANCE

For a fault to earth with resistivity of 10,000 Ωcm between points far apart with respect to penetration depth P (Fig. 14.8) of 850 m, calculate the fault impedance. The conductor is 1.27 cm ($^1/_2$ in.) in radius with a resistance of 0.54 Ω/km. Compare it with the recommended grounding resistance.

Calculation Procedure

1. Calculate Impedance

Impedance $\mathbf{Z} = r_c + 0.063 + j0.18 \log_{10}(P/\text{GMR})$, where $\mathbf{Z}$ is the complex impedance per kilometer, r_c is the conductor resistance per kilometer, 0.063 is the resistance per kilometer of the assumed 30-cm radius underground equivalent conductor (which, to engineering accuracy, is independent of earth resistivity), P is the penetration depth of the assumed earth conductor, and GMR is the geometric mean radius of the conductor in the same dimensions as P. GMR $= \sqrt{R_1R_2} = \sqrt{(0.0127)(0.30)} = 0.062$ m. Substituting in the expression for impedance, find $\mathbf{Z} = 0.6 + j0.74$ Ω.

2. Compare the Long-Line Impedance with Grounding Resistance

If the grounding resistance is kept to the recommended 5 Ω for industrial applications, then the line impedance will be on the same order as the grounding impedance. Note the need for keeping the grounding resistance below 5 and 10 Ω for lighter-duty applications.

IMPEDANCE OF SHORT-DISTANCE FAULTS COMPARED WITH GROUNDING RESISTANCE

Use the standard formula for reactance of two parallel conductors to examine the inductive reactance X of a conductor and a metallic ground return. The radius of each is 1.27 cm.

Calculation Procedure

1. Calculate **X** *for Various Spacings*

Use $X = 0.34 \log_{10}, \sqrt{D_{12}/\text{GMR}}$ Ω/km, where D_{12} is the distance between the center line of the two conductors and GMR is the geometric mean radius of the two conductors in the same dimensions as D_{12}. GMR $= \sqrt{(1.27 \text{ cm})(1.27 \text{ cm})} = 0.0127$ m. We obtain the following table.

D_{12} as multiple of 0.0127-m GMR	D_{12}, m	Ω/km
10	0.13	0.17
100	1.3	0.34
1000	13	0.51

2. Compare the Reactive Drop with Ground Resistance

Because even with the 30-m spacing between the normal conductor and metallic ground return, resistance per unit length is only 0.51 Ω/km, it is clear the ground resistance of the grounding element dominates the fault impedance for short line faults for all practical grounding resistances.

Related Calculations. Where great accuracy in impedance is desired, as in distance-relaying design, a computer approach is required. Electric Power Research Institute Reports N-1605, Vols. 1 and 2, describe the computer programming approach to solve the distributed-constant problem to the desired accuracy. In this calculation, orthogonal cubes are assumed. These cubes have front and back surfaces of equipotentials. The sides of the cubes are current sheets. By summing the currents into and out of the cube to zero and solving for voltage, the interactive process can be made to converge to a solution of the unique gradients and equipotentials.

GROUNDING ELECTRODES

Introduction

In the last few decades, a lot has been learned about the interaction between the grounding electrode and the earth. Ultimately, the resistivity of the soil determines the system design and performance. Other soil conditions such as pH and water-table fluctuations affect the life expectancy and performance of the system over time. New technology has made significant improvements to the conductivity between the grounding electrode and the surrounding soil. The grounding electrodes that are in use today are the standard driven rod, grounding plate, Ufer (concrete-encased electrode), water pipe, and the electrolytic electrode.

Sphere of Influence

An important theory about the way grounding electrodes discharge electrons into the earth is the sphere of influence. The sphere of influence is the volume of soil that the electrode will utilize to discharge any current it may carry. The depth/length of the electrode is equal to the radius of the volume of soil, measured from any point on the electrode. In other words, the volume of soil used by the electrode is proportional to its maximum depth or length. The surface area of the electrode determines the ampacity of the device and does not affect the sphere of influence. The greater the surface area, the greater the contact with the soil, and the more electrical energy that can be discharged per unit of time. The formula for calculating the volume of soil follows:

$$V = \frac{5L^3}{3}$$

where V equals volume of soil in the sphere of influence and L equals the length or depth of the electrode. A simpler version of the formula is created by modifying it by rounding π (pi) down to 3 and cross-canceling to get the formula $V = 5L^3$. Thus, a single 10 ft driven rod will utilize 5000 ft^3 of soil, whereas a single 8-ft rod will utilize only half the soil, at 2560 ft^3 (Fig. 14.9).

Driven Rod

The standard driven rod or copper-clad rod consists of a length of steel with a 5-mil coating of copper. The primary drawback of the copper-clad rod is that copper and steel are two dissimilar metals, and, when an electrical current is imposed, electrolysis will occur. Additionally, the act of driving the rod into the soil can damage the copper cladding, allowing corrosion to enter, further decreasing the life expectancy of the rod. Driven rods have a very small surface area and do not assure solid contact with the soil. This is especially true with rocky soil conditions, where the rod will only contact the edges of the surrounding rock. The environment, aging, temperature, and moisture also easily affect driven rods, giving them a typical life expectancy of 5 to 10 years, in neutral pH conditions.

Grounding Plates

Grounding plates are thin copper plates placed under poles or supplementing counterpoises. Although the surface area is greatly increased over that of a driven rod, the sphere of influence is relatively small and typically has higher resistance readings. The same environmental conditions that lead to the failure of the driven rod also plague the grounding plate.

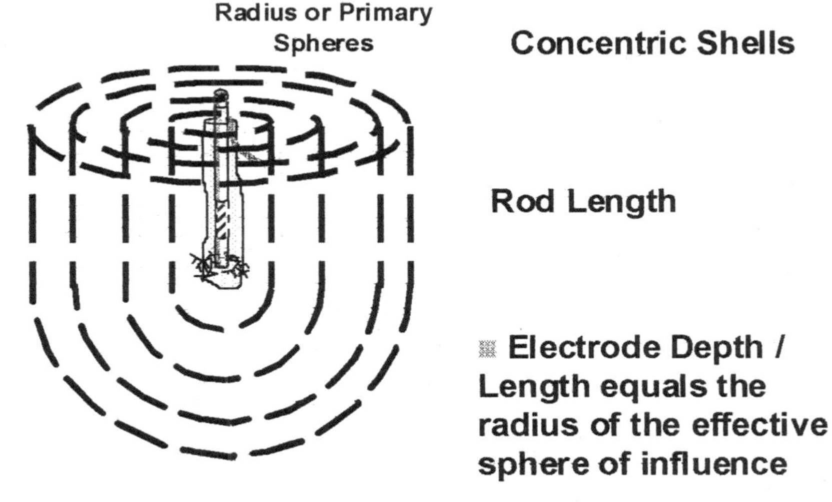

FIGURE 14.9 Sphere of influence for grounding electrodes.

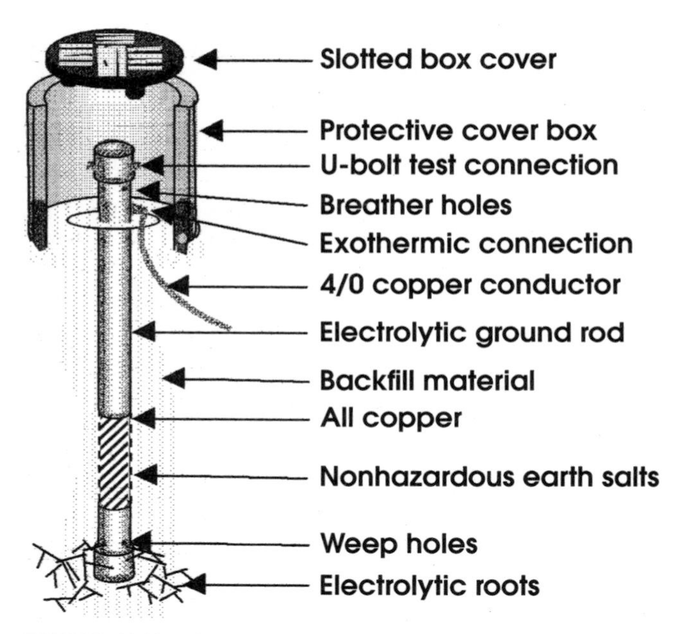

FIGURE 14.10 Electrolytic electrodes.

Ufer Ground

Originally, Ufer grounds were copper electrodes encased in the concrete surrounding ammunition bunkers. In today's terminology, Ufer grounds consist of a concrete-encased electrode in the form of the rebar in the building foundation. The concept is based on the conductivity of the concrete and the large surface area and will usually provide a grounding system that can handle very high-current loads. The primary drawback occurs during fault conditions when the moisture in the concrete superheats and rapidly expands, cracking the surrounding concrete and possibly destroying the integrity of the building foundation. Ufer grounds are not testable under normal circumstances.

Water Pipes

Water pipes have been used extensively over time as a grounding electrode. Water-pipe connections are not testable and are seldom reliable due to the use of tar coatings and plastic fittings. City water departments have begun to specifically install plastic insulators in the pipelines to prevent the flow of electrons and reduce the corrosive effects of electrolysis. Since water pipes are continuous citywide, fault conditions in adjacent neighborhoods could backfeed current into sensitive equipment, causing unintentional damage. The latest NEC code still allows this electrode when continuity and 10 ft of direct earth contact can be shown. Additionally, a supplemental electrode must be utilized.

Electrolytic Electrode

The electrolytic electrode (Fig. 14.10) was specifically engineered to eliminate the drawbacks found in other grounding systems. This active electrode consists of a hollow copper shaft filled with natural earth salts that have a hygroscopic nature to draw moisture from the air. This moisture mixes with the salts to form an electrolytic solution that continuously

TABLE 14.5 Characteristics of the Electrolytic Electrode

Features	Driven rod	Grounding plate	Ufer ground	Water pipe	Electrolytic electrode
Resistance over time	Increases	Increases	Increases	Increases	Decreases
Corrosion resistance	Low	Low	High	Low	High
UL listing	Yes	Custom application	Custom application	No	Yes
Resistance to temperature change	Low	Low	Low	Low	High
Ampacity	Low	High, dependent upon surface area	High	High, if continuous and in direct contact with earth	High
Deteriorates building foundation	No	No	Yes	No	No
Life expectancy	5–10 years	5–10 years	Unknown: difficult to test	Unknown: difficult to test	30–50 years
Manufacturer's warranty	None	None	None	None	Up to 30 years

weeps into the surrounding backfill material, keeping it moist and high in ionic content. The electrolytic electrodes are installed into an augured hole and backfilled with Lynconite™, a highly conductive clay with a neutral pH that will protect the electrode from corrosion. The electrolytic salt solution and the Lynconite™ work together to provide a solid connection between the ground conductor and the surrounding soil that is free from the effects of temperature, environment, and corrosion (Table 14.5). The electrolytic electrode is the only grounding system known to improve with age. All other electrode types will have rapidly increasing earth-to-ground resistances as the seasons change and the years pass.

SYSTEM DESIGN AND PLANNING

A grounding design starts with site analysis and collection of the geological and soil resistivity data of the area. Typically, the site engineer or equipment manufacturers specify a resistance to ground. The NEC lists 25 ohms for single-electrode installations; this does not indicate that 25 ohms is a satisfactory level. High-technology manufacturers will specify 5 or 10 ohms, depending upon the needs of the equipment. Occasionally, and only for extreme circumstances, will a 1 ohm specification be required. When designing a ground system, the difficulty level and costs increase exponentially as the required resistance to ground approaches the unobtainable goal of zero.

Once a need is established, data collection begins. Soil resistivity testing, geological surveys, and test borings provide the basis for all grounding design. Proper soil resistivity testing using the Wenner 4-point method is critical and will be discussed later. Additional data can be collected from existing ground systems located near the site, which can be tested using the 3-point fall-off-potential method (Fig. 14.11) or an Induced Frequency Test (Fig. 14.12) using a clamp-on ground resistance meter.

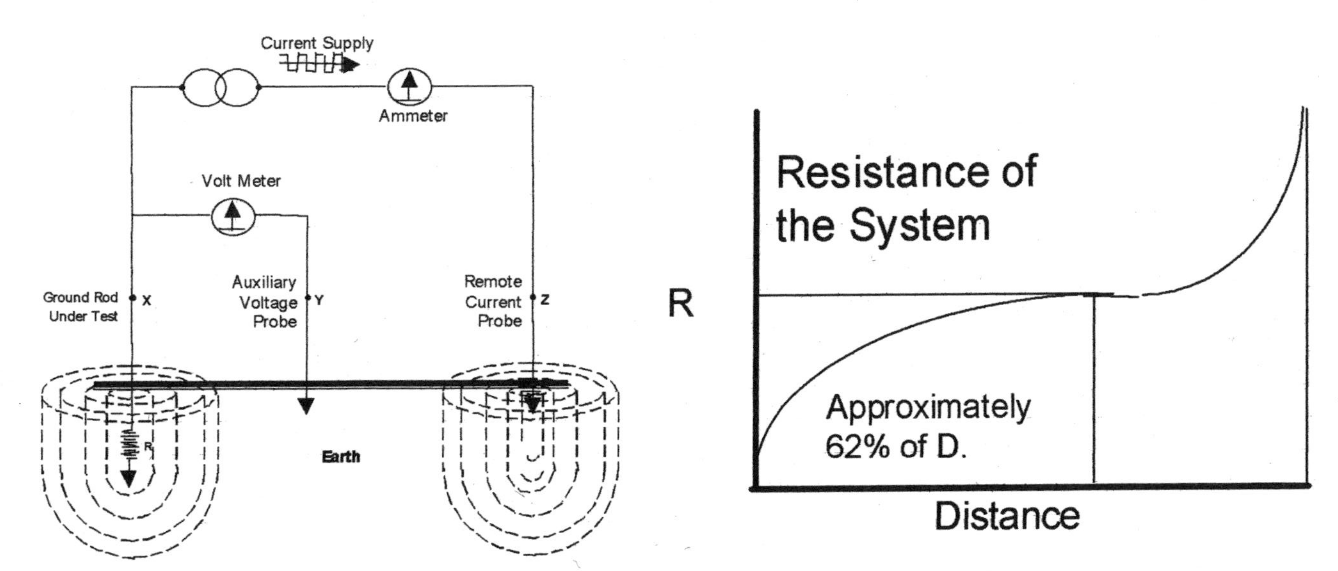

FIGURE 14.11 The three-point fall-off potential method of measuring ground resistance.

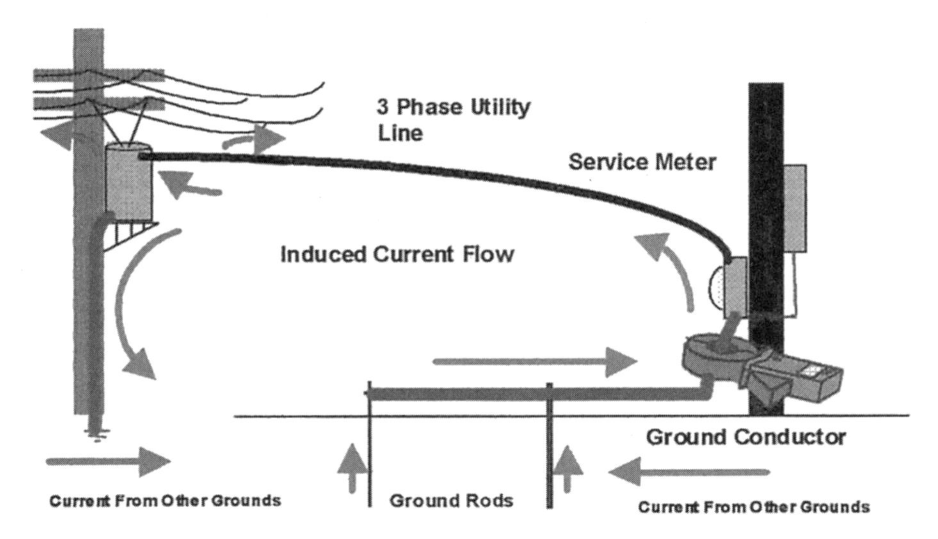

FIGURE 14.12 Example of induced frequency testing.

With all the available data, sophisticated computer programs can begin to provide a soil model showing the soil resistivity in ohmmeters at the various layer depths. Knowing at what depth the most conductive soil is located for the site will allow the design engineer to model a system to meet the needs of the application.

Soil resistivity is the key factor that determines the resistance or performance of a grounding system. It is the starting point of any grounding design. As you can see in Table 14.6, soil resistivity varies dramatically throughout the world and is heavily influenced by electrolyte content, moisture, and temperature.

TABLE 14.6 Soil Types vs. Resistivity

Soil resistivity comparison	
Soil type	Resistivity tange (ohmmeter)
Lyconite	2.4
Surface soils	2–50
Clay	2–100
Sand and gravel	50–1000
Surface limestone	100–10,000
Limestone	5–4000
Shale's	5–100
Sandstone	20–2000
Granites, basalt's, etc.	1000
Decomposed gneiss's	50–500
Slates, etc.	10–100

TESTING EXISTING GROUNDING SYSTEMS

Ground resistance monitoring is the process of automated, timed, and/or continuous ground resistance measurement. These dedicated systems use the Induced Frequency Test to continuously monitor the performance of critical grounding systems and may or may/not provide automated data reporting. As with the Induced Frequency Test, these meters test grounding systems that are in use and do not require the interruption of service to take measurements.

SOIL RESISTANCE TESTING

Four-point or Wenner Soil Resistivity Test

The Wenner four-point method is used to measure the resistivity of soil at a particular location (Fig. 14.13). Electrical resistivity is a measurement of the resistance of a unit quantity of a given material, expressed, in this case, in volume as an ohmmeter. This test is commonly performed at raw land sites, during the design and planning of grounding systems specific to the tested site. The test spaces four probes out at equal distances to approximate the depth of the soil to be tested. Typical spacing will be 5′, 10′, 15′, 20′, 30′, and 40′ intervals and will require a total of 120′ linear ft. This test is repeated five times, four times along the perimeter and once across the diagonal of the site to be measured.

In this test, a probe, C_1, is driven into the earth at the corner of the area to be measured. Probes, P_1, P_2, and C_2, are driven at 5′, 10′, and 15′, respectively, from rod C_1 in a straight line to measure the soil resistivity from 0′ to 5′ in depth. C_1 and C_2 are the outer probes; P_1 and P_2 are the inner probes. At this point, a known current is applied across probes C_1 and C_2 and the resulting voltage is measured across P_1 and P_2. Ohm's law can then be applied to calculate the measured resistance.

Probes, C_2, P_1, and P_2, can then be moved out to 10′, 20′, and 30′ spacing to measure the resistance of the earth from 0′ to 10′ in depth. Continue moving the three probes (C_2, P_1, and P_2) away from C_1 at equal intervals to approximate the depth of the soil to be measured.

Once all the resistance data is collected, the following formula can be applied to calculate the soil resistivity in ohmmeters:

$$\rho = \frac{4\pi AR}{1 + \dfrac{2A}{\sqrt{(A^2 + 4B^2)}} - \dfrac{A}{\sqrt{(A^2 + B^2)}}}$$

where,

ρ = Resistivity

B = Depth of probes

A = Spacing of probes

R = Resistance (reading from meter)

If $A > 20B$, then $\rho = 1.915\ AR$

For example, if a soil resistance was measured at 40-ft spacing and found to have a resistance of 4.5 ohms, calculate the following:

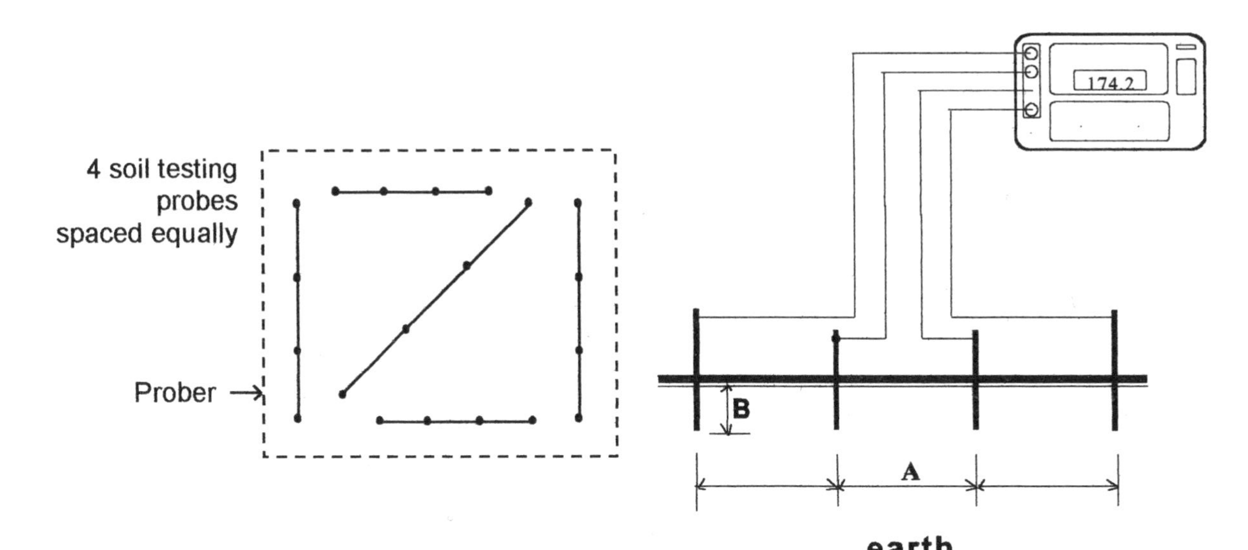

FIGURE 14.13 The four-point or Wenner soil resistivity test.

$$\rho = 1.915\,AR$$
$$\rho = 1.915\,(40)(4.5)$$
$$\rho = 344.7 \text{ ohmmeters}$$

BIBLIOGRAPHY

ANSI/IEEE Std. 142. 1982. *Green Book—Grounding of Industrial and Commercial Power Systems.*

ANSI/IEEE Std. 82. 1983. *IEEE Guide for Measuring Earth Resistivity, Ground Impedance, and Earth Surface Potentials of a Ground System.*

Fink, Donald G., and H. Wayne Beaty. 2000. *Standard Handbook for Electrical Engineers,* 14th ed. New York: McGraw-Hill.

Hartwell, Fred (ed.). 1998. *Understanding the National Electrical Code Rules on Grounding and Bonding.* Overland Park, K.S.: Intertec Publishing Co.

IEEE. 1986. *Guide for Safety in ac Substation Grounding.* New York: IEEE.

IEEE. 1991. *Recommended Practice for Grounding of Industrial and Commercial Power Systems: The Green Book.* New York: IEEE.

Jenkins, Brian D. 1993. *Touch Voltages in Electrical Installations.* Boston: Blackwell Science.

Meliopoulos, A. P. Sakis. 1999. *Power System Grounding and Transients.* New York: Marcel Dekker.

Morrison, Ralph. 1998. *Grounding and Shielding Techniques.* New York: John Wiley & Sons.

NFPA 70. 1999. *National Electric Code.*

O'Rilley, Ronald P. 1998. *Electrical Grounding: Bringing Grounding Back to Earth.* Albany, N.Y.: Delmar Publishing.

Seidman, Arthur, H. Wayne Beaty, and Haroun Mahrous. 1997. *Handbook of Electrical Power Calculations,* 2nd ed.

Sharick, Gilbert. 1999. *ABC of the Telephone Volume 13—Grounding and Bonding,* revised 2nd ed.

Watkins, William S., and Thomas M. Kovacic. 1994. *Grounding Safety: A Safety Professional's Guide to Electrical Grounding.* New York: American Society of Mechanical Engineers.

SECTION 15

POWER-SYSTEM PROTECTION

Thomas H. Ortmeyer
Clarkson University

Protective systems are designed to sense faults and initiate fault clearing in a timely manner while minimizing the affected area. Protective relays are used to sense the faults and initiate circuit breaker tripping. Alternatively, fuses are used on the distribution system to sense and clear faults.

Protective relays sense system current and voltage levels to determine both the presence and location of faults. Current transformers (CTs) and voltage transformers (VTs) are used to scale the system levels to levels suitable for the relays and measurement instruments. Voltage transformers are most commonly connected with wye secondaries, with 120Y/69 V being a standard secondary-voltage rating. They are selected through their voltage and volt-amp requirements. Current transformers carry both a continuous current rating and a burden rating. Their application is discussed in the next section.

Recently, digital protective relays have become available and are popular in many applications. These relays offer flexibility, self-checking, and ease of installation and often can provide additional functions over traditional electromechanical relays. Settings calculations for many of these relays are straightforward and are outlined in the relay's applications manual. In order to make these calculations, knowledge of peak-load current, minimum and maximum fault currents, and the CT and VT ratings is required.

This section provides a brief overview of some of the most common protective relaying applications. A much fuller discussion of these topics is available elsewhere, including Anderson (1999), Blackburn (1987), Elmore (1994), and IEEE Buff Book. Relay manufacturers commonly supply application information for their products as well. Finally, special situations can arise that will require additional or different protective measures than are discussed here.

CURRENT TRANSFORMER CONNECTION AND SIZING

Current transformers (CTs) commonly have a 5 A (continuous) secondary rating. Many current transformers are multiratio, with the user being able to select the CT turns ratio from among several available on a particular device. Current transformers are typically able to carry fault currents up to 20 times their rating for short periods of time. Their ability to do this depends on their loading being within their specified burden rating.

Example 15.1

A current transformer is specified as being 600 A:5 A class C200. Determine it's characteristics.

This designation is based on ANSI Std. C57.13-1978. 600 A is the continuous primary current rating, 5 A is the continuous secondary current rating, and the turns ratio is 600/5 = 120. C is the accuracy class, as defined in the standard. The number following the C, which in this case is 200, is the voltage that the CT will deliver to the rated burden impedance at 20 times rated current without exceeding 10 percent error. Therefore, the rated burden impedance is

$$Z_{\text{rated}} = \frac{\text{Voltage class}}{20 \cdot \text{Rated secondary current}} = \frac{200\ \text{V}}{20 \cdot 5\ \text{A}} = 2\ \Omega$$

This CT is able to deliver up to 100 A secondary current to load burdens of up to 2 Ω with less than 10 percent error. Note that the primary source of error is the saturation of the CT iron core and that 200 V will be approximately the knee voltage on the CT saturation curve. This implies that higher burden impedances can be driven by CTs which will not experience fault duties of 20 times rated current, for example.

A typical wye CT connection is shown in Fig. 15.1. The neutral points of the CTs are tied together, forming a residual point. Four wires, the three-phase leads and the residual, are taken to the relay and instrument location. The three-phase currents are fed to protective relays or meters, which are connected in series. After these, the phases are connected to form and tied back to the residual. Additional relays are often connected in the residual, as the current in this circuit is proportional to the sum of the phase currents and corresponds to the current that will eventually end up flowing through neutral or ground.

Example 15.2

The circuit of Fig. 15.1 has 600:5 class C100 CTs. The peak-load current is a balanced 475 A per phase.

1. Determine the Relay Currents for the Peak-Load Conditions.

The A phase CT secondary current is

$$I_A = \frac{475\text{A}}{120} = 3.96\ \text{A}\underline{/0^\circ}$$

Here, the A phase current is taken to be at 0°. The B and C phase currents are the same magnitude, shifted by 120°,

$$I_B = 3.96\ \text{A}\underline{/-120^\circ},\ I_C = 3.96\ \text{A}\underline{/120^\circ}$$

The residual current is

$$I_R = I_A + I_B + I_C = 3.96\ \text{A}\underline{/0^\circ} + 3.96\ \text{A}\underline{/-120^\circ} + 3.9.6\ \text{A}\underline{/120^\circ} = 0\ \text{A}$$

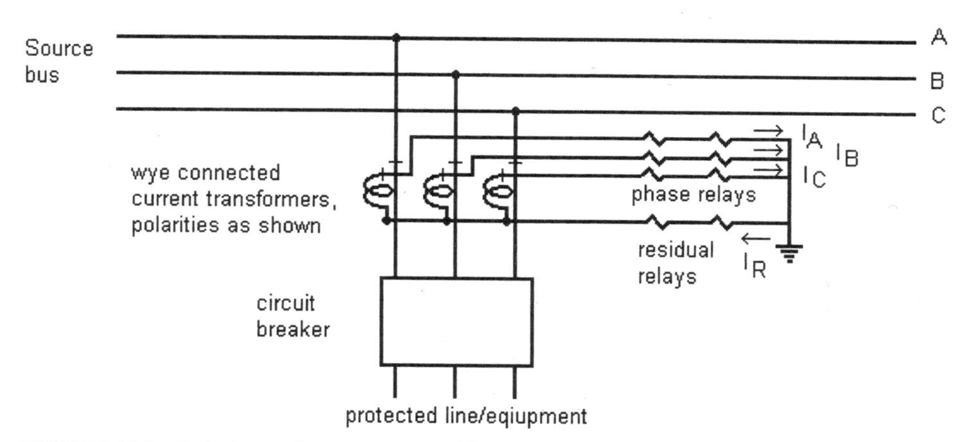

FIGURE 15.1 Typical setup for wye-connected CTs protecting a line or piece of equipment.

2. The circuit has an *A* phase to ground fault on the line, with fault current magnitude of 9000 A. Find the phase and residual relay currents. Again, assume that the *A* phase current is at 0°.

$$I_A = \frac{9000 \text{ A}}{120} = 75 \text{ A}\angle 0°$$

$$I_B = 0 \text{ A}$$

$$I_C = 0 \text{ A}$$

$$I_R = I_A + I_B + I_C = 75 \text{ A}\angle 0° + 0 \text{ A} + 0 \text{ A} = 75 \text{ A}\angle 0°$$

The current path is therefore through the *A* phase lead and back through the residual lead.

3. The circuit has a two-phase fault with 5000 amps going out *B* phase and back in on *C* phase. Choose *B* phase current to be at 0°.

$$I_A = \frac{0 \text{ A}}{120} = 0 \text{ A}$$

$$I_B = \frac{5000 \text{ A}\angle 180°}{120} = 41.7 \text{ A}\angle 0°$$

$$I_C = \frac{5000 \text{ A}\angle 180°}{120} = 41.7 \text{ A}\angle 180° = -I_B$$

$$I_R = I_A + I_B + I_C = 0 \text{ A} + 41.7 \text{ A}\angle 0° + 41.7 \text{ A}\angle 180° = 0 \text{ A}$$

This current path involves the *B* and *C* phase leads, with no current in either the *A* phase lead or residual.

4. The circuit has a three-phase fault with 8000 A per phase.

$$I_A = \frac{8000 \text{ A}\angle 0°}{120} = 66.7 \text{ A}\angle 0°$$

$$I_B = \frac{8000\ \text{A}\angle -120^\circ}{120} = 66.7\ \text{A}\angle -120^\circ$$

$$I_C = \frac{8000\ \text{A}\angle 120^\circ}{120} = 66.7\ \text{A}\angle -120^\circ$$

$$I_R = I_A + I_B + I_C = 66.7\ \text{A}\angle 0^\circ + 66.7\ \text{A}\angle -120^\circ + 66.7\ \text{A}\angle 120^\circ = 0\ \text{A}$$

The phase currents sum to zero, so no current flows in the residual for this fault.

The path of current flow for these various situations must be considered in calculating the CT excitation voltage and subsequent saturation.

Example 15.3

For part 2, 3, and 4 of Example 15.2, calculate the CT voltage if the phase relay burden is 1.2 Ω, the residual relay burden is 1.8 Ω, the lead resistance is 0.4 Ω, and the CT resistance is 0.3 Ω. Neglect CT saturation in this calculation.

1. Single-Phase Fault

The *A* phase CT will have an excitation voltage of

$$\begin{aligned} V_{exA} &= I_{A\text{sec}}(Z_{\text{CT}} + 2Z_{\text{lead}} + Z_{\text{phase}} + Z_{\text{residual}}) \\ &= 75\ \text{A}(0.3\ \Omega + 2 \cdot 0.4\ \Omega + 1.2\ \Omega + 1.8\ \Omega) \\ &= 307\ \text{V} \end{aligned}$$

The impedances are primarily resistive, and phase angle is often neglected in the voltage calculations. The impedances can be determined by tracing the path of the current through the CT secondary circuit.

2. Two-Phase Fault

$$\begin{aligned} V_{exB} &= I_{B\text{sec}}(Z_{\text{CT}} + Z_{\text{lead}} + Z_{\text{phase}}) \\ &= 41.7\ \text{A}(0.3\ \Omega + 0.4\ \Omega + 1.2\ \Omega) \\ &= 79.2\ \text{V} \end{aligned}$$

The *C* phase CT will see a similar voltage. Note that the *A* phase CT will also see a significant voltage, although it is carrying no current.

3. Three-Phase Fault

$$\begin{aligned} V_{exA} &= I_{A\text{sec}}(Z_{\text{CT}} + Z_{\text{lead}} + Z_{\text{phase}}) \\ &= 66.7\ \text{A}(0.3\ \Omega + 0.4\ \Omega + 1.2\ \Omega) \\ &= 126.7\ \text{V} \end{aligned}$$

The worst-case fault for this example is therefore the single-phase fault. It is clear that a CT with a saturation voltage of 200 V would experience substantial saturation for this fault. This saturation would cause a large reduction in the current delivered. In the other two cases, the CT remains unsaturated, so the CT will deliver the expected current at this voltage level.

Example 15.4

Multiratio Current Transformers. A current transformer has maximum load current of 650 A and maximum fault current of 10,500 A. The total burden impedance is 2.1 Ω. A

1200:5 class C200 multiratio CT is present. The available CT taps are 100, 200, 300, 400, 500, 600, 800, 900, 1000, and 1200. (These taps represent the primary current rating, with the secondary current rating being 5 A for all tap selections.)

The continuous primary current rating must be greater than 650 A. The continuous rating should also be greater than 5 percent of the maximum fault current—10,500 A/20 or 525 A. Therefore, 800, 900, 1000, and 1200 will satisfy these criteria. The other criteria involves avoiding saturation on the maximum fault. Note that the partial winding use of a CT reduces the saturation voltage in proportion the percentage of the total turns in use. Therefore, the 800:5 tap has 800/1200 = 67 percent of the turns in use. Estimating the full winding saturation voltage at 200 V, the saturation voltage for the 800 A tap is

$$V_{\text{knee}} = 0.67 \cdot 200\text{ V} = 133\text{ V}$$

Neglecting saturation, the worst-case voltage on the CT would be

$$V_e = \frac{I_{\text{fault}}}{\mathrm{N}_{\text{CT}}} Z_{\text{burden}} = \frac{10{,}500\text{ A}}{160} 2.1\ \Omega = 138\text{ V}$$

I_{fault} is the fault current flowing in the CT primary, and N_{CT} is the CT turns ratio in use. In this case, the excitation voltage is somewhat higher than the CT knee voltage, so a different tap should be considered. For the 1000:5 ratio setting,

$$V_{\text{knee}} = 0.833 \cdot 200\text{ V} = 167\text{ V}$$

and the expected maximum excitation voltage would be

$$V_e = \frac{I_{\text{fault}}}{N_{\text{CT}}} Z_{\text{burden}} = \frac{10{,}500\text{ A}}{200} 2.1\ \Omega = 110\text{ V}$$

This voltage is significantly less than the CT knee voltage, so that the CT ratio of 1000:5 is a better choice than the 800:5 ratio in this application.

TIME OVERCURRENT PROTECTION OF RADIAL PRIMARY DISTRIBUTION SYSTEMS

Time overcurrent (TOC) protection is the common protection method used on radial distribution networks. Time overcurrent relay characteristics offer fast response at high current levels, with the response time increasing as fault current level declines. With careful coordination between devices, selective coordination is possible so that the fault is sensed and cleared by downstream devices before the upstream device responds to the fault. TOC relay characteristics also allow for selective coordination between fuse and relay and between fuses. In addition to these devices, automatic circuit reclosers are self-contained devices that can be pole-mounted and will sense and clear faults to isolate feeder segments.

Example 15.5

Figure 15.2 shows an example involving a pair of fuses where the downstream fuse protects the upstream fuse for a fault at the location shown. Fuse operation is characterized by the fuse element melting, arcing, and clearing the fault. Fuses will generally coordinate if the total clearing time of the protecting fuse is less than 75 percent of the minimum

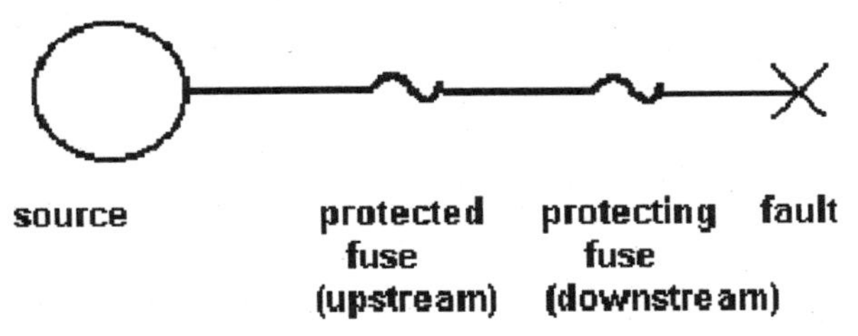

FIGURE 15.2 Protected and protecting fuse terminology.

melting time of the protected fuse for all currents up to the maximum fault current that the pair will experience. The maximum fault current is typically the bolted (zero fault impedance) fault just downstream of the protecting fuse. Figure 15.3 shows a 50-K (50-A continuous current rating, K fuse curve) fuse link protecting a 140 K fuse link for a maximum coordinating current of 5000 A. For either fuse, the lower boundary represents the minimum melting time of the fuse link, while the upper boundary represents the total clearing time of the fuse link. The curves are closest at 5000 A, and the gap between curves at this point represents the available coordinating margin. In this case, the total clearing time of the 50-K fuse link is about 80 percent of the minimum melting time of the 140-K link at 5000 A, so the margin is a bit too small to ensure coordination at 5000 A.

Time-current curves for specific devices are available from the device manufacturer. Coordination curves can be drawn by hand or on the computer using one of several software packages that are available on the market.

Example 15.6

Selection of Distribution Transformer Fuse. Transformer trf-1 is a three-phase, 900-kVA, 13.2-kV:480-V grounded wye-grounded wye transformer. Rated load current in the primary winding of the transformer is

$$I_{\text{rated}} = \frac{450 \text{ kVA}}{\sqrt{3} \cdot 13.2 \text{ kV}} = 20 \text{ A}$$

It can typically be expected that transformer inrush current on this bank will be 10 to 12 times rated current for 100 ms, with the first half-cycle of current being somewhat higher. Also, some duration of overload current can be expected due to cold load pickup. Allow 2 times rated current for 10 s due to this effect. The transformer fuse cannot operate for any of these events. On the other hand, the fuse must operate and clear the fault before transformer damage occurs during a fault on the transformer secondary. Many transformers are designed to conform to the damage points described in ANSI Std. C57.109. Figure 15.4 shows the time-current curve for these points. The figure also shows the melting and clearing curves of a 30-K fuse link. The fuse-link melting curve lies above all the load points, indicating it will not melt for these normal conditions. The fuse clearing curve lies below the transformer damage points, indicating that it will protect the transformer.

Note that the transformer fuse melting curve must also be above the transformer secondary circuit breaker total clearing curve. Also, the possibility of ferroresonance should be examined in distribution transformers with ungrounded primaries.

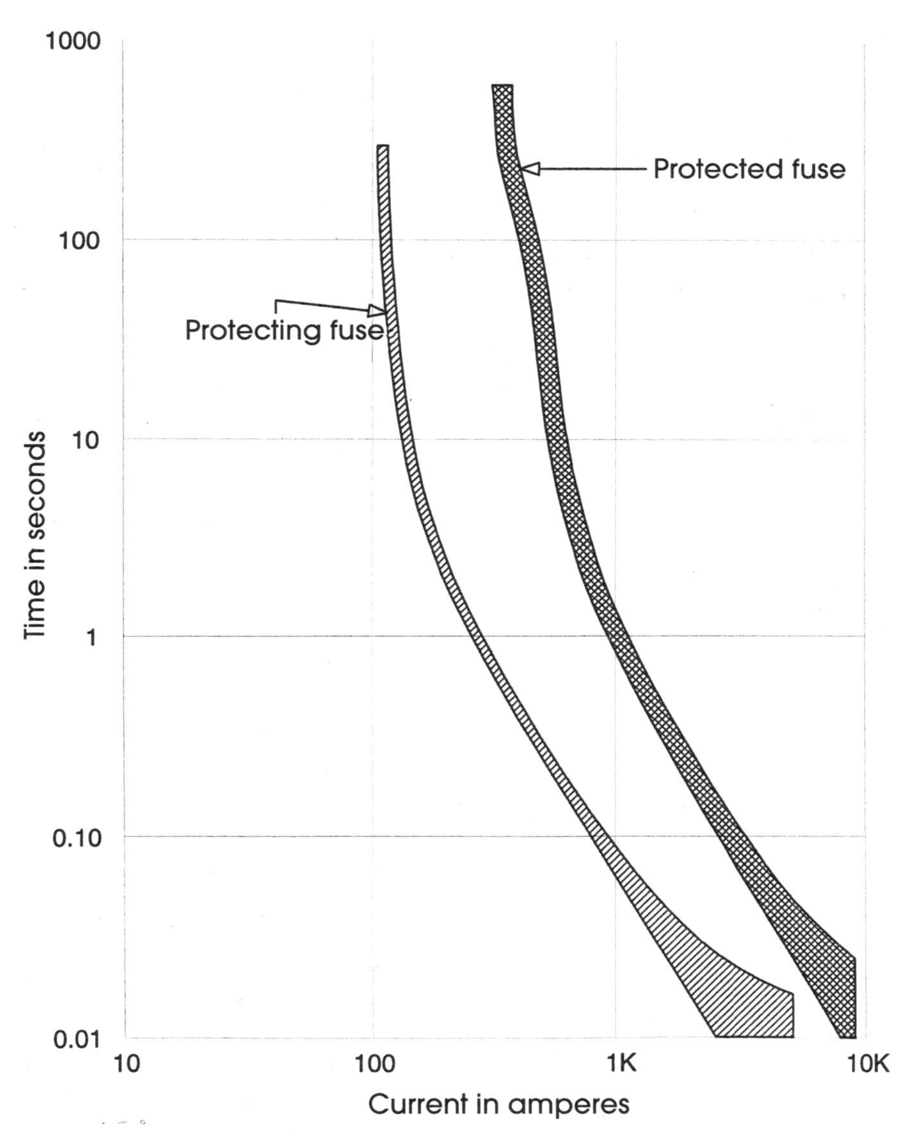

FIGURE 15.3 Time-current curve showing the coordination of a 50-K fuse link with a 140-K fuse line, for a maximum fault current of 5000 A.

Time overcurrent protection of primary radial distribution lines is accomplished in a similar fashion. The one-line diagram of a typical line is shown in Fig. 15.5. This line has a feeder-head circuit breaker that employs time overcurrent relays for sensing faults. It has a recloser located at the approximate midpoint of the line. It also has several sectionalizing fuses, which are intended to isolate feeder taps, most of which are single phase. Finally, the feeder feeds numerous distribution transformers, which are protected by fuses. There is no generation on this feeder, so that all load current and fault current flows from the source through the substation bus and out the feeder to the fault. Figure 15.6

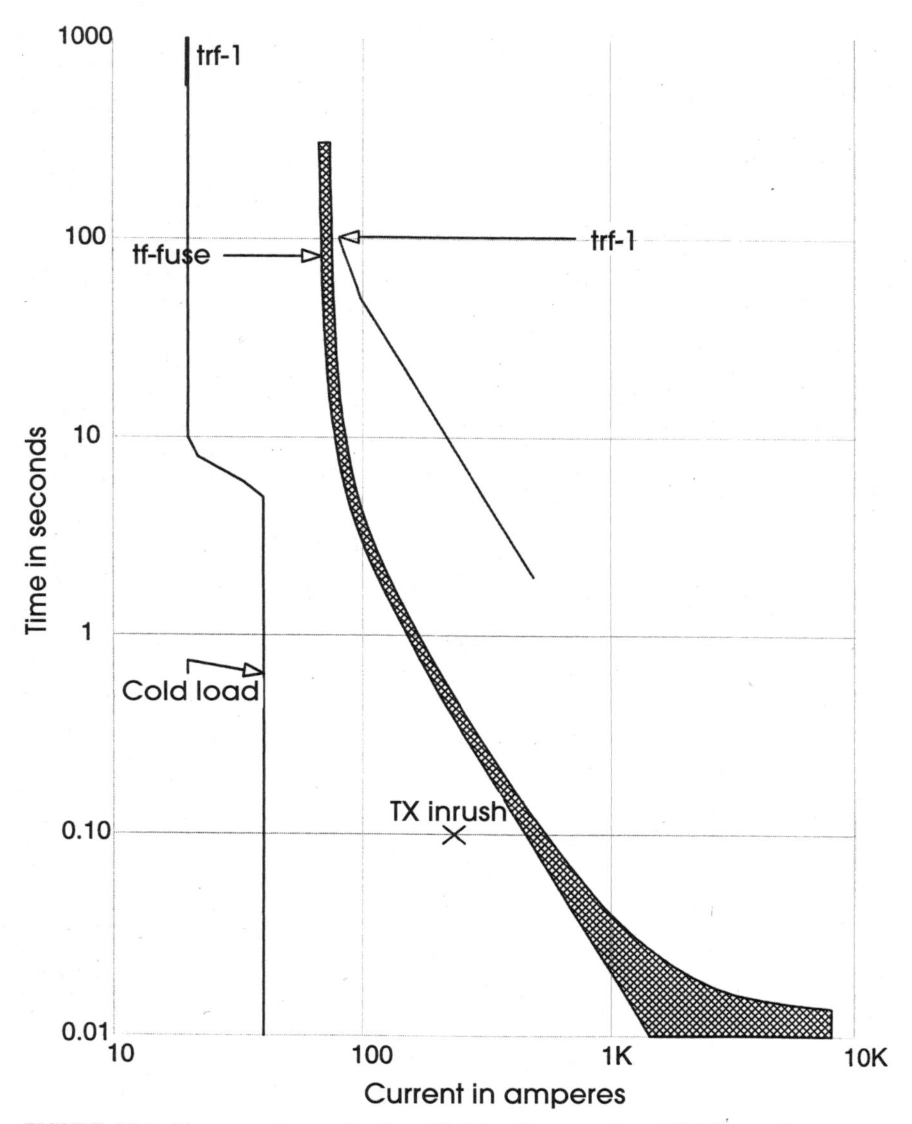

FIGURE 15.4 Time-current curve showing a 30-K fuse line protecting a 450-kVA transformer.

shows the same feeder with peak load currents noted. Figure 15.7 shows the same diagram with fault currents noted for faults at various locations on the diagram. It also shows the conductor size for the feeder.

The feeder protection must be determined to meet several objectives:

1. It must sense all faults on its section.
2. It must sense and clear faults before any equipment damage occurs.

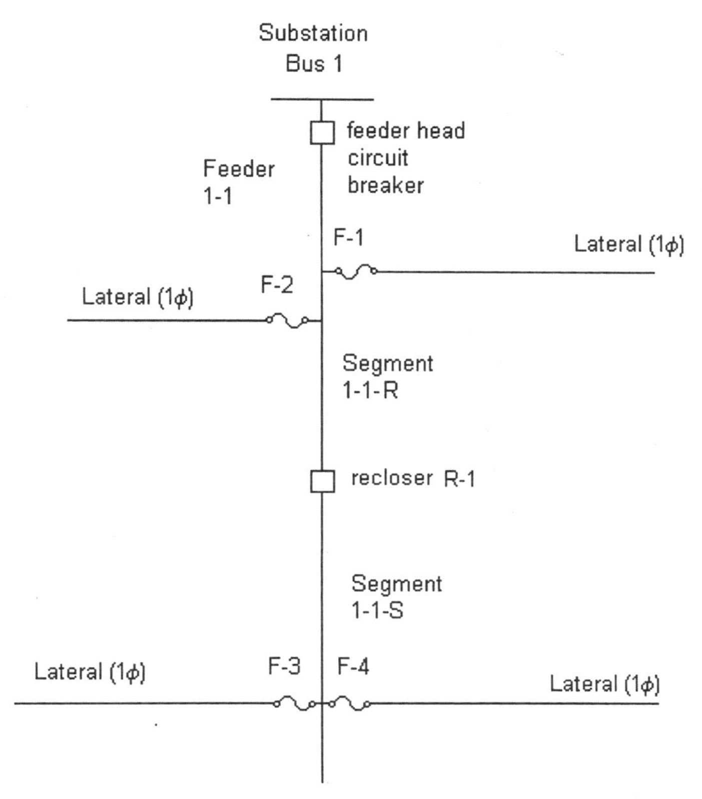

FIGURE 15.5 One-line diagram of a distribution feeder showing feeder-head circuit breaker (CB), line recloser, and sectionalizing fuses.

3. It must coordinate with upstream devices by sensing and clearing faults before the upstream device senses the fault.

4. It must avoid operating for load currents, including cold load pickup.

Finally, it must be noted that a large percentage of faults occurring on overhead lines are temporary. On temporary faults, it is possible to extinguish the arc by clearing the fault and then successfully return the line to service. This has obvious benefits for system reliability. Circuit breakers and reclosers are therefore commonly reclosed 1 to 3 times following a fault. Fuses, of course, must be replaced manually, resulting in an outage. In many cases, a fast trip is employed on a circuit breaker or recloser with the intention of having the fast trip operate and clear the fault prior to fuse melting. If the fault is temporary, the full circuit is returned to service following reclosing. If the fault is permanent, however, circuit reclosing will cause a restrike of the fault. Following one or two fast trips, the sensing device is switched to a slow trip, which allows the fuse to melt and clear before the slow-trip time elapses. This method, which is commonly referred to as fuse saving, allows service to be maintained to feeder taps for temporary faults on the tap and also allows service to be maintained on the main feeder for permanent faults on the tap.

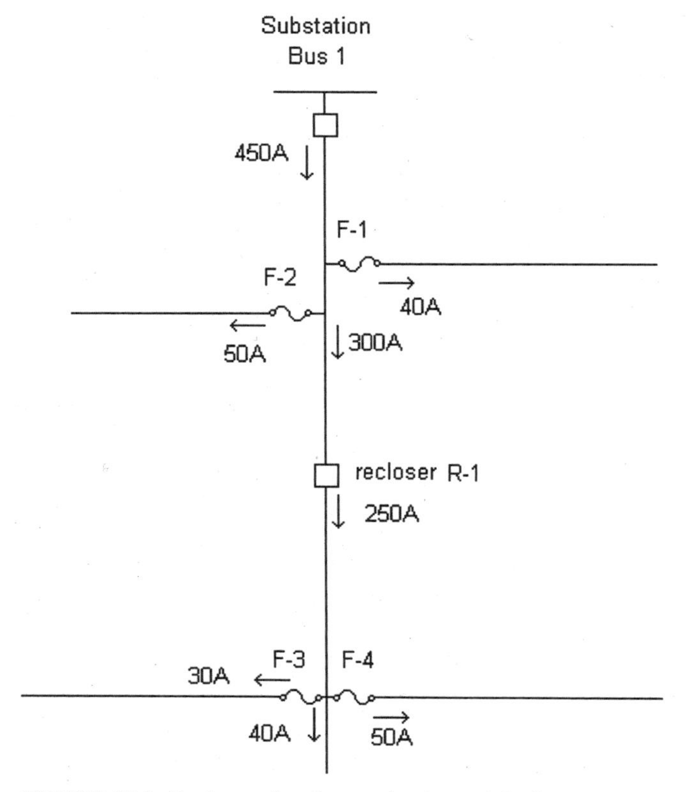

FIGURE 15.6 Feeder one-line diagram showing peak-load currents.

Example 15.7

Coordinate the Fig. 15.4 Feeder. Coordination of the various devices on the feeder is accomplished graphically through the use of time current curves (TCCs). This is generally an iterative process that is done graphically. The graphs can be drawn by hand or with one of several computer-aided engineering tools that are available. The following procedure outlines one approach to coordinating the feeder protection. Throughout this example, fuse and relay response curves and settings values are inserted into the discussion. These values are obtained from the fuse or relay manufacturer for their specific devices.

1. Determine the Feeder Head Relay Setting

The feeder head relay is fed by 600:5 current transformers, which sense the phase current in the circuit breaker. The relay setting requires a minimum pickup value, which sets the minimum fault current to which the relay will respond. The second setting is the time dial setting, which sets the time delay of the relay. A common starting point for setting the minimum pickup on primary distribution systems is twice maximum load current. Therefore, the initial setting for the circuit breaker relay should be approximately 1000 primary A. The current transformer rating is 600:5, so the TOC relay will see

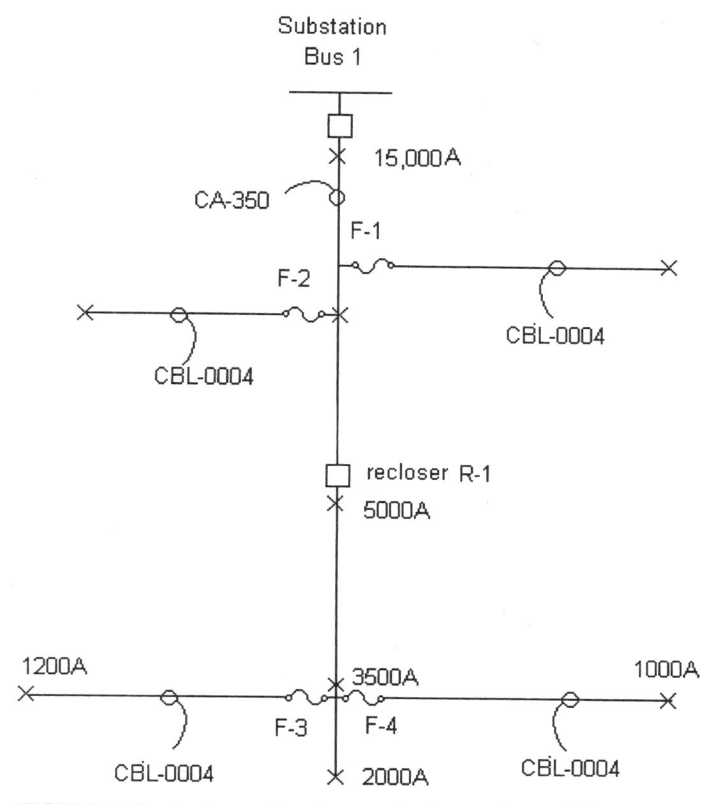

FIGURE 15.7 Feeder one-line diagram showing maximum fault currents at selected points and cable sizes.

$$I_{\text{relay}} = \frac{I_{\text{primary}}}{N_{\text{CT}}} = \frac{1000 \text{ A}}{120} = 8.33 \text{ A}$$

The closest available relay tap is 8.0 A, which corresponds to a primary setting of 960 A. The relay time lever is chosen so that the feeder circuit breaker will clear feeder faults before the substation transformer overload relay responds to the fault. Figure 15.8 shows the transformer overload relay curve. The total clearing time of the feeder circuit breaker is the relay response time plus the circuit breaker clearing time, plus auxiliary relay time if present. Therefore, the margin that must be maintained between the feeder relay curve and the bus relay curve equals the breaker time plus a reasonable margin. A typical margin is typically 0.25 to 0.3 s for electromechanical bus overload relays. Perhaps 0.10s of this margin allows for overtravel of electromechanical relays, which can be eliminated when solid-state or computer relays are used on the upstream devices (overtravel is the extra spin experienced by an electromechanical relay due to the momentum of the disk). Figure 15.8 shows the feeder-head phase relay selection with a suitable margin below the bus overload relay, up to the maximum fault current of 15,000 A. The figure also shows the damage curve for the main feeder cable, which is labeled CA-350. The feeder-head circuit breaker must clear the fault before cable damage occurs. In this case, the trans-

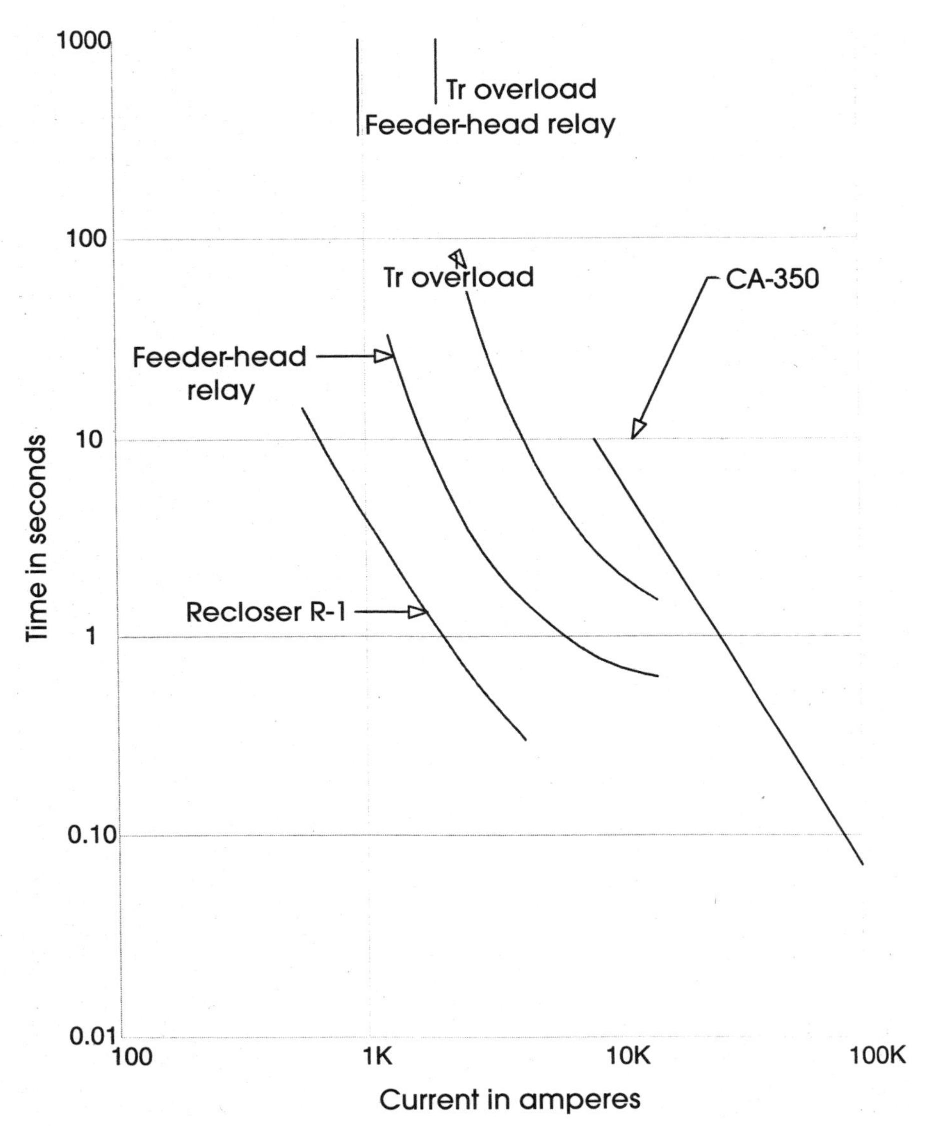

FIGURE 15.8 Time-current curve showing the coordination between the transformer overload relay, feeder-head relay, recloser slow trip, and line cable.

former overload relay will also protect the cable and will provide backup protection for the cable in the event of a failure of the feeder-head breaker.

2. Select the Recloser Minimum Pickup and Timing Curve

The recloser must sense and clear downstream faults before the feeder head relay senses the fault. The margin between the recloser clearing time and the feeder head relay sensing time is dependent on the recloser switching sequence and the type of feeder head relay. With electromechanical relays at the feeder head, the resetting time of the disk is important—during high current faults downstream from the recloser, the feeder head relay will respond with its disk traveling toward the trip position. When the recloser clears the fault, the disk on the feeder head relay will come to a stop and then reverse direction, going back toward the reset position. If the recloser closes back in on the fault before the feeder head relay resets, additional margin must be allowed between the recloser clearing time and feeder-head response curves.

Computer-based relays can be programmed to reset instantly and so are not subject to this increased margin. In a common situation, with electromechanical overcurrent relays, one fast trip on the recloser followed by two slow trips, and a 2-s delay to reclose, it has been found that coordination can be achieved by setting the recloser slow-trip clearing time to be less than 50 percent of the feeder-head-relay response time.

In this example, the peak-load current at recloser R1 is 275 A. A recloser minimum pickup setting of 560 A is therefore reasonable. The recloser would typically come with a selection of curves, which would be chosen to meet the margin requirement with the feeder head relay. Note that this margin is needed only up to a level of 5000 A, as the fault current will never exceed this value for faults beyond the recloser. This level of current marks the upper bound of the coordinating interval for this recloser-relay pair.

3. Recloser-Fuse Coordination

Any fuse downstream from the relay or recloser must melt and clear the fault before the relay or recloser senses the fault. A typical coordinating rule is that the total clearing time of the fuse must be less than 75 percent of the sensing time of the recloser or relay. The figure shows a variety of fuse curves sketched on top of the relay and recloser curves. Also shown on these curves is the largest current at which several of these pairs coordinate.

4. Fuse-Fuse Coordination

Downstream fuses must protect upstream fuses. The coordinating rule is similar to that in **(3):** The total clearing time of the downstream fuse must be less than 75 percent of the minimum melting time of the upstream fuse for coordination.

5. Fast Trip Coordination

Where it is desired to implement fuse saving through the use of a fast trip of a recloser or relay-circuit breaker, the coordinating rule is that the fast trip clearing time must be less than 75 percent of the minimum melting time of the fuse. This situation is shown in Fig. 15.9.

Items 1–5 must generally be repeated several times to achieve overall coordination on the feeder. The distribution transformer fuses place a lower bound on the coordinating curves, while the substation bus overload relays (or substation transformer overload relays) place an upper bound on these curves. Additionally, fault clearing curves must be below cable damage curves. Depending on the line configuration and loading, it is sometimes not possible to achieve coordination over the full range of coordinating intervals, which can result in two devices responding to a single fault. The design should eliminate or minimize this lack of coordination.

The preceding items discuss the response of the fuses and phase relays on the feeder. A similar procedure should be followed for the residual relay settings on the recloser and feeder-head circuit breaker. As the balanced load current does not flow through the residual relay, this relay can be set more sensitively than the phase relays. This is desirable in order to sense the low current faults that can result from one of the phases going to ground through a high fault impedance. Fuses, of course, do experience the full load current, so it is difficult to coordinate fuses with residual relays, and these relays are often set with relatively slow response times at the higher current levels. The minimum pickup of

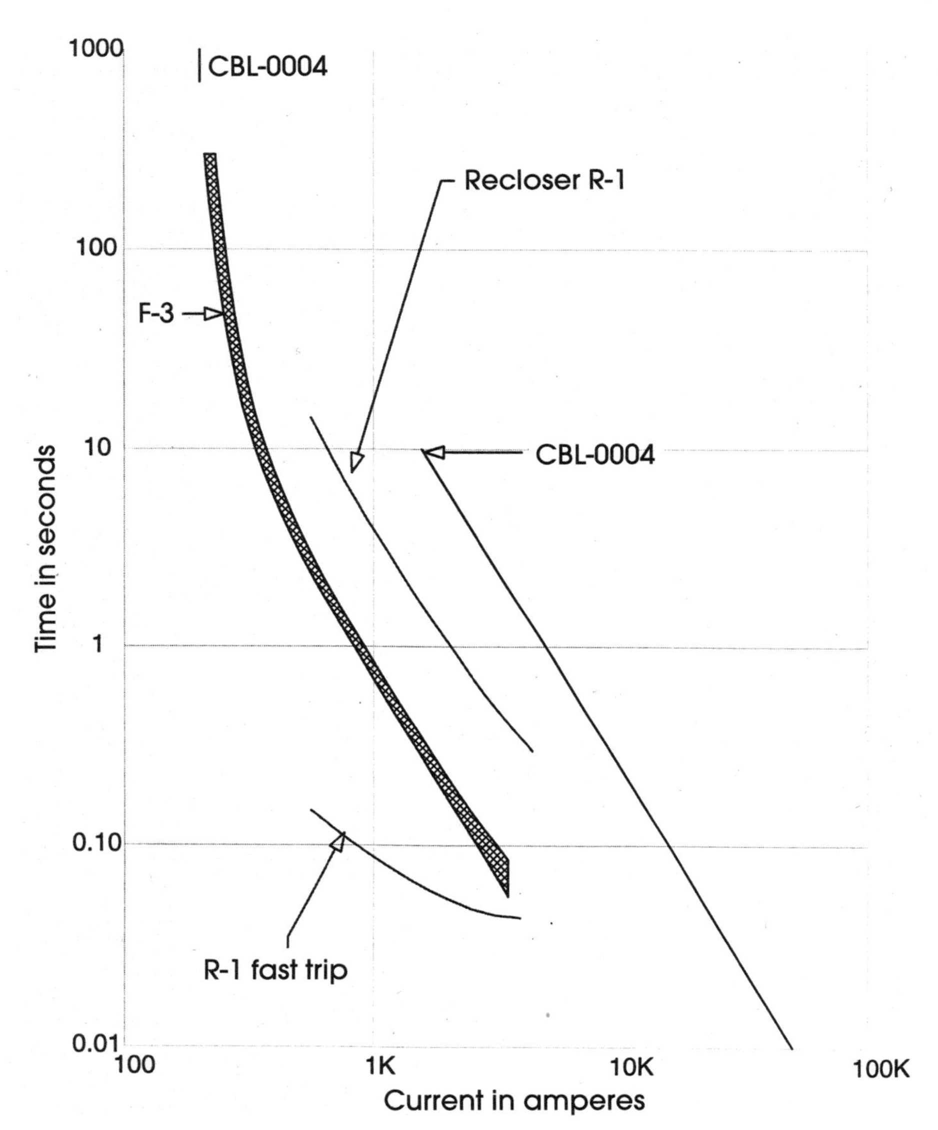

FIGURE 15.9 Time-current curve showing the coordination of the recloser fast and slow trips with sectionalizing fuse F-3.

residual elements must be above the level the relay will experience due to load imbalance, including the imbalance that is caused by the operation of a downstream fuse. A residual relay setting of approximately half of the phase load current level is reasonable in many cases. Residual relays will respond to faults that the phase relays do not sense, at the expense of some loss of selectivity. Nonetheless, there is some small but finite probability for the occurrence of high impedance faults that residual relays cannot detect. Specialized fault-detection relays for these events are under investigation.

The coordination of low-voltage distribution systems is accomplished in similar fashion. A brief consideration of low-voltage distribution is included in Section 5.

DIFFERENTIAL PROTECTION

Differential protection is applied on busses, generators, transformers, and large motors. Specialized relays exist for each of these applications, and their settings are described in the manufacturer's literature. Differential relays do require careful selection of current transformers. The full winding should be used when multiratio CTs are used in differential schemes, and other relays and meters should be fed from different CT circuits. Bus, generator, and large-motor differential schemes require matched sets of current transformers (with the same ratio and saturation characteristics) with suitable characteristics, while transformer differential protection requires CTs with limited mismatch.

Differential schemes are sometimes implemented on distribution busses with standard time overcurrent relays. An example of a single-phase differential scheme is shown in Fig. 15.10. With the current transformers connected as shown, the relay will see no current when the bus is intact (and with no CT error), as illustrated in Fig. 15.10*a* for a fault just outside the bus. During a bus fault, however, the relay will see the fault current divided by the CT ratio, as illustrated in Fig. 15.10*b*. The differential relay will see current on an external fault if one or more of the CTs saturate. In order to successfully

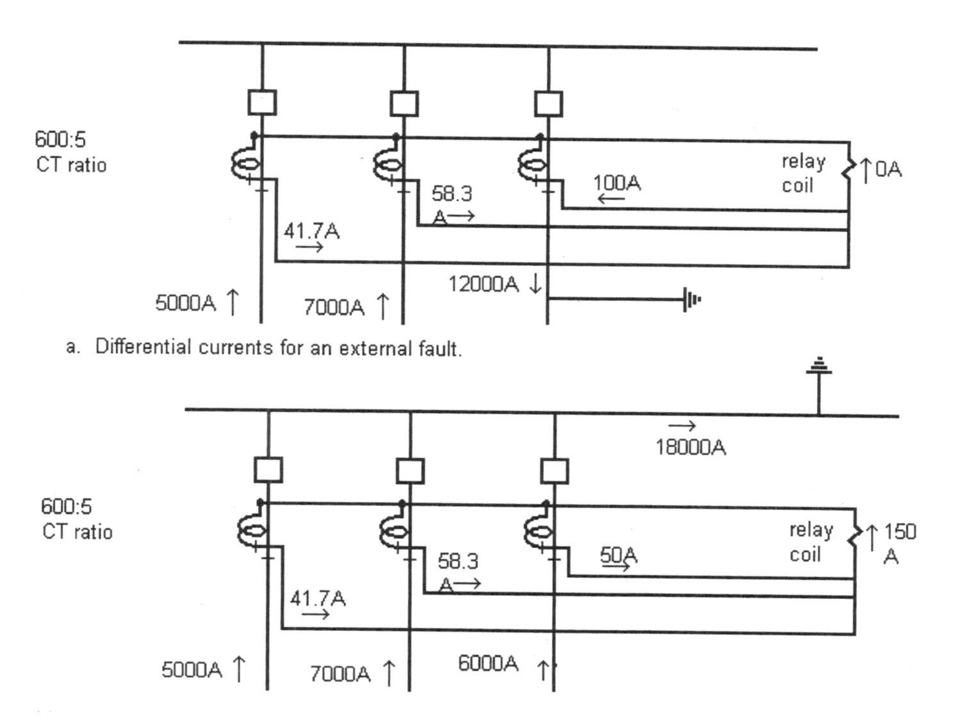

FIGURE 15.10 Example of a single-phase differential relay scheme.

implement this differential scheme, the saturation of the current transformers must not operate the relay for the maximum external fault. AC saturation of the CT will be avoided if the CT excitation voltage remains below the knee voltage of the CT. For the CT nearest the fault (and most subject to saturation), this voltage is

$$V_{CT} = (I_f/N_{CT})(R_{lead} K_p + R_{CT})$$

where

V_{CT} = the CT excitation voltage
I_f = the current in the primary of the CT nearest the fault
N_{CT} = the current transformer ratio in use
R_{lead} = the one-way lead resistance
$K_p = 2$
R_{CT} = the current transformer resistance

This scheme can be adopted to three-phase systems using wye-connected CTs. In this case, K_p is 1 for three-phase faults, and 2 for single-phase–to–ground faults.

If the excitation voltage of this CT is less than the CT knee voltage, the CT error will be no more than 10 percent. Assume no error in the other CTs. The relay current will be equal to the error current—in other words, 10 percent of the current that would be delivered with no saturation. In this example, the CT nearest the fault would ideally deliver 100 A. With 10 percent error, this current is reduced to 90 A, so 10 A will flow in the relay coil. Select a relay setting of 10 A.

The sensitivity of the differential scheme will be the relay set current times the CT ratio—10 A · 120 = 1200 A. Determine if this setting will sense the minimum bus fault. The minimum bus fault will occur when the bus is fed only by the weakest system and the fault has some resistance. In this example, the weakest system supplies 5000 A to a bolted fault. If only this breaker is closed and fault resistance reduces the fault by 75 percent, 1250 A would be the minimum expected fault current.

Select the relay time delay to avoid relay misoperation on dc saturation of the CT. Typically, this might be a two- or three-time dial setting with an inverse time characteristic, for a distribution bus with no local generation. (Note: The sensitivity of this scheme is marginal. Relays designed specifically for differential operation will offer substantial improvements in sensitivity for little additional cost. Also, the sensitivity can be improved with this scheme if the CT error is significantly less than 10 percent.)

STEP DISTANCE PROTECTION

Many transmission and subtransmission lines are protected with distance relays. These relays sense local voltage and current and calculate the effective impedance at that point. When the protected line becomes faulted, the effective impedance becomes the impedance from that point to the fault. A typical ohm distance characteristic is shown in Fig. 15.11. The maximum torque angle is set to be near the angle of the line impedance, to provide highest sensitivity for V/I ratios at that angle. The relay is inherently directional and will not sense reverse faults, which would appear in the third quadrant in Fig. 15.11. Also, the relay is less sensitive to load currents, which would be within 20 or 30° of the real axis in either direction. These desirable characteristics make distance protection popular. Distance protection is available for both phase and ground faults.

Step distance protection combines instantaneous and time delay tripping. Zone 1 is an underreaching element—any fault within Zone 1 is known to be on the protected line. When Zone 1 operates, the line is tripped instantaneously. However, Zone 1 will not oper-

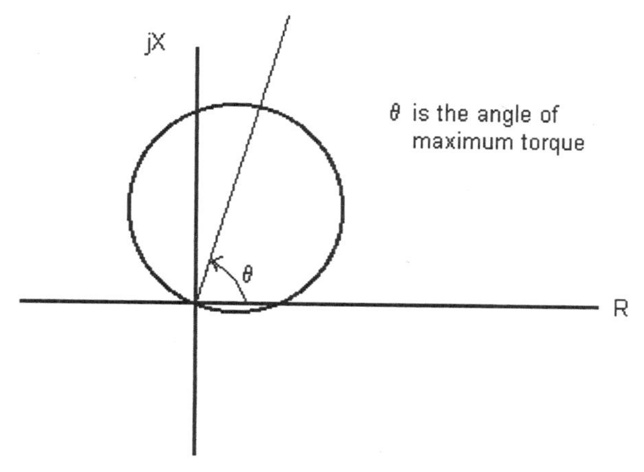

FIGURE 15.11 Distance relay ohm characteristic. Apparent impedances inside the circle will cause relay operation.

ate for all line faults. Zone 2 is an overreaching relay—it is set so that operation is guaranteed if the fault is on the line. Zone 2, however, will operate for some external faults. Selectivity is maintained by delaying Zone 2 tripping so that external faults are cleared by downstream devices—bus differential relays for a fault on the next bus, or Zone 1 relays on the next line. The Zone 2 time delay should be set for the auxiliary relay time + circuit breaker time + margin. A margin of 0.1–0.15 s is considered adequate, so that the Zone 2 delay may be 0.25 seconds for a 6-cycle circuit breaker.

Example 15.8

Figure 15.12 shows a typical example. We will consider the settings for line *PQ* at bus *P*. The impedance angle for all lines is 75°. The line length is 80 Ω. The distance relay at bus *P* is fed by current transformers rated at 2000 A:5 A and voltage transformers rated at 345 kV/200 kV Y:120 V/69 V Y. Set Zone 1 for 85 percent of this value (85–90 percent settings are typical for phase distance, slightly lower for ground distance):

$$
\begin{aligned}
\text{Zone 1 setting} &= 0.85 \cdot 80\ \Omega = 68\ \Omega, \text{ primary ohm setting} \\
\text{CT ratio} &= 2000/5 = 400 \\
\text{VT ratio} &= 200{,}000/69 = 2900 \\
\text{Relay setting} &= \text{primary setting } (\Omega) \cdot \text{CT ratio/VT ratio} \\
&= 68\ \Omega \cdot (400)/(2900) \\
&= 9.38 \text{ relay ohms}
\end{aligned}
$$

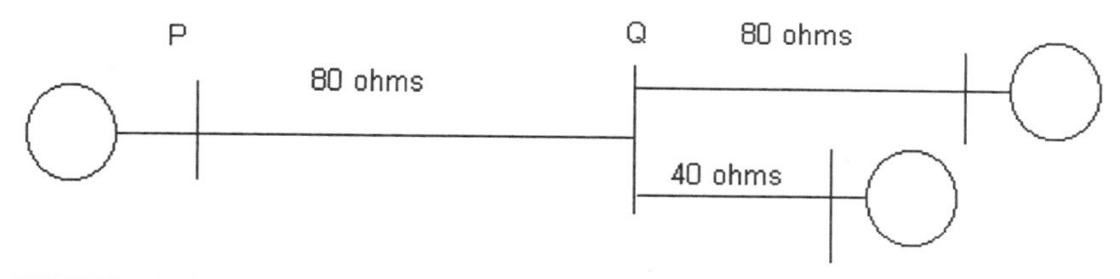

FIGURE 15.12 Example of step-distance relay setting.

$$\begin{aligned}\text{Zone 2 setting} &= \text{line length} \cdot 115 \text{ percent (minimum)}\\ &= \text{line length} + 0.5 \cdot \text{length of shortest next adjacent line (preferred)}\end{aligned}$$

The two next adjacent lines are 40 and 80 Ω, respectively. The shortest of these is 40 Ω. Half of that is 20 Ω. The setting 80 + 20 Ω = 100 Ω is greater than the minimum setting of 92 Ω (which guarantees seeing the entire line).

The relay setting is then

$$\begin{aligned}\text{Zone 2 setting} &= 100\ \Omega \text{ (primary)}\\ &= 100 \cdot 400/2900\\ &= 13.8 \text{ relay ohms}\end{aligned}$$

It must be noted that when the preferred setting is less than the minimum, the minimum must be selected. This means, however, that Zone 2 of the protected line is capable of reaching beyond Zone 1 of the short next line, so the time delay must be increased to avoid miscoordination.

Some schemes include a Zone 3 element for additional backup. Also, all bulk transmission lines include communications between line ends to provide instantaneous tripping for faults located at any point on the line.

Infeed Effect

The infeed effect shortens the reach of distance relays for relays reaching beyond a junction point. Infeed effect is illustrated in Fig. 15.13.

The apparent impedance to a distance relay at bus P is

$$Z_P = \frac{V_P}{I_P}$$

The bus voltage at P for a fault at F is

$$V_P = Z_{PQ}I_P + Z_{QF}(I_P + I_R)$$

The impedance sensed by the relay is then

$$Z_P = \frac{V_p}{I_P} = Z_{PQ} + Z_{QF}\left(1 + \frac{I_R}{I_P}\right)$$

The fault at F therefore appears to be further away than it actually is. On two terminal lines, this effect is not particularly important. On three terminal lines, however, the infeed effect must be fully considered to ensure complete coverage of the line under all conditions.

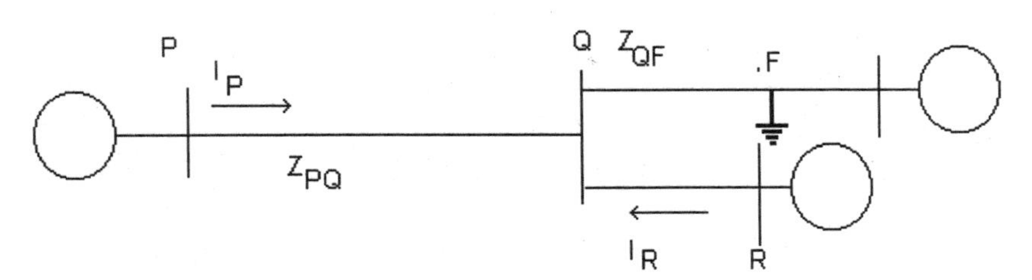

FIGURE 15.13 Example of infeed effect.

SMALL MOTOR PROTECTION

Induction motors can be damaged by a number of events, including short circuits, overloads, voltage unbalance, undervoltage, and locked rotor conditions. Small motors (up to several hundred horsepower) are commonly protected by a combination of fast protection for faults and slower protection for the other situations. The lower current protection must protect the motor from thermal damage while allowing the high currents drawn by the normal starting transient. This protection is generally provided by a combination of motor starter with overload sensing and fuses or circuit breaker. It also must be recognized that motor starters have limited current-clearing capability and must be protected from fault currents above their interrupting rating.

As motor size increases, it becomes impossible to protect against locked rotor conditions and unbalance through the sensing of stator phase current magnitude, and additional protection functions are needed. Computer-based relays are available for these situations, and the setting of these relays is relatively straightforward.

Example 15.9

Provide protection for the 100-hp induction motor shown in the one-line diagram of Fig. 15.14. System data:

Utility source: 13.2 kV, 80 MVA three-phase short-circuit capability.

Transformer XF2: 500 kVA, 13.2 kV:480 V delta-grounded wye.

Motor: 100 hp, 480 V, 0.85 pf, 85 percent efficient. Starting requires up to 5.9 · rated current for up to 8 s. Motor locked rotor thermal capability is 20 s.

1. Calculate Rated Motor Current

$$I_{\text{rated}} = \left(\frac{\text{motor horsepower} \cdot (746\text{w/hp}}{\text{power factor} \cdot \text{efficiency}}\right) \cdot \frac{1}{\sqrt{3} \cdot \text{V}_{\text{rated}}}$$

$$= \left(\frac{100 \cdot 746}{0.85 \cdot 0.85}\right) \cdot \frac{1}{\sqrt{3} \cdot 480}$$

$$= 124 \text{ A}$$

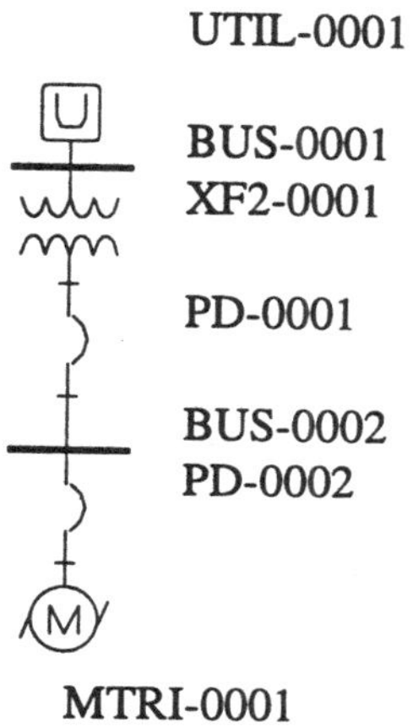

UTIL-0001

BUS-0001

XF2-0001

PD-0001

BUS-0002

PD-0002

MTRI-0001

FIGURE 15.14 One-line diagram for small motor protection.

2. Calculate the Motor Starting/Running Curve and the Locked Rotor Point

See plot on Fig. 15.15.

3. Determine Motor Protection Characteristic

The minimum pickup must be between 100 and 125 percent of the motor rated current. The delayed trip must be above the starting/running curve but below the locked rotor thermal limit. A low-voltage circuit breaker was selected with both thermal and magnetic trips, and a thermal pickup setting of 125 A, with the magnetic trip set at 6.75 times that value. The plot PD-0001 shows this characteristic.

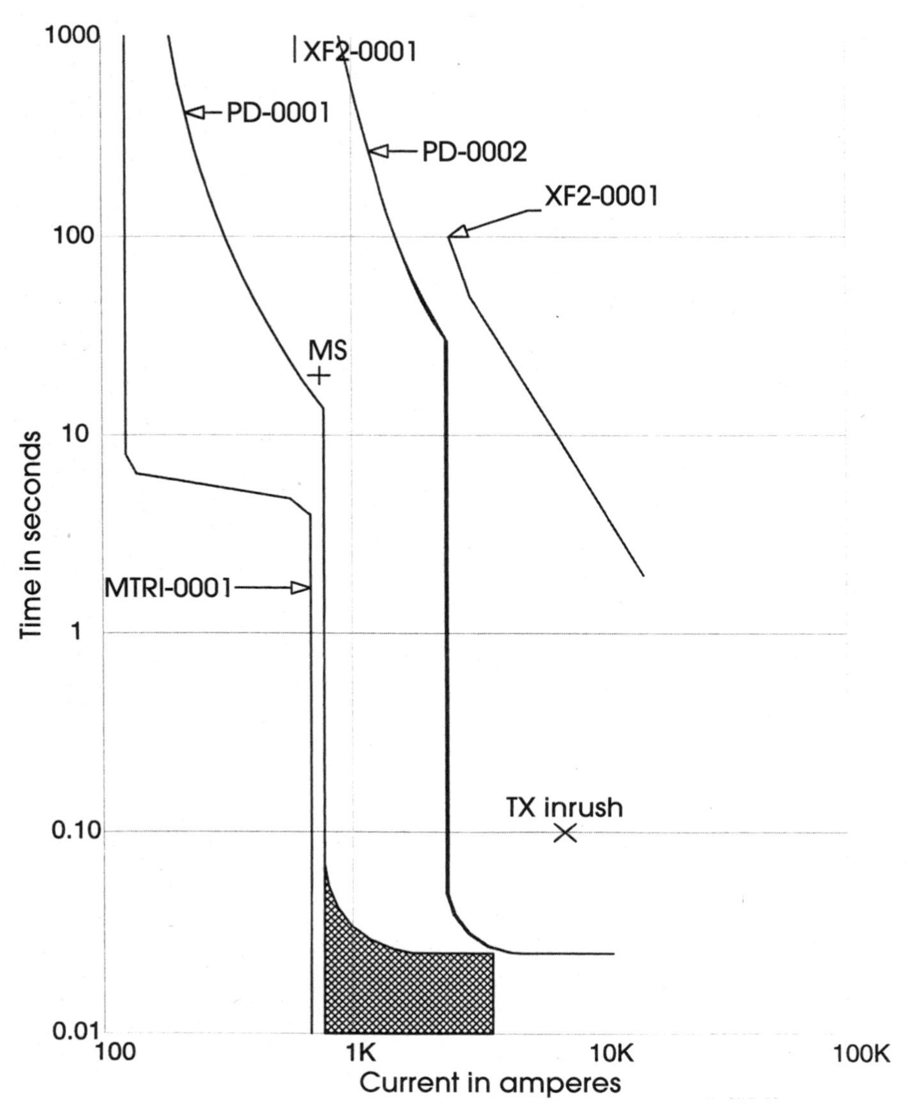

FIGURE 15.15 Time-current curve showing the coordination between motor MTRI-0001, motor protection PD-0001, transformer PD-0002, and the transformer damage characteristic.

4.

The motor protection must be faster than the transformer low-side circuit breaker, which in turn must protect the transformer from damage. The circuit-breaker characteristic is labeled PD-0002, while the curve XF-0001 shows the transformer damage characteristic.

Figure 15.16 shows similar protection provided by a motor starter with thermal overloads and fuse. The fuse in this case must protect the motor starter by interrupting currents above the starter clearing capability.

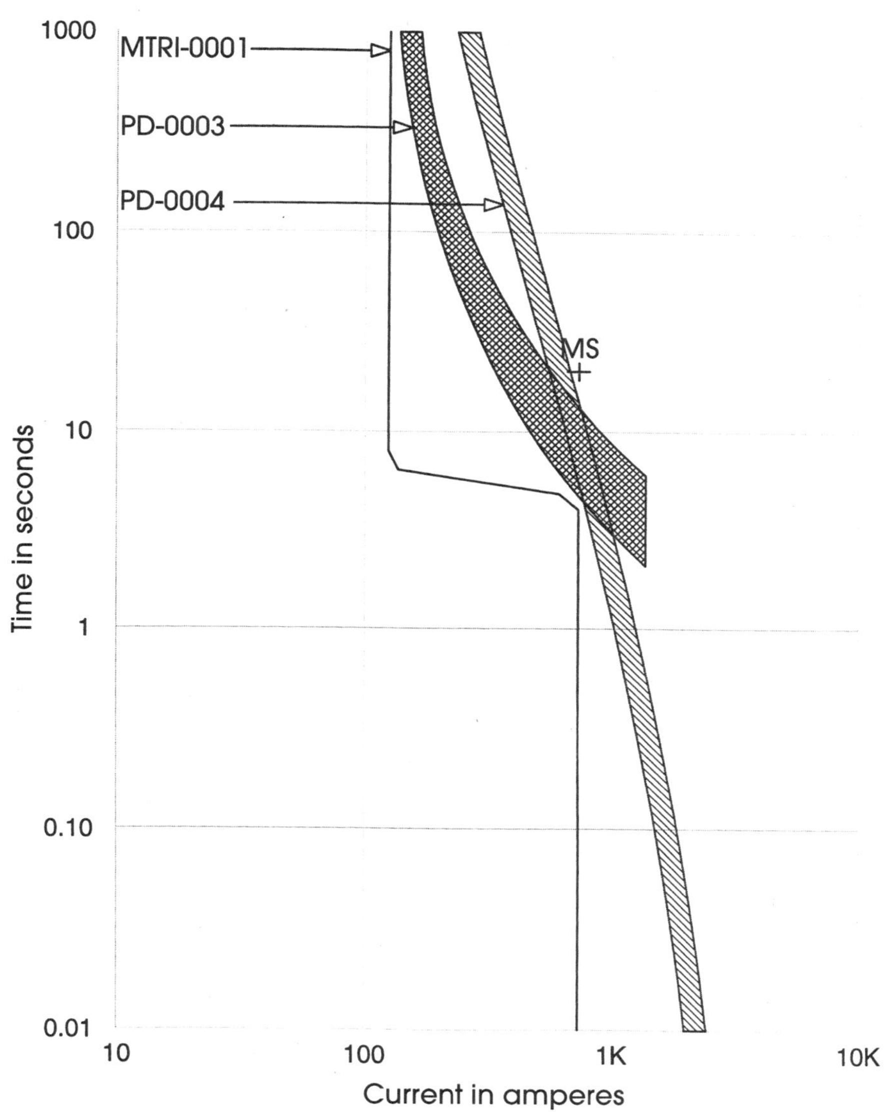

FIGURE 15.16 Time-current curve showing motor characteristics, contactor/thermal overload characteristic PD-0003, and fuse curve PD-0004.

SUMMARY

This chapter has described current and voltage transformer selection, and several common protection applications. Many protective functions exist that are not described in this chapter. Several excellent references are listed in the bibliography.

Acknowledgment

The author would like to thank SKM Systems Analysis, Inc. for allowing the use of their software in preparing the time current curves presented in this chapter.

BIBLIOGRAPHY

Anderson, P. M. 1999. *Power System Protection*. New York: IEEE Press.

Blackburn, J. Lewis. 1987. *Protective Relaying: Principles and Applications*. New York: Marcel Dekker, Inc.

Cooper Bussmann, Inc. 1998. *Bussmann Electrical Protection Handbook*. Copyright 1998 by Cooper Bussmann, Inc. http://www.bussmann.com

Elmore, Walter, ed. 1994. *Protective Relaving: Theory and Applications*. New York: Marcel Dekker, Inc.

IEEE Std. 242. 1986. *IEEE Recommended Practice for Protection and Coordination of Industrial and Commercial Power Systems* (IEEE Buff Book). New York: IEEE Press.

Manufacturer's Literature

General Electric Co. *Multi-line 139 Series Motor Protection Relay*. General Electric Co. http://www. ge.com/indsys/pm/

SEL-351 Distribution and Transmission Relay. Pullman, W.A. Schweitzer Engineering Labs. www.selinc.com

SECTION 16

POWER-SYSTEM STABILITY

Alexander W. Schneider, Jr.
Senior Engineer
Mid-America Interconnected Network

Peter W. Sauer
Professor of Electrical Engineering
University of Illinois at Urbana-Champaign

INTRODUCTION

Power-system stability calculations fall into three major categories: transient, steady-state, and voltage stability. Transient stability analysis examines the power-system response to major disturbances, such as faults or loss of generation. This analysis is typically nonlinear and focuses heavily on the ability of the system generators to maintain synchronism. Steady-state stability analysis examines the qualitative nature of steady-state equilibrium in response to infinitesimally small disturbances. This analysis uses a linear version of the detailed dynamic models and focuses on the eigenvalues of the linear system and the response of control systems. In earlier literature, steady-state stability was frequently referred to as *dynamic stability*. Voltage stability analysis examines the ability of a power system to maintain acceptable voltage levels in response to both abrupt and gradual disturbances. This analysis primarily uses steady-state modeling, although some analysis may consider dynamics of voltage controls such as tap-changing-under-load transformers.

The first procedure presented is the formulation of models that are utilized for stability analysis. Using these modeling techniques, the remaining procedures illustrate both small- and large-scale stability calculations. The small-scale procedures are provided to illustrate the mathematical steps associated with the analysis. The large-scale procedures provide the necessary considerations to perform this analysis on realistic system models.

DYNAMIC MODELING AND SIMULATION

Since power-system stability analysis is based on computations involving mathematical models, it is important to understand the formulation of these models as well as their limitations. In the dynamic analysis of the fastest transients, partial differential equations describing the properties of transmission lines are solved using approximate numerical techniques. For the analysis presented in this chapter, the fast dynamics associated with the synchronous machine stator, transmission lines, and loads are neglected. As such, they are normally modeled with algebraic equations that are essentially identical to their steady-state form. The models of the synchronous generators and their controls are normally limited to ordinary differential equations that are coupled to one another through the algebraic equations of the network. Each such differential equation expresses the derivative of a state variable, such as rotor angle, field voltage, or steam-valve position, in terms of other states and algebraic variables. This section illustrates the procedures used to create a typical model suitable for both transient and steady-state stability in a time frame that neglects the high-frequency "60-Hz" transients.

It should be noted that the differential equations are generally nonlinear. In addition to product nonlinearities, phenomena such as magnetic saturation, voltage and valve-travel limits, dead bands, and discrete time delays are inherent in many of the devices modeled, and contemporary simulation packages provide a rich assortment of features to deal with those deemed significant.

Calculation Procedure

1. Specify a Synchronous Machine Model that Captures the Dynamics of Interest

A synchronous machine has essentially three fundamental characteristics that need to be represented in the dynamic model. These are the field winding dynamics, the damper winding dynamics, and the shaft dynamics. In that order, the following model includes the field flux-linkage dynamic state variable, E'_q, a single q-axis damper winding-flux linkage state variable, E'_d, the shaft-position state variable relative to synchronous speed, δ, and the shaft-speed state variable, ω. The two flux-linkage states are shown in per unit (p.u.). The model indexes the equations for a system with m synchronous machines.

The first equation is the dynamic model that represents Faraday's law for the field winding, and E_{fd} is the field-winding input voltage in per unit. This field voltage is taken either as a constant or a dynamic state variable in the exciter model, described next. The second equation is the dynamic model that represents Faraday's law for the q-axis damper winding. The third and fourth equations are Newton's second law for the dynamics of the rotating shaft.

$$T'_{doi}\frac{dE'_{qi}}{dt} = -E'_{qi} - (X_{di} - X'_{di})I_{di} + E_{fdi} \qquad i = 1, \ldots, m$$

$$T'_{qoi}\frac{dE'_{di}}{dt} = -E'_{di} + (X_{qi} - X'_{qi})I_{qi} \qquad i = 1, \ldots, m$$

$$\frac{d\delta_i}{dt} = \omega_i - \omega_o \qquad i = 1, \ldots, m$$

$$\frac{2H_i}{\omega_o}\frac{d\omega_i}{dt} = T_{Mi} - E'_{di}I_{di} - E'_{qi}I_{qi} - (X'_{qi} - X'_{di})I_{di}I_{qi} \qquad i = 1, \ldots, m$$

The inertia constant, H, represents the scaled total inertia of all masses connected in tandem on the same shaft. The units of this constant are seconds. The X parameters are inductances that have been scaled with synchronous frequency (ω_o) to have the units of reactance in per unit. The time constants multiplying the state derivatives have the units of seconds. The algebraic variables, I_{di} and I_{qi}, are components of the generator current, which in steady-state is a fundamental-frequency phasor with the following form:

$$(I_{di} + jI_{qi})\,e^{j(\delta_i - \pi/2)} \qquad i = 1, \ldots, m$$

During transients, the generator terminal current algebraic variables, I_{di} and I_{qi}, are related to the generator terminal voltage, $Ve^{j\theta}$, and the dynamic states through the following stator algebraic equation (for $i = 1, \ldots, m$):

$$(I_{di} + jI_{qi})\,e^{j(\delta_i - \pi/2)} = \left(\frac{1}{R_{si} + jX'_{di}}\right)[E'_{di} + (X'_{qi} - X'_{di})I_{qi} + jE'_{qi}]e^{j(\delta_i - \pi/2)} - \left(\frac{1}{R_{si} + jX'_{di}}\right)V_i e^{j\theta_i}$$

The variable, T_M, is the mechanical torque that may be a dynamic state variable in the prime mover and governor model, as described next, or a constant.

2. Specify an Exciter and Voltage Regulator Model that Captures the Dynamics of Interest

Field current for a synchronous generator is normally supplied by a variable-voltage exciter whose output is controlled by a voltage regulator. The voltage regulator may be electromechanical or solid state, and the exciter may be a direct-current generator, an alternating-current generator with controlled or uncontrolled diodes to rectify the output, or a transformer with controlled rectifiers. The exciter and its voltage regulator are collectively referred to as an *excitation system*. While the exciter of a synchronous generator is normally thought of as a mechanism for changing the terminal voltage, its action also indirectly determines the generator reactive power output.

IEEE task forces have periodically issued recommendations for the modeling of excitation systems in dynamic studies. The first such effort suggested four models, of which model 1 was applicable to the largest group of machines, those with continuously acting voltage regulators and either direct-current or alternating-current rotating exciters. Although more specialized models represent these machines more rigorously, this model is still in widespread use, particularly for machines at some distance from the disturbances to be studied. An IEEE Type 1 excitation system is modeled with three or four dynamic state variables as follows:

- V_{si}: the voltage-sensing circuit output, whose time constant is generally quite small and often assumed to be zero.
- V_{Ri}: the voltage regulator output.
- E_{fdi}: the output of the exciter, which is the input to the synchronous-machine field winding.
- R_{fi}: a rate-feedback signal, which anticipates the effect of the error signal in changing the terminal voltage.

The equations for the derivatives of these state variables are

$$T_{Ri}\frac{dV_{Si}}{dt} = -V_{Si} + (V_{\text{ref}i} - V_i) \qquad i = 1, \ldots, m$$

$$T_{Ai}\frac{dV_{Ri}}{dt} = -V_{Ri} + K_{Ai}R_{fi} - \frac{K_{Ai}K_{Fi}}{T_{Fi}}E_{fdi} + K_{Ai}V_{Si} \qquad i = 1, \ldots, m$$

$$T_{Ei}\frac{dE_{fdi}}{dt} = -[K_{Ei} + S_{Ei}(E_{fdi})]E_{fdi} + V_{Ri} \qquad i = 1, \ldots, m$$

$$T_{Fi}\frac{dR_{fi}}{dt} = -R_{fi} + \frac{K_{Fi}}{T_{Fi}}E_{fdi} \qquad i = 1, \ldots, m$$

There is generally a limit constraint on the regulator output:

$$V_{Ri}^{\min} \leq V_{Ri} \leq V_{Ri}^{\max} \qquad i = 1, \ldots, m$$

The input to the voltage-sensing circuit is the error signal, $V_{\text{ref}i} - V_i$. If T_{Ri} is zero, then this difference equals V_{Si}, which becomes an algebraic variable rather than a dynamic state variable. This is the difference between the desired or reference value of terminal voltage and the measured value (typically line-to-line in per unit).

3. Specify a Prime Mover and Governor Model that Captures the Dynamics of Interest

Most modern generators have a governor, which compares their speed with synchronous speed. This governor adjusts the flow of energy into the prime mover by means of a valve. The valve regulates steam flow in the case of a steam turbine, water in the case of a hydro unit, and fuel in the case of a combustion turbine or diesel engine. The prime mover fundamentally dictates the choice of governor model, and only that for a basic steam turbine will be discussed in detail. Models of hydro turbines must consider the dynamics of fluid flow in the penstock and surge tank, if present. Models of combustion turbines must consider temperature limits as affected by ambient conditions. Under significant perturbations these involve nonlinearities, which should not be neglected and are not amenable to a generalized treatment. At this time, there is no IEEE Standard recommending governor modeling. It is not uncommon to disregard governor response and assume constant mechanical power in screening or preliminary studies, and, in general, the conclusions reached by doing so will be conservative because the mechanical power input to the unit will be higher than if the governor response was modeled.

A simple prime mover and governor model having two dynamic state variables could represent delays introduced by the steam-valve operating mechanism and the steam chest as follows. The variables are

- P_{SVi}: the steam valve position
- T_{Mi}: the mechanical torque to the shaft of the synchronous generator

The equations for the derivatives of these state variables are

$$T_{SVi}\frac{dP_{SVi}}{dt} = -P_{SVi} + P_{Ci} - \frac{1}{R_{Di}}\left(\frac{\omega_i}{\omega_o} - 1\right) \qquad i = 1, \ldots, m$$

$$T_{CHi}\frac{dT_{Mi}}{dt} = -T_{Mi} + P_{SVi} \qquad i = 1, \ldots, m$$

with the limit constraint

$$0 \leq P_{SVi} \leq P_{SVi}^{\max} \qquad i = 1, \ldots, m$$

The input to the governor is the speed error signal, which is given here as the difference between rated synchronous speed and actual speed. Notice that measuring all quantities within the model using a per-unit system permits using power and torque in the same equation, as rated speed of 1.0 is used as the scaling factor.

Large steam turbine units may have up to four stages through which the steam passes before it returns to the condenser. If there are two or three stages, steam from the boiler initially enters a high-pressure turbine stage. Steam leaving the first stage returns to a reheater section within the boiler. After reheating, the steam enters the second turbine stage. The dynamic response of such "reheat" units is generally dominated by the time constant of this reheater, typically 4 to 10 s. Units that are at or near the maximum size for the vintage when they were built are often built as cross-compound units with two separate turbines and two generators. Steam from the boiler initially enters a high-pressure turbine stage, returns to the reheater, passes through an intermediate-pressure stage on the same shaft as the high-pressure stage, and then crosses over to a low-pressure stage on the other shaft. The high-pressure and low-pressure machine rotor angles will vary considerably from each other, because the former tend to have lower inertia and the low-pressure unit input steam flow is subject to reheater and crossover piping lags.

4. Specify a Network/Load Model that Captures the Dynamics of Interest

In order to complete the dynamic model, it is necessary to connect all of the synchronous generators through the network to the loads. In most studies all fast network and load dynamics are neglected, so this model consists of Kirchhoff's power law at each node (also called *bus*). It is common to characterize the initial real and reactive load at each bus as a particular mix of constant power, constant current, and constant impedance load with respect to voltage changes at that bus. The proportions will depend on the nature of the loads (residential, agricultural, commercial, or industrial) anticipated under the conditions being modeled. The proportions of interest are the short-time responses of load to a voltage change; longer-term dynamics such as load tap-changer operation and reduced cycling of thermostatically controlled heating or cooling loads are not normally modeled. If single, large synchronous or induction motors comprise a significant proportion of the load at a bus near a disturbance being simulated, their dynamics should be modeled in a manner similar to generators.

The following algebraic equations give these constraints both for the synchronous machine stator and the interconnected network. The variable indexing assumes that the m synchronous machines are each at a distinct network node and that there are a total of n nodes in the network. (Loads are "injected" notation.) The i-kth entry of the bus admittance matrix has magnitude Y and angle α.

$$0 = V_i e^{j\theta_i} + (R_{si} + jX'_{di})(I_{di} + jI_{qi})\, e^{j(\delta_i - \pi/2)} - [E'_{di} + (X'_{qi} - X'_{di})I_{qi} + jE'_{qi}]\, e^{j(\delta_i - \pi/2)} \qquad i = 1, \ldots, m$$

$$V_i e^{j\theta_i}(I_{di} - jI_{qi})\, e^{-j(\delta_i - \pi/2)} + P_{Li}(V_i) + jQ_{Li}(V_i) = \sum_{k=1}^{n} V_i V_k Y_{ik}\, e^{j(\theta_i - \theta_k - \alpha_{ik})} \qquad i = 1, \ldots, m$$

$$P_{Li}(V_i) + jQ_{Li}(V_i) = \sum_{k=1}^{n} V_i V_k Y_{ik} e^{j(\theta_i - \theta_k - \alpha_{ik})} \qquad i = m + 1, \ldots, n$$

For given functions $P_{Li}(V_i)$ and $Q_{Li}(V_i)$, the $n + m$ complex algebraic equations must be solved for V_i, θ_i ($i = 1, \ldots, n$), and I_{di}, I_{qi} ($i = 1, \ldots, m$) in terms of states, δ_i, E'_{di}, and E'_{qi} ($i = 1, \ldots, n$). The currents can clearly be explicitly eliminated by solving the m stator algebraic equations (linear in currents) and substituting into the differential equations and remaining algebraic equations. This leaves only the n complex network algebraic equations to be solved for the n complex voltages, $V_i e^{j\theta_i}$.

5. *Compute Initial Conditions for the Dynamic Model*

In dynamic analysis it is customary to assume that the system is initially in sinusoidal constant-speed steady-state. This means that the initial conditions can be found by setting the time derivatives in the preceding model equal to zero. The steady-state condition that forms the basis for the initial conditions begins with the solution of a standard power flow. From this, the steps to compute the initial conditions of the two-axis model for machine i are as follows:

Step 1. From an initial power flow, compute the generator terminal currents, $I_{Gi}e^{j\gamma_i}$, as $(P_{Gi} - jQ_{Gi})/V_i e^{-j\theta_i}$.

Step 2. Compute δ_i as angle of $[V_i e^{j\theta_i} + jX_{qi}I_{Gi}e^{j\gamma_i}]$.

Step 3. Compute $\omega_i = \omega_o$.

Step 4. Compute I_{di}, I_{qi} from $(I_{di} + jI_{qi}) = I_{Gi}e^{j(\gamma_i - \delta_i + \pi/2)}$.

Step 5. Compute E'_{qi} as $E'_{qi} = V_i \cos(\delta_i - \theta_i) + X'_{di}I_{di} + R_{si}I_{qi}$.

Step 6. Compute E_{fdi} as $E_{fdi} = E'_{qi} + (X_{di} - X'_{di})I_{di}$.

Step 7. Compute R_{fi} as $R_{fi} = \dfrac{K_{Fi}}{T_{Fi}} E_{fdi}$.

Step 8. Compute V_{Ri} as $V_{Ri} = [K_{Ei} + S_{Ei}(E_{fdi})]E_{fdi}$.

Step 9. Compute E'_{di} as $E'_{di} = (X_{qi} - X'_{qi})I_{qi}$.

Step 10. Compute V_{Si} as $V_{Si} = \dfrac{V_{Ri}}{K_{Ai}}$.

Step 11. Compute $V_{\text{ref}i}$ as $V_{\text{ref}i} = V_{Si} + V_i$.

Step 12. Compute T_{Mi} as $T_{Mi} = E'_{di}I_{di} + E'_{qi}I_{qi} + (X'_{qi} - X'_{di})I_{di}I_{qi}$.

Step 13. Compute P_{ci} and P_{svi} as $P_{ci} = P_{svi} = T_{Mi}$.

Step 14. As an additional check, verify that all dynamic model states defined by these initial conditions are within relevant limits and that all state derivatives are zero. This ensures that the specified initial condition is attainable and is in equilibrium.

This computational procedure is unique in that the control inputs for generator terminal voltage and turbine power are not specified. Alternatively, the generator terminal voltage and terminal power are specified in the power flow, and the resulting current is used to compute the required control inputs.

6. *Specify the Disturbance(s) to Be Simulated*

The purpose of a dynamic simulation is to predict the behavior of the system in response to some disturbance or perturbation. This disturbance could be any possible change in the conditions that were used to compute the initial steady state. Transient stability studies normally consider disturbances such as a fault of a specified type (e.g., single-phase to ground) and duration, the loss of a generating unit, the false trip of a line, the sudden addition of a load, or the change of a control input. This disturbance is normally modeled by a change in one or more of the constants specified in the dynamic model. Steady-state stability studies consider response to smaller but more frequent perturbations, such as the ramping up and down of system loads over the operating day. Voltage stability studies consider both such small perturbations and system changes such as the loss of a generating unit or line, but generally do not consider conditions during faults. Additional information on this step of the procedure is discussed next in the description of the various stability analyses.

TRANSIENT STABILITY ANALYSIS

Transient stability analysis is concerned with simulating the response of the power system to major disturbances. The primary objective of the analysis is to determine if all synchronous machines remain in synchronism after a fault or other disturbance. A secondary objective is to determine if the nonlinear dynamic response is acceptable. For example, the bus voltages during the transient must remain above acceptable transient levels. While these transient levels are considerably below steady-state acceptable levels, they must be monitored at each instant of the simulation. This analysis is normally done through dynamic simulation using a detailed dynamic model and numerical integration. The output of this simulation includes synchronous-machine rotor angle and bus voltage trajectories as a function of time. For a typical case where the disturbance is a fault, the analysis includes the following procedures.

Calculation Procedure

1. Compute the Prefault Initial Conditions

Prepare the case data and solve a power-flow problem that gives all of the prefault bus voltages and generator currents. This involves the solution of the network power-balance equations for specified values of generator voltage magnitudes and real-power outputs. From this power-flow solution, the initial conditions for all dynamic state variables are computed using the procedure provided earlier.

2. Modify the System Model to Reflect the Faulted Condition

The faulted condition is usually modeled as a modification to the system bus-admittance matrix which makes up a portion of the algebraic equations of the dynamic model given earlier. The fault impedance, computed using symmetrical component theory, is added as a shunt admittance to the diagonal element of the faulted bus.

3. Compute the System Trajectories while the Fault Is On

With the modification of the admittance matrix, the solution of the algebraic-differential equation model will now produce changes in algebraic variables such as voltage and current. These changes disturb the equilibrium, so the derivatives of the state variables are no longer zero. Solution for the time series of values of each state, termed its *trajectory*, is obtained using a numerical integration technique. The solution continues until the prespecified fault clearing time.

Changes in the states over time will in turn impact the algebraic variables describing the remainder of the power system. There are basically two approaches used in power-system simulation packages: Simultaneous-Implicit (SI) and Partitioned-Explicit (PE). The SI method solves both the differential and algebraic equations simultaneously using an implicit integration routine. A typical integration routine is the trapezoidal method, which results in the need to solve a set of nonlinear algebraic equations at each time step for the new dynamic states—hence the name *implicit.* Since the algebraic equations of a power system are also nonlinear, it is efficient to solve these together—hence the name *simultaneous.* The solution of the model at each time step is performed using Newton's method, with an initial guess equal to the solution obtained at the converged solution of the previous time step.

The PE method solves the differential and algebraic equations in separate steps. The differential equations are first solved using an explicit integration scheme (such as Runge-Kutta) that directly gives the new dynamic states in terms of the old dynamic states and the old algebraic states. Since this method considers the algebraic variables to be constant at their previous time-step value, this method is noniterative—hence, the name *explicit.* After the dynamic states are computed for the new time, the nonlinear algebraic equations are solved iteratively using the dynamic states as constants. This gives the algebraic states for the new time. Since the solution of the dynamic and algebraic equations is done separately, this is called a *partitioned solution.* This PE method is typically less numerically stable than the SI method. Since the algebraic equations are nonlinear, the PE method is still iterative at each time step even though an explicit integration routine is used.

4. Remove the Fault at the Prespecified Fault Clearing Time

When the numerical integration routine attains the prespecified clearing time, the system admittance matrix is modified again to reflect the specified postfault condition. This could be simply the removal of the fault itself or the removal (or change in degree, as from three-phase to phase-to-ground) of the fault plus the removal of a line or generating unit—simulating the action of protective relaying and circuit breakers.

5. Compute the Postfault System Trajectories

With the modified system-admittance matrix, the numerical integration routine continues to compute the solution of the differential-algebraic equations. This solution is usually continued to a specified time, unless the system is deemed unstable because one or more machine rotor angles have exceeded a specified criterion, such as 180°, with respect to either a synchronous reference or to other, less-perturbed machines. An unstable system normally contains trajectories that increase without bound. However, if all of the trajectories of the synchronous machine angles increase without bound, the system may be considered stable in the sense that all machines are remaining in synchronism, although the final speed may not be rated synchronous speed. Alternatively, a solution may be terminated because one or more of the trajectories is unacceptable, that is, voltage dips below acceptable transient levels.

Related Calculations. If the system is found to be stable in Step 5, Steps 3 to 5 can be repeated with a larger prespecified clearing time. When an unstable system is encountered, the clearing time has exceeded the "critical clearing time." Determining the exact value of the critical clearing time can be done by repeated simulations with various prespecified clearing times.

SINGLE MACHINE-INFINITE BUS ILLUSTRATION

To illustrate several of these stability calculations, several major simplifying assumptions are made to the machine dynamic model. In the dynamic model given earlier, assume

1. $T'_{do} = T'_{qo} = \infty$. This makes the machine dynamic states, E'_q and E'_d, constants and removes the exciter from consideration.

2. $X'_q = X'_d$. This greatly simplifies the machine torque and circuit model.
3. $T_{CH} = T_{SV} = \infty$. This makes the governor dynamic states, T_M and P_{SV}, constants.
4. $R_s = 0$. This greatly simplifies the machine power model.

With these assumptions, the dynamic model of a synchronous machine is greatly simplified to be

$$\frac{d\delta}{dt} = \omega - \omega_o$$

$$\frac{2H}{\omega_o}\frac{d\omega}{dt} = P_m - P_e$$

where P_e is the real power out of the machine and P_m is the constant input power from the turbine. The algebraic equations for the machine simplify to

$$[E'_d + jE'_q]e^{j(\delta - \pi/2)} = +jX'_d(I_d + jI_q)e^{j(\delta - \pi/2)} + Ve^{j\theta}$$

Subtracting the appropriate constant from each angle, this can be written as the classical circuit model:

$$V'e^{j\delta} = jX'_d Ie^{j\theta_I} + Ve^{j\theta_V}$$

where V' is the constant voltage magnitude behind transient reactance and $Ve^{j\theta_V}$ is the terminal voltage. This classical dynamic model is used in the following illustration.[1]

The equivalent network of a synchronous machine connected to an infinite bus is shown in Fig. 16.1. All quantities in Fig. 16.1 are given per unit, and the impedance values are expressed on a 1000-MVA base. The machine operates initially at synchronous speed (377 rad/s on a 60-Hz system). A three-phase fault is simulated by closing switch S1 at time $t = 0$. S1 connects the fault bus F to ground through a fault reactance X_f. Compute the machine angle and frequency for 0.1 s after S1 is closed. Compute the machine angle and frequency with $X_f = 0$ and 0.15 p.u.

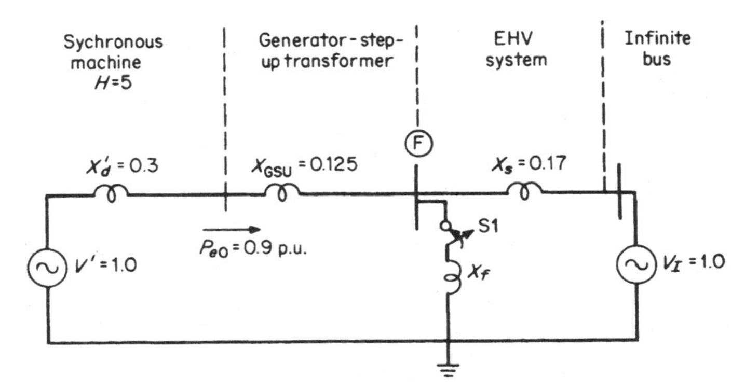

FIGURE 16.1 Equivalent network of single machine connected to an infinite bus. All quantities are per unit on common 1000-MVA base.

[1] Originally presented in a previous edition of this chapter by Cyrus Cox and Norbert Podwoiski.

Calculation Procedure

1. *Compute the Initial Internal Machine Angle*

Use $\delta_0 = \sin^{-1}(P_{e0}X_{tot}/V'V_l)$, where δ_0 = initial internal machine angle with respect to the infinite bus (for this machine, δ_0 is the angle of the voltage V'), P_{e0} = initial machine electrical power output, V' = voltage behind transient reactance X'_d, V_l = infinite-bus voltage, and X_{tot} = total reactance between the voltages, V' and V_l. Thus, $\delta_0 = \sin^{-1}[(0.9)(0.3 + 0.125 + 0.17)/(1.0)(1.0)] = 32.4°$.

2. *Determine Solution Method*

Computing synchronous-machine angle and frequency changes as a function of time requires the solution of the swing equation $d\delta^2/dt^2 = (\omega_0/2H)(P_m - P_e)$, where δ = internal machine angle with respect to a synchronously rotating reference (infinite bus), ω_0 = synchronous speed in rad/s, H = per-unit inertia constant in s, t = time in s, P_m = per-unit machine mechanical shaft power, and P_e = per-unit machine electric-power output. The term $(P_m - P_e)$ is referred to as the *machine accelerating power* and is represented by the symbol, P_a.

If P_a can be assumed constant or expressed explicitly as a function of time, then the swing equation will have a direct analytical solution. If P_a varies as a function of δ, then numerical integration techniques are required to solve the swing equation. In the network of Fig. 16.1, for $X_f = 0$ the voltages at bus F and P_e during the fault are zero. Thus, $P_a = P_m$ = constant, and therefore with $X_f = 0$, there is a direct analytical solution to the swing equation. For the case in which $X_f = 0.15$ (or $X_f \neq 0$), the voltages at bus F and P_e during the fault are greater than zero. Thus, $P_a = P_m - P_e$ = a function of δ. Therefore, with $X_f = 0.15$, numerical integration techniques are required for solution of the swing equation.

3. *Solve Swing Equation with* $X_f = 0$

Use the following procedure:

a. Compute the machine accelerating power during the fault. Use $P_a = P_m - P_e$; for this machine, $P_m = P_{e0} = 0.90$ p.u.; $P_e = 0$. Thus, $P_a = 0.90$.

b. Compute the new machine angle at time, $t = 0.1$ s. The solution of the swing equation with constant P_a is $\delta = \delta_0 + (\omega_0/4H)P_a t^2$, where the angles are expressed in radians and all other values are on a per-unit basis. Thus,

$$\delta = \left(\frac{32.4°}{57.3°/\text{rad}}\right) + \left(\frac{377}{4 \times 5}\right) 0.90(0.1)^2$$

$$= 0.735 \text{ rad, or } 42.1°$$

c. Compute the new machine frequency at time, $t = 0.1$ s. The machine frequency is obtained from the relation, $\omega = d\delta/dt + \omega_0$, where $d\delta/dt = (\omega_0 P_a t)/2H$. Thus, $\omega = [(377)(0.90)(0.1)/(2)(5)] + 377 = 380.4$ rad/s, or 60.5 Hz.

It is obvious from the equation for δ in Step b that the rotor angle will increase indefinitely as long as the fault is on, indicating instability.

4. *Solve Swing Equation with* $X_f = 0.15$ *p.u.*

Use the following procedure.

a. Select a numerical integration method and time step. There are many numerical integration techniques for solving differential equations including Euler, modified

Euler, and Runge-Kutta methods. The Euler method is selected here. Solution by Euler's method requires expressing the second-order swing equation as two first-order differential equations. These are: $d\delta/dt = \omega(t) - \omega_0$ and $d\omega/dt = (\omega_0 P_a(t))/2H$. Euler's method involves computing the rate of change of each variable at the beginning of a time step. Then, on the assumption that the rate of change of each variable remains constant over the time step, a new value for the variable is computed at the end of the step. The following general expression is used: $y(t + \Delta t) = y(t) + (dy/dt)\Delta t$, where y corresponds to δ or ω and Δt = time step and $y(t)$ and dy/dt are computed at the beginning of the time step. A time step of 1 cycle (0.0167 s) is selected.

b. From Fig. 16.2, determine the expression for the electrical power output during this time step. For this network, Case 3 is used, where $X_G = X'_d + X_{GSU} = 0.3 + 0.125 = 0.425$ p.u., $X_s = 0.17$, and $X_f = 0.15$. Thus,

$$P_e = [(1.0)(1.0)\sin\delta]/\{0.424 + 0.17 + [(0.425)(0.17)/0.15]\}$$
$$= 0.93 \sin\delta$$

c. Compute $P_a(t)$ at the beginning of this time step ($t = 0$ s). Use $P_a(t) = P_m - P_e(t)$ for $t = 0$, where $P_m = P_{eo0} = 0.90$ p.u.; $P_e(0) = 0.930 \sin 32.4°$. Thus, $P_a(t = 0) = 0.90 - 0.93 \sin 35.2° = 0.40$ p.u.

d. Compute the rate of change in machine variables at the beginning of the time step ($t = 0$). The rate of change in the machine phase angle $d\delta/dt = \omega(t) - \omega_0$ for $t = 0$, where $\omega(0) = 377$ rad/s. Thus, $d\delta/dt = 377 - 377 = 0$ rad/s. The rate of change in machine frequency is $d\omega/dt = \omega_o P_a(t)/2H$ for $t = 0$, where $P_a(0) = 0.40$ p.u. as computed in **c**. Thus, $d\omega/dt = (377)(0.40)/(2)(5) = 15.08$ rad/s^2.

e. Compute the new machine variables at the end of the time step ($t = 0.0167$ s). The new machine phase angle is $\delta(0.0167) = \delta(0) + (d\delta/dt)\Delta t$, where the angles are

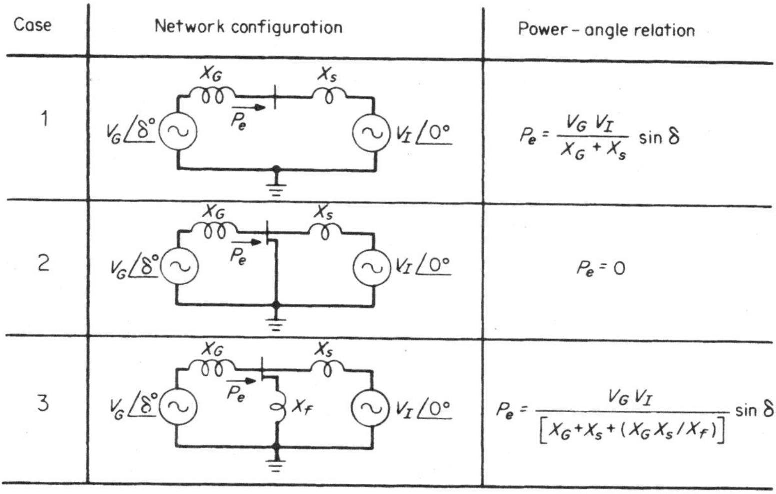

Case	Network configuration	Power-angle relation
1	$V_G\angle\delta°$, X_G, P_e, X_s, $V_I\angle 0°$	$P_e = \dfrac{V_G V_I}{X_G + X_s} \sin\delta$
2	$V_G\angle\delta°$, X_G, P_e, X_s, $V_I\angle 0°$	$P_e = 0$
3	$V_G\angle\delta°$, X_G, P_e, X_s, X_f, $V_I\angle 0°$	$P_e = \dfrac{V_G V_I}{[X_G + X_s + (X_G X_s/X_f)]} \sin\delta$

FIGURE 16.2 Power-angle relations for general network configurations; resistances are neglected. $V_G\angle\delta$ = internal machine angle, X_G = machine reactance, X_s = system reactance, P_e = machine electrical power output, $V_I\angle 0°$ = infinite bus voltage, and X_f = fault reactance.

TABLE 16.1 Computations for Solution of Swing Equation by Euler's Method

Time	Frequency		Angle		Accel.	Derivatives		Increments	
	Radians/		Radians		power	Angle	Speed	$d\delta/dt$	$d\omega/dt$
(s)	second-ω	Hz	δ	Degrees	$P_a(t)$	$d\delta/dt$	$d\omega/dt$	Δt	Δt
0	377	60	0.564	32.4	0.402	0	15.14	0	0.253
0.0167	377.25	60.04	0.564	32.4	0.402	0.253	15.14	0.0042	0.253
0.0334	377.51	60.08	0.569	32.6	0.398	0.506	15.02	0.0084	0.251
0.0501	377.76	60.12	0.577	33.1	0.392	0.757	14.77	0.0126	0.247
0.0668	378.00	60.16	0.590	33.8	0.382	1.003	14.40	0.0168	0.241
0.0835	378.24	60.20	0.606	34.8	0.369	1.244	13.91	0.0208	0.232
0.1000	378.48	60.24	0.627	36.0	0.353	1.476	13.32	0.0247	0.222

$P_a = 0.93 \sin\delta$, $d\delta/dt = \omega(t) - \omega_0$
$d\omega/dt = (\omega_0 P_a(t))/2H$, $\Delta t = 0.0167$ s, $\delta(t + \Delta t) = \delta(t) + d\delta/dt\, \Delta t$
$\omega(t + \Delta t) = \omega(t) + d\omega/dt\, \Delta t$

expressed in radians and Δt = time step = 0.0167 s. Thus, $\delta(0.0167) = 32.4°/(57.3°/\text{rad}) + (0)(0.0167) = 0.565$ rad, or 32.4°. The new machine frequency is $\omega(0.0167) = \omega(0) + (d\omega/dt)\Delta t = 377 + (15.08)(0.0167) = 377.25$ rad/s, or 60.04 Hz.

f. Repeat **c**, **d** and **e** for the desired number of time steps. Table 16.1 displays the remaining calculations for the 0.1-s solution time. The machine frequency at 0.1 s = 378.48 rad/s, or 60.24 Hz; the corresponding machine angle is 36.0°.

Related Calculations. The preceding illustration simply computes the fault-on trajectories for a time of 0.1 s. In an actual study, the simulation would need to be performed well beyond the time that the fault is cleared to observe the responses to determine if the unit is stable for that fault-clearing time. If it is not stable, the calculation is repeated with successively smaller clearing times until a stable response is observed. The longest clearing time showing a stable response is called the *critical* clearing time. The procedure can readily be generalized to consider the case of a second switching step—for instance, if only two of three breaker poles operate to clear a three-phase fault in primary clearing time and the remaining phase-to-ground fault is cleared later by backup circuit breaker operation.

The preceding example can readily be set up on a spreadsheet program such as Excel® and the calculation extended sufficiently to demonstrate whether the machine will remain stable. The curves shown in Fig. 16.3 were produced by such an application, which is included as file swingcurve.xls on the diskette enclosed with this *Handbook.** Selected additional cases are also presented; by changing the parameters in shaded cells, the user can observe their effect on the plotted rotor angle curves.

There are direct methods that seek to determine this critical clearing time with less numerical integration effort and without repeated analysis with various clearing times. These methods are called *equal area criteria*, for single-machine systems, and *direct methods* for large-scale systems.

* See above-referenced file on accompanying CD for curves shown in Fig.16.3 and other selected cases.

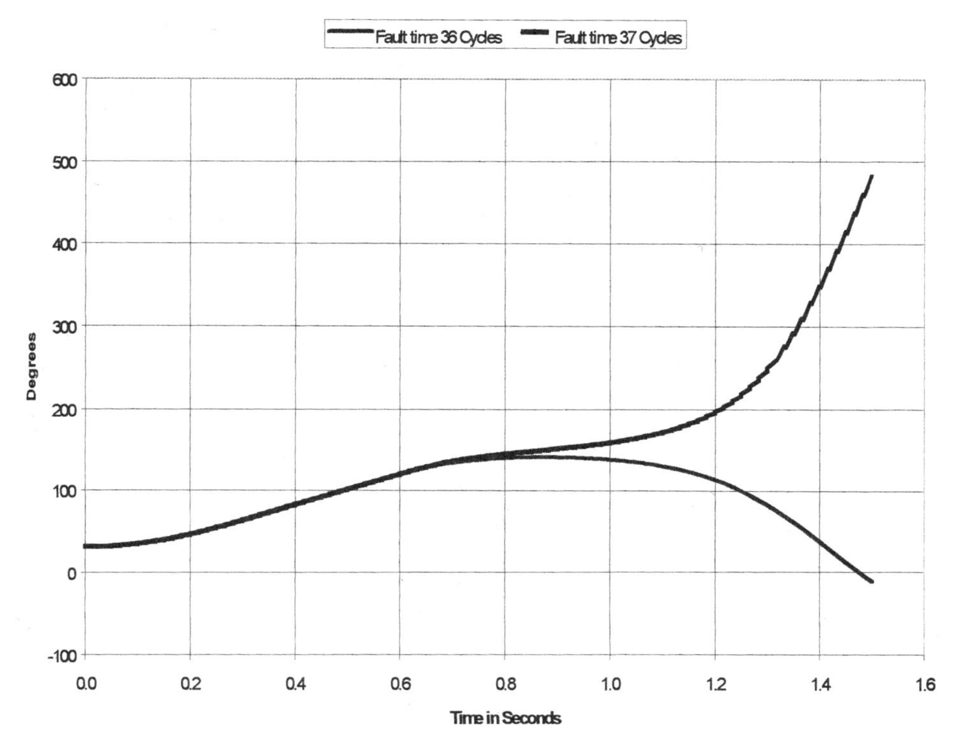

FIGURE 16.3 Rotor angle curves for stable and unstable cases with system as shown in Fig. 16.1. Fault duration is 36 cycles for the stable case and 37 cycles for the unstable case.

SELECTING TRANSIENT STABILITY DESIGN CRITERIA

Stability design criteria may be deterministic or probabilistic. A deterministic criterion requires that the generating unit be demonstrated to be stable for any disturbance of a specified level of severity—for instance, any three-phase fault tripping a single transmission system element in six cycles. Events of greater severity—say, faults cleared by backup protection schemes in a greater time following initiation—are disregarded. A probabilistic criterion requires that the total annual frequency of events leading to instability be less than a specified number or, equivalently, that the mean time to instability be greater than a certain number of years.[2]

Stability design criteria are to be selected for the generating facility shown in Fig. 16.4. The design basis for the plant is that the mean time to instability (MTTI-failure due to unstable operation) is greater than 500 years.

[2] Example of application of probabilistic design criteria by Cyrus Cox and Norbert Podwoiski.

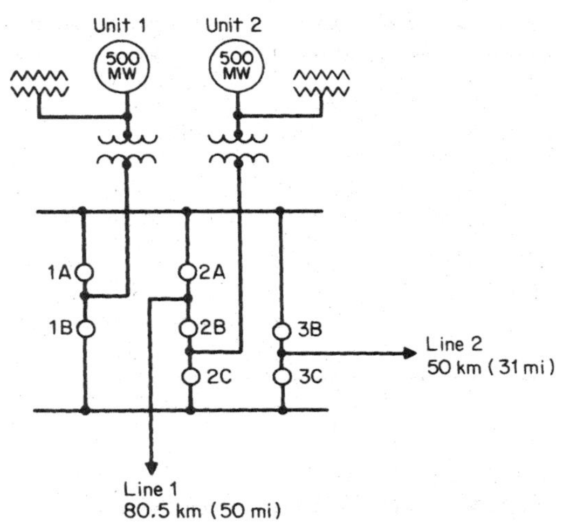

FIGURE 16.4 A two-unit generating facility, switchyard, and out-of-plant transmission.

Calculation Procedure

1. Specify a Fault-Exposure Zone

A fault-exposure zone defines a boundary beyond which faults can be neglected in the selection of the stability criteria. For example, on long out-of-plant transmission lines it is typically unnecessary to consider line faults at the remote end. Specifying a fault-exposure zone is generally based on the fault frequency and/or the magnitude of the reduction in the plant electric output during the fault.

For this plant, specify a fault-exposure zone of 50 km (31 mi); that is, it will be assumed that the plant will maintain transient stability to faults beyond this distance, even if breaker failure delays fault clearing. In addition, faults in the switchyard are considered outside the fault-exposure zone. This is done because switchyard-fault frequency is generally much less than line-fault frequency.

2. Compute Fault Frequency within the Exposure Zone

Use Table 16.2 to estimate the fault frequency per line. The values given are very general and based on composite data from numerous power-industry sources. Line-fault frequency varies with factors such as line design, voltage level, soil conditions, air pollution, storm frequency, etc.

For this plant, use the typical line-fault frequency rates. Thus, the fault frequency within the exposure zone is f = (number of lines) (fault frequency) (exposure zone/

TABLE 16.2 Typical Extra-High Voltage (EHV) Probability Data

Data description	Optimistic	Typical	Pessimistic
Line-fault frequency, faults/100 km, yr	0.62	1.55	3.1
Conditional probability of relay failure to sense or stuck breaker*	0.0002	0.003	0.01

* Assumes nonindependent pole tripping. For independent pole tripping, probabilities are for a 1 Φ failure.

TABLE 16.3 Composition of EHV Transmission-Line Faults

Type of fault	Percent of total faults		
	765 kV	EHV composite	115 kV
Phase to ground	99	80	70
Double-phase to ground	1	7	15
Three-phase to ground	0	3	4
Phase to phase	0	2	3
Unknown	0	8	8
Total	100	100	100

100 km), where the fault frequency is in faults/100 km-yr and exposure zone is in km. Then $f = (2)(1.55)(50/100) = 1.55$ faults/yr within the exposure zone.

Use Table 16.3 to estimate the fault frequency on the basis of the number of phases involved. For this plant, use the composite values and assume that the unknown fault types consist entirely of single-phase faults. Thus, the three-phase fault frequency $f_{3\phi}$ = (percent of total faults that involve $(3\phi)(f) = (0.03)(1.55) = 0.465 - 3\phi$ faults/yr, or $1/f_{3\phi} = -\text{MTBF}_{3\phi}$ = mean time between 3ϕ faults $= 1/0.0465 = 21.5$ yr. Similar calculations are made for the other fault types and are tabulated next.

Fault type	Faults per year	MTBF, yr
Three-phase	0.0465	21.5
Double-phase	0.140	7.17
Single-phase (+ unknown)	1.36	0.733

Note that in these calculations $f \cong 1/\text{MTBF}$. The precise relation for the frequency is $f = 1/(\text{MTBF} + \text{MTTR})$, where MTTR = mean time to repair (i.e., time to clear the fault and restore the system to its original state). The term MTTR is required in the precise relation because a line cannot fail while it is out of service. However, for stability calculations, MTTR $\ll$ MTBF, and, thus, the approximation, $f \cong 1/\text{MTBF}$, is valid.

3. Compute Breaker-Failure Frequencies

Only breaker failures in which both generating units in Fig. 16.4 remain connected to the transmission system are considered. Tripping one unit at a multiunit plant substantially improves the stability of the remaining units because they can utilize the tripped unit's portion of the available power transfer. Intentional tripping of a generating unit in the process of clearing a "stuck breaker" is sometimes used as a stability aid. For this plant and transmission system, the failure of breaker 2B for a fault on line 1 will result in tripping unit 2. Thus, only three breaker failures affect stability: 2A, 3B, and 3C.

From Table 16.2, the conditional probability that a relay/breaker will fail to sense or open for a line fault is $p_{bf} = 0.003$ (typical value for nonindependent pole tripping). The frequency of a three-phase fault plus a breaker failure is $f_{3\phi,bf} = f_{3\phi}\, p_{bf}\, B$, where B = number of breaker failures that impact stability. Thus, $f_{3\phi,bf} = (0.465)(0.003)(3) = 4.19 \times 10^{-4}$ three-phase faults plus breaker failures per year, or $1/f_{3\phi,bf} = 2390$ yr. Similar calculations are made for the other fault types and are tabulated next.

Fault type	Fault plus breaker failures per year	Mean time between faults plus breaker failures, yr
Three-phase	0.000419	2390
Double-phase	0.00126	793
Single-phase (+ unknown)	0.122	81.7

4. Select the Minimum Criteria

As a design basis, the plant must remain stable for at least those fault types for which the MTBF is less than 500 years. Thus, from the fault-type frequencies computed in Steps 2 and 3, only a three-phase or two-phase fault plus breaker failure could be eliminated to select the minimum criteria. A check is now made to ensure that neglecting three- and two-phase faults plus breaker failures is within the design basis.

5. Compute the Frequency of Plant Instability

Use $f_I = \Sigma$ frequencies of fault types eliminated in Step 4, where f_I = frequency of plant instability. Thus, $f_I = f_{3\phi,\mathrm{bf}} + f_{2\phi,\mathrm{bf}} = 0.000419 + 0.00126 = 0.00168$ occurrences of instability per yr, or $1/f_I = -$mean time to instability = 595 yr. Thus, neglecting three-phase and two-phase faults plus breaker failures in the criteria, the value is within the original design basis of MTTI $>$ 500 yr. If the MTTI computed in this step was less than the design basis, then additional fault types would be added to the minimum criteria and the MTTI recomputed.

6. Select the Stability Criteria

The following stability criteria are selected:

a. Three-phase fault near the plant's high-voltage bus cleared normally

b. Single-phase fault near the plant's high-voltage bus plus a breaker failure

If the plant remains stable for these two tests, the original design basis has been satisfied.

Related Calculations. Specifying a design basis, although not a step in the procedure, is critical in selecting stability criteria. No general guidelines can be given for specifying a design basis. However, most utilities within the United States specify some type of breaker failure test in their criteria.

In addition to neglecting low-probability fault types, low-probability operating conditions can also be neglected in the selection of stability criteria. As an example, suppose that for this plant it was determined that during leading power-factor operation, the plant could not remain stable for a single-phase fault plus a breaker failure. Further assume that the plant transition rate into leading power-factor operation is $\lambda = 2$ yr and the mean duration of each occurrence is $r = 8$ h. Use the following general procedure to compute the mean time to instability.

1. Sketch a series-parallel event diagram as shown in Fig. 16.5*a*. Any break in the continuity of the diagram causes instability. A break in continuity could be caused by Event 3 or 4, or the simultaneous occurrence of Events 1 and 2.

2. Reduce the series-parallel event diagram to a single equivalent event. By recursively applying the relations in Fig. 16.6, the diagram can be reduced to an equivalent MTTI. To

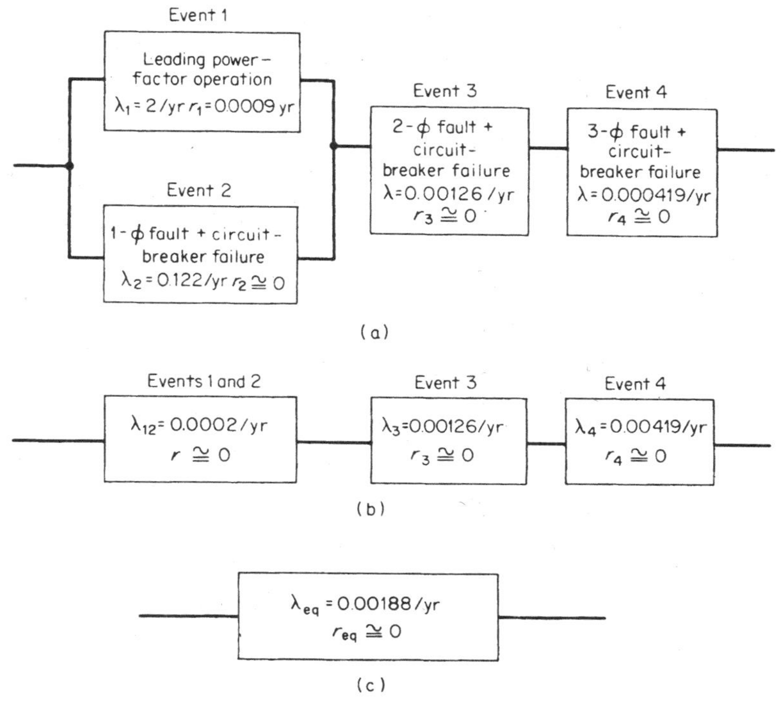

FIGURE 16.5 Event diagrams. (*a*) Series-parallel event diagram. (*b*) Events 1 and 2 reduced to equivalent event. (*c*) All four events reduced to a single event.

reduce the parallel combination of Events 1 and 2, use $\lambda_{12} = \lambda_1\lambda_2(r_1 + r_2)(1 + \lambda_1 r_1 + \lambda_2 r_2)$, where λ_{12} = equivalent transition rate for Events 1 and 2, λ_1 = transition rate for Event 1 in occurrences per year, λ_2 = transition rate for Event 2 in occurrences per year (note that for this case, $f_2 \cong \lambda_2 = 0.122$ occurrences per year as computed in Step 3), r_1 = mean duration of Event 1 in years = 8 h (8760 h/yr) = 0.0009 yr, r_2 = mean duration of Event 2 in years (the duration of a single-phase fault plus breaker failure = 0 yr). Thus, $\lambda_{12} = (2)(0.122)(0.0009)/[1+(2)(0.0009)] = 0.0002$ transitions/yr. The diagram is now reduced to Fig. 16.5*b*. Note that the equivalent duration of Events 1 and 2 is $r_{12} \cong 0$ yr. The equivalent transition rate for the three series events is $\lambda_{eq} = \lambda_{12} + \lambda_3 + \lambda_4 = 0.0002 + 0.00126 + 0.000419 = 0.00188$ occurrences/yr. Since $r_{eq} \cong 0$ yr, we have $\lambda_{eq} \cong f_{eq}$ = frequency of instability f_I, or $1/f_I$ = MTTI = 1/0.00188 = 532 yr (Fig. 16.5*c*).

This general procedure can be used either to select a criterion or to determine the adequacy of final stability design.

TRANSIENT STABILITY AIDS[3]

If a generating facility such as that of Fig. 16.4 cannot meet the stability design criteria and the number and configuration of transmission circuits is fixed, the following procedure is used to select a set of stability aids to meet the criteria.

[3] This section was originally written by Cyrus Cox and Norbert Podwoiski for a previous edition.

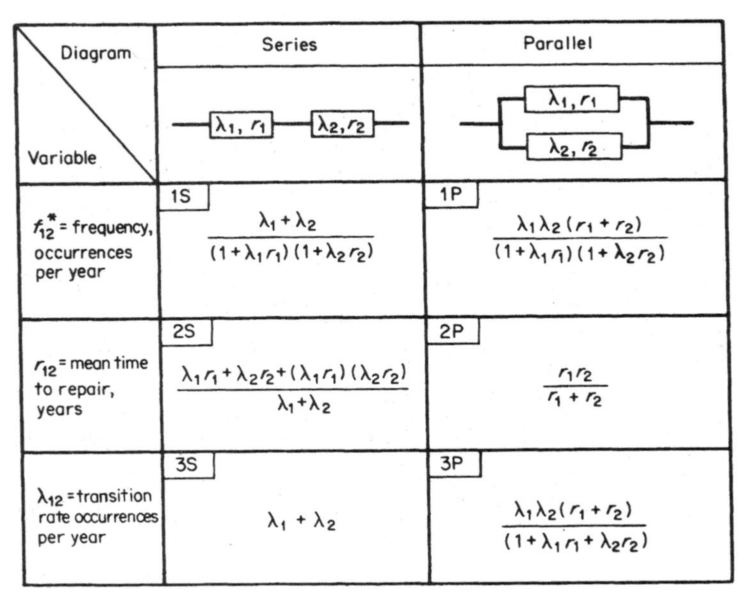

FIGURE 16.6 Equivalent frequency-duration formulas for occurrence of events in series or parallel. *Although outage rate and frequency have same units of outage per year, they are not equivalent; λ is the reciprocal of mean time to failure (MTTF); f is the reciprocal of the sum of the MTTF and mean time to repair (MTTR). If MTTF $\gg$ MTTR, then $f \cong \lambda$.

Calculation Procedure

1. Determine Critical Fault-Clearing Times for the System as Specified

Compute or use computer simulation to determine the critical fault-clearing times for each contingency specified in the stability criteria.

2. Compute the Attainable Fault-Clearing Time

Table 16.4 displays the range of typical EHV (extra-high voltage) relay-breaker clearing times. Most components of the total clearing times are limited by the type of

TABLE 16.4 Range of Typical EHV Relay-Breaker Clearing Times

Function	Time in cycles (60 Hz)		
	Fast	Average	Slow
Primary relay	0.25 – 0.5	1.0 – 1.5	2.0
Breaker clearing	1.0	3.0	3.0 – 5.0
Total normal clearing time	1.3 – 1.5	3.0 – 3.5	5.0 – 7.0
Breaker-failure detection	0.25 – 0.5	0.5 – 1.5	1.0 – 2.0
Relay coordination time	3.0	3.0 – 5.0	5.0 – 6.0
Auxiliary relay	0.25 – 0.5	0.5 – 1.0	1.0
Backup breaker clearing	1.0	2.0	3.0 – 5.0
Total backup clearing	5.75 – 6.5	9.0 – 13.0	15.0 – 20.0

TABLE 16.5 Typical Range of Machine Transient Reactance and Inertia Constant for Modern Turbine-Generators

		Transient reactance X'_d, percent on nameplate MVA base		Inertia constant H MWs/MVA	
Turbine type	Speed, rpm	Low	High	Low	High
Steam	3600	30	50	2.5	4.0
	1800	20	40	1.75	3.5
Hydro	600 or less	20	35	2.5	6.0

equipment used. The function most susceptible to reduction is the relay-coordination time for backup clearing. However, reduction in the coordination time can result in erroneous backup clearing. The minimum achievable times displayed in Table 16.4 are associated with state-of-the-art equipment (1-cycle breakers and ultra-high-speed relaying).

3. Modify Machine and GSU Transformer Parameters

A decrease in the machine transient reactance and/or an increase in the inertia constant can provide increases in CFC time. Table 16.5 displays typical ranges of these parameters for modem turbine generators. Table 16.6 displays the typical range of standard impedances for GSU transformers. The most common way to achieve CFC increases in the generation system is to reduce the generator step-up transformer (GSU) impedance. GSU transformer impedances below the minimum standard can be obtained at a cost premium. Reductions in GSU transformer impedance may be limited by the fault-interrupting capabilities of the circuit breakers in the EHV switch-yard.

4. Survey Transient Stability Aids

Transient stability aids fall into three general categories: machine controls, relay system enhancements, and network modifications. The most commonly used aids are tabulated next.

TABLE 16.6 Typical Range of Impedance for Generator Step-Up Transformers

	Standard impedance in percent on GSU MVA base	
Nominal system voltage, kV	Minimum	Maximum
765	10	21
500	9	18
345	8	17
230	7.5	15
138–161	7.0	12
115	5.0	10

Transient stability aids	
Category	Stability aid
Machine control	High initial response excitation
	Turbine fast-valve control
Relay control	Independent pole tripping
	Selective pole tripping
	Unit rejection schemes
Network control	Series capacitors
	Braking resistors

Table 16.7 is a summary of typical improvements in CFC time and typical applications of the more commonly used stability aids. In general, stability aids based on machine control provide CFC increases of up to 2 cycles for delayed clearing and only marginal increases for normal clearing. Transmission relay control methods focus on

TABLE 16.7 Summary of Commonly Used Supplementary Transient Stability Aids

		Maximum improvement in CFC time, cycles		
	Stability aid	Normal clearing	Delayed clearing	Remarks
Machine controls	High initial response excitation	1/2	2	Most modern excitation and voltage regulation systems have the capability for high initial response.
	Turbine fast-valve control	1/2	2	Not applicable at hydro plants. Generally involves fast closing and opening of turbine intercept valves. Available on most steam turbines manufactured today.
Relay enhancements	Independent pole tripping	N.A.	5	Reduces multiphase faults to single-phase faults for a breaker failure (delayed clearing). Increases relay costs.
	Selective pole tripping	N.A.	5	Opens faulted phase only for single-phase faults. Generally only used at plants with one or two transmission lines. Increases relay cost and complexity.
	Unit rejection schemes	Limited by amount generated that can safely be rejected		System must be capable of sustaining loss of the unit(s).
Network configuration	Series capacitors	Limited by amount of series compensation that can be added		May be required for steady-state power transfer. High cost; typically only economical for plants greater than 80 km (50 mi) from the load centers.
	Braking resistors	Limited by the size of the resistor		High cost, typically only economical for plants greater than 80 km (50 mi) from the load centers.

CFC improvements associated with multiphase faults cleared in backup time. Network configuration control methods provide relatively large increases in CFC time for both primary and backup clearing.

5. Select Potential Stability Aids

In general, breaker-failure criteria have the most severe stability requirements. Thus, selection of stability aids is generally based on the CFC time improvements required to meet breaker-failure criteria. It should be noted that CFC time improvements associated with stability aids are not necessarily cumulative.

6. Conduct Detailed Computer Simulations

Evaluation of the effects on stability of the stability aids outlined in this procedure is a nonlinear and complex analysis. Detailed computer simulation is the most effective method to determine the CFC time improvements associated with the stability aids.

7. Evaluate Potential Problems

Each transient-stability control aid presents potential problems, which can be evaluated only through detailed computer analysis. Potential problems fall into three categories: (a) unique problems associated with a particular stability control method, (b) misoperation (i.e., operation when not required), and (c) failure to operate when required or as expected. Table 16.8 briefly summarizes the potential problems associated with the stability aids and actions that can be taken to reduce the risk.

Related Calculations. Evaluation of stability controls is a complex and vast subject area. For more detailed discussions of stability controls, see Byerly and Kimbark (1974).

TABLE 16.8 Potential Problems Associated with Stability Controls

Stability aid	Potential problems	Action to reduce risk
High initial response excitation systems	Dynamic instability	Reduce response or add power system stabilizer (PSS)
	Overexcitation (misoperation)	Overexcitation relay protection
Independent pole tripping	Unbalanced generator operation (misoperation)	Generator negative-phase-sequence relay protection
Selective pole tripping	Sustain faults from energized phases	Add shunt reactive compensation
	Unbalanced generator operation (misoperation)	Generator negative-phase-sequence relay protection
Turbine fast valving	Unintentional generator trip	Maintain safety valves
Unit rejection schemes	Reduced generator reliability	Provide unit with capability to carry just the plant load (fast load runback)
Series capacitors	Subsynchronous resonance (SSR)	SSR filter
	Torque amplification	Static machine-frequency relay
	Self-excitation	Supplementary damping signals
Braking resistors	Instability (misoperation)	High-reliability relay schemes

SELECTION OF AN UNDERFREQUENCY LOAD-SHEDDING SCHEME[4]

An underfrequency load-shedding scheme is to be selected for a portion of a power system shown in Fig. 16.7. The scheme should protect this portion of the system from a total blackout in the event the two external power lines are lost. Load in the area varies 2 percent for each 1 percent change in frequency.

Calculation Procedure

1. Select Maximum Generation Deficiency

Selection of the maximum initial generation deficiency for which the load-shedding scheme should provide protection is an arbitrary decision. For the system of Fig. 16.7, an initial maximum generation deficiency of 450 MW is selected. This value is based on the difference between the peak load and the rated generator output.

2. Compute Corresponding System Overload

It is convenient to define a system overload as $\text{OL} = (L - P_m)/P_m$, where L = initial load and P_m = initial generator output. Thus, the maximum initial generation deficiency can be expressed in terms of a system overload as $\text{OL} = (900 - 450)/450 = 1.0$ p.u., or a 100 percent system overload.

3. Select a Minimum Frequency

The minimum frequency is the lowest allowable frequency that the system should settle to after all load shedding has occurred. In general, the minimum frequency should be above the frequency at which the generating unit will be separated from the system. Assume the generator trip frequency is 57 Hz and select a minimum frequency of 57.5 Hz.

4. Compute the Maximum Amount of Load to Be Shed

The maximum load shed is that required to allow the system frequency to decay to the minimum frequency for the maximum system overload. Use $L_m = [\text{OL}/(\text{OL} + 1) - \alpha]/(1 - \alpha)$, where L_m = maximum load shed; OL = maximum initial system overload as

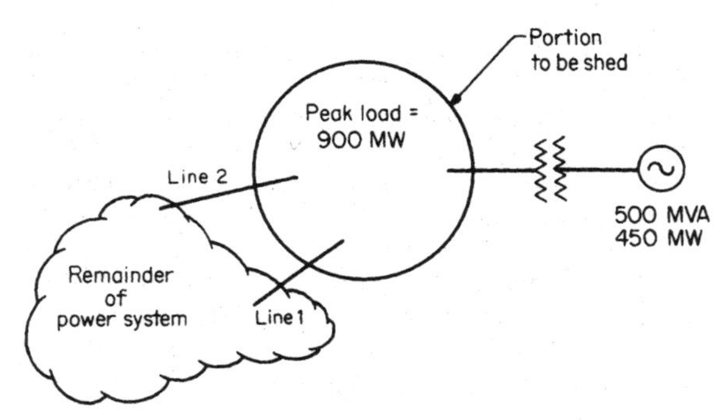

FIGURE 16.7 Simple system for underfrequency load-shedding example.

[4] Cyrus Cox and Norbert Podwoiski originally wrote this section for a previous edition.

selected in Step 1; $\alpha = d(1 - \omega_m/60)$, where d = system-load damping factor (given as 2); ω_m = minimum frequency. Thus, $2(1 - 57.5/60) = 0.833$ and $L_m = [I/(I + 1)\ 0.08331]/(1 - 0.0833) = 0.453$ p.u., or 45.3 percent of initial system load.

5. Select Load-Shedding Scheme

Selection of the load-shedding scheme involves specification of the number of load-shedding steps and the frequency set points. In general, the more load-shedding steps, the better. However, as the number of steps increases, the cost may also increase. In addition, too many steps may create relay coordination problems. Load should be shed gradually; that is, each step should drop progressively more load. Frequency set points can be divided up in equal intervals from the maximum to minimum frequency set points. The following table displays the load-shedding scheme selected.

	Frequency set points	Percent of initial load to be shed
Step 1	59.5	5
Step 2	59.0	15
Step 3	58.5	25
Total		45

6. Check Relay Coordination

Relay coordination checks between adjacent steps are required to assure the minimum amount of load is shed for various initial system overloads. The reason for this check is that there may be a significant amount of time delay between the time the system frequency decays to a set point and the time the actual load shedding occurs. The time delays include relay pickup time, any intentional time delay, and breaker-opening times. This time delay may cause the frequency to decay through two adjacent steps unnecessarily. Use the following procedure to check relay coordination.

a. For load-shedding Steps 1 and 2, compute the initial system overload that will result in the frequency decaying to Step 2. Use $OL_2 = [L_d + (\alpha/(1 - \alpha))]/(1 - L_d)$, where OL_2 = initial system overload for frequency to decay Step 2; L_d = per-unit load that should be shed prior to Step 2; $\alpha = d(I - \omega_2/60)$, where ω_2 is the relay set point for Step 2. Thus, $\alpha = (2)(1 - 59.0/60) = 0.033$ p.u., and $OL_2 = [0.05 + 0.033/(1 - 0.033)]/(1 - 0.05) = 0.088$ p.u.

b. Compute the corresponding initial system load. Use $L_i = P_m(OL_2 + 1)$, where L_i is the initial system load that results in a 0.088 p.u. system overload. Thus, $L_i = 450(0.088 + 1) = 0.979$ p.u. or, in MW, L_i = (per-unit load)(base MVA) = (0.979) (500) = 489 MW.

c. Conduct computer simulation. A digital computer simulation is made with an initial generation of 450 MW and an initial load of 489 MW. If relay coordination is adequate, load-shedding Step 2 should not shed any load for this initial system overload. If load-shedding Step 2 picks up and/or sheds load, then the relay set points can be moved apart, time delay may be reduced, or the load shed in each step may be revised.

d. Repeat steps **a** through **c** for the remaining adjacent load-shedding steps.

Related Calculations. This procedure is easily expanded for use with more than one generator. Digital computer simulations are required in the relay coordination steps to take into account the effects of automatic voltage regulation, automatic governor control, and load dependence on voltage and frequency.

STEADY-STATE STABILITY ANALYSIS

Steady-state stability analysis is concerned with the ability of a system to remain in an equilibrium condition after a small disturbance. By definition, this analysis is based solely on the characteristics of a system when it is analyzed in a linear form. The steps for a typical analysis are as follows.

Calculation Procedure

1. Compute an Equilibrium Condition

In most systems, the equilibrium condition is computed by setting all derivatives to zero in the time-invariant dynamic model. In power systems, the equilibrium is normally computed using a standard power-flow routine. From this, the equilibrium condition is computed as identified earlier for computing initial conditions. In either case, the equilibrium condition is typically some steady-state case of interest. Since nonlinear systems can have multiple equilibrium conditions, it may also be necessary to ensure that the solution is the "condition of interest." In power systems, these multiple solutions arise because the given data is typically given in terms of power. In a circuit, power is an ambiguous piece of information. For example, if the power absorbed by a load is given as zero, the possible solutions are twofold: (1) voltage equals zero (short circuit) and (2) current equals zero (open circuit). Both of these conditions result in zero power for the load and both are physically reasonable. The same is true for cases where power is specified other than zero.

2. Linearize the Dynamics about the Equilibrium Condition

In order to perform steady-state stability analysis, the nonlinear system must be "linearized." This means that the nonlinear functions that appear on the right-hand side of the state derivatives and the algebraic equations must be made linear. This is done by expanding these nonlinear functions in a Taylor series about the equilibrium condition and retaining only the linear portion.

3. Compute the Eigenvalues of the Linearized Model

Once the dynamic model has been linearized, the stability computation can be done directly by simply computing the eigenvalues of the system dynamic matrix. If the system has algebraic equations, the algebraic variables must be eliminated by matrix operation to obtain the dynamic matrix. If the real parts of all the eigenvalues of the dynamic matrix are less than zero, the system is declared steady-state stable. If all machine angles are retained and there is no infinite bus in the system for angle reference, then one of the eigenvalues of the dynamic matrix will be zero. The zero eigenvalue does not imply that the system is unstable. It is an indication of the fact that the machines in the system could remain in synchronism at a speed different from the original synchronous speed. According to traditional power-system stability, this is considered stable.

VOLTAGE STABILITY ANALYSIS

Voltage stability analysis of power systems means different things to different people. It can be loosely defined as the state of a power system when its voltage levels do not respond in an expected manner when changes are made. This could mean a dynamic

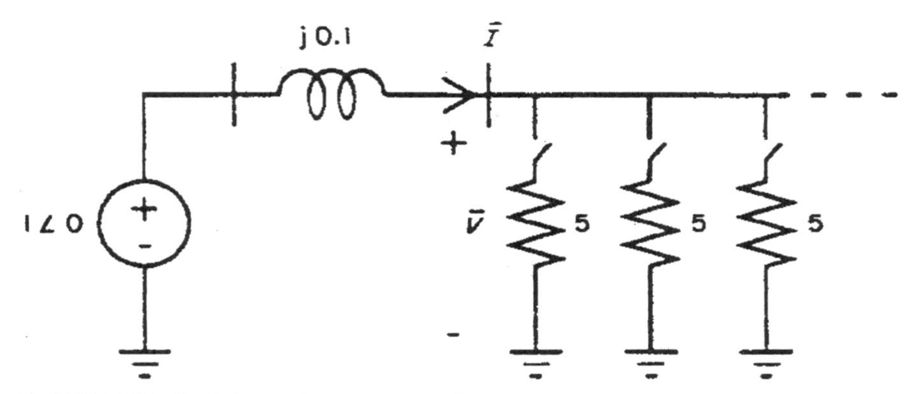

FIGURE 16.8 Single-bus maximum power transfer.

instability caused by tap-changing-under-load (TCUL) transformer controls that change according to a preset criteria that is not appropriate for the existing condition (increase of a tap position results in a lowering of a voltage level), or it could simply mean that the switching in of a "load" causes the power consumed to decrease (turning on another light makes the room get darker). This latter definition has been interpreted in terms of power-flow solutions. When the load power at a bus (or interchange schedule) is specified too high, the power-flow algorithm may fail to converge, or may converge to a "low-voltage" solution. If a failure to converge is due to the lack of an actual solution, this is considered by some to indicate a "voltage collapse" in the actual system. As such, the voltage-collapse point has been interpreted as a point where the Jacobian of the power-flow algorithm becomes singular.

Figure 16.8 illustrates one of these phenomena in terms of a simple model. The 1.0 p.u. voltage source serves parallel resistive loads of resistance 5.0 p.u. each through a line having a reactance of 0.1 p.u. When no resistors are switched in, the load voltage is equal to 1.0. When 10 resistors are switched in, the parallel load voltage drops to 0.981 and the power delivered increases to 1.924. When 50 resistors are switched in, the load voltage is equal to 0.707 (unacceptably low, but that is not the point in this case) and the power level reaches 5.0. When more than 50 resistors are switched in, the total power consumption drops below this power level of 5.0—termed the *nose* of the curve of load voltage vs. power consumed, as given in Fig. 16.9. This plot is known as a PV curve. For this simple system, a PV curve can be calculated by hand as shown.

Number of loads	Voltage	Current	Power
0	1.000	0.000	0.000
5	0.995	0.995	0.990
10	0.981	1.961	1.924
15	0.958	2.873	2.753
30	0.857	5.145	4.409
50	0.707	7.071	5.000
160	0.298	9.540	2.844
400	0.124	9.923	1.230

In large-scale systems, the procedure for computing a PV curve is given next.

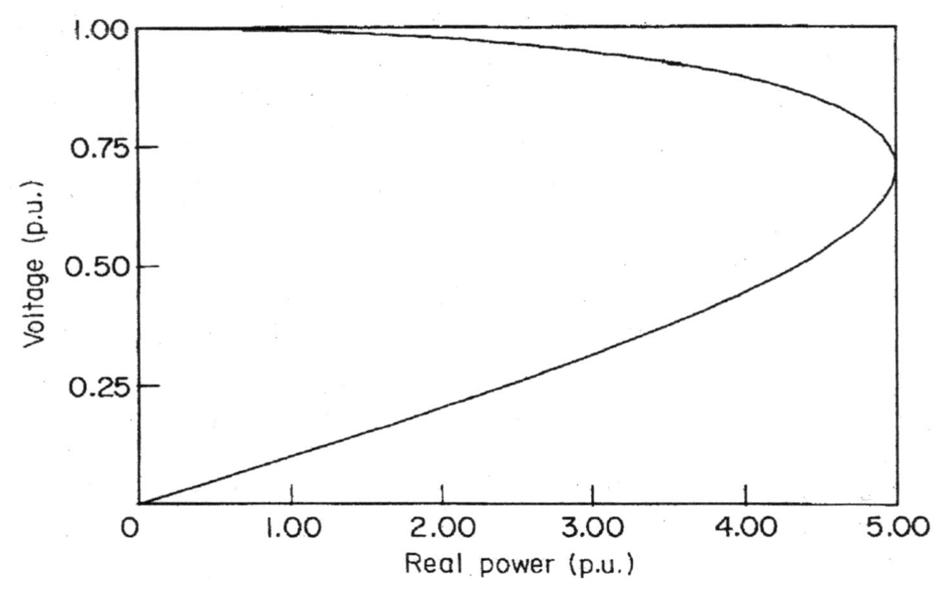

FIGURE 16.9 PV curve.

Calculation Procedure

1. Specify a Base-Case Condition

The base-case condition is normally some future scenario with forecasted loads and/or transactions. For this base-case condition, the network power-flow equations are solved for the network voltages.

2. Specify a Direction for Power Increase

This type of voltage stability analysis considers the increase of power in some direction. This power could be total load, the load at a particular bus, or a wheeling transaction from one or more import locations to one or more export locations. This power level increase will make up the horizontal axis of the PV curve.

3. Increment the Power Increase and Solve for System Voltages

Using power-flow analysis, the power level is increased by some amount and the power flow is solved for the new system condition. If the power flow fails to solve, the increase is reduced until convergence is reached. If the power flow solves, the power level is increased further and the power flow is solved again. This process is repeated until the nose of the PV curve is reached. Any bus voltage of interest may be plotted on the vertical axis against the power level on the horizontal. When the power-flow solution gets near the nose of the PV curve, the convergence may become a problem due to the ill-conditioning of the Jacobian matrix.

Related Calculations. The "continuation power flow" was created to solve the problems of ill-conditioned power-flow Jacobian matrices near the nose of the PV curve. This type of power flow uses more robust solution techniques that actually allow the precise determination of the nose and the lower half of the PV curve.

DATA PREPARATION FOR LARGE-SCALE DYNAMIC SIMULATION

The North American interconnected power system network is probably the largest network of interconnected systems whose dynamic performance is routinely simulated under postulated disturbances. The generating units, transmission elements, and loads comprising this network are owned and operated by hundreds of utility companies and millions of customers. Because they are interconnected, the consequences of a disturbance on one system may be felt across ownership and regulatory boundaries, affecting the operation of interconnected systems. Advances in modeling of system elements can be fully exploited only if the parameters of those models can be efficiently obtained for a study. To achieve this, in November 1994, the North American Electric Reliability Council (NERC) created the System Dynamics Database Working Group to develop and maintain an integrated system dynamics database and associated dynamics simulation cases for the Eastern Interconnection. This database is maintained in the Microsoft Access® database format. Similar databases have been established by the Electric Reliability Council of Texas (ERCOT) and the Western Systems Coordinating Council (WSCC), as these regions are electrically isolated from the Eastern Interconnection, except for asynchronous (dc) ties. Preparation and submission of data to the appropriate database is now mandatory for all control areas under NERC Planning Standard II.A.I, Measurement 4.

Data Requirements and Guidelines

The following requirements are taken from the NERC System Dynamics Database Working Group *Procedural Manual*, dated October 5, 1999, "Appendix II: Dynamics Data Submittal Requirements and Guidelines." A current copy may be obtained from the NERC Web site, www.nerc.com. It should be noted that the data-submission format utilized is that of the PSS/E® program of Power Technologies, Inc., Schenectady, N.Y. The level of modeling detail required may be considered as the current industry consensus even in the absence of regulatory requirements. MMWG refers to the Multi-Regional Modeling Working Group, which annually prepares a series of power-flow cases representing anticipated conditions at intervals over the planning horizon.

Requirements

I. Power Flow Modeling Requirements

1) All power flow generators, including synchronous condensers and SVCs, shall be identified by a bus name and unit id. All other dynamic devices, such as switched shunts, relays, and HVDC terminals, shall be identified by a bus name and base kV field. The bus name shall consist of eight characters and shall be unique within the Eastern Interconnection. Any changes to these identifiers shall be minimized.

2) Where the step-up transformer of a synchronous or induction generator or synchronous condenser is not represented as a transformer branch in the MMWG power flow cases, the step-up transformer shall be represented in the power flow generator data record. Where the step-up transformer of the generator or condenser is represented as a branch in the MMWG power flow cases, the step-up transformer impedance data fields in the power flow generator data record shall be zero and the tap ratio unity. The mode of step-up transformer representation, whether in the power flow or the generator data record, shall be consistent from year to year.

3) Where the step-up transformer of a generator, condenser, or other dynamic device is represented in the power flow generator data record, the resistance and reactance shall be given in per unit on the generator or dynamic device nameplate MVA. The tap ratio shall reflect the actual step-up transformer turns ratio considering the base kV of each winding and the base kV of the generator, condenser or dynamic device.

4) In accordance with PTI PSSE requirements, the Xsource value in the power flow generator data record shall be as follows:

   a) Xsource = X'' di for detailed synchronous machine modeling
   b) Xsource = X' di for non-detailed synchronous machine modeling
   c) Xsource = 1.0 p.u. or larger for all other devices

II. Dynamic Modeling Requirements

1) All synchronous generator and synchronous condenser modeling and associated data shall be detailed except as permitted below. Detailed generator models consist of at least two d-axis and one q-axis equivalent circuits. The PTI PSSE dynamic model types classified as detailed are GENROU, GENSAL, GENROE, GENSAE, and GENDCO.
   The use of non-detailed synchronous generator or condenser modeling shall be permitted for units with nameplate ratings less than or equal to 50 MVA under the following circumstances:

   a) Detailed data is not available because manufacturer no longer in business.
   b) Detailed data is not available because unit is older than 1970.

   The use of non-detailed synchronous generator or condenser modeling shall also be permitted for units of any nameplate rating under the following circumstances only:

   c) Unit is a phantom or undesignated unit in a future year MMWG case.
   d) Unit is on standby or mothballed and not carrying load in MMWG cases.

   The non-detailed PTI PSSE model types are GENCLS and GENTRA. When complete detailed data are not available, and the above circumstances do not apply, typical detailed data shall be used to the extent necessary to provide complete detailed modeling.

2) All synchronous generator and condensers modeled in detail per Requirement II.A shall also include representations of the excitation system, turbine-governor, power system stabilizer, and reactive line drop compensating circuitry. The following exceptions apply:

   a) Excitation system representation shall be omitted if unit is operated under manual excitation control.
   b) Turbine-governor representation shall be omitted for units that do not regulate frequency such as base load nuclear units and synchronous condensers.
   c) Power system stabilizer representation shall be omitted for units where such device is not installed or not in continuous operation.
   d) Representation of reactive line drop compensation shall be omitted where such device is not installed or not in continuous operation.

3) All other types of generating units and dynamic devices including induction generators, static var controls (SVC), high-voltage direct current (HVDC) system,

and static compensators (STATCOM), shall be represented by the appropriate PTI PSSE dynamic model(s).

4) Standard PTI PSSE dynamic models shall be used for the representation of all generating units and other dynamic devices unless both of the following conditions apply:

   a) The specific performance features of the user-defined modeling are necessary for proper representation and simulation of inter-regional dynamics, and
   b) The specific performance features of the dynamic device being modeled cannot be adequately approximated by standard PSSE dynamic models.

5) Netting of small generating units, synchronous condensers or dynamic devices with bus load shall be permitted only when the unit or device nameplate rating is less than or equal to 20 MVA. (Note: any unit or device which is already netted with bus load in the MMWG cases need not be represented by a dynamic model.)

6) Lumping of similar or identical generating units at the same plant shall be permitted only when the nameplate ratings of the units being lumped are less than or equal to 50 MVA. A lumped unit shall not exceed 300 MVA. Such lumping shall be consistent from year to year.

7) Where per unit data is required by a dynamic model, all such data shall be provided in per unit on the generator or device nameplate MVA rating as given in the power flow generator data record. This requirement also applies to excitation system and turbine-governor models, the per unit data of which shall be provided on the nameplate MVA of the associated generator.

III. Dynamics Data Validation Requirements

1) All dynamics modeling data shall be screened according to the SDDB data screening checks.

   All data items not passing these screening tests shall be resolved with the generator or dynamic device owner and corrected.

2) All regional data submittals to the SDDWG coordinator shall have previously undergone satisfactory initialization and 20 second no-disturbance simulation checks for each dynamics case to be developed. The procedures outlined in Section III.H* of this manual (*yet to be written) may be applied for this purpose.

Guidelines

1) Dynamics data submittals containing typical data should include documentation, which identifies those models containing typical data. When typical data is provided for existing devices, the additional documentation should give the equipment manufacturer, nameplate MVA and kV, and unit type (coal, nuclear, combustion turbine, hydro, etc.).
2) When user-defined modeling is used in the SDDWG cases, written justification shall be supplied explaining the dynamic device performance characteristics that necessitate use of a user-defined model. The justifications for all SDDWG user-defined modeling shall be posted on the SDDWG Internet site as a separate document. (Note: Guideline 2 above shall become a requirement in January 2001 and thereafter.)
3) The voltage dependency of loads should be represented as a mixture of constant impedance, constant current, and constant power components (referred to as the ZIP model). The Regions should provide parameters for representing loads via the PTI PSSE CONL activity. These parameters may be specified by area, zone, or bus. Other types of load modeling should be provided to SDDWG when it becomes evident that accurate representation of interregional dynamic performance requires it.

MODEL VALIDATION

The response of excitation systems and governors, like other control systems, to a change in set point should be rapid but well damped. A modest overshoot of the intended final output may be acceptable if it permits more rapid attainment of the target, but sustained or poorly damped oscillations are not. A technique that has been applied to such models is to consider the exciter or governor isolated from the unit and the rest of the power system and simulate variation of the field voltage or mechanical torque in response to step change in set point.

An excitation system model is validated by simulating its response for a two-to-five-s interval following an increase in its voltage set point of perhaps 5 percent of the nominal value (e.g., from 1.0 to 1.05 p.u.). Governor models are similarly validated by simulating a change in the desired power input, but governor response tests must be run for a considerably longer time, particularly for reheat units, as discussed above.

In making such simulations it is important that both initial and final output values be within applicable limits $V_{R\max}$, $V_{R\min}$, $P_{\max}$, and $P_{\min}$.

Figure 16.10 shows responses from a well-tuned IEEE Type 1 excitation system model and two with different parameter values that are not well tuned. For comparison, the response of IEEE Type 3 (GE SCPT) and IEEE Type 4 (noncontinuous regulator) excitation systems are also shown in Fig. 16.11. The latter have not been installed in recent years but are still found on older machines. These have a rapid and a slow responding mechanism in parallel controlling the dc exciter, hence the name *noncontinuous*. The rapid-responding mechanism has a significant dead band, typically 2 or 3 percent of nominal voltage, in its response.

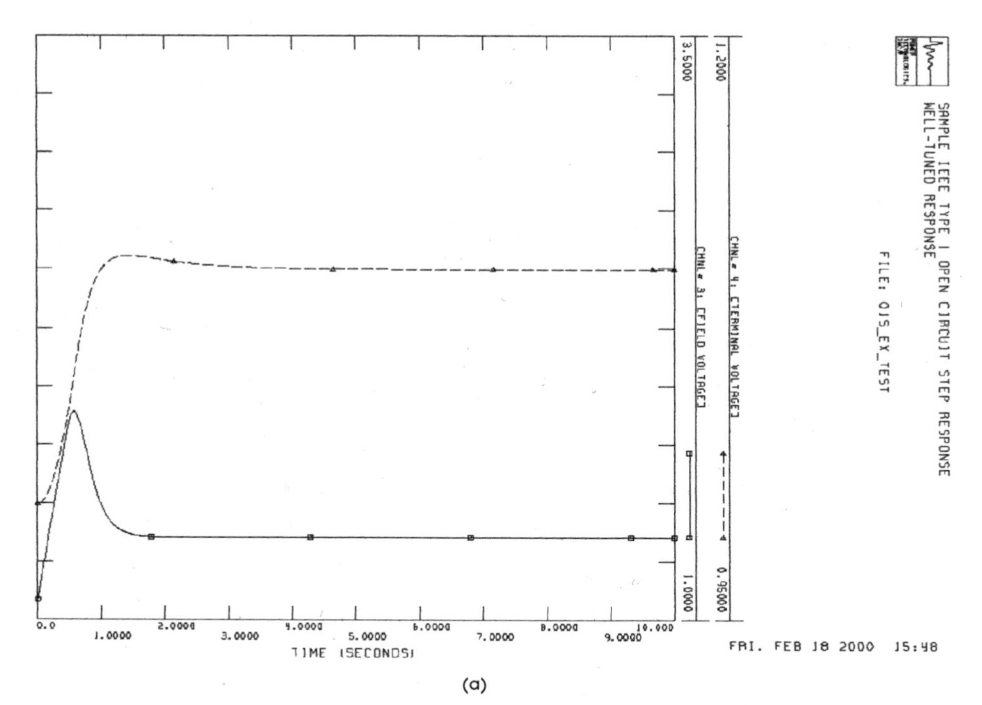

(a)

FIGURE 16.10 Sample IEEE Type 1 excitation system simulations of open-circuit step response. (*a*) Well-tuned response. (*b*) Gain too high; oscillatory response. (*c*) Gain too low; sluggish response.

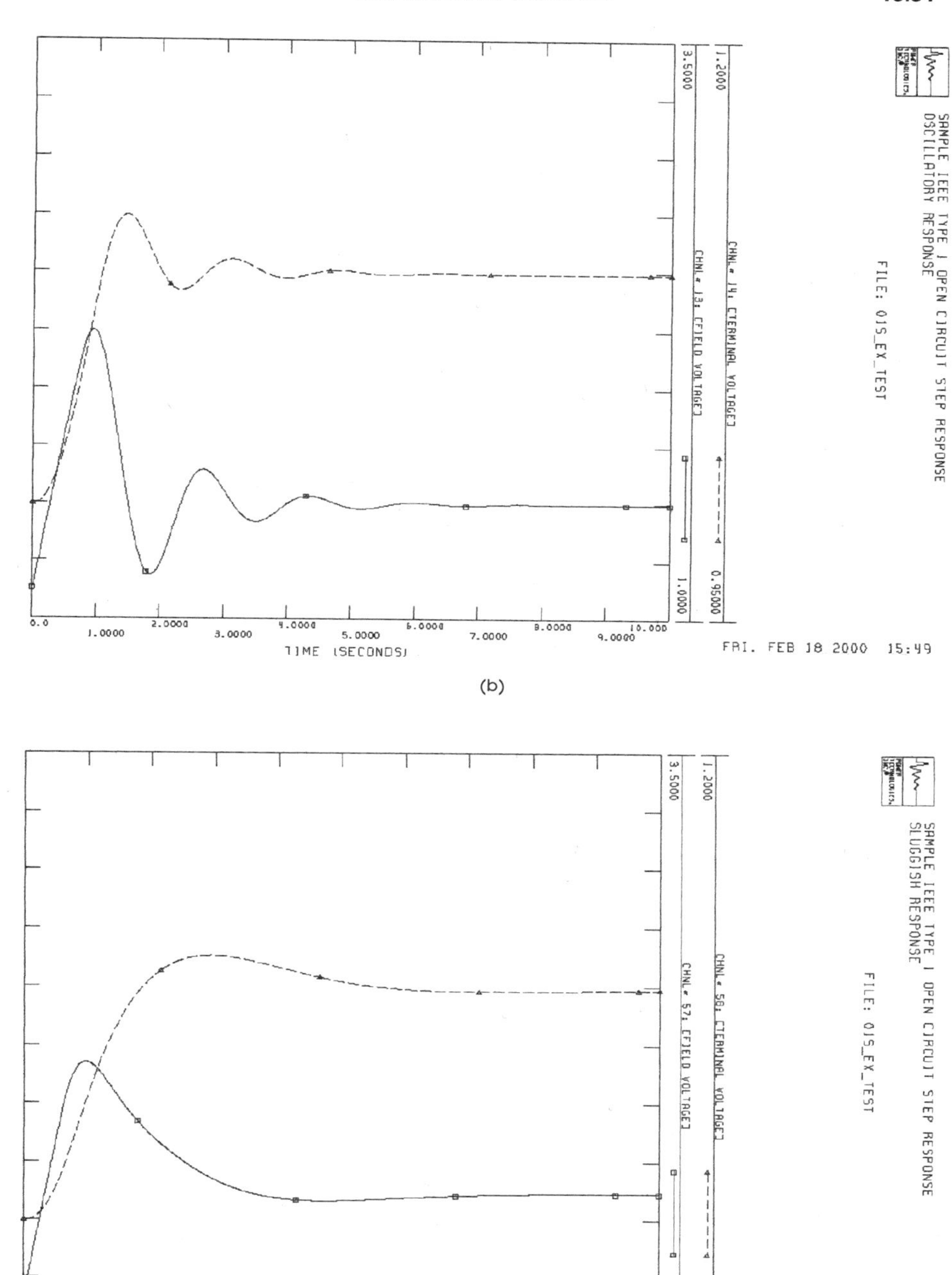

FIGURE 16.10 (*Continued*). Sample IEEE Type 1 excitation system simulations of open-circuit step response. (*a*) Well-tuned response. (*b*) Gain too high; oscillatory response. (*c*) Gain too low; sluggish response.

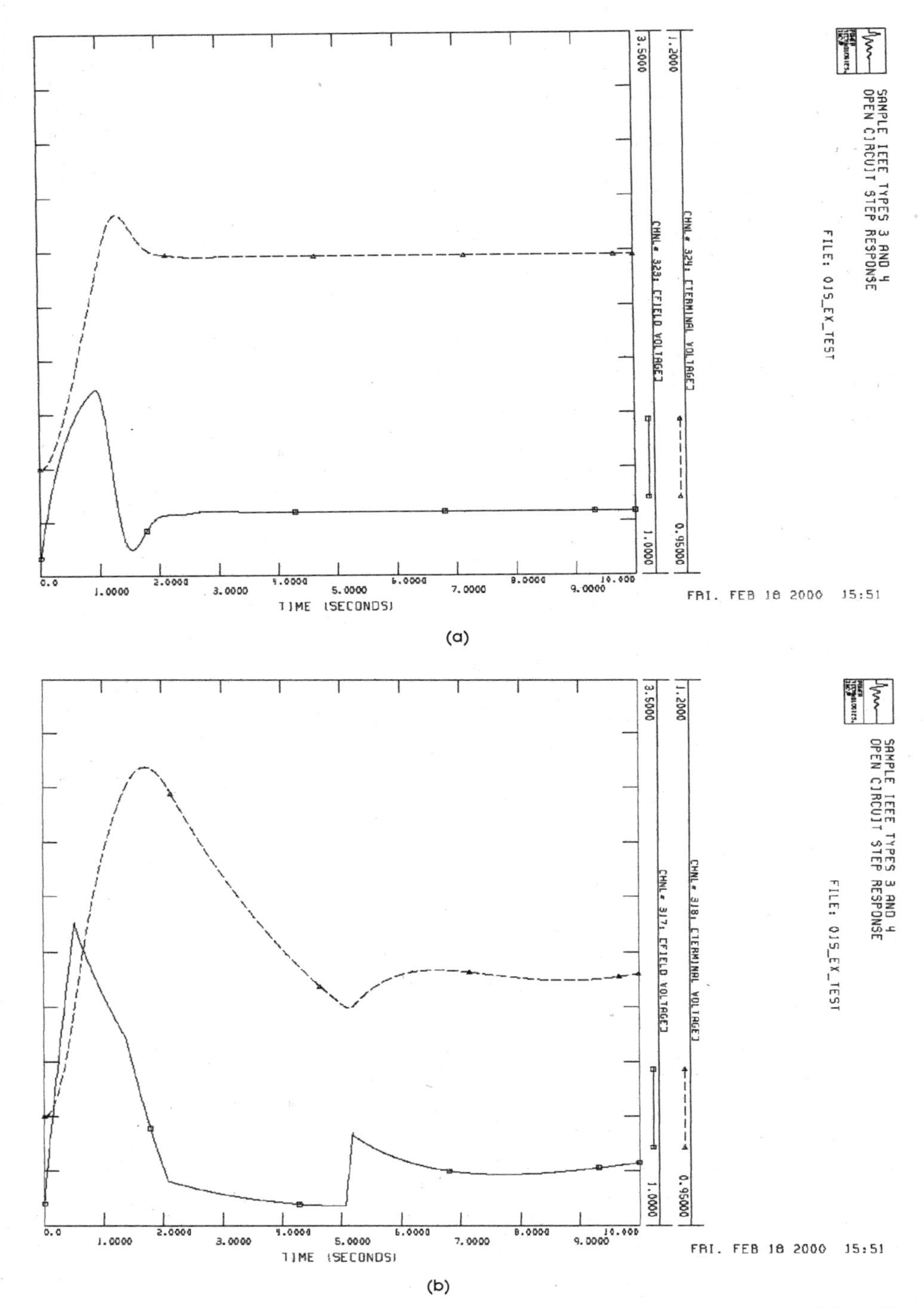

FIGURE 16.11 IEEE Type 3 (SCPT) and IEEE Type 4 (noncontinuous) excitation system simulations of open-circuit step response.

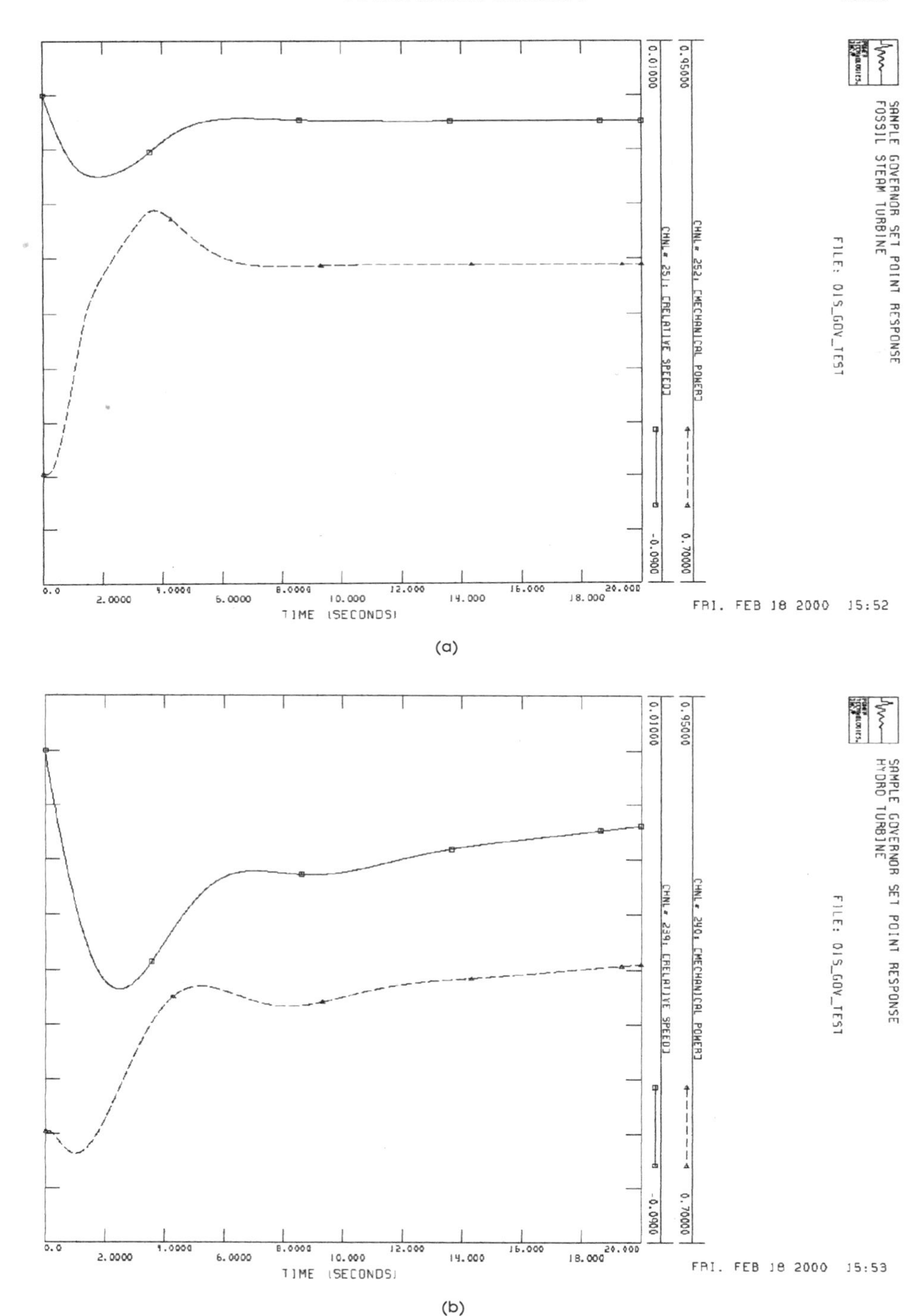

FIGURE 16.12 Governor response simulations. (*a*) Steam turbine units. (*b*) Hydro unit.

Figure 16.12*a* shows governor responses to a 10 percent change in load set point from a reheat (tandem compound) steam turbine. (Response of a nonreheat steam turbine is similar.) Figure 16.12*b* shows the governor response of a hydro unit. Note that an initial reduction in power output from an increased valve setting is typical, due to the inertia of the water flowing through the penstock. When the gate valve is opened some, the head is initially used to accelerate the water column and the head available at the turbine to generate power is reduced.

BIBLIOGRAPHY

Anderson, Paul M., and Aziz A. Fouad. 1994. *Power System Stability and Control (Revised Printing)*. New York: IEEE Press.

Bergen, Arthur R., and Vijay Vittal. 2000. *Power System Analysis*. Upper Saddle River, N.J.: Prentice Hall.

Byerly, Richard T., and Edward W. Kimbark. 1974. *Stability of Large Electric Power Systems*. New York: IEEE Press.

Elgerd, Olle I. 1982. *Electric Energy Systems Theory: An Introduction*. New York: McGraw-Hill.

Energy Development and Power Generating Committee of the Power Engineering Society. 1992. IEEE Standard 421.5-1992, *IEEE Recommended Practice for Excitation System Models for Power System Stability Studies*. New York: Institute of Electrical and Electronics Engineers, Inc.

Kundur, Prabha. 1994. *Power System and Stability and Control*. New York: McGraw-Hill.

Machowski, Jan, Janusz Bialek, and James R. Bumby. 1997. *Power System Dynamics and Stability*. Chichester, U.K.: Wiley.

Pai, M. A. 1981. *Power System Stability—Analysis by the Direct Method of Lyapunov*. New York: North Holland.

Power System Engineering and Electric Machinery Committees of the IEEE Power Engineering Society. 1991. IEEE Standard 1110-1991, *IEEE Guide for Synchronous Generator Modeling Practices in Stability Analyses*. New York: Institute of Electrical and Electronics Engineers, Inc.

Sauer, Peter W., and M. A. Pai. 1998. *Power System Dynamics and Stability*. Upper Saddle River, N.J.: Prentice Hall.

Taylor, Carson W. 1994. *Power System Voltage Stability*. New York: McGraw-Hill.

SECTION 17

COGENERATION

Hesham Shaalan, Ph.D.

Assistant Professor
Georgia Southern University

INTRODUCTION

During the 1970s, energy issues became prominent, and one of those issues was to improve the efficiency of energy utilization. The major piece of legislation to focus on this issue was the 1978 Public Utilities Regulatory Act (PURPA). This act required that utilities purchase power and pay as much as the utility's avoided cost per kilowatt-hour for the cogenerated electricity. Cogeneration involves the simultaneous production of process steam and electricity. PURPA gives various legal rights to nonutility owners of cogeneration facilities meeting certain standards. The effect of these rights is to make on-site generation a viable alternative for large industrial users of steam and to open up the generation sector of the electric industry to competition.

During the 1960s and 1970s, power generated by industrial companies averaged 500 MW of additions per year. These capacity additions were very small in comparison to the 15,000 MW per year of utility additions during the same time period. After 1980, however, nonutility capacity additions rose significantly as a major source of electric capacity additions. The key incentive for a company to begin cogeneration is that it reduces overall industrial process steam and electricity costs. Cogeneration facilities use significantly less fuel to produce electricity and steam than would be needed to produce the two separately. Thus, by using fuels more efficiently, cogeneration can also make a significant contribution to the nation's efforts to conserve energy resources.

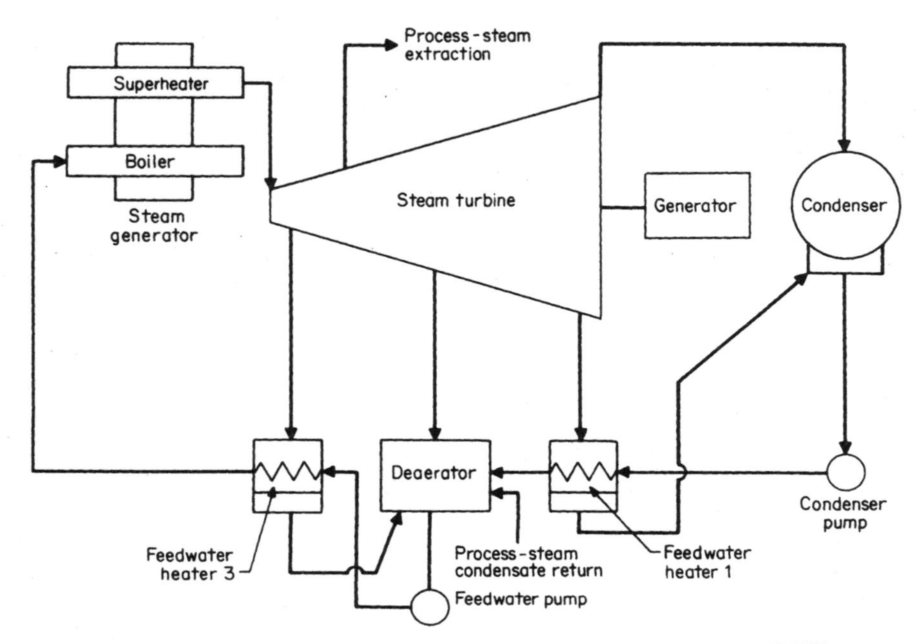

FIGURE 17.1 Cogeneration-plant cycle based on a steam turbine as the prime mover (STCP).

A cogeneration plant is a power plant that produces electrical power along with heat output, in the form of steam flow or water flow (through installation of steam-water heat exchangers), for industrial or residential consumption. The ratio of electric power to heat load varies, depending on the type of power plant. If the plant is located in an industrial complex, its main objective is often the supply of steam or hot water for industrial consumption. In this case, electric power output is considered as a byproduct and is relatively small. The public utility companies have the inverse ratio of electric output to heat load, because heat output is considered as a byproduct.

There are two conceptually different generation plant types. One is the steam-turbine–based cogeneration plant (STCP) consisting of a steam turbine with the usual controlled steam extraction(s) for process steam supply (Fig. 17.1). The other is a gas-turbine–based cogeneration plant (GTCP) consisting of one or more gas turbines exhausting products of combustion through one or more heat-recovery steam generators (HRSGs), which produce steam for the heat supply (Fig. 17.2).

A cogeneration plant has the following major operational features: electric output in kWh, kJ; heat output in kJ (kcal, Btu); and heat rate in kJ/kWh (Btu/kWh). The cogeneration-plant heat rate requires special definition. For a conventional power plant, heat rate represents the heat consumption in kJ (kcal, Btu) per kilowatthour of electric output. However, this definition of heat rate is not applicable for cogeneration plants because it does not account for heat output.

The heat rate calculations presented in this section are based on the following definition: heat rate = $(Q_1 - Q_2)/P$, where Q_1 = cogeneration plant heat input in kJ (Btu, kcal); Q_2 = conditional heat input with fuel to conventional steam generator to produce heat output equal to that produced by a cogeneration plant, in kJ (Btu, kcal); and P = electrical

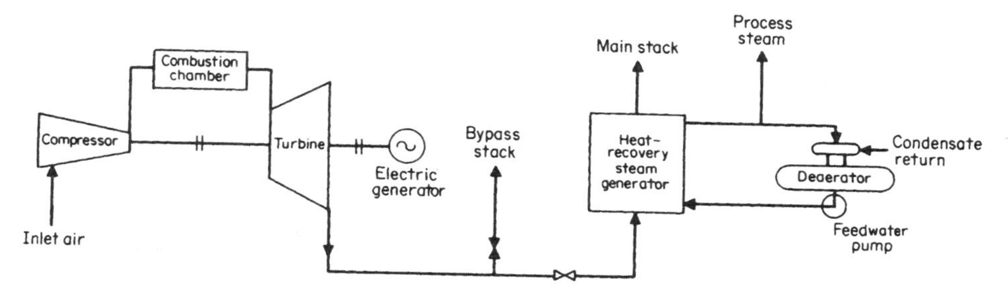

FIGURE 17.2 Cogeneration-plant cycle based on a gas turbine as the prime mover (GTCP).

output in kWh. This heat-rate definition assigns all the benefits from the combined power and steam generation to power production.

The thermodynamic efficiency of a cogeneration plant shall be evaluated using the expression: efficiency $= (P + H)/Q_1$, where P and H are the power and heat outputs of the cogeneration plant, respectively. They are expressed in the same heat units as Q_1.

POWER OUTPUT DEVELOPED BY TURBINE STAGES

Calculate the power output for the power plant of Fig. 17.3. The flows of steam through the various parts of the turbine are different as a result of steam extractions for the feedwater heating and process-steam supply.

Calculation Procedure

1. Divide Turbine into Sections

To calculate the power output, the turbine is divided into sections (Fig. 17.4) which have constant steam flow and no heat addition or extraction.

2. Calculate Total Power Output in Each Section

The power output of each section is determined by multiplying the flow through that section by the enthalpy drop across the section (the difference between the steam enthalpy entering and that leaving the section). Hence, the section power output $= w_i \Delta H_i / 3600$, where w_i = steam flow through section in kg/h (lb/h) and ΔH_i = enthalpy drop across section in kJ/kg (Btu/lb).

3. Calculate Total Power Output

Use:

$$\text{Total power output} = \sum_{t=1}^{n} w_i \Delta H_i / 3600$$

where n is the number of sections. The calculated power output for each section of the turbine and the total power output of 96,000 kW is summarized in Table 17.1.

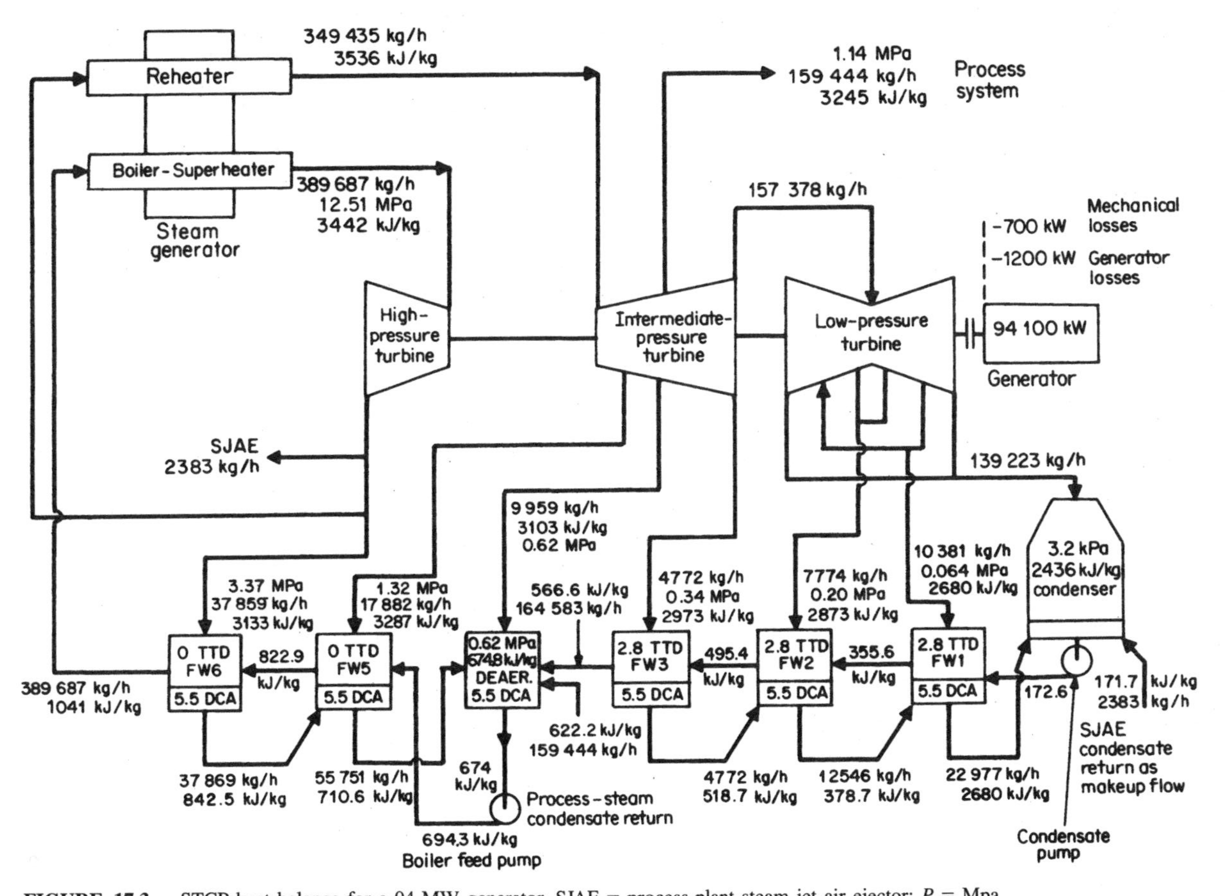

FIGURE 17.3 STCP heat balance for a 94-MW generator. SJAE = process-plant steam jet air ejector; P = Mpa measured as absolute pressure; H = kJ/kg; W = kg/h.

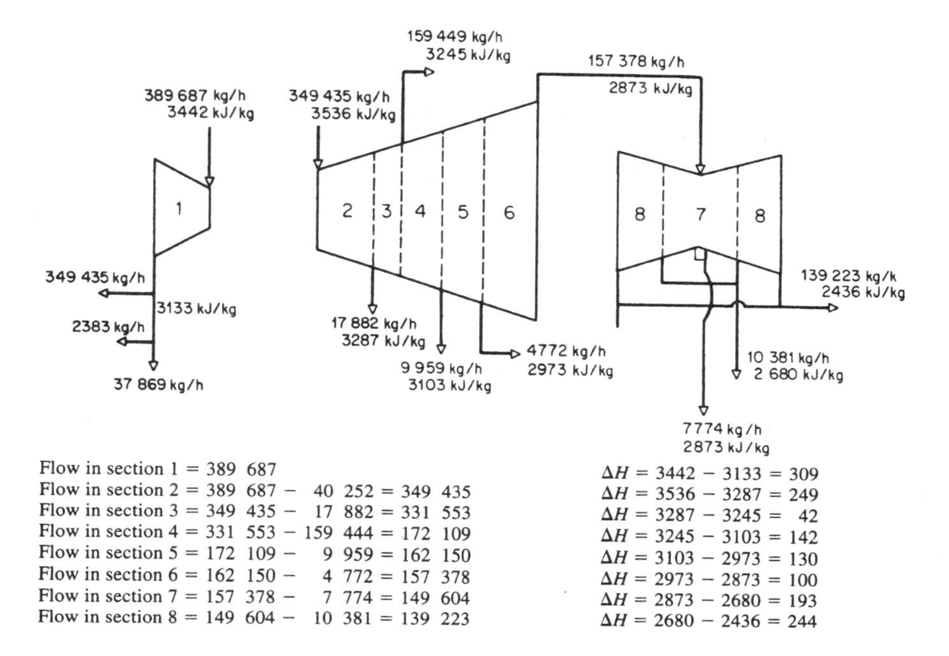

FIGURE 17.4 Turbines in Fig. 17.3 divided into sections.

GENERATOR AND MECHANICAL LOSSES

Assume the power factor is 0.85 and the electric generator rating equals the 100 percent operating load (in kVA) plus 10 percent. Use a trial-and-error method in which the power output of the generator is assumed and the mechanical and generator losses are obtained from Figs. 17.5, 17.6, and 17.7, which give the respective losses as a function of the

TABLE 17.1 Gross Power Output Calculations*

Steam turbine section number	Steam flow through section, w_i, kg/h	Enthalpy drop across section, ΔH, kJ/kg	Power output, kW
1	389,687	309	33,460
2	349,435	249	24,170
3	331,553	42	3870
4	172,108	142	6790
5	162,150	130	5870
6	157,378	100	4370
7	149,604	193	8020
8	139,223	244	9450
Sum of turbine-sections power outputs			96,000

*See Fig. 17.4.

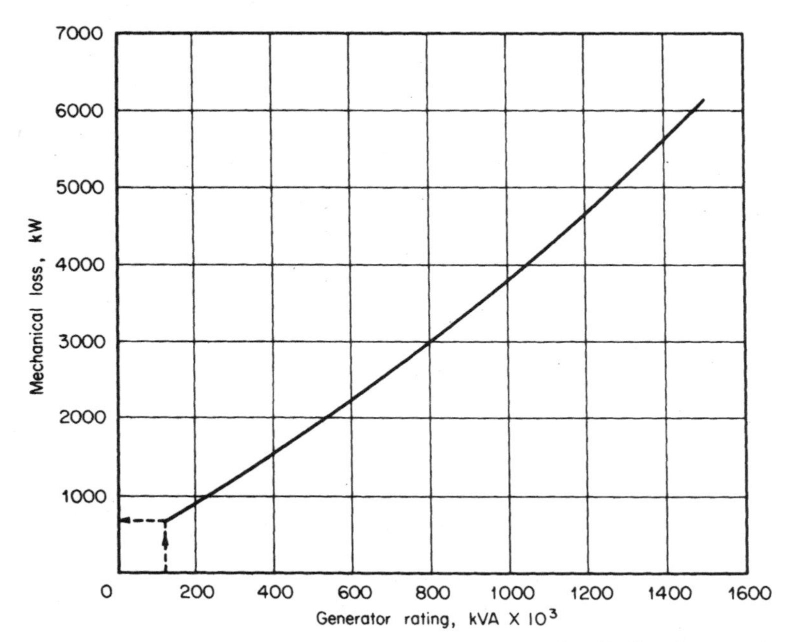

FIGURE 17.5 Mechanical loss as a function of generator rating. If oil coolers are utilized in the condensate line, the recoverable loss = 0.85 mechanical loss.

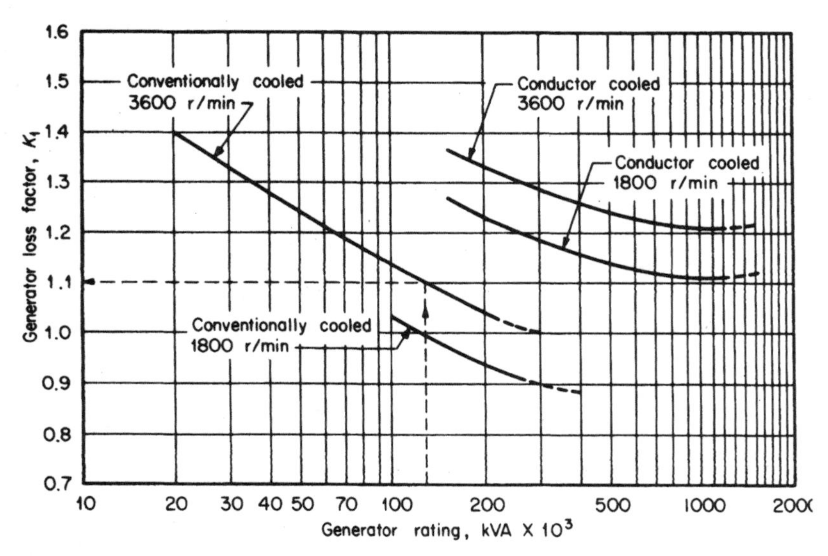

FIGURE 17.6 Generator loss factor K_1 as a function of generator rating. Generator loss [at rated H_2 pressure = operating kVA $(K_1/100)$ K_2; K_1 from this figure, K_2 from Fig. 17.7]. If hydrogen coolers are utilized in the condensate line, the recoverable loss at any kVA value = generator loss − 0.75 generator loss at the rated kVA value.

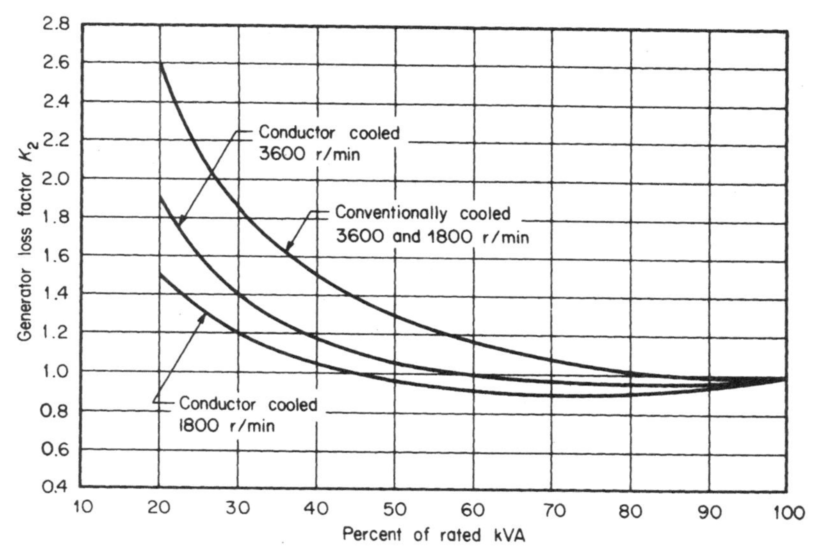

FIGURE 17.7 Generator loss factor K_2 as a function of generator rating.

operating load in kVA. The generator power output plus the mechanical and generator losses shall be equal to the sum of the turbine sections power outputs. Determine the generator and mechanical losses.

Calculation Procedure

1. Assume Generator Power Output = 94,100 kW

Operating kVA = assumed power output/power factor = 94,100 kW/0.85 = 111,000 kVA. Generator rating = operating kVA × 1.1 = (111,000 kVA)(1.1) = 122,000 kVA.

From Fig. 17.5, the mechanical losses = 700 kW and the generator losses = (operating kVA)(K_1)/100K_2 = (111,000 kVA)(1.1)/(100)(1.0) = 1200 kW, where $K_1 = 1.1$ from Fig. 17.6 and $K_2 = 1.0$ from Fig. 17.7.

2. Check the Assumed Value of Generator Power Output

The generator power output (94,100 kW) + mechanical losses (700 kW) + generator losses (1200 kW) = 96,000 kW, which equals the sum of the power of the turbine sections (Table 17.1).

BOILER-FEED AND CONDENSATE PUMP POWER CONSUMPTION

Using the values given in Fig. 17.3, calculate the boiler-feed pump (BFP) and condensate pump (CP) power consumption.

Calculation Procedure

1. Calculate Pump Power Consumption

Use: pump power consumption (kW) = enthalpy increase across the pump (kJ/kg) × mass flow (kg/s). Based on the values given in Fig. 17.4 for the boiler-feed pump, $\Delta H = 694.3 - 674.8 = 19.5$ kJ/kg and $w = 9959 + 164{,}532 + 159{,}445 + 55{,}751 = 389{,}687$ kg/h. Therefore, BFP = (19.5 kJ/kg)(389,687 kg/h)/(3600 kJ/kWh) = 2111 kW. For the condensate pump, $\Delta H = 172.6 - 171.7 = 0.9$ kJ/kg and $w = 139{,}223 + 22{,}977 + 2383 = 164{,}583$ kg/h. Therefore, CP = (0.9 kJ/kg)(164,583 kg/h)/(3600 kJ/kWh) = 41 kW.

2. Determine Power Consumption of Electric Motors

In order to evaluate the power consumption of the electric motors of these pumps, assume a 90 percent motor efficiency for BFP and 85 percent for CP. Hence, BFP electric-motor power consumption = 2111 kW/0.9 = 2345 kW and CP electric-motor consumption = 41 kW/0.85 = 48.5 kW. The total CP and BFP motor power consumption is 2345 + 48.5, or approximately 2400 kW.

GROSS AND NET POWER OUTPUT

Calculate the gross and net power output for the power plant in Fig. 17.3.

Calculation Procedure

1. Calculate Gross Power Output

Gross power output = sum of turbine-section power outputs − mechanical losses − generator losses = 96,000 − 700 − 1200 = 94,100 kW.

2. Calculate Net Power Output

Net power output = gross power output − internal plant power consumption (for simplicity this is assumed to be only BFP and CP power consumption) = 94,100 kW − 2400 kW = 91,700 kW.

HEAT AND FUEL CONSUMPTION

Steam flow and enthalpy for characteristic cycle points for the cogeneration plant are given in Fig. 17.3. Assume the steam generator efficiency is 86 percent and no. 4 fuel oil having a heating value of 43,000 kJ/kg is used. Determine the heat and fuel consumption for the power plant.

Calculation Procedure

The step-by-step calculations are summarized in Table 17.2.

TABLE 17.2 Heat and Fuel Consumption Calculations

Item no.	Defined item	Source	Value
1	Main steam flow	Heat balance, Fig. 17.3	389,687 kg/h
2	Main steam-flow enthalpy	Heat balance, Fig. 17.3	3442 kJ/kg
3	Final feedwater enthalpy	Heat balance, Fig. 17.3	1041 kJ/kg
4	Enthalpy change across steam generator	Line (2) − Line (3)	2401 kJ/kg
5	Heat added to main flow	Line (1) × Line (4)	0.936×10^9 kJ/h
6	Hot reheat steam flow	Heat balance, Fig. 17.3	349,435 kg/h
7	Hot reheat enthalpy	Heat balance, Fig. 17.3	3536 kJ/kg
8	Cold reheat enthalpy	Heat balance, Fig. 17.3	3133 kJ/kg
9	Enthalpy change across reheater	Line (7) − Line (8)	403 kJ/kg
10	Reheater heat added	Line (6) × Line (9)	0.140×10^9 kJ/h
11	Total heat added in steam generator	Line (5) + Line (10)	1.076×10^9 kJ/h
12	Steam generator efficiency	Assumed	86 percent
13	Total heat added with fuel	$\frac{\text{Line (11)}}{\text{Line (12)}} \times 100$	1.25×10^9 kJ/h
14	No. 4 fuel-oil heating value	Assumed	43,000 kJ/kg
15	No. 4 fuel-oil consumption	Line (13): Line (14)	29,100 kg/h
16	Process-steam flow	Heat balance, Fig. 17.3	159,445 kg/h
17	Process-steam enthalpy	Heat balance, Fig. 17.3	3245 kJ/kg
18	Condensate return enthalpy	Heat balance, Fig. 17.3	622.2 kJ/kg
19	Enthalpy change across steam generator for process-steam flow	Line (17) − Line (18)	2622.8 kJ/kg
20	Heat input for process-steam supply	$\frac{\text{Line (16)} \times \text{Line (19)}}{\text{Line (12)}}$	0.49×10^9 kJ/h

HEAT RATE

The major parameters indicating the efficiency of a power plant are gross and net power-plant heat rates, which represent heat expenditure to produce 1 kWh of electrical energy. Using appropriate data in Table 17.2, determine the gross and net power-plant heat rates.

Calculation Procedure

1. Calculate Gross Cogeneration Plant Heat Rate

Use: gross plant heat rate = $(Q_1 - Q_2)$/gross power output, where Q_1 = total heat added with fuel in cogeneration plant and Q_2 = heat in conventional steam generator. From Table 17.2, $Q_1 = 1.25 \times 10^9$ kJ/h and $Q_2 = 0.49 \times 10^9$ kJ/h. For gross power output = 94,100 kW, the gross plant heat rate = $(1.25 \times 10^9 - 0.49 \times 10^9)/94{,}100 = 8077$ kJ/kWh.

2. Calculate Net Cogeneration-Plant Heat Rate

Use: net plant heat rate = $(Q_1 - Q_2)$/(net power output). For net power output = 91,700 kW, the net plant heat rate = $(1.25 \times 10^9 - 0.49 \times 10^9)/91{,}700 = 8288$ kJ/kWh.

FEEDWATER-HEATER HEAT BALANCE

The heat balance of any heat exchanger is based on the law of conservation of energy; that is, heat input minus heat losses is equal to heat output. A heat balance helps to determine any unknown flow or parameter if all other flows or parameters are known. Perform heat-balance calculations to determine the required steam flow (assumed to be unknown) to high-pressure feedwater heater 5 (Fig. 17.3).

Calculation Procedure

1. Write Heat Balance Equation

Denoting the unknown steam extraction flow as X and referring to Table 17.3, write the following heat balance for heater 5: $X \times$ Line (4) $\times$ [100 − Line (8)]/100 + Line (6) $\times$ Line (5) + [Line (1) $\times$ Line (2)] = Line (1) $\times$ Line (3) + [X + Line (6)] $\times$ Line (7), where the left-hand side of the equation represents heat flow into feedwater heater 5 and the right-hand side represents exiting heat flows.

2. Determine Steam Flow

The procedure is summarized in Table 17.3; X = 17,888 kg/h [Line (9)].

GAS-TURBINE–BASED COGENERATION PLANT

A cogeneration-plant cycle with a gas turbine as the prime mover and the corresponding heat balance are shown in Fig. 17.8. The gas-turbine–based cogeneration plant consists of two major components: gas-turbine generator(s), which produce electric power, and heat-recovery steam generator(s), which produce process steam by recovering heat from the gas-turbine exhaust gases.

TABLE 17.3 Feedwater-Heater 5 Heat-Balance Calculations

Item no.	Defined item	Source	Value
1	Feedwater flow	Heat balance, Fig. 17.3	389,687 kg/h
2	Enthalpy of feedwater flow entering heater	Heat balance, Fig. 17.3	694.3 kJ/kg
3	Enthalpy of feedwater flow leaving heater	Heat balance, Fig. 17.3	822.9 kJ/kg
4	Enthalpy of extraction steam	Heat balance, Fig. 17.3	3287 kJ/kg
5	Enthalpy of drain flow from feedwater heater 6	Heat balance, Fig. 17.3	842.5 kJ/kg
6	Drain flow from feedwater heater 6	Heat balance, Fig. 17.3	37,869 kg/h
7	Enthalpy of drain flow from feedwater heater 5	Heat balance, Fig. 17.3	710.6 kJ/kg
8	Radiation losses in feedwater heater	Assumed	1.5 percent
9	Steam flow to feedwater heater 5	Feedwater-heater heat balance	17,883 kg/h

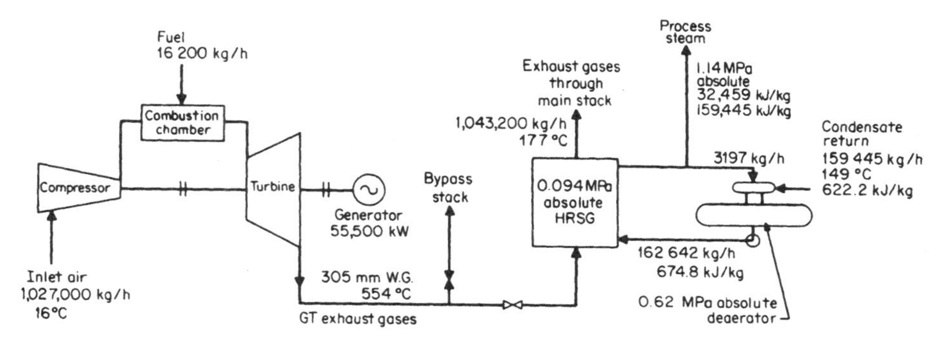

FIGURE 17.8 GTCP heat balance.

Gas-turbine performance data for a site temperature of 16°C, simple cycle application, and base load are assumed as follows: net generator output = 56,170 kW, net turbine heat rate HHV = 12,895 kJ/kWh, efficiency = 27.9 percent, airflow = 1,027,000 kg/h, turbine-exhaust temperature = 554°C, exhaust-gas flow = 1,043,200 kg/h, and light fuel oil HHV = 44,956 kJ/kg. (HHV ≡ high heating value of a fuel.)

The process-steam supply requirements are assumed as follows: process-steam flow = 159,445 kg/h, steam pressure from HRSG = 1.14 MPa (absolute pressure), steam enthalpy = 3245.9 kJ/kg, assumed pressure drop across HRSG = 305 mmH_2O gauge, condensate return temperature = 149°C, and condensate return enthalpy = 622.1 kJ/kg. Analyze the performance of the system.

GAS-TURBINE OUTPUT AND HEAT RATE IN COGENERATION-PLANT MODE

A gas turbine operating in the simple cycle mode (exhaust gases being vented to atmosphere and not into HRSG) has the following parameters at 16°C ambient air temperature: net power output = 56,170 kW and net turbine heat rate LHV = 12,985 kJ/kWh.

A gas turbine operating in a cogeneration plant has an increase in back pressure because of the additional pressure drop in the flue gases across the HRSG, which is assumed to be 305 mmH_2O gauge.

Calculation Procedure

1. Calculate Corrected Output and Heat Rate of Gas Turbine

From manufacturer's information the effect of the pressure drop across HRSG on the gas-turbine power output and heat rate is linear, and approximate recommendations are presented in Table 17.4.

Corrected output and heat rate of gas turbine in cogeneration-plant application area: power output = (56,170 kW)[1.00 − (0.004)(305 mmH_2O gauge)/(102 mmH_2O gauge)] = 55,500 kW and the heat rate = (12,985 kJ/kWh)[1.00 + (0.004)(305 mmH_2O gauge)/(102 mmH_2O gauge)] = 13,140 kJ/kWh.

TABLE 17.4 Effect of Pressure Losses on Gas-Turbine Performance

	Effect on output (percent)	Effect on heat rate (percent)	Increased exhaust temperature (°C)
102 mmH_2O inlet	−1.4	+0.4	+1.1
102 mmH_2O exhaust	−0.4	+0.4	+1.1

2. Calculate Compressor-Inlet Air-Temperature Effects on Plant Output and Gas-Turbine Heat Rate

The influence of the compressor-inlet air-temperature on the gas-turbine power and heat rate is shown in Fig. 17.9. For 32°C ambient temperature (16°C is the design temperature), the correction factor for the heat rate is 1.025 and for the power output is 0.895. Therefore, the corrected power output = (55,496 kW)(0.895) = 49,670 kW and the corrected heat rate = (13,140 kJ/kWh)(1.025) = 13,470 kJ/kWh.

3. Calculate Steam Productivity of HRSG Recovering Heat from Gas-Turbine Exhaust Gases

The calculation of steam productivity is based on the heat balance for HRSG. Heat transferred from gas turbine-generator exhaust gases less heat losses will be equal to the heat received by the HRSG medium (condensate steam). In calculation of HRSG steam productivity, the following is assumed: HRSG exhaust gas temperature is 177°C, which provides sufficient margin above the dew-point temperature of the products of combustion of no. 2 fuel oil (distillate) in a gas turbine. For conceptual calculations with acceptable accuracy, the specific heat of the exhaust gases can be obtained from air tables at average temperature, since the products of combustion in gas turbines have an air/fuel ratio much higher than theoretically required. Heat loss in an HRSG is assumed to be 2 percent. The HRSG heat-balance equation is: $W_gC_p(T_{g1} - T_{g2})(0.98) = W_{st}(H_1 - H_2)$, where W_g =

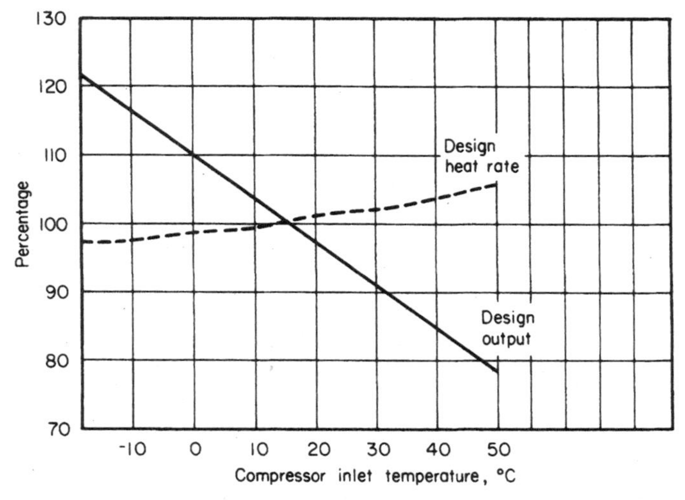

FIGURE 17.9 Effect of compressor inlet air/temperature on gas-turbine power output and heat rate. The fuel used was natural gas distillate oil.

exhaust-gas flow, 1,043,100 kg/h; T_{g1} and T_{g2} = temperatures of exhaust gases entering and leaving the HRSG, 554° and 177°C, respectively; C_p = specific heat of air at average exhaust-gas temperature; and W_{st} = produced steam flow, kg/h. The average temperature = (554°C + 177°C)/2 = 366°C. From air tables, C_p = 1.063 kJ/kg°C and H_1 and H_2 = enthalpy of produced steam and entering HRSG feedwater, 3245.9 kJ/kg and 674.8 kJ/kg, respectively. From the solution of the heat-balance equation with the preceding data, the produced steam flow is W_{st} = 162,642 kg/h.

The process steam is the difference between the produced steam flow (162,642 kg/h) and the steam flow required by the deaerator (3197 kg/h, obtained from Fig. 17.8). Thus, process steam = 162,642 − 3197 = 159,445 kg/h.

4. Calculate Heat and Fuel Consumption

Heat consumption (kJ/h) = corrected gas-turbine heat rate (kJ/kWh) × corrected gas-turbine power (kW). Hence, (13,049 kJ/kWh)(55,496 kW) = 0.73 × 10^9 kJ/h.

Fuel consumption = (heat consumption: 0.73 × 10^9 kJ/h)/(no. 2 fuel oil HHV: 45,124 kJ/kg) = 16,200 kg/h.

5. Calculate Cogeneration-Plant Heat Rate

Cogeneration-plant heat rate = $(Q_1 - Q_2)$/net power output. Heat consumption Q_1 = 0.73 × 10^9 kJ/h and heat input for process steam supply, Q_2 = 0.49 × 10^9 kJ/h (from Table 17.2, line 20). Corrected gas-turbine power output = 55,496 kW. Therefore, cogeneration-plant heat rate = (0.73 × 10^9 kJ/h − 0.49 × 10^9 kJ/h)/(55,500 kW) = 4325 kJ/kWh.

COMPARATIVE ANALYSIS OF STCP AND GTCP

The selection of the most economical cogeneration-plant type—steam-turbine–based cogeneration plant vs. gas-turbine–based cogeneration plant—for special power and heat consumption requirements is the most important problem for the conceptual definition of a power plant. There are two approaches to an optimization of the cogeneration plant:

1. Comparative thermodynamic analysis of STCP and GTCP, which represents a comparison of the cost of a fuel and may be critical for the selection of the cycle in regions with high fuel cost and when the number of operating hours per year exceeds 6000.
2. Comparative economic analysis, which is based on the evaluation of the present-worth dollar values of the cogeneration plant capital and operating costs for two cogeneration-plant designs. This analysis requires, in addition to performance characteristics for both cogeneration plants, information relevant to the costs of equipment and other features that are not always available at the conceptual design phase.

Calculation Procedure

Calculation procedures in this section represent a first approach for the conceptual selection of a cogeneration-plant cycle. Final selection is based upon more detailed calculations, along with other considerations such as water availability, environmental conditions, operational personnel priorities, etc.

The conceptual analysis of the STCP and GTCP cycles, as well as a number of cogeneration plants' cycle optimizations, shows that the main criterion for efficiency comparison of both cogeneration plants is the ratio of required heat output to power output, Q/P. The following calculation procedure and results, presented as heat rate (HR) vs. Q/P curves for both cogeneration plant arrangements, will help the engineer to select a more efficient cogeneration plant cycle for special Q/P requirements.

Heat-rate calculations for STCP and GTCP for various Q/P ratios are presented in Table 17.5. The calculations for STCP were done for three Q/P ratios and the respective heat balances presented in Figs. 17.3, 17.10, and 17.11. The heat-rate calculations for GTCP were done for two Q/P ratios because of the evident linear character of this function. HR curves for both cogeneration plants are presented in Fig. 17.12.

Analysis of the HR vs. Q/P curves shows that the break-even point for STCP and GTCP is at Q/P = 4200 kJ/kWh, or 1.6 kg/h per kilowatt. When the Q/P ratio is less than 4200 kJ/kWh, STCP is more economical (HR for STCP is less than for GTCP); where Q/P is higher than 4200 kJ/kWh, GTCP is more economical.

These results are based upon evaluation of selected STCP and GTCP cycle parameters. The calculations show that the conclusions reached with acceptable accuracy are applicable to various cycles; that is, break-even points between STCP and GTCP exist and are located in the vicinity of the preceding Q/P ratio.

It is emphasized that these results can be used for preliminary conceptual selection of the cogeneration-plant arrangement in regions with high fuel costs, where capital costs of the equipment represent a small share of total evaluated costs. The final selection of the optimum cogeneration-plant arrangement for particular design and operational conditions is based on minimum present worth of total evaluated costs.

An economic analysis, based upon average equipment, fuel, and maintenance costs and current economic factors, shows that the break-even points for STCP and GTCP are

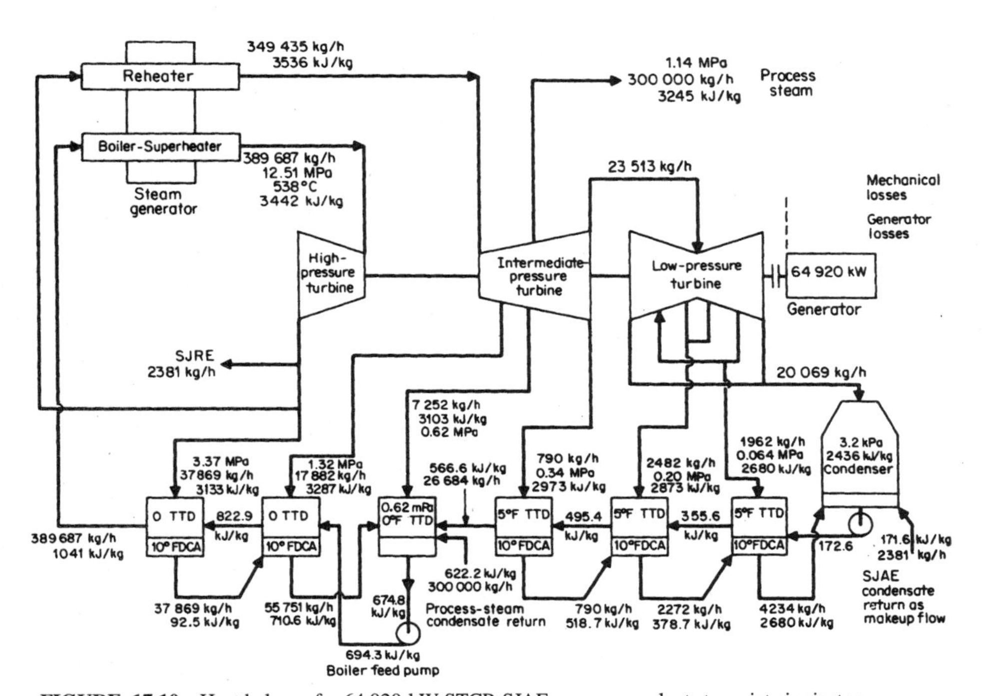

FIGURE 17.10 Heat balance for 64,920-kW STCP. SJAE = process-plant steam jet air ejector.

TABLE 17.5 Heat-Rate Calculations for Various STCP and GTCP Cycles

			STCP			GTCP	
Item no.	Defined item	Source	Case 1	Case 2	Case 3	Case 1	Case 2
1	Exporting steam	Assumed	159,445 kg/h	250,000 kg/h	300,000 kg/h	0	159,445 kg/h
2	Power output	Heat balances, Figs. 17.3, 17.10, and 17.11	91,700 kW	72,630 kW	62,520 kW	55,500 kW	55,500 kW
3	Heat output	Line (1) × 2622.8, kJ/h	0.42×10^9 kJ/h	0.655×10^9 kJ/h	0.786×10^9 kJ/h	0	0.42×10^9 kJ/h
4	Heat output/power output ratio	Line (3)/Line (2), kJ/kWh	4580 kJ/kWh	9018 kJ/kWh	12,680 kJ/kWh	0	7636 kJ/kWh
		Line (1)/Line (2), kg/h per kilowatt	1740 kg/h per kilowatt	3.40 kg/h per kilowatt	4.8 kg/h per kilowatt	0	2.9 kg/h per kilowatt
5	Heat input for process-steam supply*	Line (3)/0.86	0.49×10^9 kJ/h	0.76×10^9 kJ/h	0.91×10^9 kJ/h	0	0.49×10^9 kJ/h
6	Total heat input	Heat balances, Figs. 17.3, 17.10, and 17.11	1.25×10^9 kJ/h	1.25×10^9 kJ/h	1.25×10^9 kJ/h	0.72×10^9 kJ/h	0.72×10^9 kJ/h
7	Heat rate	$\frac{\text{Line (6)} - \text{Line (5)}}{\text{Line (2)}}$	8353 kJ/kWh	6747 kJ/kWh	5406 kJ/kWh	13,050 kJ/kWh	4289 kJ/kWh

*In conventional steam generator with assumed efficiency 0.86.

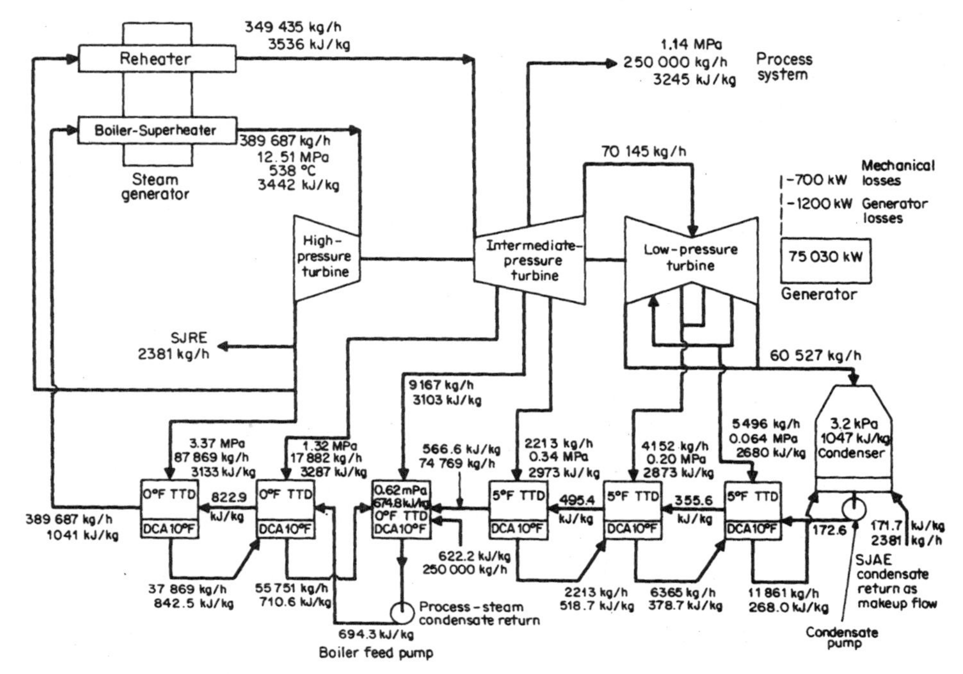

FIGURE 17.11 Heat balance for 75,030-kW STCP. SJAE = process-plant steam jet air ejector.

located in the region of a smaller Q/P ratio. Thus it shortens even more the Q/P ratio's span (between zero and break-even Q/P), where the STCP is more economical than the GTCP. These results can be explained by the considerably lower installed costs of the GTCP, which compensate for the higher heat rate.

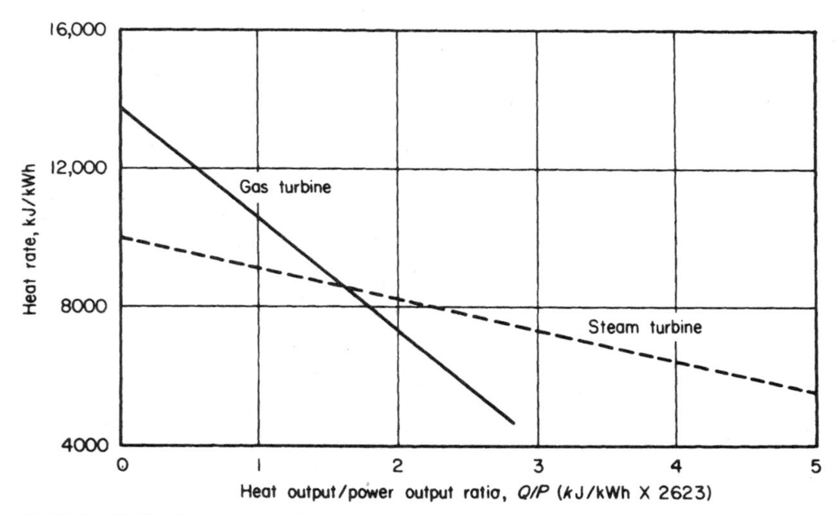

FIGURE 17.12 Heat rate as a function of Q/P ratios for STCP and GTCP.

The number of expected operating hours per year for the cogeneration plant is essential information in determining the applicable break-even Q/P ratio. The fewer annual operating hours, the lower the break-even Q/P ratio. Economic calculations show that if the number of operating hours per year is less than 2500, the break-even Q/P ratio is based only upon heat-rate considerations.

The obtained results prove that for a considerable number of feasible power-steam-supply requirement ratios, the GTCP is more economical than the STCP. This is an important conclusion as it is generally assumed that because of its high heat rate, gas-turbine cogeneration cycles are inherently less economic.

BIBLIOGRAPHY

Baumeister, Theodore, and Eugene A. Avallone (eds.). 1996. *Marks' Standard Handbook for Mechanical Engineers.* New York: McGraw-Hill.

Fink, Donald G., and H. Wayne Beaty. 2000. *Standard Handbook for Electrical Engineers.* New York: McGraw-Hill.

Lockerby, Robert W. 1981. *Cogeneration: Power Combined with Heat.* Monticello, Ill.: Vance Bibliographies.

Orlando, Joseph A. 1991. *Cogeneration Planner's Handbook.* Lilburn, Ga.: The Fairmont Press.

Potter, Phillip J. 1988. *Power Plant Theory and Design.* Malabar, Fla.: R. E. Kreiger Press.

Spiewak, Scott A. and Larry Weiss. 1997. *Cogeneration & Small Power Production Manual.* Lilburn, Ga.: The Fairmont Press.

Stoll, Harry G. 1989. *Least-Cost Electric Utility Planning.* New York: John Wiley & Sons, Inc.

Weaver, Rose. 1982. *Industrial Cogeneration of Energy: A Brief Bibliography.* Monticello, Ill.: Vance Bibliographies.

World Energy Council Report. 1991. *District Heating, Combined with Heat and Power: Decisive Factors for a Successful Use.* London: World Energy Council.

SECTION 18

STATIONARY BATTERIES

Marco W Migliaro, P.E., Fellow IEEE
Chief Electrical and I&C Engineer
Florida Power & Light
Nuclear Division

SELECTION

Batteries may be classified into two broad categories, *primary* batteries, which are not rechargeable, and *secondary* batteries, which are rechargeable. Secondary batteries are also referred to as *storage* batteries. They too, may be classified into two broad categories. The first is *SLI* (starting, lighting, and ignition) batteries and the second is *industrial* batteries.

SLI batteries include those used in automobiles, trucks, farm equipment, small internal-combustion generators, pleasure boats, etc. These batteries provide power for cranking (i.e., starting) and/or loads such as lights, controllers, and small motors (e.g., for a bilge pump).

Industrial batteries include motive power (e.g., fork lifts and aircraft tugs), railcar, locomotive, electric vehicle, missile, submarine, stationary, etc. Of these types, it is the *stationary* battery that will be discussed in this section, and it is the calculations associated with stationary batteries that will be addressed. Stationary batteries are used in standby service and derive their name from the fact that once they are installed, they remain stationary in one location. Stationary batteries are used in many services, including telecommunications, electric power (substations and generating stations), industrial control, uninterruptible power supply (UPS) systems, emergency lighting, photovoltaic systems, and energy storage systems. Table 18.1 provides relevant data on some of the many applications in which the stationary battery is used. Photovoltaic systems may be stand-alone power systems in applications such as marine buoys, remote railway crossings, billboards, highway tunnel lighting, etc. These systems have additional considerations, such as the battery remaining in a partial state-of-discharge during periods of insolation, that are discussed in documents included in the bibliography.

Selecting a stationary battery requires an evaluation of the application and factors that could affect it. The user has to make decisions on which battery technology to use as well as the plate type and cell design. The two commonly used technologies are lead-acid (nominal 2.0 Vdc per cell) and nickel-cadmium (nominal 1.2 Vdc per cell). Table 18.2 lists many of the factors that need to be considered before a battery type for an application can be selected and sized.

Both lead-acid and nickel-cadmium batteries are available in *vented* and *valve-regulated designs*. Nickel-cadmium batteries are also available in sealed designs. The valve-regulated lead-acid (VRLA) battery immobilizes the electrolyte in the cell. This is accomplished by absorbing the electrolyte in a fine *glass mat* used as a separator in the cell or by adding a gelling agent to the cell that causes the electrolyte to become a *thixotropic gel*, which has the consistency and feel of white petroleum jelly. The *absorbed glass mat* VRLA designs are sometimes referred to as AGM cells.

When selecting a cell an important consideration could be the plate type and, in the case of lead-acid cells, the plate alloy. Various positive plate types are available for lead-acid cells. These include Planté, modified Planté, pasted (also known as flat or Fauré) and tubular in rectangular plates and flat in round plates. All negative plates for lead-acid cells are the flat type. Nickel-cadmium plates include the pocket, fiber, and the sintered plastic-bonded types.

Vented lead-acid cells are available in long-duration, general-purpose, and high-performance designs. Similarly, nickel-cadmium cells are available in L, M, and H designs. Valve-regulated cell designs have no classification of design types; however, absorbed glass mat VRLAs are typically high-performance cells due to their thin plates, high-density electrolyte, and low-resistance separator. Gelled electrolyte VRLA cells could be classified as general purpose. Long-duration cells are normally applied in telecommunications, while general-purpose cells are used in substation, generating sta-

TABLE 18.1 Typical Applications for Stationary Batteries

Application	Typical back-up time	Representative capacity range	Nominal voltage (Vdc)	Typical loads
Traditional telecommunications	3 to 5 h	to 3500 Ah per string*	48	Telecommunications equipment (e.g., switch)
Microwave telecommunications	3 to 5 h	to 500 Ah per string*	24	Microwave equipment
Cellular telecommunications	3 to 5 h	to 500 Ah per string*	24	Telecommunications equipment
Electric substations and switchyards	5 to 8 h	to 1800 Ah	48 to 240	Relays, indicator lamps, circuit-breaker control, motor operators, SCADA, event recorders
Electric-power-generating stations	2 to 8 h	to 2400 Ah	48 to 240	Inverters, emergency motors, event recorders, relays, PLCs, annunciators, indicating lamps, circuit-breaker control, plant computers, valve operators, solenoids
Industrial control	1 to 5 h	to 2000 Ah	48 to 120	Circuit-breaker control, motor operators, process control
UPS systems	5 to 15 m	5 to 1500 kW	120 to 520	Inverters
Central emergency lighting systems	1.5 h	to 1000 Ah	120	Lighting fixtures
Self-contained emergency lighting packs	1.5 h	to 100 Ah	6 to 12	Lamps
Security systems	2 to 5 h	to 2500 Ah	84 to 480	Turnstiles, metal detectors, explosive detectors, x-ray machines, card readers
Fire-protection systems	8 h	to 75 Ah	84	Detectors, deluge systems, solenoids
Meteorological stations	24 h	to 100 Ah	84	Instruments
Railroad crossings	periodic	to 250 Ah	2 to 40	Signals, gate motors, lights
Evacuation sirens	short duration	to 100 Ah	12 to 24	Siren, small motor, radio/receiver
Energy storage systems	10 to 40 m	to 40 MW	to 2000	Electric power system

*Parallel strings are often used to ensure availability and provide for ease of maintenance.

TABLE 18.2 Factors to Consider When Selecting and Sizing Stationary Batteries

• New or replacement	• Capacity at the time of installation
• Real estate (room size)	• Wet vs. dry charged (vented cells)
• Load changes	• Flame-retardant jars and covers
• Voltage	• Ventilation
• Duty cycle	• Flame arrestor vents
• Temperature correction	• Ambient temperature
• Design margin	• Mounting requirements
• Aging factor	• Seismic requirements
• Float correction (NiCd)	• Expected life
• Frequency of discharge	• Cycle life
• Depth of discharge	• Available short-circuit current
• Cell/monobloc size and weight	• Method of charging
• Lead-acid vs. nickel-cadmium	• Constant potential
• Single string vs. parallel strings	• Constant current
• Alloy (lead-acid)	• Temperature compensation
• Plate type	• Float correction factor (NiCd)
• Available system voltage window	• Effects of ac ripple
• Maximum voltage	• Current
• Minimum voltage	• Voltage
• Float voltage requirements	• Maintenance requirements
• Recharge/equalize voltage	• Provisions for monitoring
• Number of cells	• Provisions for capacity testing

Source: From Migliaro, *Stationary Battery Workshop—Sizing Module.*

tion, and industrial control. High-performance cells are used with UPS systems, although some battery manufacturers are promoting their use in switchgear tripping applications.

Related Topics

Although there are no calculations associated with selecting a cell type, the selection process is critical to the success of the installation. Each of the factors to be considered, along with the suitability of a specific battery technology, plate type, or plate-alloy, needs to be thoroughly evaluated. In practice, a few cell types would be chosen for evaluation before making the final cell selection. The books, standards, and papers included in the Bibliography should be consulted for additional guidance during the process of cell selection. Battery manufacturers can provide additional information on their cell types.

RATINGS

Stationary batteries are rated in terms of capacity of a cell expressed in *ampere-hours* (Ah) or *Watt-hours* (Wh); however, this rating, when stated, is dependent upon a number of conditions. These are the *discharge rate*, *end-of-discharge voltage* (also referred to as the end or final voltage), *cell temperature* (also referred to as electrolyte temperature), and, for lead-acid cells, the *full-charge density of the electrolyte*. The standard discharge rate for lead-acid stationary batteries is 8 h in North America and 10 h in other parts of the world. There are some exceptions to this. UPS batteries are often rated at the 15-min rate and small batteries for applications such as emergency lighting are rated at the 20-h rate. Nickel-cadmium batteries are generally rated at the 5-h rate; however, the 8- and 10-h rates are sometimes used. The rating of a battery at rates other than its standard rate will vary based on the parameter used to rate the battery. So, for example, if the end voltage is held constant as discharge current is decreased (i.e., the discharge rate in hours is increased), the capacity of the cell will increase. Similarly, as the discharge current is increased (i.e., the discharge rate in hours is decreased), the capacity of the cell will decrease. These are nonlinear, so for example, the 4-h rate of a battery rated at the 8-h rate *is not* simply one-half the 8-h rate.

The standard end-of-discharge voltage used for lead-acid cells is 1.75 *volts per cell* (Vpc) in North America. In other parts of the world, 1.80 or 1.81 Vpc is frequently used. These end voltages typically apply to cells rated at discharge rates equal to or greater than 1 h. The use of these end voltages ensures that the potential for damaging a lead-acid battery due to overdischarge (i.e., caused by overexpansion of the cell's plates) is minimized. For discharge rates less than 1 h, a lower end voltage may be used without causing any damage to the cell's plates. For example, high-performance lead-acid cells rated at the 15-min rate may use 1.67, 1.64, or 1.60 Vpc as an end voltage; however, 1.67 Vpc is becoming the industry standard.

The standard end-of-discharge voltage used for nickel-cadmium cells is 1.0 Vpc, although 1.14 Vpc is sometimes used in North America. Nickel-cadmium cells may be overdischarged without any negative effects to the cell.

The standard for cell temperature is 25°C in North America, Japan, Australia, and New Zealand. In other parts of the world, 20°C is used.

The ratings for lead-acid cells include the density of the electrolyte, since the electrolyte enters into the chemical reactions of the cell. Changing the nominal full-charge electrolyte density will change the performance of the cell. In North America, the standard for electrolyte density for vented cells is 1215 $\pm$ 10 kg/m^3 when measured at 25°C; however, the standard for UPS cells is becoming 1250 $\pm$ 10 kg/m^3 at 25°C. In other parts of the world, the standard electrolyte density for vented cells is 1230 or 1240, $\pm$5 or $\pm$10 kg/m^3 at 20°C (25°C in Japan, Australia, and New Zealand). VRLA cells throughout the world use nominal electrolyte densities of 1260 to 1310, with those below 1280 typically used in gelled-electrolyte cells.

From this discussion, it is apparent that for the purposes of comparison, the stated battery ratings may not be comparable without putting them on the same base. For example, a 1000 Ah cell rated at the 8-h rate is not equivalent to a 1000 Ah battery rated at the 10-h rate.

Rating of a Battery with Cells Connected in Series

Determine the rating of a battery that has 24 lead-acid cells connected in series, if the rating of a single cell is 200 Ah at the 8-h rate to 1.75 Vpc, with a nominal electrolyte density of 1215 at 25°C.

Calculation Procedure

The capacity of cells connected in series remains the same; however, the nominal voltage increases.

1. Compute the Nominal Voltage

Since the nominal voltage of a lead-acid cell is 2.0 Vdc, the battery voltage would be 48 Vdc (i.e., 24 cells $\times$ 2 Vpc), but its capacity would still be 200 Ah at the 8-h rate to 1.75 Vpc, with a nominal electrolyte density of 1215 at 25°C.

Rating of a Battery with Cells Connected in Parallel

Determine the rating of a battery that is made up of two nominal 48-Vdc lead-acid batteries connected in parallel if the rating of a single cell is 200 Ah at the 8-h rate to 1.75 Vpc, with a nominal electrolyte density of 1215 at 25°C.

Calculation Procedure

The capacity of cells connected in parallel is the sum of the capacities of the individual cells that are connected in parallel; however, the nominal voltage of the battery will remain the same. Although it is possible to connect cells of different capacities in parallel, as long as the nominal voltage of the cells is equal, normal design practice for a battery made up of parallel strings is to use cells of the same capacity in parallel.

1. Compute the Nominal Voltage

The nominal capacity of each 48-Vdc battery is 200 Ah at the 8-h rate to 1.75 Vpc, with a nominal electrolyte density of 1215 at 25°C; therefore, two strings connected in parallel would have a capacity of 400 Ah (i.e., 200 Ah + 200 Ah) at the 8-h rate to 1.75 Vpc, with a nominal electrolyte density of 1215 at 25°C. The nominal voltage of the battery would remain 48 Vdc.

C RATE

Based on the discussion on battery ratings, it should be easy to see that simply stating a battery's capacity in Ah or Wh can be confusing. Outside of North America, the capacity of a cell (or battery) is generally expressed in terms of a *C rate*. This is a term that clearly expresses a cell's (or battery's) capacity (at specified references). It is also used to express rates for charge and discharge of the cell (or battery). When used to indicate capacity, the symbol C is followed by a number (normally shown as a subscript) that represents the rate of discharge, in hours. When used to represent charge or discharge current, the current is referenced to a specific cell capacity. It is usually expressed as the symbol C preceded by a number that represents the current as a multiple of the cell (or battery) capacity and followed by a number (normally shown as a subscript) representing the rate on which the current is based.

Express the rate of a lead-acid cell as a C rate, assuming the cell is rated 200 Ah at the 8-h rate to 1.75 Vpc, with a nominal electrolyte density of 1215 at 25°C.

Calculation Procedure

1. Express the Rate as a C Rate

The C rate is simply the letter C followed by the discharge rate, expressed in hours, as a subscript, or C_8.

Related Calculations. The 10-h, 3-h, 1-h, 30-min, and 15-min capacities for the same cell could be expressed as C_{10}, C_3, C_1, $C_{0.5}$, and $C_{0.25}$, respectively. Similarly, a nickel-cadmium cell rated 38 Ah at the 5-h rate to 1.0 Vdc at 25°C would be expressed as C_5.

The C_{10} rating of a battery is 60 Ah at the 10-h rate to 1.75 Vpc, with a nominal electrolyte density of 1215 at 25°C. The battery is capable of delivering 6 A for 10 h at rated conditions. Express this discharge rate as a function of the C rate.

Calculation Procedure

1. Express the Current as a Multiple of the C_{10} Rate

This is done by dividing the current by the C_{10} capacity in Ah, or 6/60 = 0.1.

2. Express the Rate as a C Rate

0.1 C_{10}.

Related Calculations. If the same cell was to be discharged at 162 A or 42 A, these discharge rates could be expressed as 2.7 C_{10} and 0.7 C_{10}, respectively.

Determine the charging current in A if a battery manufacturer states that the charging current should be limited to 0.25 C_8, and the C_8 rate, at standard conditions, for the cell to be charged is 96 Ah.

Calculation Procedure

1. Use the C Rate as Expressed to Determine the Charging Current Limit

$0.25\ C_8 = 0.25 \times 96 = 24$ Adc.

CALCULATING BATTERY END VOLTAGE FOR A DC SYSTEM

Although this might seem straightforward, since the standard ratings for individual cells include an end voltage, the requirements of a dc system often dictate the use of other than the standard end voltage. The *voltage window* (i.e., the voltages from the minimum to maximum system voltage) for applications varies. Telecommunications applications often have a narrow voltage window due to the limitation of acceptable equipment voltages; however, this is changing with new equipment that can tolerate lower input voltages being introduced in the market. Electric utility and industrial control-voltage windows generally use the standard cell end voltage but are sometimes limited by the minimum system-voltage limit. UPS applications can offer the largest voltage window, since the inverter input can be designed to accept a wide range of voltage without affecting the inverter's output voltage.

End voltage is important in battery sizing because the lower the end voltage, the more energy that can be removed from the battery. This results in a smaller battery for the application. Conversely, higher end voltages increase the size of the battery needed to perform the desired design function.

When determining end voltage at the battery terminals for the purpose of sizing the cell, do not forget to factor in the voltage drop to the loads. This may require an analysis of each load to determine which represents the worst case.

Determine the cell end voltage for a 24-cell, 48-Vdc telecommunications battery if the minimum system voltage is 45 V.

Calculation Procedure

1. Calculate the End Voltage per Cell Using the Minimum System Voltage

45 Vdc/24 cells = 1.88 Vpc

Determine the cell end voltage for a 60-cell, 120-Vdc substation battery if the minimum system voltage is 105 V.

Calculation Procedure

1. Calculate the End Voltage per Cell Using the Minimum System Voltage

105 Vdc/60 cells = 1.75 Vpc

Related Calculations. If the number of cells in the battery had been 58 due to limitations on maximum system voltage during recharge, then the end voltage per cell would have been 1.81 Vpc.

SIZING METHODS—CONSTANT CURRENT

Two methods are used to size stationary batteries, constant current and constant power. Constant current is used for applications such as generating stations, substations, and industrial control. This method is particularly useful when loads are switched on and off at various times during the battery discharge. When using this method, the current of each load at the nominal battery voltage is determined, and it is assumed to remain constant over the period the load is on the battery. All loads on the battery are normally tabulated and a load profile of current vs. time is drawn. A more detailed discussion of constant current along with a sizing worksheet is included in IEEE Standard 485-1997.

SIZING METHODS—CONSTANT POWER

Constant-power sizing is used almost exclusively for UPS system applications. It may also be used for telecommunications applications; however, other methods, including constant current and those involving the number of busy hours of an exchange, may be used for this application. Unlike constant-current loads, the current of a constant-power load (e.g., an inverter or dc motor) increases as the battery terminal voltage decreases during a discharge, so that $P = E \times I$ remains constant. Load profiles are not normally drawn for constant-power sizing since the load, once applied, remains the same throughout the battery discharge. A more detailed discussion of constant-power sizing is included in IEEE Standard 1184-1994.

DETERMINING THE CURRENT OF A CONSTANT-POWER LOAD

There are instances where a constant-power load must be included when constant-current sizing is used. This may occur when a constant-power load such as an inverter is one load on a dc system (e.g., at a generating station). When determining the current to be used for sizing, it is calculated based on the average voltage during the time the constant-power load is on the battery. In each case, the voltage at the start of the discharge is taken to be the open-circuit voltage of the cell (or battery). For lead-acid cells, this may be taken as 2.0 Vdc, or it may be calculated since it varies with cell electrolyte specific gravity. The open-circuit voltage of nickel-cadmium cells is taken as 1.2 Vdc.

Determine the current to be used for constant-current sizing for a constant-power load rated 10 kW that will be on a 116-cell, lead-acid battery with a nominal 1215 density electrolyte for the entire duty cycle. The end voltage is 1.81 Vpc.

Calculation Procedure

1. Determine the Open-Circuit Voltage at the Start of Discharge

The open-circuit voltage of a lead-acid cell can be approximated by the equation $E_{oc} = 0.84 +$ specific gravity. Thus, $E_{oc} = 0.84 + (1215/1000) = 2.055$, or 2.05 Vdc.

2. Determine the Average Voltage per Cell During Discharge

Average voltage over the discharge $= (2.05 + 1.81)/2 = 1.93$ Vpc.

3. Determine the Average Battery Voltage

Average battery voltage $=$ 1.93 Vpc $\times$ 116 cells $=$ 223.9 Vdc.

4. Compute the Current at the Average Voltage

$I =$ 10,000 W/223.9 V $=$ 44.7 Adc

Related Calculations. When making calculations, make certain that the dc input kW is used. For example, an inverter may be rated 100 kVA at 0.8 power factor (pf) and 0.91 efficiency, which is its output rating. The dc input kW required to be used for sizing would be kVA $\times$ pf/efficiency, or 87.9 kW.

In lieu of calculating the current based on the average voltage, a more conservative approach is sometimes taken: calculating the current based on the minimum voltage during discharge. For the example above, this would result in a current of $10{,}000/(1.81 \times 116) = 47.6$ Adc.

NUMBER OF POSITIVE PLATES

When sizing batteries, some data are based on one positive plate and other data are based on one cell. It will become necessary to convert these data; therefore, the number of positive plates in a cell must be determined. Lead-acid battery manufacturers normally include the total number of plates in a cell on the cell catalog sheet or in its model number (e.g., an XYZ-17 would have 17 plates). Lead-acid batteries typically have one more negative plate than positive plate, while nickel-cadmium cells may have an equal number of plates.

Determine the number of positive plates in a model number XYZ-25 lead-acid cell.

Calculation Procedure

1. Determine the Number of Positive Plates by Applying the Formula

No. of positive plates $=$ (number of plates $-$ 1)/2 $= (25 - 1)/2 = 12$

DISCHARGE CHARACTERISTICS

Battery manufacturers publish typical discharge data for the cells they manufacture. The cell discharge rates are expressed in amperes, ampere-hours, or watts for various discharge times to various end voltages at the standard cell temperature. Most often, the data is a part of the catalog cut for the cell and may be in the form of a table or curve. Data in tables are typically for a single cell at the discharge times commonly used in applications the cell is designed for, while cell discharge curves cover all discharge times. In addition to discharge data, the cell catalog cuts will normally include information such as plate alloy, cell dimensions and weights, nominal full-charge specific gravity, rack selection data, jar and cover materials, and expected life.

When sizing a battery using constant current, the sizing may be based on positive plates or ampere-hours. Sizing using positive plates requires the use of a capacity rating factor, R_T, and sizing using Ah uses a capacity rating factor of K_T. In North America, lead-acid battery manufacturers typically publish values of R_T, and nickel-cadmium manufacturers publish values of K_T.

Discharge curves are developed by a battery manufacturer by discharging a cell(s) at various currents and measuring the voltage at a number of points during the discharge. These data are plotted as cell voltage vs. time at each of the discharge currents. A typical plot for a lead-acid cell type is shown in Fig. 18.1. Note the initial voltage dip as the cell is discharged, particularly at higher discharge currents. This phenomenon in lead-acid cells is known as the *coup de fouet*. (There is no similar voltage dip in nickel-cadmium cells.) Once these data are obtained, they are expressed on a common base of amperes per positive plate. In addition, for some discharge curves, the ampere-hours removed from the cell at various times are also computed and expressed on a base of Ah per positive plate. These data are then plotted as a curve known as a discharge curve. Two commonly used curves are the *fan curve* and an *S curve*. Although these curves look very different, either can be used to determine battery characteristics.

All of the data on a discharge curve is based on the standard cell voltage and, in the case of a lead-acid cell, the nominal electrolyte density. Discharge data at other cell temperatures can be calculated using temperature correction factors in industry standards or obtained from the battery manufacturer.

USING FAN CURVES

Figure 18.2 is an example of a fan curve. The curve is for a single cell. The x-axis is labeled amperes per positive plate, while the y-axis is labeled ampere-hours per positive plate. There are a number of radial lines from the origin that represent discharge times (in minutes or hours) and curves for various end-of-discharge voltages. At the top of the curve, there is a line labeled "initial voltage." Fan curves can be used for many purposes, including determination of the capacity rating factor, R_T, for constant-current sizing. Fan curves are also available for ratings in watts or kW (see Fig. 18.3).

Determine the capacity rating factor, R_T, for the 60-min rate to 1.75 Vpc at 25°C, for the cell type shown in Fig. 18.2.

Calculation Procedure

1. Graphically Determine Where the 60-Min Line Intersects the 1.75 Vpc Line

Referring to Fig. 18.4, find the point at which the radial line for 60 min intersects the curve for 1.75 FV. This is indicated on the figure as point *A*.

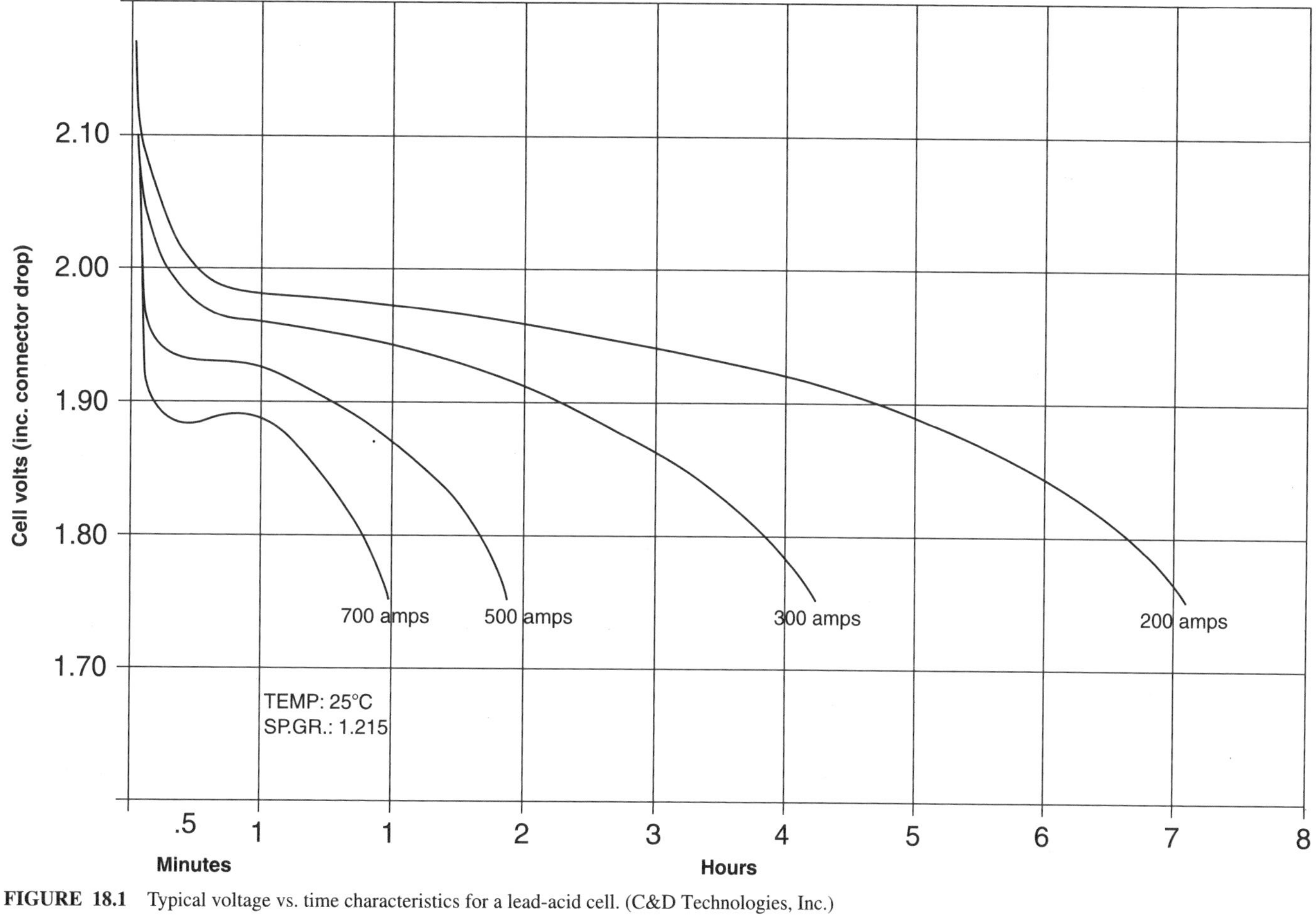

FIGURE 18.1 Typical voltage vs. time characteristics for a lead-acid cell. (C&D Technologies, Inc.)

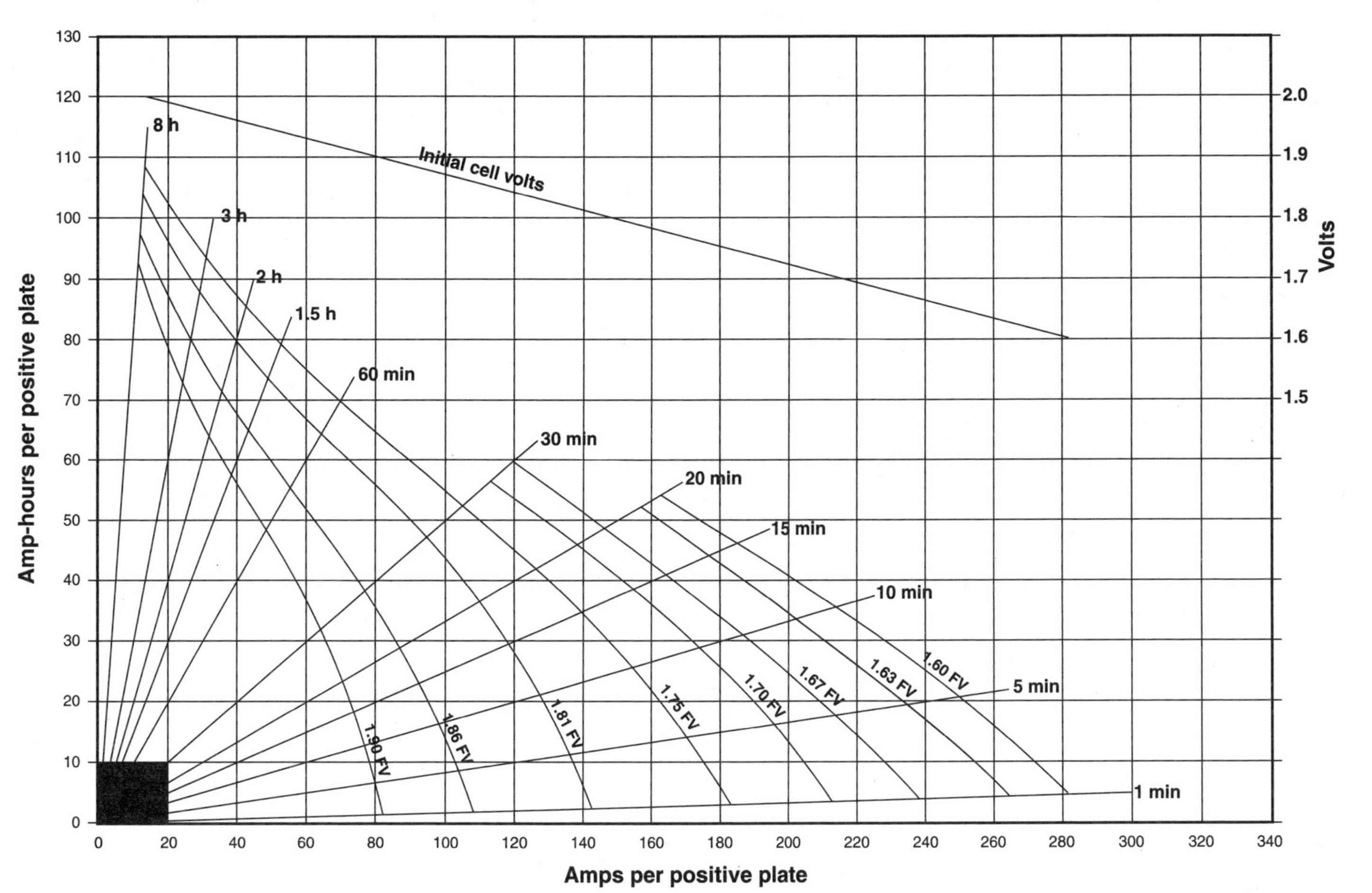

FIGURE 18.2 Typical fan curve—amperes. (C&D Technologies, Inc.)

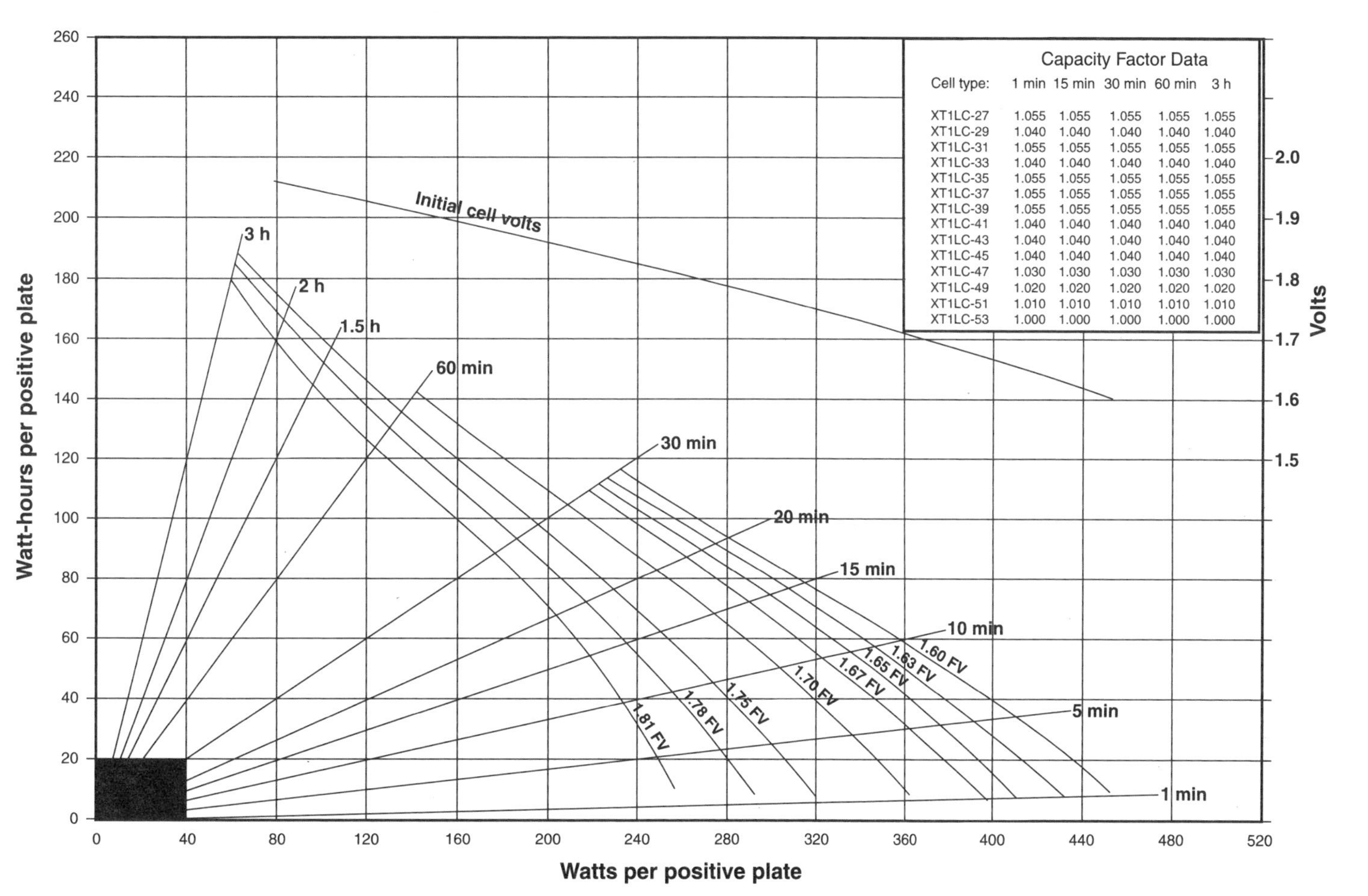

Cell type:	1 min	15 min	30 min	60 min	3 h
XT1LC-27	1.055	1.055	1.055	1.055	1.055
XT1LC-29	1.040	1.040	1.040	1.040	1.040
XT1LC-31	1.055	1.055	1.055	1.055	1.055
XT1LC-33	1.040	1.040	1.040	1.040	1.040
XT1LC-35	1.055	1.055	1.055	1.055	1.055
XT1LC-37	1.055	1.055	1.055	1.055	1.055
XT1LC-39	1.055	1.055	1.055	1.055	1.055
XT1LC-41	1.040	1.040	1.040	1.040	1.040
XT1LC-43	1.040	1.040	1.040	1.040	1.040
XT1LC-45	1.040	1.040	1.040	1.040	1.040
XT1LC-47	1.030	1.030	1.030	1.030	1.030
XT1LC-49	1.020	1.020	1.020	1.020	1.020
XT1LC-51	1.010	1.010	1.010	1.010	1.010
XT1LC-53	1.000	1.000	1.000	1.000	1.000

FIGURE 18.3 Typical fan curve—watts. (C&D Technologies, Inc.)

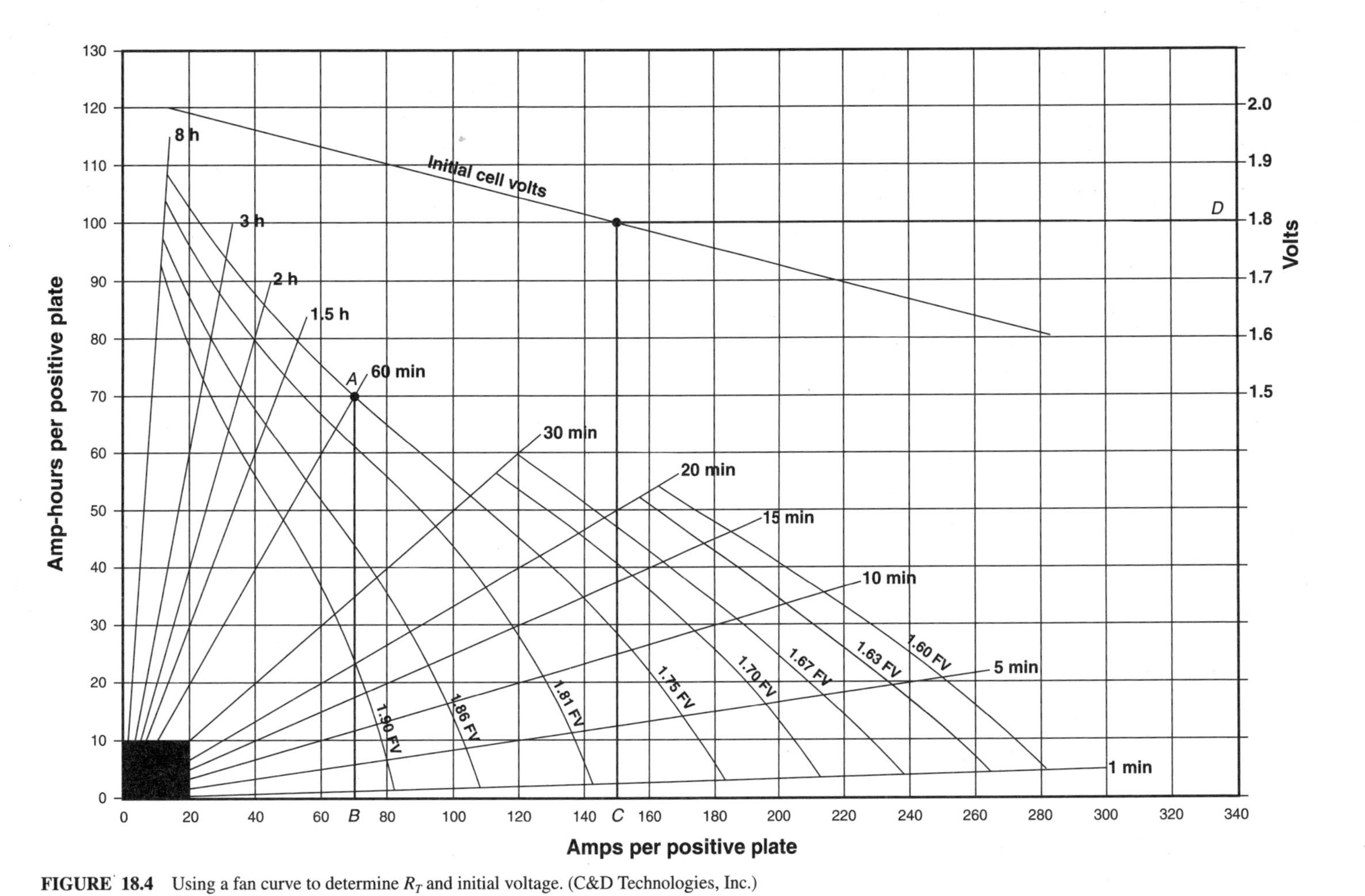

FIGURE 18.4 Using a fan curve to determine R_T and initial voltage. (C&D Technologies, Inc.)

2. Determine the $\mathbf{R_T}$ ***in Amperes/Positive Plate***

From point *A*, drop a vertical line that intersects the *x*-axis and read the R_T. This is identified as point *B* and is 70 amperes/positive plate.

Related Calculations. If this cell had 9 plates, then the discharge capability of one cell to 1.75 V at 25°C would be 4 positive plates × 70 amperes/positive plate = 280 Adc.

The fan curve may also be used to determine the voltage profile of a battery during discharge and data pertaining to testing a battery. Details on the process may be found in industry standards (e.g., IEEE Standard 485-1997).

Determine the initial voltage of the cell used in the previous example if a load of 600 A was placed on it.

1. Determine the Current per Positive Plate

Current per positive plate = 600 A/4 positive plates = 150 A/positive plate.

2. Graphically Determine Initial Voltage

Referring to Fig. 18.4, find 150 A/positive on the *x*-axis and draw a vertical line up to where it intersects the initial voltage line. This is shown as point *C*. Draw a horizontal line to the right-hand *y*-axis and read the initial voltage, which is at point *D* and is equal to 1.8 Vdc. (Note: This is the initial cell voltage even though the data was in A/positive plate. Although there are 4 positive plates in the cell, they are in parallel so the voltage of a single plate and of the cell would be the same.)

USING S CURVES

Figure 18.5 is an example of an S curve. It too is for a single cell and can be used for many purposes, including determination of R_T capacity rating factors for constant-current sizing. The curve has two distinct sections. The lower section (which is log-log) shows the cell discharge characteristics for various discharge times (in minutes) to various end-of-discharge (or final) voltages in relation to amperes per positive plate. The upper section (the *x*-axis is logarithmic) shows cell voltages during discharge in relation to amperes per positive plate.

Determine the capacity rating factor, R_T, for the 10-min rate to 1.84 Vpc at 25°C, for the cell type shown in Fig. 18.5.

Calculation Procedure

1. Graphically Determine Where the 10-Min Line Intersects the 1. 84 Vpc Line

Referring to Fig. 18.6, draw a horizontal line from 10 min on the *y*-axis until it intersects the 1.84 curve. This is indicated on the figure as point *A*.

2. Determine the $\mathbf{R_T}$ ***in Amperes/Positive Plate***

From point *A*, drop a vertical line that intersects the *x*-axis and read the R_T. This is identified as point *B* and is 65 amperes/positive plate.

Related Calculations. The S curve can be used to determine initial voltage in a manner similar to that for the fan curve. It can essentially be used to determine any data that can be determined with a fan curve.

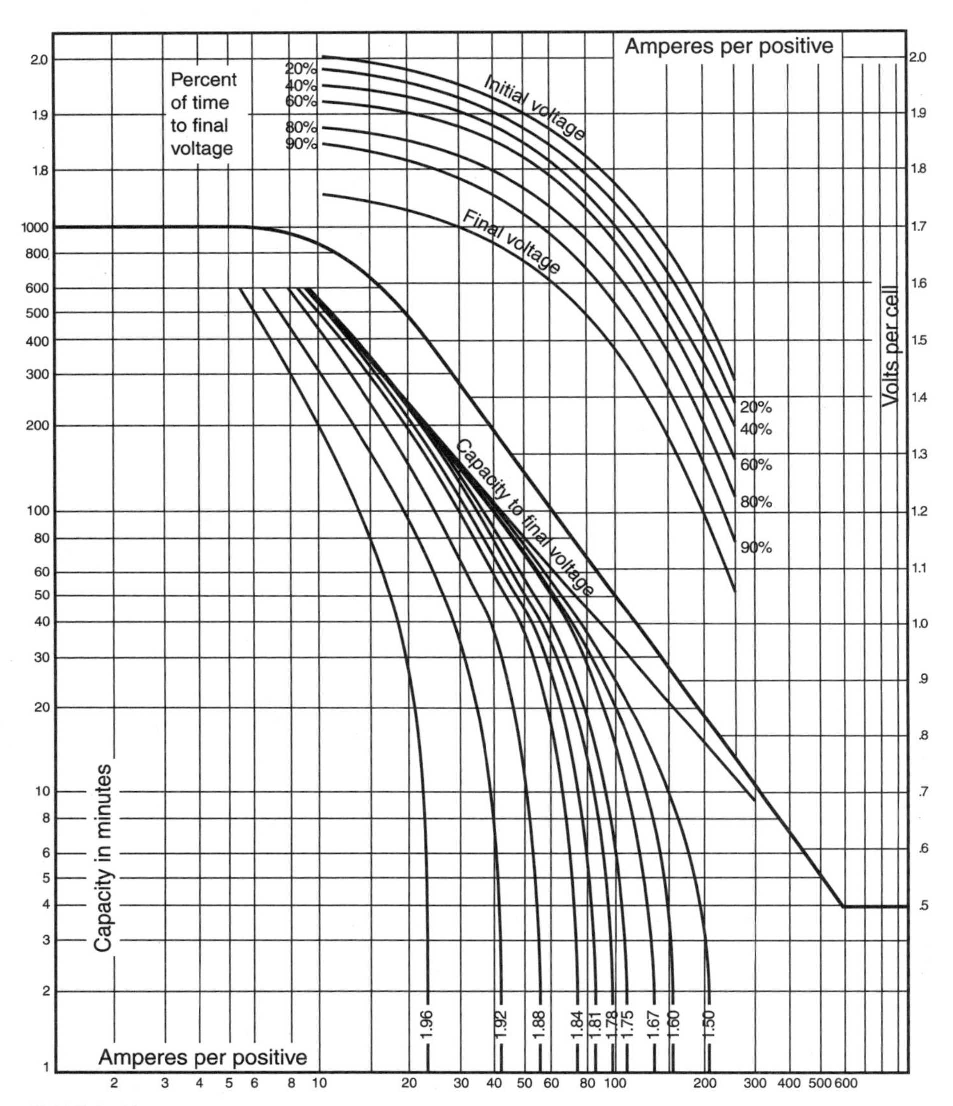

FIGURE 18.5 Typical S curve. (Yuasa, Inc.)

As a reminder: when interpolating between values on the S curve, the axes are a logarithmic scale.

USING K FACTORS

If ampere-hour sizing will be used, the K_T capacity rating factors for the cell must be obtained. Some manufacturers have tables or curves (Fig. 18.7) available; however, they may be calculated directly from the cell rating and discharge data, using the following formula $K_T = C/I_T$, where C is the rated capacity of the cell in Ah, I is the discharge current, and T is the discharge time period.

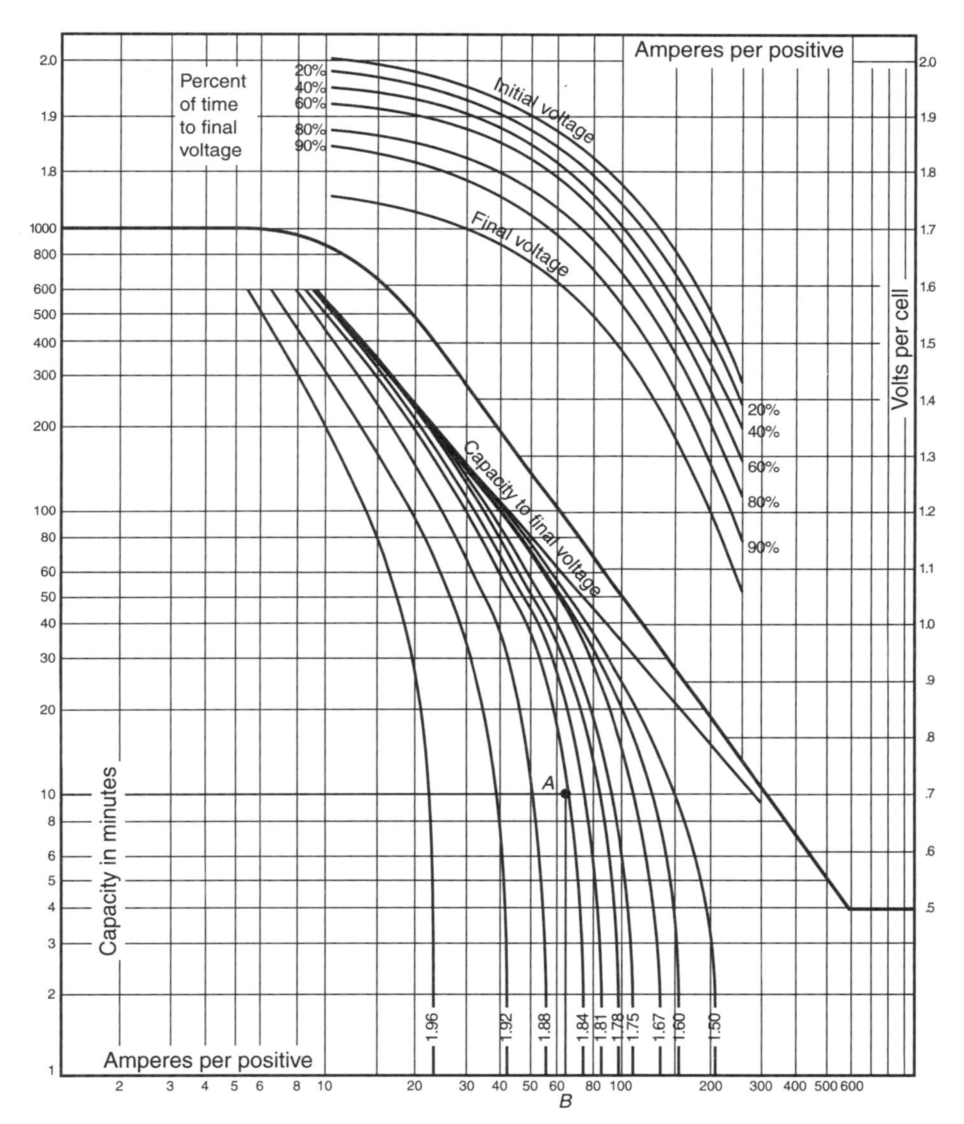

FIGURE 18.6 Using an S curve to determine R_T. (Yuasa, Inc.)

Determine the K_T capacity rating factor for the 3-hour discharge rate to 1.75 Vpc, at 25°C, for a cell that is rated 410 Ah, C_8. The 3-h discharge current is 108.7 Adc to 1.75 V.

Calculation Procedure

1. Determine $\mathbf{K_T}$ ***by Applying the Formula***

$K_T = 410/108.7 = 3.77.$

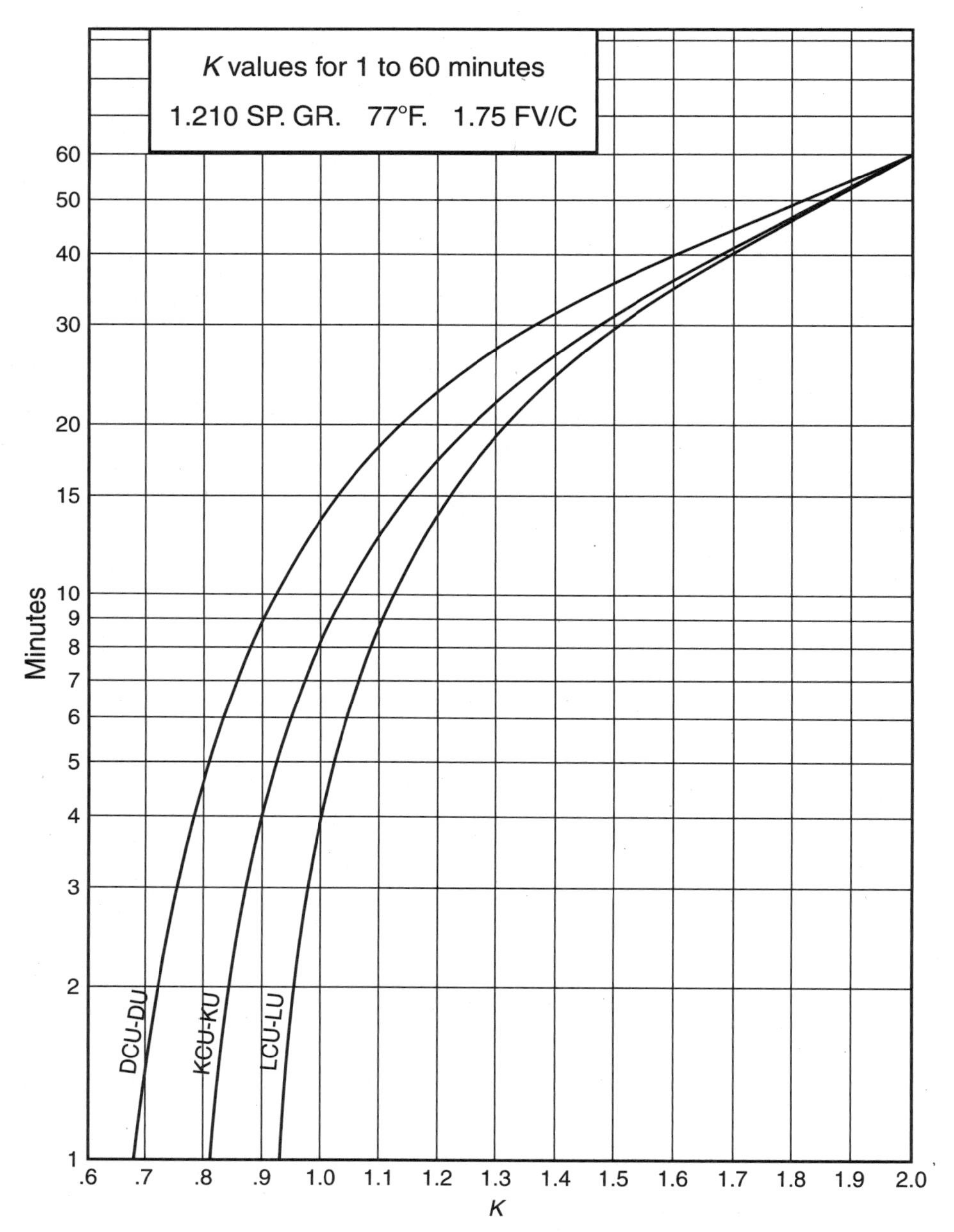

FIGURE 18.7 Typical K_T capacity rating factors. (C&D Technologies, Inc.)

Related Calculations. If the discharge currents for the same cell for 1 min, 30 min, 1 h, and 8 h to 1.75 Vpc were 500, 285, 204, and 51.25 A, respectively, then the corresponding K_T for each, in the same order, would be 0.82, 1.44, 2.01, and 8.00.

VOLTAGE DEPRESSION—NICKEL-CADMIUM ON FLOAT CHARGE

IEC 60622-1988 and IEC 60623-1990 rate nickel-cadmium cells. In the mid-1980s, testing conducted in North America found differences in published vs. tested capacities. A study revealed that unlike nickel-cadmium batteries in other parts of the world where constant-current charging is used, batteries in North America were being float-charged at constant potential. The effect of a prolonged constant potential charge on nickel-cadmium batteries is to cause a lowering of their average voltage on discharge. This voltage depression must be accounted for when sizing if the battery will be constant potential charged and the battery manufacturer has not corrected its published capacity rating factors (i.e., R_T or K_T) to include the effect of voltage depression.

Related Calculations. If the published capacity rating factors for a nickel-cadmium cell that will be float-charged have not been corrected by the manufacturer, they must be multiplied by a correction factor, normally available in a table published by the manufacturer. For example, if the published 60-min K_T for a cell to 1.0 Vpc is 1.44, based on constant-current charging, and the constant potential correction factor is 0.93, the K_T, corrected for use in sizing the battery, would be $1.44 \times 0.93 = 1.34$.

NUMBER OF CELLS FOR A 48-VOLT SYSTEM

Determine the number of lead-acid cells required of a battery for a nominal 48-Vdc system (42-Vdc minimum to 56-Vdc maximum).

Calculation Procedure

1. Compute Number of Cells

The nominal voltage of a lead-acid cell is 2.0 Vdc; therefore, the number of cells = 48/2 = 24 cells.

2. Check the Minimum Voltage Limit

Minimum volts/cell = (min. volts)/(number of cells) = 42 V/24 cells = 1.75 V/cell. This is an accepted end-of-discharge voltage for a lead-acid cell.

3. Check Maximum Voltage Limit

Maximum volts/cell = (max. volts)/(number of cells) = 56 V/24 cells = 2.33 V/cell. This is an acceptable maximum voltage for a lead-acid cell. Therefore, select 24 cells of the lead-acid type.

NUMBER OF CELLS FOR 125- AND 250-VOLT SYSTEMS

Determine the number of lead-acid cells required of a battery for a nominal 125-Vdc system (105 = Vdc minimum to 140 = Vdc maximum) and for a nominal 250-Vdc system (210 = Vdc minimum to 280 = Vdc maximum).

Calculation Procedure

1. Compute Number of Cells

If 1.75 V/cell is the minimum voltage of the 125-Vdc system, the number of cells = min. volts/(min. volts/cell) = 105/1.75 = 60 cells.

2. Check Maximum Voltage

Using 2.33 V/cell, the maximum voltage = (number of cells)(max. volts/cell) = (60)(2.33) = 140 Vdc. Therefore, for a 125-Vdc system, select 60 cells of the lead-acid type.

3. Calculate Number of Cells for 250-V System

Number of cells = 210 V/(1.75/cell) = 120 cells.

4. Check Maximum Voltage

Max. voltage = (number of cells)(max. volts/cell) = (120)(2.33) = 280 Vdc. Therefore, for a 250-Vdc system, select 120 cells of the lead-acid type.

SELECTING NICKEL-CADMIUM CELLS

Select the number of nickel-cadmium (NiCd) cells required for a 125-Vdc system with limits of 105 to 140 Vdc. Assume minimum voltage per cell is 1.14 Vdc for the NiCd cell.

Calculation Procedure

1. Determine Number of Cells

Number of cells = min. volts/(min. V/cell) = 105/1.14 = 92.1 cells (use 92 cells).

2. Check Maximum Voltage per Cell

Max. volts/cell = max. volts/number of cells = 140/92 = 1.52 V/cell. This is an acceptable value for a NiCd cell.

LOAD PROFILES

Determine the worst-case load profile of a dc system for a generating station consisting of a 125-Vdc nominal, stationary-type, lead-acid battery and a constant-voltage charger in full-float operation. The loads on the system and their load classification are as follows:

Load description	Rating	Classification
Emergency oil pump motor	10 kW	Noncontinuous*
Controls	3 kW	Continuous
Two inverters (each 5 kW)	10 kW	Continuous
Emergency lighting	5 kW	Noncontinuous
Breaker tripping (20 at 5 A)	100 A	Momentary (l-s duration)†

*It will be assumed here that the equipment manufacturer requires the emergency oil pump to run continuously for 45 min following a unit trip.

†Occurs immediately following unit trip.

Calculation Procedure

1. Determine Load Conditions for Battery

The first step in determining the worst-case load profile is to develop the conditions under which the battery is required to serve the dc system load. These conditions will vary according to the specific design criteria used for the plant. Load profiles will be determined for the following three conditions:

a. Supply of the emergency oil pump for 3 h (with charger supplying continuous load).
b. Supply of the dc system for 1 h upon charger failure.
c. Supply of the dc system for 1 h following a plant trip concurrent with loss of the auxiliary system's ac supply.

2. Develop Load List for Each Condition

A load list for each condition is required to determine the time duration of each load. Once this is done, the load profile may be plotted for each condition.

Condition a. The load list is as follows:

Load	Current	Duration
Emergency oil pump (inrush current)	250 A	0–1 min*
Emergency oil pump (full-load current)	80 A	1–180 min

*Although the inrush current lasts for a fraction of a second, it is customary to use a duration of 1 min for lead-acid batteries because the instantaneous voltage drop for the battery for the period of inrush is the same as the voltage drop after 1 min.

The load profile is plotted in Fig. 18.8.

Condition b. The load list is as follows:

Load	Current	Duration
Controls	24 A	0–60 min
Inverters (two)	80 A	0–60 min

The load profile is shown in Fig. 18.9.

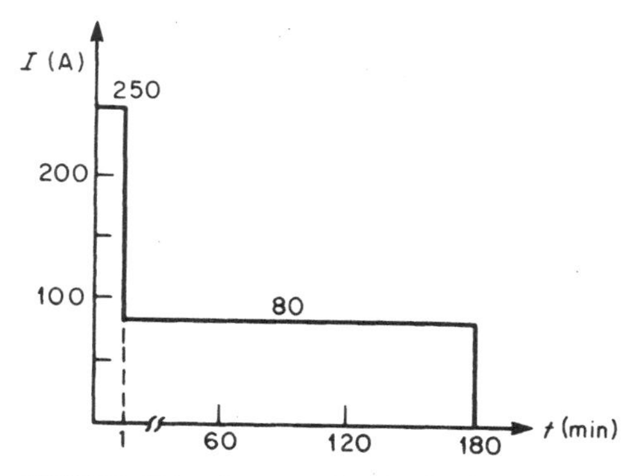

FIGURE 18.8 An example of a load profile for a battery.

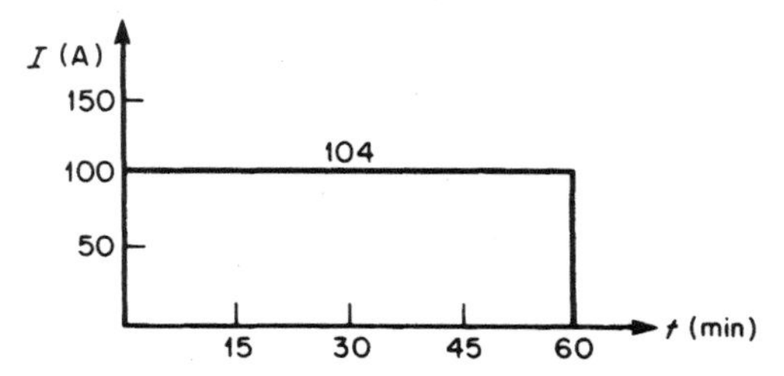

FIGURE 18.9 Load profile for controls and two inverters.

Condition c. The load list is as follows:

Load	Current	Duration
Circuit-breaker tripping (20)	100 A	0–1 min*
Emergency oil pump (inrush)	250	0–1 min*
Emergency oil pump (full load)	80	1–45 min
Controls	24	0–60 min
Inverters (two)	80	0–60 min
Emergency lighting	40	0–60 min

*As stated previously for Condition *a*, the duration of 1 min is used for lead-acid cells even though the load lasts for a fraction of a second. *Note:* It is assumed that breaker tripping and emergency oil pump inrush occur simultaneously.

The load profile is plotted in Fig. 18.10.

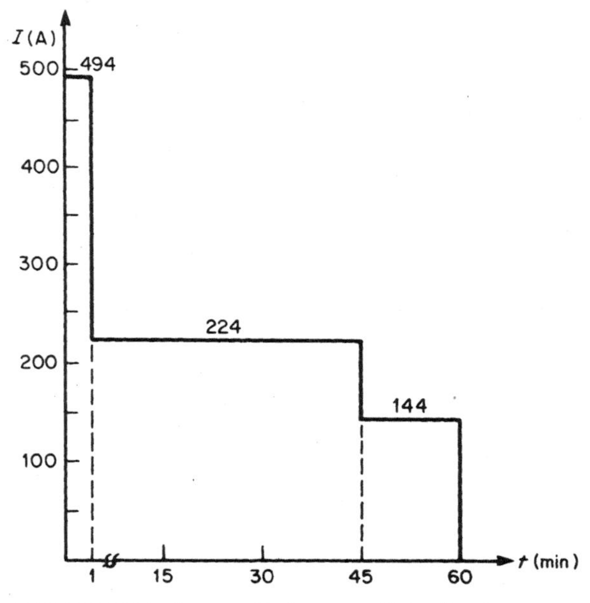

FIGURE 18.10 Load profile for various kinds of equipment.

Related Calculations. Typically, a battery may have a number of load profiles (i.e., one for each set of specified conditions); for sizing the battery, however, the worst-case profile must be used. In some instances, the worst-case profile is apparent and may be used directly for battery sizing. In other cases, each profile should be used to calculate a battery size.

PROFILE FOR RANDOM LOAD

Determine the load profile of a lead-acid cell for the following load list, which includes a random load (i.e., a load that may be imposed on the battery at any time during the duty cycle).

Load	Current, A	Duration
Circuit-breaker tripping (30)	150	0–1 s
Control	15	0–180 min
Fire-protection components	10	0–180 min
Emergency lighting	30	0–180 min
Sequence-of-events recorder	8	0–60 min
Oscillograph	17	0–1 min
	9	1–60 min
Emergency oil pump (inrush)	88	0–1 s
(full load)	25	1–15 min (cycle repeats at the 60th and 120th minute)
Random load	45	1 min, occurring at any time from 0–180 min

Calculation Procedure

1. Develop Method for Constructing Load Profile

The load profile can be constructed in a manner similar to that presented in the previous example, except that the random load must be considered separately. Because it is not known when the load occurs, the normal procedure is to develop a load profile without considering the random load. The battery is then sized on the basis of the profile and the effect of the random load is added to the portion of the load profile that is found to control the battery size.

2. Consider the First Minute

If the first minute of the load profile is examined separately, discrete loads may be identified to develop the 1-min profile, as shown in Figs. 18.11 and 18.12. The load selected for the first minute for the final sizing is the maximum load that occurs at any point in the 60-s period.

3. Construct Complete Load Profile

The complete load profile is shown in Fig. 18.13.

Related Calculations. The load profile of Fig. 18.13 could be modified for use with NiCd cells by recognizing the availability of discharge rates under 1 min.

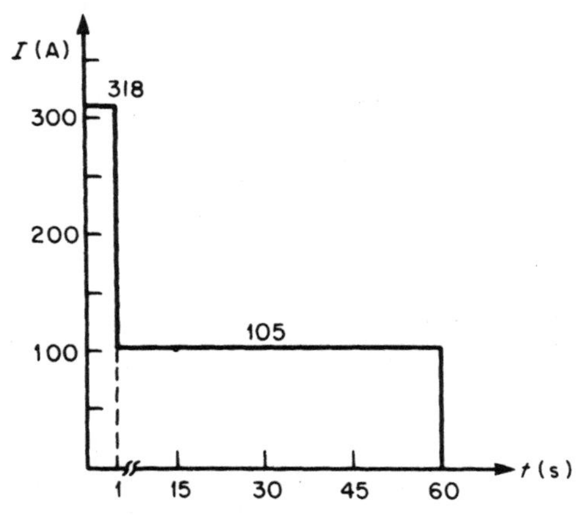

FIGURE 18.11 Dealing with a random load.

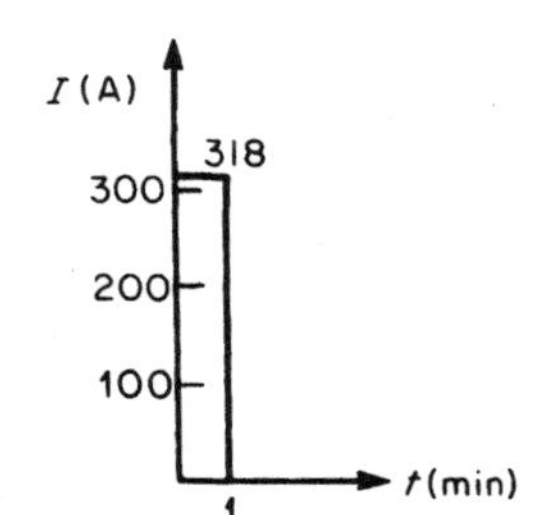

FIGURE 18.12 Load profile for the first minute.

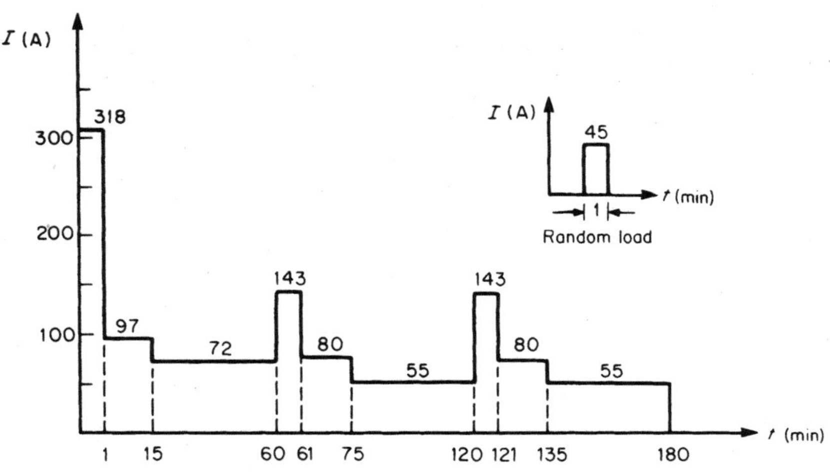

FIGURE 18.13 Complete load profile.

LOADS OCCURRING DURING FIRST MINUTE

Each of the following loads occur during the first minute; however, a discrete sequence of the loads cannot be established. Draw the load profile for the first minute, assuming a lead-acid cell is used.

Load	Current, A	Duration
Breaker closing	40	1 s
Motor inrush	110	1 s
Control	15	1 min
Miscellaneous	30	30 s

Calculation Procedure

1. Develop Method for Drawing Load Profile

Because a discrete sequence cannot be established, the common practice is to assume that all of the loads occur simultaneously.

2. Draw Load Profile

The load profile for the first minute is drawn in Fig. 18.14.

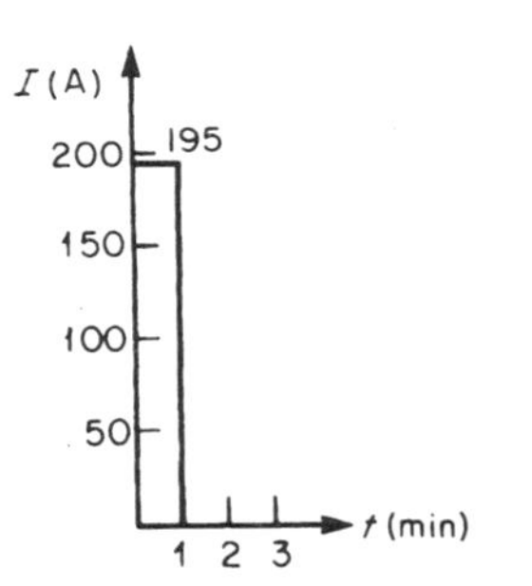

FIGURE 18.14 Load profile for the first minute.

Related Calculations. If NiCd cells are used, manufacturers publish rates for discharge times less than 1 min. Since these rates exist, assumptions can be made that will enable a number of profiles to be developed that may lead to a reduction in battery size, while still being able to support the load requirements. For example, one could assume all loads occur simultaneously for 1 s, followed by control and miscellaneous loads for another 29 s, followed by control load for 30 s. It can similarly be reasoned that the control load could occur for 30 s, followed by control and miscellaneous loads for 1 s.

These possibilities are illustrated in Figs. 18.15 and 18.16. Each profile is then analyzed to determine which represents the worst case. The reader will find that determining

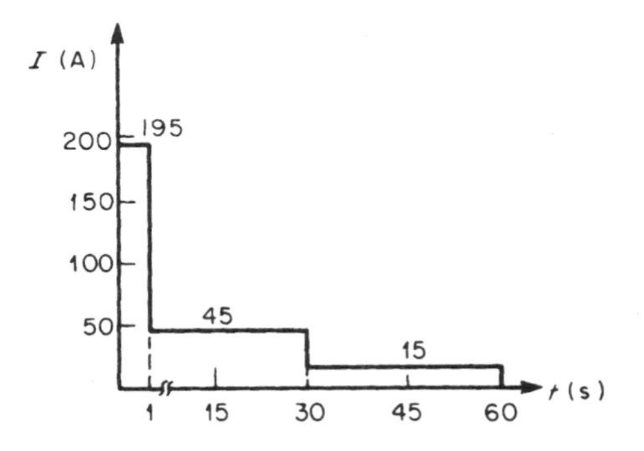

FIGURE 18.15 A load profile to be analyzed for worst-case conditions.

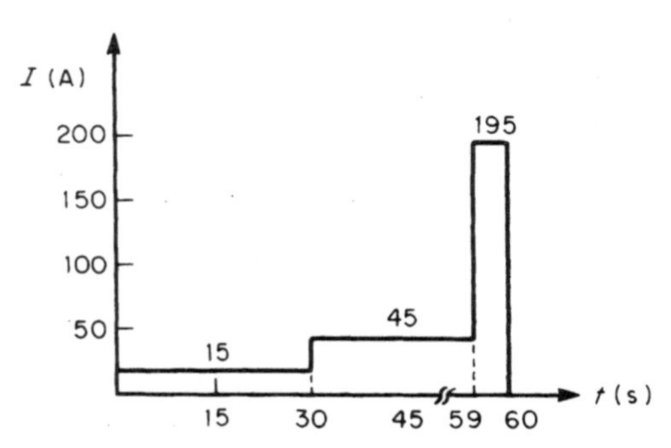

FIGURE 18.16 Another load profile to be analyzed for worst-case conditions.

which profiles represent the worst case will be intuitive, once a number of battery-sizing calculations have been performed.

SIZING BATTERY FOR SINGLE-LOAD PROFILE

Given the load profile of Fig. 18.17, calculate the number of positive plates of lead-acid cell type X required to supply the load. Assume: design margin = 10 percent, lowest electrolyte temperature is 10°C (50°F), 125-Vdc system, 60 cells, 105-Vdc minimum (i.e., 1.75-V/cell end-of-discharge voltage), and age factor = 25 percent. (Note: for these calculations, two decimal places provide sufficient accuracy.)

Calculation Procedure

1. Calculate Uncorrected Cell Size

Use*:

$$\text{Cell size (positive plates)} = \max_{S=1}^{S=N} \sum_{P=1}^{P=S} \frac{A_P - A_{(P-1)}}{R_T}$$

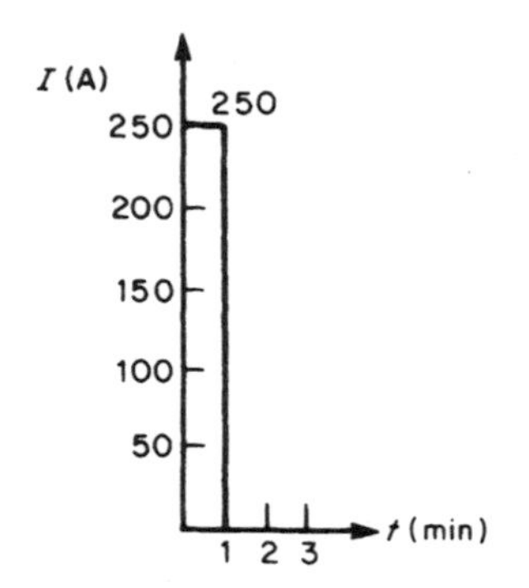

FIGURE 18.17 Sizing battery for shown profile.

where S = section of load profile being analyzed, N = number of periods in the load profile, P = period being analyzed, A_P = amperes required for period P (note that $A_0 = 0$), T = time in minutes (may also be seconds for NiCd cells) from the beginning of period P through the end of section S, and R_T = capacity rating factor representing the number of amperes that each positive plate can supply for T minutes at 25°C (77°F) to a specified end-of-discharge voltage.

*Formula from IEEE Standard 485-1997.

Because there is only one section in the load profile, only one calculation need be made: number of positive plates (uncorrected) = $(A_1 - A_0)/R_T$. Because $T = 1$ min, R_T can be found for $T = 1$ min to 1.75 V/cell from manufacturer's data. Assume $R_T = 75$ A per positive plate; hence, number of positive plates (uncorrected) = (250 − 0)/75 = 3.33.

2. Determine Required Size

Required size = (uncorrected size)(temperature correction factor)(design margin)(age factor). From IEEE 485-1997, the temperature correction for 10°C (50°F) = 1.19. Therefore, required size = (3.33)(1.19)(1.10)(1.25) = 5.45 positive plates. Because it is impossible to get a fraction of a positive plate, it is normal practice to round the answer to the next-higher whole number. Thus, for this example, 6 positive plates of cell type *X* would be required.

Related Calculations. In determining the size of a battery for a specific application, adequate margin should be included for load growth of the dc system. Typically, a generating station battery may be sized a number of times before it is purchased (e.g., conceptual sizing, followed by periodic resizing as load requirements are firmed up during the generating station design) followed by final sizing checks before it is placed in service. Each of these calculations require that a design margin be included; however, the margin will vary according to the type of calculation. For example, a design margin for a conceptual sizing might be 25 to 50 percent, but for the sizing calculation for purchasing the battery it might be only 10 to 20 percent. The design margin to be included is, therefore, dependent upon the specific battery installation. (IEEE Standard 485-1997 recommends the use of 10 to 15 percent design margin as a minimum.)

As another example, consider a distribution substation with a single 138-kV line in and two 12-kV feeders out. If there were plans to expand the station to include an additional 138-kV line and four more 12-kV feeders within 5 yr, the battery would most likely be sized with enough design margin to carry the future load. However, if the expansion was not to take place for 15 years, the design margin might not include the future loads if, by economic analysis, it was determined to be more economical to size only for present load requirements and replace the battery with a larger one at the time the future load is added.

A similar calculation could be performed for a nickel-cadmium cell. IEEE Standard 1115-1992 provides additional information. Temperature correction factors and float-charging correction factors (if applicable) should be obtained from the battery manufacturer.

SIZING BATTERY FOR MULTIPLE-LOAD PROFILE

Calculate the number of positive plates of a lead-acid cell type required to supply the load of Fig. 18.18. Assume: design margin = 10 percent, lowest electrolyte temperature is 21.1°C (70°F), age factor = 25 percent, 250-Vdc system, 120 cells, 210 Vdc minimum (i.e., 1.75 V/cell end-of-discharge voltage).

The load profile has four sections, each of which must be analyzed to determine which section controls the battery size, because the current in each period decreases with time. If the current in any period increased over that in the previous period, the section ending with the period just before the period of increased current would not have to be analyzed.

Calculation Procedure

1. Calculate Uncorrected Cell Size for Section 1 (Fig. 18.19)

Assume $R_1 = 125$ A per positive plate ($T_1 = 1$ min). Hence, the number of positive plates (uncorrected) = $(A_1 - A_0)/R_1$ = (176 − 0)/125 = 1.41 positive plates.

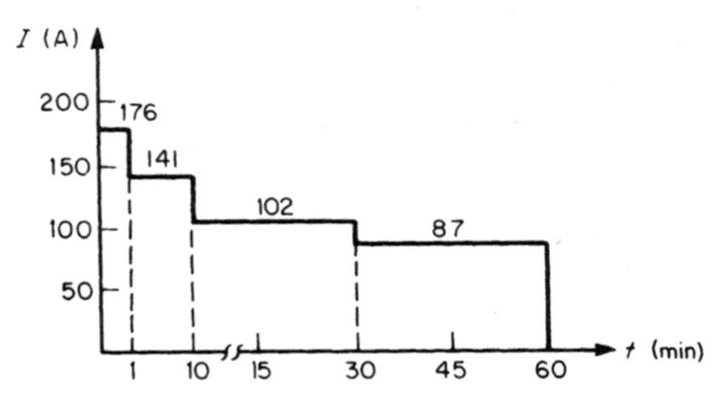

FIGURE 18.18 Sizing battery for a multiple-load profile.

2. Calculate Uncorrected Cell Size for Section 2 (Fig. 18.20)

Assume R_1 = 110 A per positive plate (T_1 = 10 min) and R_2 = 112 A per positive plate (T_2 = 9 min). Number of positive plates (uncorrected) = $(A_1 - A_0)/R_1 + (A_2 - A_1)/R_2$ = (176 − 0)/110 + (141 − 176)/112 = 1.6 − 0.31 = 1.29 positive plates.

3. Calculate Uncorrected Cell Size for Section 3 (Fig. 18.21)

Assume R_1 = 93 (T_1 = 30 min), R_2 = 94 (T_2 = 29 min), and R_3 = 100 (T_3 = 20 min) A per positive plate. Number of positive plates (uncorrected) = $(A_1 - A_0)/R_1 + (A_2 - A_1)/R_2 + (A_3 - A_2)/R_3$ = (176 − 0)/93 + (141 − 176)/94 + (102 − 141)/100 = 1.89 − 0.37 − 0.39 = 1.13 positive plates.

4. Calculate Uncorrected Cell Size for Section 4 (Fig. 18.18)

Assume R_1 = 75 (T_1 = 60 min), R_2 = 76 (T_2 = 59 min), R_3 = 80 (T_3 = 50 min), and R_4 = 93 (T_4 = 30 min) A per positive plate. Number of positive plates (uncorrected) = $(A_1 - A_0)/R_1 + (A_2 - A_1)/R_2 + (A_3 - A_2)/R_3 + (A_4 - A_3)/R_4$ = (176 − 0)/75 + (141 − 176)/76 + (102 − 141)/80 + (87 − 102)/93 = 2.35 − 0.46 − 0.49 − 0.16 = 1.24 positive plates.

5. Determine Controlling Section

Reviewing the positive plates required for each section, one finds that section 1 requires the most positive plates (i.e., 1.41) and is thus the controlling section.

6. Determine Required Size

With the correction factors and margin applied, required size = (max. uncorrected size)(temp. correction)(design margin)(age factor) = (1.41)(1.04)(1.10)(1.25) = 2.02 positive plates.

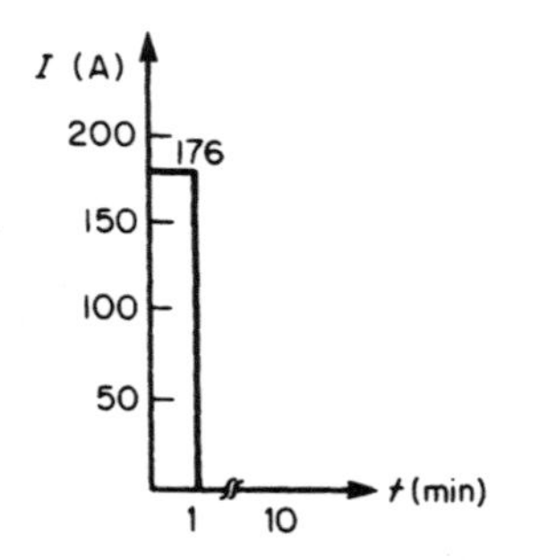

FIGURE 18.19 Considering section 1 of a load profile.

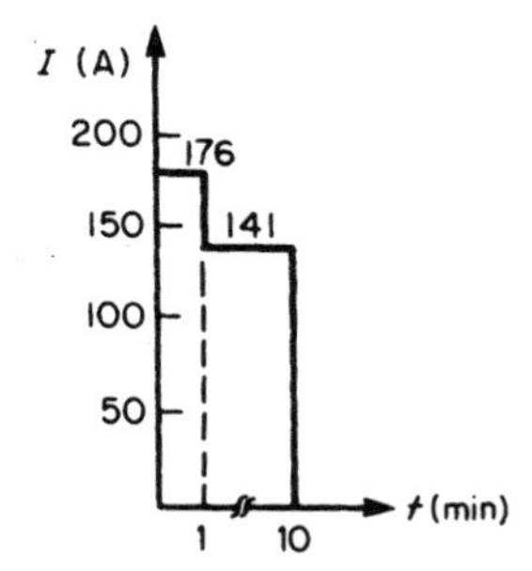

FIGURE 18.20 Considering section 2 of a load profile.

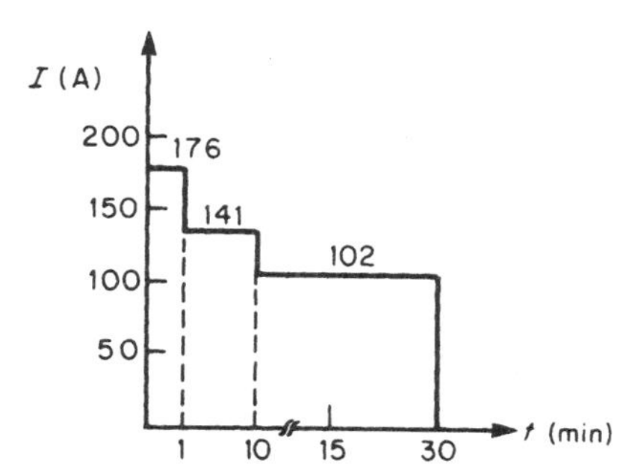

FIGURE 18.21 Considering section 3 of a load profile.

7. Select Cell

Select a type-Z cell having three positive plates. This hypothetical cell would have a capacity of approximately 500 Ah at an 8-h rate, 25°C (77°F), to 1.75 V/cell, 1.210 specific-gravity electrolyte. Even though the cell requires only a fraction of a plate above 2, the selected cell is still rounded up to the next-higher whole number.

Related Calculations. If, for example, the given load profile appears as shown in Fig. 18.22, period 2 does not have to be analyzed because it is exceeded by the current in period 3.

The effect of a random load, if any, would be added to the controlling section of the battery size.

AMPERE-HOUR CAPACITY

Calculate the ampere-hour capacity of a lead-acid cell required to satisfy the load profile of Fig. 18.22. Assume: design margin = 10 percent, lowest electrolyte temperature is 26.7°C (80°F), age factor = 25 percent, 1.75 V/cell end-of-discharge voltage, 60 cells, 125-Vdc system, 105 Vdc minimum voltage.

Calculation Procedure

1. Compute Uncorrected Ampere Hours for Section 1 (Fig. 18.23)

Use*:

$$\text{Cell size (Ah)} = \max_{S=1}^{S=N} \sum_{P=1}^{P=S} [A_P - A_{(P-1)}]K_T$$

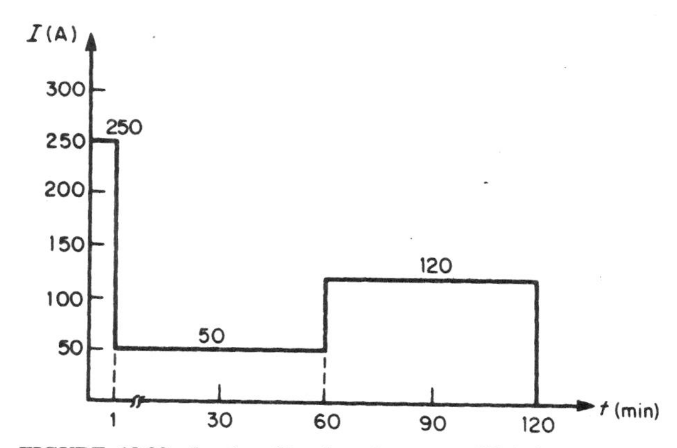

FIGURE 18.22 Load profile where the current (50 A) in section 2 is exceeded by the current (120 A) in section 3.

*Formula from IEEE Standard 485-1997.

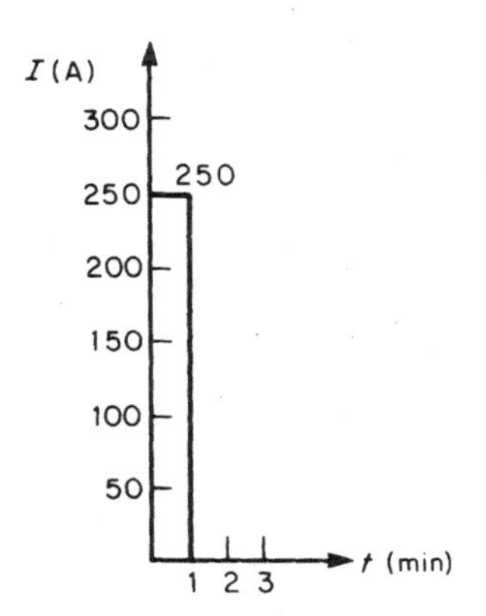

FIGURE 18.23 Considering only section 1 of a load profile.`

where K_T = a capacity rating factor representing the ratio of rated ampere-hour capacity at a standard time rate, at 25°C (77°F), and to a standard end-of-discharge voltage, to the amperes which can be supplied by the cell for T minutes at 25°C (77°F), and to a given end-of-discharge voltage. (For this example, the standard time rate is the 8-h rate.) The other terms have the same meanings as in the previous equations.

Assume $K_{T1} = 0.93$ ($T_1 = 1$ min); then, ampere-hours (uncorrected) = $(A_1 - A_0)K_{T1} = (250 - 0)0.93 = 232.5$ Ah.

2. Compute Uncorrected Capacity in Ampere-Hours for Section 3 (Fig. 18.22)

Assume $K_{T1} = 3$ ($T_1 = 120$ min), $K_{T2} = 2.95$ ($T_2 = 119$ min), and $K_{T3} = 2$ ($T_3 = 60$ min). Then, ampere-hours (uncorrected) = $(A_1 - A_0)K_{T1} + (A_2 - A_1)K_{T2} + (A_3 - A_2)K_{T3} = (250 - 0)3 + (50 - 250)2.95 + (120 - 50)2 = 750 - 590 + 140 = 300$ Ah.

3. Determine Controlling Section

Reviewing the capacity in ampere-hours required for each section, one finds section 3 requires the most capacity and is thus the controlling section.

4. Determine Required Size

Applying the correction factors and margin, one finds: required size = (max. uncorrected size)(temp. correction)(design margin)(age factor) = (300)(1.0)(1.10)(1.25) = 412.5 Ah. Therefore, select a standard type-Z cell with an ampere-hour rating at the 8-h rate greater than or equal to 412.5 Ah; for example, a standard cell of 450 Ah might be selected.

Related Calculations. Common practice for temperatures above 25°C (77°F) is to use a factor of 1.0 rather than to take credit for the extra capacity available.

A similar calculation could be performed for a nickel-cadmium cell. Temperature-correction factors and float-charging correction factors (if applicable) can be obtained from the battery manufacturer.

SIZING BATTERY FOR CONSTANT POWER

Determine the battery required to supply a UPS system rated 600 kVA, 0.8 pf, for 15 min to an end voltage of 1.67 Vpc. The battery has 200 lead-acid cells in 50, 4-cell monoblocs. Inverter efficiency is 0.92. Assume temperature correction and design margin are equal to 1.00 and that an aging factor of 1.25 will be used.

Calculation Procedure

1. Determine the Watts/Cell Required to Serve the Load

Use: Cell Size (watts) = (UPS dc input/number of cells) × design margin × temp corr × aging factor.

2. Compute the dc Input Watts Required by the Inverter

W = (VA × pf)/efficiency = 600 000 × 0.8/0.92 = 521,739 W. Round to 522 kW.

3. Compute the Cell Size

Cell size = (522 kW/200) × 1.00 × 1.00 × 1.25 = 3.26 kW/cell.

4. Select a Standard Cell

Using a battery manufacturer's discharge tables or curve, in watts, select a cell that has a minimum discharge capability of 3.26 kW/cell at the 15-min rate to 1.67 Vpc.

Related Calculations. If the UPS inverter had a low-voltage disconnect set for 330 Vdc and there was a 6 Vdc drop in the main leads to the inverter, the sizing would not be adequate since the battery terminal voltage at the end of discharge would have to be 336 Vdc. It would actually be 1.67 Vpc × 200 cells = 334 Vdc. In this case, the watts per cell required would not change; however, the end-of-discharge voltage would become 336 Vdc/200 cells = 1.68 Vpc. The cell selected would then have to have a minimum discharge capability of 3.261 kW/cell at the 15-min rate to 1.68 Vpc.

If the battery manufacturer did not have a cell large enough to meet the required kW, two strings in parallel could be used. Each string would be able to supply 1.63 kW for 15 min to the desired end voltage.

CALCULATING BATTERY SHORT-CIRCUIT CURRENT

A battery can provide a substantial current if it is shorted. If the internal resistance of a cell (or battery) is known, the short-circuit current can be calculated by dividing the open-circuit voltage of the cell by its internal resistance, that is, $I_{sc} = E_{oc}/R_c$. The internal resistance of a cell may be obtained from the battery manufacturer, but it may also be calculated.

Determine the available short-circuit current for a 9-plate lead-acid cell with 1250 density. The fan curve for the cell is shown in Fig. 18.24.

Calculation Procedure

1. Graphically Determine the Internal Resistance of the Cell Using the Initial Voltage Line to Obtain Two Points That Will Allow Computation of ΔV and ΔI

Referring to Fig. 18.24, pick two points on the initial voltage line and read the voltage and current/positive plate for each. Point *A* is 1.96 V and 60 A/positive. Point *B* is 1.7 V and 244 A/positive.

2. Determine ΔV

$\Delta V = 1.96 - 1.7 = 0.26$ V

3. Determine ΔI

$\Delta I = 244 - 60 = 184$ A

4. Determine the Resistance of One Positive Plate

$R = \Delta V/\Delta I = 0.26/184 = 1.41$ mΩ

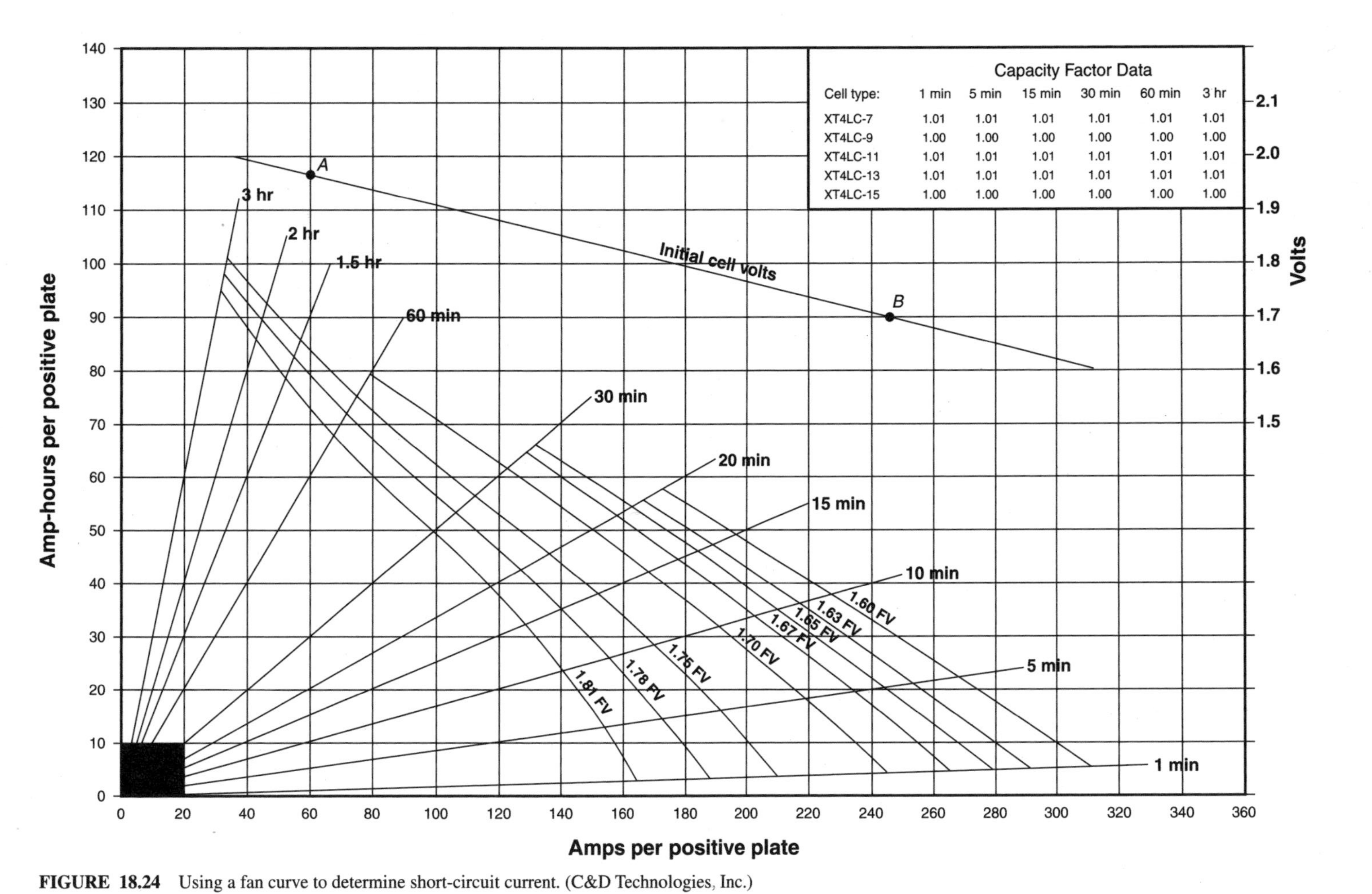

Cell type:	1 min	5 min	15 min	30 min	60 min	3 hr
XT4LC-7	1.01	1.01	1.01	1.01	1.01	1.01
XT4LC-9	1.00	1.00	1.00	1.00	1.00	1.00
XT4LC-11	1.01	1.01	1.01	1.01	1.01	1.01
XT4LC-13	1.01	1.01	1.01	1.01	1.01	1.01
XT4LC-15	1.00	1.00	1.00	1.00	1.00	1.00

FIGURE 18.24 Using a fan curve to determine short-circuit current. (C&D Technologies, Inc.)

5. Determine the Resistance of the Cell

The cell has 4 positive plates connected in parallel; therefore $R_c = 1.41 \times 10^{-3}/4 = 0.35\ \text{m}\Omega$

6. Determine the Short-Circuit Current of the Cell

$I_{sc} = 2.09/0.35 \times 10^{-3} = 5971$ Adc

Related Calculations. There are also rules of thumb used in the industry to compute short-circuit current. Some are based on the short-circuit current being a multiple of the cell's 1-min rate to a standard end-of-discharge voltage (e.g., IEEE 946-1992 uses 10 times the 1-min rate to 1.75 Vpc for a lead-acid cell), while others are based on a multiple of the cell's C rate (AS 2676.1-1992 uses 20 times the C_3 rate). Most of these rules of thumb produce conservative results. For example, the 1-min rate to 1.75 V of the cell in the example is 840 A, and 10 times its 1-min rate would be 8400 A. Similarly, the C_3 capacity of the cell is 403 Ah, and 20 times that would be 8060 A. (Note: these data can found using the fan curve in Fig. 18.24.)

CHARGER SIZE

Determine the required charger output to recharge a lead-acid battery in 16 h while serving a dc system load of 20 A. The battery has the following characteristics: 500 Ah at the 8-h rate, 25°C (77°F), 1.210 specific-gravity electrolyte to 1.75 V/cell.

Calculation Procedure

1. Compute Charger Current

The capacity of the required charger is calculated from the equation charger A = $[(\text{Ah})K/T] + L$, where charger A = required charger output amperes, Ah = ampere-hours removed from the battery, K = a constant to compensate for losses during the charge (normally 1.1 for lead-acid cells), T = desired recharge time in hours, and L = steady-state dc load, in amperes, to be served by the charger while the battery is being recharged.

Substitution of given values in the equation yields: charger A = $[(500)(1.1)/16] + 20 = 54.38$ A.

2. Select Charger

Select the next standard-rated charger with an output current that is greater than 54.38 A.

Related Calculations. Variables T and L are specific to the system under consideration and must be determined by the designer, although T is also a function of charging voltage. The higher the recharge voltage, the faster the recharge, within limits (consult the battery manufacturer).

Sometimes, however, the maximum dc system voltage may limit the recharge voltage. In these instances, a longer recharge time is required unless the system design allows the battery to be isolated from the system during recharge. (Some non-U.S. chargers have an internal dropping diode that regulates the dc bus voltage while allowing the charger voltage to increase above the system limits for recharging the battery.)

The formula in this example is applicable to nickel-cadmium cells; however, the battery manufacturer would have to provide the constant K. For nickel-cadmium cells, it is generally in the range of 1.2 to 1.4, depending on plate type.

RECHARGING NICKEL-CADMIUM BATTERIES—CONSTANT CURRENT

Calculate the required charger size to recharge a 340-Ah (at the 8-h rate) NiCd block battery having eight NC modules, using the two-rate charging method. It is desired to have 70 to 80 percent of the Ah capacity (previously discharged) replaced within 10 h.

The battery is used on a 125-Vdc system, maximum allowable voltage is 140 Vdc, and there are 92 cells in the battery. If the high-rate charge can be accomplished in 10 h or less, how many additional hours will be required to replace the additional 20 to 30 percent of capacity? Also, assume the charger must carry a 20-A steady-state dc system load.

Calculation Procedure

1. Check Manufacturer's Data

From these data, it is determined that the battery can be recharged (first period) in 9 h if the recharge voltage is 1.50 V/cell or in 10 h at 1.55, 1.60, or 1.65 V/cell. In all cases, it is assumed that the charger output divided by the cell capacity during the first period is 0.1. All these values meet the specified criteria for recharge.

2. Check Maximum System Voltage

Maximum voltage per cell = max. system voltage/number of cells = 140/92 = 1.52 V/cell. From this result, the only acceptable recharge rate is 1.50 V/cell.

3. Calculate High-Rate Current

High-rate current = (0.1)(cell capacity) = (0.1)(340) = 34 A.

4. Determine Charger Size

Charger A = high-rate current + steady-state dc load = 34 + 20 = 54 A.

5. Select Charger

Select the next standard-rated charger with an output greater than 54 A. (From the manufacturer's data, an additional 170 h is required to restore the remaining 20 to 30 percent capacity.)

Related Calculations. NiCd batteries normally use two-rate charging. This type of charge accomplishes recharge over two time periods, each with its own charging rate.

The battery manufacturer has data available to aid in the selection of the required charger. These data provide the charging times required to charge the battery assuming a specific voltage per cell recharge voltage and specific charger output as related to cell capacity during the first charge period.

For two-rate charging, the first period (or high-rate period) is the time required to replace 70 to 80 percent of the Ah capacity discharged. The second period (or finish-rate period) is the time required to replace the remaining 20 to 30 percent of the discharge capacity.

TEMPERATURE AND ALTITUDE DERATING FOR CHARGERS

If a battery charger is installed in locations above 1000 m or where ambient temperature can be greater than 40°C, a derating is generally required. Derating factors can be obtained from the charger manufacturers.

Determine the required charger size for the charger in a previous calculation that had a minimum rating of 54.38 Adc, if it is installed in a location that is at 1500 m and the ambient temperature may be as high as 55°C in the summer. The battery charger manufacturer has provided a derating factor for altitude of 1.11 and 1.25 for temperature.

Calculation Procedure

1. Determine the Minimum Charger Rating by Multiplying the Original Current Required by the Derating Factors

Charger A = 54.38 × 1.25 × 1.11 = 75.45 Adc.

CALCULATION OF THE SULFURIC ACID (H_2SO_4) CONTENT OF ELECTROLYTE

Applicable codes and standards, such as the Uniform Fire Code, require reporting of the amount of H_2SO_4 in a battery so that emergency responders know the hazards present in the event of an emergency. The electrolyte in a lead-acid battery is a solution of water and H_2SO_4. The percentage, by weight, of H_2SO_4 for different densities is provided in Table 18.3.

Determine the quantity of H_2SO_4 in a 180-cell battery in which each cell contains 27.6 kg of a nominal 1215 density electrolyte at 25°C.

Calculation Procedure

1. Determine the Percentage of H_2SO_4 by Weight in the Electrolyte Using Table 18.3

For 1215 density electrolyte, percentage of H_2SO_4 = 30.

2. Compute the Weight of H_2SO_4 Using the Weight of Electrolyte in the Cell (Normally Shown on the Battery Manufacturer's Catalog Sheet for the Cell) and the Percentage from Table 18.3

Weight of H_2SO_4 = 27.6 kg/cell × 0.3 = 8.3 kg/cell.

3. Compute the Total Weight of H_2SO_4 for the Battery

Total H_2SO_4 = 8.3 kg/cell × 180 cells = 1494 kg.

TABLE 18.3 Weight Percentage of Sulfuric Acid in Lead-Acid Battery Electrolyte

Electrolyte density (kg/m^3)	Weight of H_2SO_4 by percent @ 15°C	@ 20°C	@ 25°C
1215	29.2	29.6	30.0
1225	30.4	30.8	31.2
1230	31.0	31.4	31.8
1240	32.2	32.6	33.0
1250	33.4	33.8	34.2
1265	35.2	35.6	36.0
1280	37.0	37.4	37.8
1300	39.3	39.7	40.1

Related Calculations. If the volume of electrolyte is known rather than the weight, the weight of H_2SO_4 can still be computed, since the weight of 1 liter of electrolyte (in kg) is approximately equal to its specific gravity (SG) at standard temperature. SG is density/1000, since the density of water is 1000 kg/m^3. So, for example, a cell with 20 L of 1250 density electrolyte at 25°C, would contain 25 kg of electrolyte and 8.6 kg of H_2SO_4.

The tolerance for density is typically ± 10 kg/m^3. When computing the weight of H_2SO_4, the normal electrolyte density may be increased by $+10$ kg/m^3 to determine a maximum weight. Alternately, the ± 10 kg/m^3 may be used to provide a range of weights.

BIBLIOGRAPHY

AS 2191. 1978. *Stationary Batteries of the Lead-Acid Planté Positive Plate Type, with Amendment 1, 1989*. Sydney, NSW: SAA.

AS 2676.1. 1992. *Guide to the Installation, Maintenance, Testing and Replacement of Secondary Batteries in Buildings—Part 1: Vented Type*. Sydney, NSW: SAA.

AS 2676.2. 1992. *Guide to the Installation, Maintenance, Testing and Replacement of Secondary Batteries in Buildings—Part 2: Sealed Type*. Sydney, NSW: SAA.

AS 4029.1. 1994. *Stationary Batteries—Lead-Acid Part 1: Vented Type*. Sydney, NSW: SAA.

AS 4029.2. 1992. *Stationary Batteries—Lead-Acid Part 2: Valve-Regulated Sealed Type*. Sydney, NSW: SAA.

AS 4029.3. 1993. *Stationary Batteries—Lead-Acid Part 3: Pure Lead Positive Pasted Plate Type Supersedes AS 1981, with Amendment 1, 1995*. Sydney, NSW: SAA.

AS 4086. 1993. *Secondary Batteries for Use with Stand-Alone Power Systems Part 1: General Requirements*. Sydney, NSW: SAA.

Belmont, Thomas K. 1979. "A Calculated Guide for Selecting Stand-by Batteries," *Specifying Engineer*, pp. 85–89, May.

Berndt, D. 1997. *Maintenance Free Batteries: A Handbook of Battery Technology*, 2nd ed. Somerset, England: Research Studies Press Ltd.

Craig, D. Norman, and W. J. Hamer. 1954. "Some Aspects of the Charge and Discharge Processes in Lead-Acid Storage Batteries," *AIEE Transactions (Applications and Industry)*, Vol. 73, pp. 22–34.

Crompton, T. R. 1996. *Battery Reference Book*, 2nd ed. Oxford, England: Reed Educational and Professional Publishing Ltd.

Fink, Donald G., and H. Wayne Beaty. 1999. *Standard Handbook for Electrical Engineers*, 14th ed. New York: McGraw-Hill.

Hoxie, E. A. 1954. "Some Discharge Characteristics of Lead-Acid Batteries," *AIEE Transactions (Applications and Industry)*, Vol. 73, pp. 17–22.

Hughes, Charles J. 1979. "Duty Cycle Is Key to Selecting Batteries: Ampere-Hours Should Never Be Used Alone to Specify a Battery," *Electrical Consultant*, pp. 32, 34, 36, May/June.

IEC 60622. 1988. *Sealed Nickel-Cadmium Prismatic Rechargeable Single Cells*. Geneva, Switzerland: IEC.

IEC 60623. 1990. *Vented Nickel-Cadmium Prismatic Rechargeable Single Cells*. Geneva, Switzerland: IEC.

IEC 60896-1. 1987. *Stationary Lead-Acid Batteries—General Requirements and Methods of Test, Part 1: Vented Types*. Geneva, Switzerland: IEC.

IEC 60896-2. 1995. *Stationary Lead-Acid Batteries—General Requirements and Test Methods, Part 2: Valve Regulated Types*. Geneva, Switzerland: IEC.

IEEE. 1997. *IEEE Standards Reference Database on CD-ROM: An Enhanced Version of the IEEE Standard Dictionary of Electrical and Electronics Terms*. SE105. [CD-ROM]. New York: IEEE.

IEEE Standard 446. 1995. *IEEE Recommended Practice for Emergency and Standby Power Systems for Industrial and Commercial Power Systems: Chapter 5—Stored Energy Systems, The Orange Book*. New York: IEEE.

IEEE Standard 450. 1995. *IEEE Recommended Practice for Maintenance, Testing and Replacement of Vented Lead-Acid Batteries for Stationary Applications*. New York: IEEE.

IEEE Standard 485. 1997. *IEEE Recommended Practice for Sizing Lead-Acid Batteries for Stationary Applications*. New York: IEEE.

IEEE Standard 946. 1992. *IEEE Recommended Practice for the Design of DC Auxiliary Power Systems for Generating Stations*. New York: IEEE.

IEEE Standard 928. 1986. *IEEE Recommended Criteria for Terrestrial Photovoltaic Power Systems*. New York: IEEE.

IEEE Standard 1013. 1990. *IEEE Recommended Practice for Sizing Lead-Acid Batteries for Photovoltaic (PV) Systems*. New York: IEEE.

IEEE Standard 1115. 1992. *IEEE Recommended Practice for Sizing Nickel-Cadmium Batteries for Lead-Acid Applications*. New York: IEEE.

IEEE Standard 1144. 1996. *IEEE Recommended Practice for Sizing Nickel-Cadmium Batteries for Photovoltaic (PV) Systems*. New York: IEEE.

IEEE Standard 1184. 1994. *IEEE Guide for the Selection and Sizing of Batteries for Uninterruptible Power Systems*. New York: IEEE.

IEEE Standard 1188. 1996. *IEEE Recommended Practice for Maintenance, Testing and Replacement of Valve-Regulated Lead-Acid (VRLA) Batteries for Stationary Applications*. New York: IEEE.

IEEE Standard 1189. 1996. *IEEE Guide for the Selection of Valve-Regulated Lead-Acid (VRLA) Batteries for Stationary Applications*. New York: IEEE.

Keegan, Jack W., Jr. 1980. "Factors to be Considered When Specifying Engine/Turbine Starting Batteries," *Electrical Consultant*, pp. 35–36, 38–40, January/February.

Linden, David. 1995. *Handbook of Batteries*, 2nd ed. New York: McGraw-Hill.

Migliaro, M. W. 1987. "Considerations for Selecting and Sizing Batteries," *IEEE Transactions on Industry Applications*, Vol. IA-23, No. 1, pp. 134–143.

Migliaro, M. W. 1988. "Application of Valve-Regulated Sealed Lead-Acid Batteries in Generating Stations and Substations," *Proceedings of the American Power Conference*, Vol. 50, pp. 486–492.

Migliaro, M. W. 1991. "Determining the Short-Circuit Current from a Storage Battery," *Electrical Construction & Maintenance*, pp. 20, 22, July.

Migliaro, M. W. 1991. "Specifying Batteries for UPS Systems," *Plant Engineering*, pp. 100–102, 21 March.

Migliaro, M. W. 1993. "Stationary Batteries—Selected Topics," *Proceedings of the American Power Conference*, Vol. 55-I, pp. 23–33.

Migliaro, M. W. 1995. *Stationary Battery Workshop—Sizing Module*. Jupiter, F.L.: ESA Consulting Engineers, PA.

Migliaro, M. W. 1995. *Battery Calculations Associated with Maintenance and Testing*. Jupiter, F.L.: ESA Consulting Engineers, PA.

Migliaro, M. W. 1998. "In the Battery World, Do You Get Exactly What You Ordered?," *Batteries International Magazine*, pp. 45, 47–48, 50–51, July.

Migliaro, M. W., Ed. 1993. *IEEE Sourcebook on Lead-Acid Batteries*. New York: IEEE Standards Press.

Migliaro, M. W., and G. Albér. 1991. "Sizing UPS Batteries," *Proceedings of the Fourth International Power Quality Conference*, pp. 285–290.

Montalbano, J. F., and H. L. Bush. 1982. "Selection of System Voltage for Power Plant DC Systems," *IEEE Transactions on Power Apparatus and Systems*, Vol. 101, pp. 3820–3829.

Murugesamoorthi, K. A., R. Landwehrle, and M. W. Migliaro. 1993. "Short Circuit Current Test Results on AT&T Round Cells at Different Temperatures," *In Conference Record 1993 International Telecommunications Energy Conference (INTELEC '93)*, 93CH3411-6, Vol. 2, pp. 369–373.

RS-1219. 1992. *Standby Battery Sizing Manual*. Plymouth Meeting. P.A.: C&D Technologies, Inc.

SAND81-7135. 1981. *Handbook of Secondary Storage Batteries and Charge Regulators in Photovoltaic Systems—Final Report.* Albuquerque, N.M.: Sandia National Laboratories.

Section 12-300. 1963. *C&D Switchgear Control Batteries and Chargers—Simplified Method for Selecting Control Batteries*. Plymouth Meeting, P.A.: C&D Technologies, Inc.

Section 50.50. 1990. *How to Use "S" Curves*. Reading, P.A.: Yuasa Inc.

Smith, Frank W. 1976. "Control Voltages for Power Switchgear," *IEEE Transactions on Power Apparatus and Systems*, Vol. 96, pp. 969–977.

Uniform Building Code. 1997. Whittier, C.A.: International Conference of Building Officials.

Uniform Fire Code. 1997. Whittier, C.A.: International Conference of Building Officials.

Vigerstol, Ole K. 1988. Nickel Cadmium Batteries for Substation Applications. Saft, Inc.

SECTION 19

ELECTRIC ENERGY ECONOMIC METHODS

Gerald B. Sheblé*

Professor

Electrical and Computer Engineering

Iowa State University

INTRODUCTION

This work is divided into the following sections: business structure, general economic theory, production costing, opportunity costs and resources, project selection, and uncertainty. The first section outlines the division of the previously vertically structured electric power industry into a horizontally structured industry to foster competition without the presence of market power. The second section outlines different competitive economic conditions as the industry reregulates. The third section defines how to calculate the production cost for short-term financial statements. The fourth section discusses opportunity costs and demonstrates some deterministic approximations commonly used. The fifth section reviews the basic techniques to select projects for capital expansion. It is assumed that the potential projects are simulated for each year of operation. However, the examples are simplified for the sake of presentation. The sixth section introduces the concept of uncertainty and how to more accurately account for the risks of operation and of capital expansion.

*Some of the material originally by Bjorn M. Kaupang, General Electric Company.

REREGULATED ENVIRONMENT

New legislation calls for competition in the power industry, from the wholesale level to the retail level. New market structures are sought to search for a good business environment that would ultimately satisfy the regulatory bodies, customers, and suppliers. One approach is the application of brokerage systems to the power industry to promote competition. To accomplish competitive markets, the vertically integrated utilities shall be broken up. Figure 19.1 shows the framework of the energy market.

Government regulation is the top tier of the structure. There are many such government commissions in the United States. The federal level is assigned to the Federal Energy Regulatory Commission (FERC). Each state has a corresponding commission typically called State Public Utilities Commission (SPUC). The National Reliability Standards should be de-

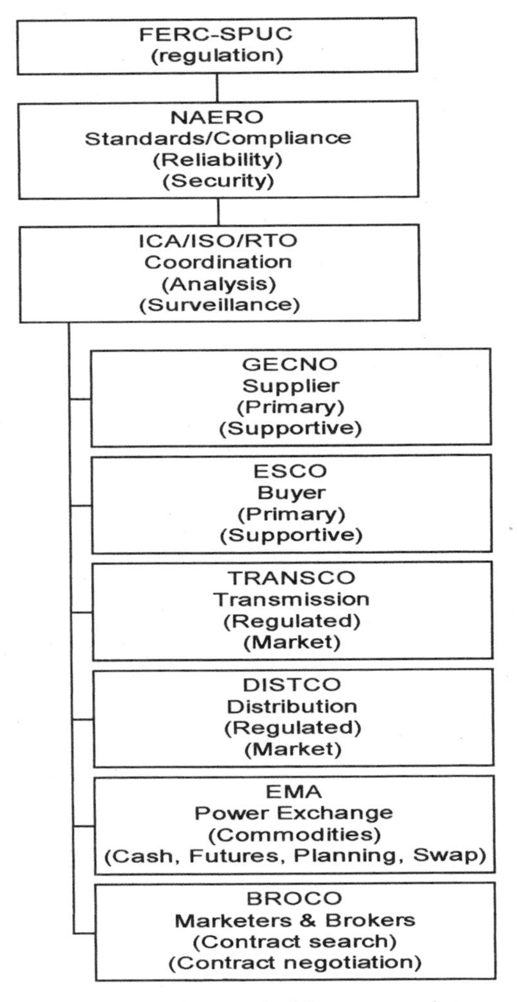

FIGURE 19.1 Framework of the energy market.

fined and standardized by the reorganized North American Electric Reliability Organization (NAERO) at the national level (formerly National Electric Reliability Council, NERC). The generation companies primarily produce electric energy (GENCOs). The transmission companies (TRANSCOs) own and operate transmission lines. The distribution companies (DISTCOs) own and operate distribution lines. As of this writing, TRANSCOs and DISTCOs may not be forced to divest from each other, as they are both regulated companies. The energy mercantile association (EMA) is a generic financial domain name for a power exchange. The brokers (BROCOs) buy and sell electricity for profit. The energy service companies (ESCOs) purchase electricity acting as agents for consumers.

The primary entity that will have direct dealings with the end-users under the new market structure is the energy service company (ESCO). The ESCO collects its revenue from the distribution customers for the energy and ancillary services it has provided. It can also act as a wholesaler, purchasing the electric energy through the auction market and reselling it to the other ESCOs, GENCOs, etc. To purchase the desired electric energy to serve its purpose, the ESCO may purchase through an auction market or direct contracts or utilize the energy reserves that it has accumulated through load-management programs or the ownership of generation units. Additionally, all supportive services (e.g., load following, regulation, etc.) have to be procured to enable delivery of the energy across the transmission grid.

In the deregulated environment, every customer is free to choose any ESCO to serve energy demand. In addition, the energy purchased from the auction market bears the risk of market price fluctuation. These, from the demand factors to the supply factors, are the risks that the ESCO has to undertake in the new market structure. Since deregulation will render governmental protection (financially) obsolete, risk management and assessment tools should be considered and applied.

This work introduces economic models for risk management and assessment tools to assist GENCO or ESCO operation.

THEORY OF THE FIRM

Each company has a multifaceted optimization problem of using scarce capital resources, scarce operating resources, and scarce labor resources. Each GENCO and ESCO has a fixed and a variable operating cost based on the selected mix of capital and operating expenditures. The GENCO sees a demand curve by the buyers (ESCOs). The GENCO is a demand curve seen by the fuel suppliers. The ESCOs see the GENCOs through the supply curve. The ESCOs are a supply curve to all customers. This section works from the market model to exemplify the solution of the supply-demand curves. Then the basic models of the supply and demand curves are generated from basic cost components. The supply curve of the marketplace is the sum of the individual supply curves of all suppliers (GENCOs). GENCOs pool suppliers. The demand curve of the marketplace is the sum of the individual demand curves of the buyers (ESCOs). ESCOs pool purchasers (consumers). These curves are assumed in this work to be the actual cost curves, including profit to pay for capital contracts (stocks, bonds, mortgages, etc.). These curves may be altered if the supplier or the buyer can implant a perceived differentiation of product or if there is a differentiation of product. Such details are beyond the scope of this work.

Supplier Solutions

Assume that the demand curve is known in a perfectly competitive environment. Thus, the marketplace fixes the price (P) since any individual company cannot change the price.

Once the quantity (Q) is selected (known), then the price is easily found:

$$P = f(Q)|_{\text{static equilibrium}}$$

Note that the evaluation assumes that all other factors are constant, as this is a static equilibrium. Almost all markets are dynamic. Such dynamic analysis techniques are beyond the scope of this work. However, most markets may be sufficiently explained from a static analysis.

The revenue (R) for the company to use for capital expenses and operating expenses is easily found:

$$R = P \cdot Q$$

Supplier Costs

Costs can be segmented into the following categories: capital, revenue, and residual. Capital costs (K) consist of equipment design and procurement. Such costs include equipment specifications, design, development, manufacturing or procurement, installation, training for operations and for maintenance, as well as spare parts for repair. Once these expenditures have been committed, the firm can produce a product for consumers. However, these expenditures are balanced against their impact on asset reliability, asset maintainability, and asset availability as an impact on production factors. The production factors include output quantity, product quality, material usage, labor usage, and asset utilization. Revenue costs (Cr) include operating and maintenance costs. Operating costs include cost for direct materials, direct labor, and overheads (direct expenses). Operating costs also include the indirect costs of materials, labor, and establishment overheads. However, these expenditures are balanced against their impact on production factors. Maintenance costs include spare inventory costs, labor costs, facilities and equipment costs, establishment overhead, and revenue lost due to downtime. However, these expenditures (preventative repair) are balanced against their impact on production factors. The residual costs are the disposal value and the disposal costs. However, these expenditures are balanced against their impact on the production factors. A detailed discussion of each cost component is beyond the scope of this work. Residual costs are assumed to be negligible for this work.

Individual Supplier Curves

This work uses the following model for the production cost (Cp) given the capital cost (K):

$$Cp = g(K) + h(Cr)$$

Most introductory economic texts use linear approximations to this function. This work assumes that the function can be segmented into a piecewise linear representation as additional accuracy is required.

The marginal cost of production (MC_p) is the first derivative of the cost of production, including all cost components:

$$MC_p = dCp/dQ$$

The average cost of production (AC_p) is the cost of production divided by the quantity sold:

$$AC_p = P/Q$$

The total production cost includes indirect costs (e.g., capital costs, as discussed below) as well as the direct production cost discussed above. When the total cost is referenced in this work, the subscript p is removed. The firm's profit, Π, is the revenue minus the cost of production:

$$\Pi = R - Cp$$

Marginal costs are separated into short-run and into long-run components. Average costs are separated into short-run and into long-run components. Short-run does not normally include the possibility of capital expenditures. Long-run does include the possibility of capital expenditures as well as other major efficiency improvement projects.

Competitive Environments

Note that the cost of production does not set the price in a perfectly competitive environment. Thus, it is natural to understand why all firms endeavor to not engage in a perfectly competitive environment. Many buyers and sellers typify a perfectly competitive market, each firm produces a homogenous product, buyers and sellers have perfect information, there are no transaction costs, and there is free entry and exit. Under perfect competition the optimization conditions require that the first derivative of the revenue must be equal to the first derivative of the cost of production:

$$MR = MC = P$$

This is most easily visualized through supply vs. demand curves. Under the assumption of perfect competition, the curves do not interact. The producer surplus is that area between the supply curve and the perfectly competitive price line bounded by the actual demand. The consumer surplus is that area between the supply curve and the perfectly competitive price line bounded by the actual demand. These surpluses, the actual price, as well as the competitive price, are shown in Fig. 19.2. As the actual price approaches the

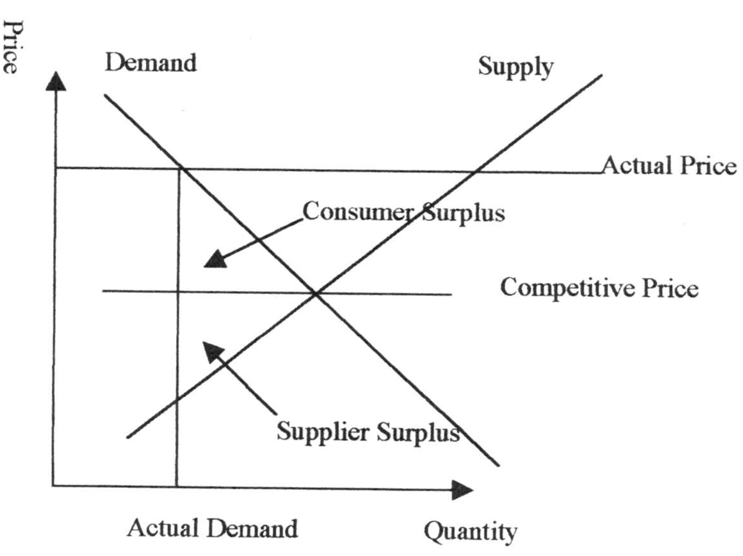

FIGURE 19.2 Supply vs. demand curve.

competitive price, note that the surpluses decrease to zero. When the price is above the competitive price, the supplier is collecting the consumer surplus as well as the producer surplus. Also, the quantity produced is less than the quantity ideally desired by society.

Consumers may alter the demand curve by substitution with alternative products through multimarkets. Use of natural gas instead of electricity is such as substitution. Such substitutions do not necessarily eliminate the consumption of a product like electric energy. However, the slope of the curve may dramatically change. Also, the demand curve (Fig. 19.1) may change based on budget constraints.

The elasticity of demand is due to budget constraints or to substitutes. The elasticity is the normalized derivative of the demand curve expressed as a percentage change when the change in P and Q is small. Elasticity of demand is defined as:

$$E_d = -(P \cdot \partial Q)/(\partial P \cdot Q)$$

When $E_d > 1$, the demand is elastic. When $E_d = 1$, the demand is constant. When $E_d < 1$, the demand is inelastic.

Instead of elasticity, indifference curves may be used to model the use of substitutes.

The solution of the supply-demand curve is simply the intersection of the two curves under perfect competition.

Imperfect Competition

Other market environments are more common than perfect competition. A single firm serving an entire market for products that have no close substitutes typifies monopoly markets. Alternatively, monopoly power is achieved by economies of scale, economies of scope, cost complementarity, patents, and legal barriers. Since the price is set by the market conditions and equating marginal revenue to marginal cost sets quantity, a supply curve does not technically exist. However, drawing a perfectly competitive supply curve and then calculating the monopoly rents to increase the marginal cost to the price can reveal the impact of monopoly power.

Monopolistic competition is a very common condition where there are many buyers and sellers, each firm produces a differentiated product, and there is free entry and exit. Differentiated products may be established in reality or in perception. Customer selection may be based on time of day, reliability, and quality to differentiate electric-energy products from different providers. Firms engage in comparative advertising, brand equity, niche marketing, and green marketing to virtually differentiate products.

Other market types are defined under separate categories, such as oligopoly. Only a few firms supplying the product denote an oligopoly market, each firm being large with respect to the total industry. Several types of oligopoly markets are found in the literature: Cournot, Bertrand, Stackelberg, and Sweezy. Such market structures are beyond this work.

Oligopolies also spawn the most research due to tacit and explicit collusion. Collusion reduces the competitive market to a monopolistic market. Tacit collusion occurs when there is sufficient publicly disclosed information for the firms to faithfully predict the competitive actions or reactions. Explicit collusion occurs when two or more firms engage in price fixing.

Other factors, such as the utility of the product instead of the cost, are beyond the scope of this work.

Time-domain solutions are needed for accurate managing of a firm as the time-dependent cash flow is the most critical component of operational management. Such time analyses can be performed by application of state-variable modern control models. Such applications are beyond the scope of this work.

Supply curves are inelastic in the short run as production cannot change quickly. If demand increases, then price increases until a new equilibrium is found. As the price increases, then more fuel can be procured to meet the increased demand. Continuous pressure on supplies with increased price will entice new entry into the market to meet the new demand. Short-run response assumes that equipment is not improved, expanded, or augmented. Long-run response assumes that new equipment may be added or production efficiencies can be implemented. Technology changes are not normally included in the long-run response, as technology changes are fundamental production changes.

A perfectly competitive market reaches the following equilibrium:

$$SRMC = SRAC = LRMC = LRAC$$

It is interesting to note that under such conditions, the following observations are found. There is no need for marketing, as all products are the same at the same price. All products have the same quality; thus no firm can differentiate its products. There is no need for research, as profit is fixed. Such conditions are ideal for the consumer in the short run. However, the lack of product improvement would be one of the constant social concerns.

Other Factors

There are other exogenous components that can significantly change the supply–demand solution. Electric-production processes create costs for a society who are not buyers or sellers of the product. Such external costs are negative externalities. Pollution is one such negative externality. The proper economic handling of such externalities requires government intervention either directly or indirectly. Direct intervention implies taxes to provide cost reimbursements. Indirect intervention implies some type of market solution like the exchange of pollution permits (SO_2, NO_x, etc.). Then the firms causing pollution as part of their business strategy can efficiently seek the cost of reimbursement. However, there have to be well-defined property rights for such markets to be complete and effective.

The presence of regulations to fix the price provides a ceiling on the solution. A ceiling is a maximum price that can be charged for a product. The presence of a ceiling increases the scarcity of the product. The price increases as a result. Note that price ceiling must be lower than the competitive price to be effective. Alternatively, floors can be introduced through regulation. A floor is a minimum price that can be charged for a product. Floors must be greater than the competitive price to be effective. Floors increase the quantity as the price increases beyond the quantity that would have been produced at the competitive price. The excess quantity has to be purchased by a regulatory body (government) and stored, given away, or destroyed.

Import quotas can lower the supply, thus raising prices. As always, such indirect taxes increase the cost of production. Such taxes lower the quantity of the product as the supply curve shifts upward. Lump-sum or per-unit (excise) tariffs limit foreign competition and thus increase the cost of production. Public goods are often provided from electrical production for goods that are nonrival and nonexclusionary in consumption. Streetlights are a common example.

The threat of regulatory changes, based on customer requirements, is a constant problem for the electric-power industry. Deregulation is actually reregulation. The industry is being forced into a competitive market. The establishment of such markets has taken centuries in other industries. Many government approaches have been taken to best unify the goal of competition with the unique delivery and structure of the electric-energy industry.

There are many other markets that significantly influence the electric-energy market. All fuel markets (coal, natural gas, oil, and uranium) determine the major component of operational expenses. The labor market for engineers has recently become a major component of corporate operations and planning. The financial markets have always been the significant components for planning and operations. The interaction of all of these markets establishes the price for electricity both directly and indirectly.

PRODUCTION COSTING

The following gives examples of the primary techniques for calculating the total production cost, including capital, revenue, and residual costs. Residual costs are folded into the capital cost of equipment. The end result is a complete financial description of operation for the year.

Cost of Money

The cost of money is defined as the income that would be produced if the money were invested for the given period of time. Since a firm has at least three generic capital accounts, a weighted sum of all accounts is normally used. There are three major capital sources: stock, bonds, and bank contracts (e.g., mortgages). Find the weighted cost of money when given the bond interest rate $i_B = 10$ percent, the preferred stock interest rate $i_P = 12$ percent, and the return on common stock $i_C = 15$ percent. Also, the debt ratio (fraction of bonds) $DR = 50$ percent, the common-stock ratio (fraction of common stock) $CR = 35$ percent, and the preferred-stock ratio (fraction of preferred stock) $PR = 15$ percent.

Calculation Procedure

1. Compute the Cost of Money

The weighted cost of money (i) is given by

$$i = i_B DR + i_P PR + i_C CR.$$

Substitution of given values in the equation yields

$$i = (10)(0.5) + (12)(0.15) + (15)(0.35) = 12.05 \text{ percent.}$$

Related Calculations. The basic sources of capital, other than capital surplus from operations, are the bond and stock markets for long-term capital needs and banking institutions for short-term (less than a year) borrowings. Bonds are the most common long-term debt instrument in the utility industry. The first-mortgage bond is the most senior and, therefore, has the first claim on a company's assets. This security is reflected in the relatively low interest rate for first-mortgage bonds.

Stocks could be issued as preferred, with a fixed dividend rate, or as common. In most companies, the only voting stock is the common stock. It also carries the highest risk and, therefore, the highest return.

Capitalization is an accounting term for total outstanding bonds and stock. The relationship between the bonds and stock making up the capitalization may be expressed as capitalization ratios that can be used for estimating the weighted cost of money.

Equivalent and Compound Interest

Cash flows are often depicted as cash payments as discrete points of time, as shown in Fig. 19.3. Each income payment is shown as a positive (drawn to the top of the page) arrow. Each outgoing payment is shown as a negative (drawn to the bottom of the page) arrow.

The value of such payments is brought back to present values using the conversion function for future values (f) to present values (p) given a rate of inflation (i_i) for each future period (j) and a future period for the payment (n):

$$p = f/[(1 + i_i) \cdot \cdot \cdot (1 + i_n)]$$

This simplifies if the inflation is constant for each future period:

$$p = f/[(1 + i)^n]$$

The ratio of p/f is called the present value factor for a future single payment.

An annuity is a special cash flow consisting of a debt (mortgage) repaid by a string of equal annual payments. All cash flows are equal in this case. The present value of such an annuity (a) is:

$$p = a(1/i)[(1 + i)^n - 1]/[(1 + i)^n]$$

The ratio of p/a is called the present value factor for an annuity.

It is often convenient to convert interest rates to an equivalent rate for a given period. Consider converting a monthly rate to an annual rate. If the number of discounting periods $= m$, the interest for every period $= i_m$, then the effective annual rate, i_{eff}, can be found using the nominal interest rate, i_{nom}:

$$i_{\text{nom}} = (1 + i_{\text{eff}})^{1/m} - 1$$

When discounting is continuous, the natural exponential (e = exp) is used based on the limit of m approaching infinity:

$$i_{\text{eff}} = \exp(i_{\text{nom}}) - 1$$

Sum-of-Years Digit Depreciation

One common method used for a faster accumulation of depreciation expenses than the straight-line method is the sum-of-years digit method. The annual depreciation expense with this method is equal to the initial investment times the remaining life divided by the sum-of-years digits.

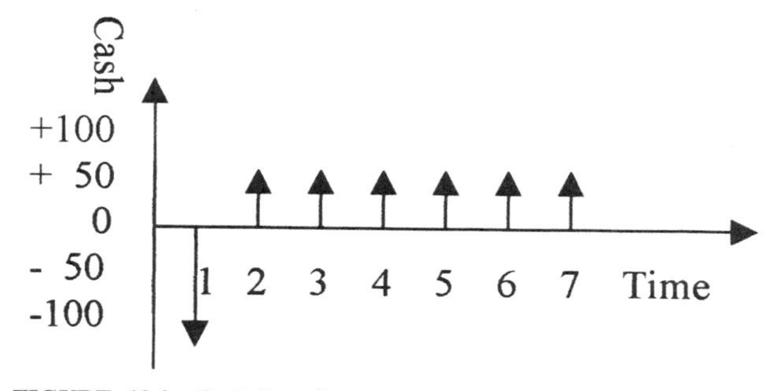

FIGURE 19.3 Cash-flow diagram.

TABLE 19.1 Data for Calculating Sum-of-Years Digit Depreciation

Year	Remaining life	Remaining life sum-of-years digit	Accumulated depreciation
1	5	0.33333	0.33333
2	4	0.26667	0.60000
3	3	0.20000	0.80000
4	2	0.13333	0.93333
5	1	0.06667	1.00000

Determine the accumulated depreciation using the sum-of-years digit method for $1 investment over a 5-yr life.

Calculation Procedure

1. Find the Sum-of-Years Digit

The sum-of-years digit $= 1 + 2 + 3 + 4 + 5 = 15$.

2. Compute Annual Depreciation in Each Year

Use $d_n = (N - n + 1)/[N(N + 1)/2]$, where d_n = annual depreciation. For example, if $n = 2$, $d_n = (5 - 2 + 1)/[5(5 + 1)/2] = 0.26667$. This and other values for n are tabulated in Table. 19.1.

3. Determine Accumulated Depreciation for Each Year

The accumulated depreciation for each year may be obtained by accumulating the annual values from Step 2. For example, if $n = 2$, the accumulated depreciation $= 0.33333 + 0.26667 = 0.6000$. This and other values for n are tabulated in Table 19.1.

Declining-Balance Method

Another method, which yields a faster depreciation than straight-line accumulation of depreciation expense, is the declining-balance method. The rate of depreciation is applied to the remaining balance of initial investment minus accumulated depreciation. The rate is normally higher than the straight-line method. If the rate is double the straight-line method, we have double-declining-balance depreciation.

Find the annual depreciation expenses for a $1 investment depreciated at 40 percent for a 5-yr depreciation life.

Calculation Procedure

1. Compute the Annual Depreciation

The annual depreciation may be computed by $d_n = r(1 - r)^{n-1}$, where r = rate of depreciation. For example, if $n = 2$, $d_n = 0.4(1 - 0.4)^{2-1} = 0.24$. This and other values for n are tabulated in Table 19.2.

2. Compute the Accumulated Depreciation

Accumulate the values found in Step 1. For example, if $n = 2$, the accumulated depreciation $= 0.40 + 0.24 = 0.64$. This and other values for n are tabulated in Table 19.2.

TABLE 19.2 Data for Declining-Balance Method Calculations

Year	Remaining balance	Annual depreciation	Accumulated depreciation
1	1.0	0.40	0.40
2	0.6	0.24	0.64
3	0.36	0.144	0.784
4	0.216	0.0864	0.8704
5	0.1296	0.0518	0.9222

Related Calculations. Because the accumulated depreciation does not reach 1.0 before n goes to infinity, it is common to transfer from the declining-balance method to another depreciation method at some point during the depreciation life.

Transfer of Depreciation Methods

Calculate the accumulated depreciation when a transfer is made to the straight-line depreciation method in the last 2 yr of the previous example.

Calculation Procedure

1. Find Remaining Balance to Be Depreciated in Last 2 Years

From Table 19.3, the value is 0.216.

2. Find Annual Straight-Line Depreciation for Last 2 Years

The value is 0.216/2 = 0.108.

3. Determine Accumulated Depreciation

The values are given in Table 19.3.

Related Calculations. The most common depreciation method is the straight-line method over the useful life of the equipment. The useful life should be based on knowledge of mechanical life as well as economic life. Technical improvements might make a device obsolete before its mechanical life is up.

The fast depreciation methods, like the sum-of-years digit and declining-balance, are commonly used for calculating deductible expenses for income tax purposes. The

TABLE 19.3 Values of Annual and Accumulated Depreciation

Year	Remaining balance	Annual depreciation	Accumulated depreciation
1	1.0	0.40	0.4
2	0.6	0.24	0.64
3	0.36	0.144	0.784
4	0.216	0.108	0.8704
5	0.108	0.108	1.0

effect is to delay, not to reduce, paying income taxes, thereby reducing external financing. It should be noted that most tax regulations have been rewritten to allow only a selected depreciation each year. This is called asset depreciation range in the United States (ADR).

Taxes

Find the income tax and the resulting net income to the shareholders given the following information. The corporation has $2500 in revenues and $350 in annual operating expenses. The original investment was $3000 and the annual interest expense is $300. The depreciation life is 5 yr. The sum-of-years digit depreciation will be used for tax purposes and the straight-line method for book depreciation. The tax rate is 50 percent and the corporation is in its first year of operation.

Calculation Procedure

1. Find Total Deductible Expenses

These values are given directly in the example except for depreciation. From $d_n = (N - n + 1)/[N(N + 1)/2]$ with $n = 1$, the first-year depreciation expense is $1000. Total deductible expenses, therefore, for income tax purposes are: 350 + 300 + 1000 = $1650.

2. Compute Taxable Income

The revenues are given as $2500, which gives 2500 − 1650 = $850 in taxable income.

3. Determine Income Tax

With a taxable income of $850 and a tax rate of 50 percent, the income tax is (850) (0.50) = $425.

4. Calculate Income Statement for Shareholders

The only value not yet calculated is the book depreciation, which is $3,000/5 = $600 yr. The resulting income statement is provided in Table 19.4.

Related Calculations. For an industry or a utility, the common taxes are property, or ad valorem, taxes and income taxes. The ad valorem taxes are normally simple percentages of the assessed value of property and therefore are simple to calculate. Federal and most state governments normally levy income taxes by applying a fixed rate to a taxable income.

TABLE 19.4 Income Statement for Shareholders

Revenues		$2500
Operating expenses	$350	
Interest	300	
Depreciation	600	
Income taxes	425	
Total expenses		$1675
Net income		$825

TABLE 19.5 Modified Tax Statement, Including Investment Tax Credit

Revenues	$2500
Total deductions	1650
Taxable income	850
Income tax	425
Investment tax credit, 10% of $3000	300
Income tax payable	$125

As a rule, taxable income is calculated differently from the income reported by the owners of the company. The major difference in the calculation comes from the treatment of depreciation. It is typical for a corporation to have two sets of books, one for income tax purposes and one for the management and shareholders.

Investment Tax Credit

This is a method used by government to encourage investment in new production facilities. The tax credit is normally calculated as a fixed percentage of new investment in a year and then subtracted directly from the income tax.

Find the effect of a 10 percent investment tax credit when applied to the previous income tax example.

Calculation Procedure

1. Find Investment Tax Credit

The tax credit is ($3000)(0.10) = $300.

2. Compute Income Taxes

Table 19.5 shows the values from the income tax example modified by the investment tax credit.

3. Generate Income Statement to Shareholders

The new income statement is given in Table 19.6.

TABLE 19.6 New Income Statement for Shareholders

Revenues		$2500
Operating expenses	$350	
Interest	300	
Depreciation	600	
Income tax payable	125	
Total expenses		$1475
Net income		$1025

OPPORTUNITY COSTS AND RESOURCES

The opportunity cost of a project is the saving that will accrue if the project is not implemented. After projects are selected for implementation, it is hoped that the resources will be used to their capacity. Unfortunately, this is not always true. As discussed in the following paragraphs, the units are not always run at capacity. It is hoped that each unit will be run at the most optimum point of performance. However, the cyclical demand does not permit such operation. Indeed, even forecasting what the demand will be is yet a black art. Such cyclical operation is not unique to the power industry, but is also present in all industries. The operations research literature is full of techniques to optimize the production utilization of machines, as indicated by all of the literature on job shop scheduling.

The following sections discuss how to optimize the operation of units within an hourly period (economic dispatch). Shorter time periods may be selected for optimizing production. However, the industry has used the hourly period for convenience. It should be noted that some of the reregulated countries now use the half-hour as a convenient time period. Then the accumulation of the hourly periods is emulated by approximate analysis by daily factors. The next period of scheduling is normally the week. This is the unit commitment problem.

The planning problem aggregates the daily or the weekly simulations into yearly summaries. The annual demand factor is an approximation to explain how to evaluate the impact of alternative scheduling, control, and project investments.

System Economic Dispatch

Find the loading schedule for Units A, B, and C given the data in Table 19.7. Linear interpolation is assumed between data points.

Calculation Procedure

1. Find Breakeven Loading between Units

The values are calculated from the table values. For example, $17.55/MWh is found at a 40-MW loading on Unit A, and $18.00/MWh is found at a 6-MW loading level on Unit C.

TABLE 19.7 Data for Units A, B, and C

Data	Unit A	Unit B	Unit C
Full load, MW	50	35	16
Heat rate, kJ/kWh	12,000	12,500	13,000
Btu/kWh	11,375	11,849	12,323
Fuel price, $/GJ	1.50	2.00	1.50
$/MBtu	1.58	2.11	1.58
Minimum load, MW	13	10	4
Incremental HR			
Minimum load, kJ/kWh	10,560	11,100	11,700
50% load, kJ/kWh	11,280	11,870	12,480
100% load, kJ/kWh	12,000	12,500	13,000
Incremental fuel cost			
Minimum load, $/MWh	15.84	22.4	17.55
50% load, $/MWh	16.92	23.74	18.72
100% load, $/MWh	18.00	25.00	19.50

TABLE 19.8 Loading Schedule for Units A, B, and C

		Unit loading (MW)		
System load	Fuel cost, $/MWh	Unit A	Unit B	Unit C
27	15.84	13	10	4
39	16.92	25	10	4
50	17.55	40	10	4
66	18.00	50	10	6
68	18.72	50	10	8
76	19.50/22.2	50	10	16
84	23.74	50	18	16
102	25.00	50	36	16

2. Find Loading Schedule

From the given data and Step 1, the results are tabulated in Table 19.8. The loading schedule shows all possible combinations of unit loadings listed in the order of increasing incremental fuel cost.

Related Calculations. The process of dispatching the committed generation is based on minimizing the cost of fuel. This results in an incremental generation loading, beyond minimum generation, in the order of increasing incremental fuel cost. The minimization of emissions may also be treated in a similar fashion if the incremental cost curve above is replaced by an incremental emission curve.

Daily Load Factor

The daily load factor is defined as the ratio of the load energy in the day to the energy represented by the daily peak demand multiplied by 24 h. Find the daily load factor, LF_D, when the daily-load energy is 21 GWh and the daily peak demand is 1000 MW.

Calculation Procedure

1. Multiply the Peak Demand by 24 h

The multiplication is (1000)(24) = 24,000 MWh = 24 GWh.

2. Determine the Daily Load Factor

$LF_D = 21 \text{ GWh}/24 \text{ GWh} = 0.875$

Related Calculations. The shape of the daily, seasonal, and annual load curves are important characteristics for operation and expansion of generation systems to meet the system load. Utilities record the chronological hourly loads on a continuous basis. A typical hourly load curve for a day is shown in Fig. 19.4. Note that the daily curve is an aggregate of the hourly curves.

System Unit Commitment

The unit commitment problem traditionally has been to schedule the operation of the units such that the production costs are minimized. A reregulated environment requires maximization of profit. If the price is fixed, then maximization of profit is the same as

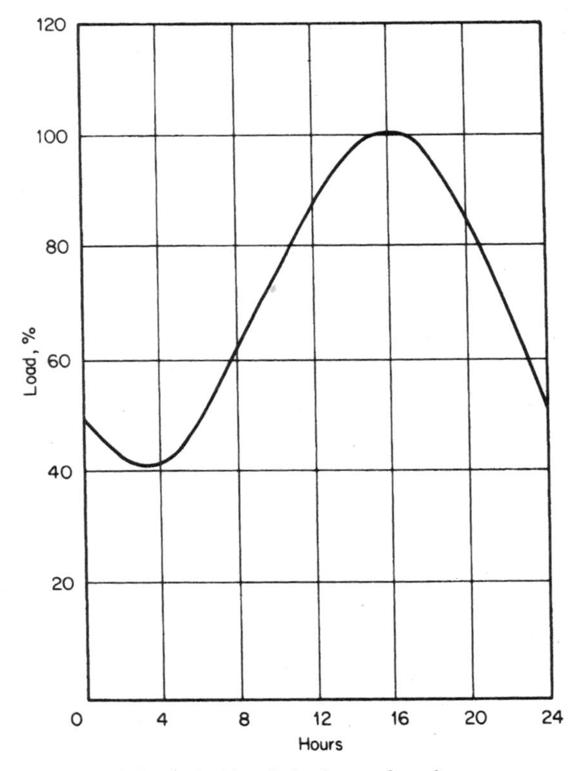

FIGURE 19.4 Typical hourly load curve for a day.

minimization of cost. The typical unit commitment solution arranges the starting and the stopping of the units such that the weekly demand cycle is followed subject to minimum generation time, to minimum downtime, startup costs, startup cycles, crew constraints, etc. Unit commitment is typically associated with the operation of thermal units. However, the same problem exists with hydro units.

Annual Load Factor

The annual load factor is defined as the ratio of annual load energy to the energy represented by the annual peak demand multiplied by 8760 h. It is possible to estimate the annual load factor from the average daily load factor by using typical daily and monthly peak-load variations.

Find the annual load factor LF_A when the average daily load factor $LF_D = 0.875$, the ratio of average daily peak load to monthly peak load $R_{WM} = 0.85$, and the ratio of average monthly peak load to annual peak load $R_{MA} = 0.8$.

Calculation Procedure

***1. Compute* $\mathbf{LF_A}$**

$LF_A = LF_D \cdot R_{WM} \cdot R_{MA} = (0.875)(0.85)(0.8) = 0.595.$

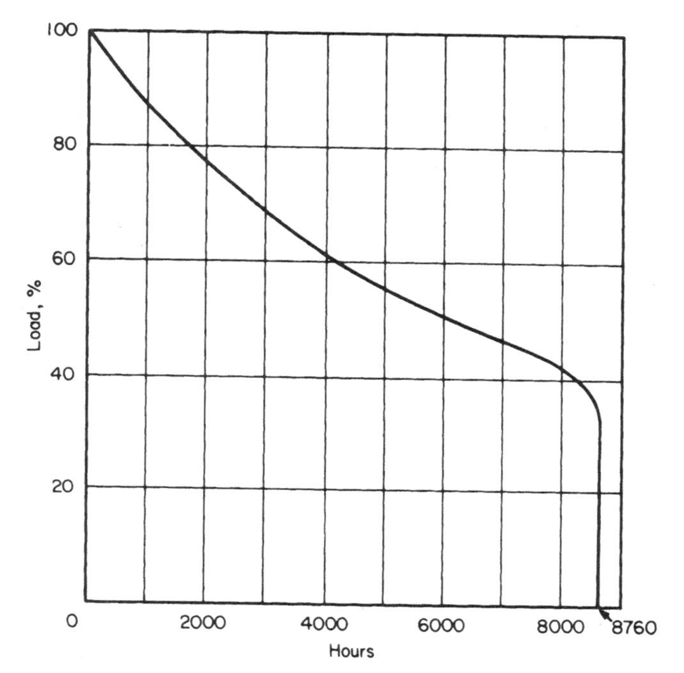

FIGURE 19.5 Annual load-duration curve.

Related Calculations. Load-duration curves are curves in which all the hourly loads in a time period (commonly a year) are arranged in descending order. This curve is not chronological. Because the area under the curve is the period-load energy, the curve is often useful in simplified utility economics calculations. A typical load-duration curve is provided in Fig. 19.5.

Load Management

The purpose of load management is to direct energy usage away from peak-load periods. Methods for load management include peak-sensitive rate structures and automatic control of power consumption to increase load diversity.

Find the annual load factor LF_A when the average daily peak load is reduced 5 percent from 1000 MW to 950 MW. Assume $R_{WM} = 0.85$ and $R_{MA} = 0.8$.

Calculation Procedure

1. Find Energy Associated with 950 MW for 24 h

The energy is $(950)(24) = 22{,}800$ MWh.

2. Find Resulting Annual Load Factor with a Daily Load Energy of 21,000 MWh

$LF_A = LF_D \cdot R_{WM} \cdot R_{MA} = (21{,}000/22{,}800)(0.85)(0.8) = 0.626.$

Related Calculations. The major problem with evaluating load-management devices is the difficulty in predicting the effect on demand and energy consumption based on actual operation of such devices. The economic value of the reductions is highly location-specific and time-specific. However, such devices provide dramatic relief when properly combined with risk-management techniques.

System Storage

Find the cost saving realized by operating a storage device with conversion efficiency $Os = 70$ percent on the utility system described by the daily load curve of Fig. 19.6. The storage reservoir has an additional storage capacity equivalent to 6 h charging at 50 MW as this day starts.

Calculation Procedure

1. Find Cost of 6-h Charging at 50 MW

The values on the right-hand vertical axis in Fig. 19.6 are the variable fuel, operations, and maintenance costs in mills/kWh for the generating plants operating this day.

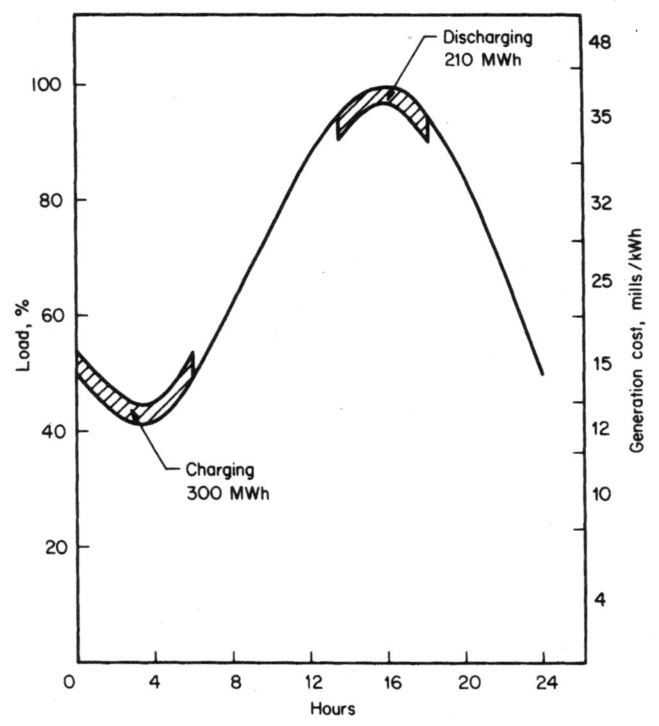

FIGURE 19.6 Example of economic operation of utility system storage.

(Cost of electricity is often expressed in mills/kWh or $MWh). The cost proportion for 6-h charging is estimated to be 60 percent at 12 mills/kWh, 40 percent at 15 mills/kWh. The cost of 6 h of charging (C_{CH}), then, is $C_{CH} = (0.60)(12)(6)(50) + (0.40)(15)(6)(50) = \$3960$.

2. Determine Cost of Stored Energy, in mills/kWh, Available for Discharging When Needed

Because the energy available for discharge is less than the energy used for charging by the conversion efficiency of 70 percent, the unit cost of the energy for discharge (C_{DC}) is: $C_{DC} = 39{,}601(6)(50)(0.70) = 18.9$ mills/kWh.

3. Find Cost of Existing Generation during the Daily Peak

From Fig. 19.6, it is seen that existing generation would cost 35 mills/kWh.

4. Find Cost Saving When Discharging Storage Device during Peak Period, Displacing Energy from Existing Generation

The energy available from the storage device is (6)(50)(0.7) = 210 MWh. The savings from operating the storage (S_{ST}) for 18.9 mills/kWh displacing energy at 35 mills/kWh is $S_{ST} = 210\text{ J}(35 - 18.9) = \$3381$.

Related Calculations. These savings are energy savings only. Potential capacity savings in conventional generation are dependent on storage capacity and the cost of the storage system. These costs must be included for a total system economic analysis. The economic value of the reductions is highly location-specific and time-specific. However, such devices provide dramatic relief when properly combined with risk-management techniques.

Forced-Outage Rate

Find the forced-outage rate, *FOR*, in percent for a generating unit that operated 6650 h in one year, with 350 h on forced outage and 1860 h on scheduled shutdown.

Calculation Procedure

***1. Compute* FOR**

Use $FOR = FOH/(FOH + SH)$, where FOH = forced-outage hours and SH = service hours. Substituting values, find $FOR = 350/(350 + 6650) = 0.05 = 5$ percent.

Related Calculations. Total utility economic analysis includes the analysis of both the cost of reliability and the cost of production of all the generating units and the utility system loads they serve. The study time period should be long enough to adequately include at least the major economic effects of the alternatives studied. In some cases, this could mean as much as 20 yr.

Cost of Electric Energy Production

Find the cost of electricity (CE) from a steam turbine-generator plant for the parameters in Table 19.9.

TABLE 19.9 Data for Cost of Electricity Calculation

Plant cost	\$1000/kW
Fixed charge rate	20 percent
Fixed operation and maintenance costs	\$200/kW per yr
Variable operation and maintenance costs	8 mills/kWh
Heat rate	10,000 kJ/kWh (9479 Btu/kWh)
Fuel cost	\$1.30/GJ (\$1.37/MBtu)
Capacity factor	0.70 p.u.
Assumed inflation	8 percent per yr
Assumed discount rate	12 percent per yr
Study period	30 yr

Calculation Procedure

1. Find Plant Cost

Use the given values: annual plant cost (1000)(0.20) = \$200/kW per yr and annual operating hours (8760)(0.70) = 6132 h/yr. Therefore, plant cost $(200 \times 10^3)/6132 =$ 32.6 mills/kWh.

2. Find Fixed Operation and Maintenance Costs

Fixed operation and maintenance costs $= (15 \times 10^3)/6132 = 2.4$ mills/kWh.

3. Compute Levelized Fuel Cost

The levelizing factor is found from $P = S/(1 + i)^N$ and $CRF = 0.12414$. Hence, fuel cost $= (10{,}000)(1.3)(2.06 \times 10^{-3}) = 26.8$ mills/kWh.

4. Find Levelized Variable Operation and Maintenance Costs

Operation and maintenance costs = (8)(2.06) = 16.5 mills/kWh.

5. Determine Cost of Electricity from Plant

The cost is found by adding the cost components found in Steps 1, 2, 3, and 4: cost of electricity = 32.6 + 2.4 + 26.8 + 16.5 = 78.3 mills/kWh.

Related Calculations. It should be noted that the result is a levelized value and, as such, should only be used for comparison with other similar generation alternatives. This value is not directly comparable to the cost of electricity in any one year. Note that risk factors (e.g., forced-outage rates for each unit) are not included properly. It is appropriate for quick evaluation of the ideal situation from a generation perspective when reliability does not include a financial penalty.

Life-cycle cost calculations may be made by estimating the changing operational costs as system economic conditions during the estimated lifetime of the unit. The life-cycle cost should be expressed in \$/kW per year and may include differences in unit rating and reliability, as well as the changes in unit capacity factor over the lifetime of the alternatives. Because the total lifetime must be included in the calculation, it is common to use levelized values for the variable components.

The effective capacities of the alternatives, as well as their capacity factors, may be different. It is, therefore, necessary to assume values for system replacement capacity and system replacement energy costs.

PROJECT SELECTION

The most frequent use of economics for engineers is to recommend the implementation of a project. The following techniques are used to select projects based on the economic factors, assuming that all other project characteristics are the same. If the characteristics are different, then the differences have to be converted to economic benefits and costs for a complete comparison. One or more of the following techniques may be used. The following gives examples of the primary techniques to calculate the total production cost, including capital, revenue, and residual costs for production expansion projects. Residual costs are folded into the capital cost of equipment. The end result should include a complete financial description of operation for each year with the new equipment in operation. Actual aggregation of each year for the life of the project is beyond this work.

Rate of Return / Minimum Attractive Rate of Return

The rate of return for a given cash flow is a method to estimate the interest rate equivalent for the cash loan. Assume that a new efficiency unit could be purchased for $1500 with a $500 down payment. Assume that your friendly banker offered a 2-yr "10 percent" business loan. The payments were calculated in the following way:

Amount borrowed	$1500
10 percent interest for 2 yr	$300
Bank's charge for arranging the loan	$20
Total	$1800
Monthly payment	$1800/24 = $75.83

What is the effective interest rate?

1. Find the Cash Flow Stream

The cash flow is +$1500 at the end of month 0, and −$75.83 at the end of each month 1 to 24.

2. Solve for the Interest Rate Using the Present Value for an Annuity Formula

The (p/a) ratio is found from the definition above:

$$75.83(p/a)_{24}^{i} = 75.83(1/i)[(1+i)^{24} - 1]/[(1+i)^{24}] = 1500$$

$$(p/a)_{24}^{i} = 1500/75.83 = 19.78$$

Assume an interest rate:

$$i = 2 \text{ percent}$$

Find the error for the assumed interest rate:

$$\varepsilon = 19.78 - 18.91 = 0.87$$

Update the interest rate based on the error using interpolation of an interest table or any other optimization technique:

$$i = 1.5 \text{ percent}$$

Find the error for the assumed interest rate:

$$\varepsilon = 19.78 - 20.03 = -0.25$$

Update the interest rate based on the error using interpolation of previous values:

$$0.87i - 1.305 = 0.5 - 0.25i$$

$$i = 1.612 \text{ percent}$$

Find the error for the assumed interest rate:

$$\varepsilon = 19.78 - 19.77 = 0.01$$

Repeat finding the error and updating the interest rate until the error is sufficiently small.

Note that a monthly rate of 1.612 percent is nominally 21.155 percent per yr.

Levelized Annual Cost

Levelizing of a nonuniform series of fixed and variable costs is often used in economic evaluation to compare the economic value of a cost series different in timing and magnitude. Find the levelized cost for the 5-yr cost series in Table 19.10. Assume a 10 percent discount rate (weighted cost of money) and a capital recovery factor $CRF = 0.2638$.

Calculation Procedure

1. Determine Sum of Present Values of Each Annual Cost

Use p/f from above, where f = future worth, p = present value, n = number of years, and i = interest rate. The calculated results are given in Table 19.10.

2. Compute Levelized Cost

The levelized annual cost is equal to the product of the present value and the CRF. Hence, the levelized annual cost = (\$2888.60)(0.2638) = \$762.01. A uniform series of \$762.01 per yr for 5 yr has the same present value as the actual cost series in the table.

Sinking Fund Depreciation

Find the annual depreciation expense for a \$1 investment with no salvage value after 5 yr. The discount rate is 10 percent and the sinking fund factor is 0.16380.

TABLE 19.10 Data for Calculating Levelized Costs

Year	Annual cost ($)	Present-value factor $[1/(1+t)^N]$	Present value P($)
1	400	0.9091	363.64
2	600	0.8264	495.84
3	800	0.7513	601.04
4	1000	0.6830	683.00
5	1200	0.6209	745.08
Present value			2888.60

Calculation Procedure

1. Find Accumulated Depreciation for Each Year

The accumulated depreciation at the end of year, n, may be calculated by

$$\sum_j d = [(1 + i)^n - 1]/[(1 + i)^N - 1] \qquad \forall_j = 1, \ldots, n$$

For example, if $n = 2$, then the accumulated depreciation is $[(1 + 0.1)^2 - 1]/[(1 + 0.1)^5 - 1] = 0.34398$. This and other values for n are tabulated in Table 19.11.

2. Find Interest on Last Year's Accumulated Depreciation

Multiply the accumulated depreciation by 0.1. The resulting values are given in Table 19.11.

3. Find Annual Depreciation Expense

This value may be found by adding the results from Steps 1 and 2 or by using the equation $d_n = [i(1 + i)^{n-1}]/[(1 + i)^N - 1]$. For example, if $n = 2$, $d_n = [0.1(1.1)^{2-1}]/[(1.1)^5 - 1] = 0.18018$. This and other values for n are tabulated in Table 19.11.

Related Calculations. Depreciation is a method of accounting to ensure recovery of the initial capital investment, adjusted for a possible salvage value. The depreciation expense is typically based on a yearly basis, but the yearly amount does not necessarily reflect the actual loss of value of the investment.

There are several common methods of calculating annual depreciation expenses. All methods calculate an annual series of expenses that adds up to the initial capital expense less the salvage value at the end of useful life.

The most common method for calculating annual depreciation expense is the straight-line method. The annual expense is equal to the initial capital investment minus the end-of-life salvage value divided by the depreciation life in years. For example, an investment of $1 with no salvage value after 5 yr would have $0.20/yr in depreciation.

The sinking fund depreciation method accumulates depreciation expenses more slowly than the straight-line method. The annual depreciation expense is equal to the sinking fund factor times the initial investment plus the interest charges on the previous accumulated depreciation.

Fixed-Charge Rate

Assume that the tax rate is 50 percent, the return R multiplied by the weighted cost of money in equity is 0.045, accounting depreciation is 0.2, the levelized tax depreciation is 0.215, and the cost of money is 12 percent. Calculate the fixed-charge rate.

TABLE 19.11 Data for Sinking Fund Depreciation Calculation

Year n	Sinking fund factor (SFF)	Accumulated depreciation	Interest on last year's accumulated depreciation	Total annual depreciation expense d
1	0.16380	0.16380	0	0.16380
2	0.16380	0.34398	0.01638	0.18018
3	0.16380	0.54218	0.03440	0.19820
4	0.16380	0.76020	0.05422	0.21802
5	0.16380	1.00000	0.07602	0.23982

Calculation Procedure

1. Find Income Tax Effect

Use

$$T = [t/(1 - t)](R_e - d_T - d_A),$$

where t is the tax rate, R_e is the return (R) multiplied by the weighted cost of money in equity, d_T is the levelized tax depreciation based on sum-of-years digits, and d_A is the accounting depreciation. Substituting values yields

$$T = (0.5/0.5)(0.045 - 0.2150 - 0.2) = 0.03.$$

2. Determine the Fixed-Charge Rate

The fixed-charge rate (FCR) is given by

$$FCR = [R + d_A + T + T_A + I],$$

where T_A is the ad valorem tax, R is the return needed to cover the cost of money and depreciation, and I is the insurance. Assume

$$T_A + I = 0.025;$$

hence,

$$FCR = 0.0774 + 0.2 + 0.03 + 0.025 = 0.362.$$

Related Calculations. Investment decisions often involve comparing annual operating expenses with the cost of capital needed for the investment and the revenues generated from a new project. For regulated electric utilities, the revenues are regulated and it is therefore convenient to relate investment decisions to revenue requirements. The revenue requirements are equal to the annual expenses, such as fuel costs and operations and maintenance (O&M) costs, plus the annual fixed charges on the investment. The fuel, operation, and maintenance costs are obtained from estimates of production needs.

The assumed fixed charges are normally estimated by a levelized fixed-charge rate applied to the initial investment. The fixed charges should include a return to the shareholder, interest payments on debt, depreciation expenses, income tax effects, property taxes, and insurance.

In the first example, the cost of money was approximately 12 percent. For a 5-yr plant life and 12 percent cost of money, the annual charge necessary to recover the capital is the interest plus the sinking fund factor equal to $0.12 + 0.1574 = 0.2774$. The sinking fund could be regarded as depreciation. It is common, however, to use straight-line depreciation where the depreciation rate is $1/5 = 0.2$. An excess of $0.2 - 0.1574 = 0.0426$ is thereby obtained. This gives a revenue requirement resulting from the cost of money and depreciation method equal to $0.12 - 0.0426 = 0.0774$.

The equation for the return R needed to cover the cost of money and depreciation is

$$R = i - (d - SFF)$$

where i is the weighted cost of money, d is the depreciation used for accounting purposes, and SFF is the sinking fund factor.

Rate of return (ROR), also called return on investment (ROI), is a more common measure in an industrial environment. This measure often refers to the discounted cash-flow method of economic evaluation. Because most industries attempt to maximize the ROI,

TABLE 19.12 Annual Operating Costs of Alternatives A and B

Year	Annual operating costs	
	Alternative A	Alternative B
1	$6000	$6500
2	$5800	$6600
3	$5600	$6700
4	$5400	$6800
5	$5200	$7000

the approach is to estimate revenues and expenditures and then find the interest rate that, when used to discount the cash flows over the life of the project, will make the present-worth sum of the cash flows equal to the initial investment. An example of *ROI* and discounted cash-flow analysis will be considered later.

Revenue-Requirements Method

Alternatives A and B have operational characteristics that allow both to perform the same service. The capital investment required is $50,000 and $48,000 for Alternatives A and B, respectively. The useful life for both alternatives is 5 yr. Annual operating costs are given in Table 19.12. The weighted cost of money is 12 percent and the fixed-charge rate is 36.2 percent as calculated in the previous example. Make an economic choice between Alternatives A and B.

Calculation Procedure

1. Determine the Sum of Present Values of Revenue Requirements to Cover Annual Operating Costs

The calculated results are summarized in Table 19.13.

2. Find the Present Value of Annual Revenue Requirements for the Capital Investment

The levelized annual fixed charges are found by multiplying the investment by the fixed-charge rate. For Alternative A, ($50,000)(0.362) = $18,100/yr; for Alternative B, ($48,000)(0.362) = $17,376/yr.

TABLE 19.13 Present Values of Revenue Requirement; 12 Percent

Year	Alternative A	Alternative B
1	$5357	$5804
2	$4624	$5261
3	$3986	$4769
4	$3432	$4322
5	$2951	$3972
Sum of present values	$20,350	$24,128

The present values are found by dividing the levelized annual fixed charges by the capital recovery factor, $CRF = [i(1 + i)^N]/[(1 + i)^N - 1]$, where i = interest rate and n = number of years. For $n = 5$ yr and $i = 12$ percent, $CRF = 0.12(1.12)^5/[(1.12)^5 - 1] = 0.2774$, and the present values of the annual revenue requirements for the capital investment are: Alternative A, \$18,100/0.2774 = \$65,249 and Alternative B, \$17,376/0.2774 = \$62,639.

3. Find the Present Value of Total Revenue Requirements for the Two Alternatives

This is the sum of the results obtained in Steps 1 and 2: Alternative A, 20,350 + 65,249 = \$85,599 and Alternative B, 24,128 + 62,639 = \$86,767. Alternative A, having the lowest present value of total revenue requirements, is the economic choice.

Related Calculations. The economic evaluation of engineering options is an important factor in equipment application. The capital cost of each alternative must be combined with its operating cost to develop a base for comparison. It is most likely that the capital costs differ for the alternatives, as do the operating costs.

The time period chosen for a study should include at least a major portion of the expected useful life of the equipment. For a utility, this would mean 15 to 20 yr, while study periods of only a few years could be appropriate for an industry.

Often, economic comparisons are made without considering the effects on the operation of the interconnected systems. This is acceptable if the alternatives have similar operational characteristics, that is, equal annual energy production for generating units or operating hours for electric motors. If the effects on the interconnected system are uncertain, total system cost evaluations must be performed. The revenue-requirements method is typical of the utility industry, while the discounted cash-flow and payback-period methods are typical in industrial evaluation.

Discounted Cash-Flow Method

Using the data in the previous example, choose between Alternatives A and B by the discounted cash-flow method.

1. Find Difference in Annual Cash Flows for Alternatives

The annual cash flows are provided in Table 19.14 and include operating expenses, a tax depreciation based on a sum-of-years digit, and an ad valorem tax at a rate of 0.025.

TABLE 19.14 Annual Cash Flows for Alternatives A and B

	Year 1	Year 2	Year 3	Year 4	Year 5
Incremental investment	−2000				
Operating cost B	6500	6600	6700	6800	7000
Operating cost A	6000	5800	5600	5400	5200
Operating saving for incremental investment	500	800	1100	1400	1800
Tax depreciation	−667	−533	−400	−267	−133
Ad valorem tax	−50	−50	−50	−50	−50
Taxable income	−217	217	517	1083	1617
Tax at 50 percent	109	−109	−259	−542	−809
Net income	−108	108	258	541	808
Tax depreciation	667	533	400	267	133
Cash flow	559	641	658	808	941

TABLE 19.15 Annual Cash Flows for Alternative A

	Year 1	Year 2	Year 3	Year 4	Year 5
Cash flow	−1441	−644	685	888	+945

Calculation Procedure

2. Find Discount Rate Using the Sum of Present Values Equal to the Investment Difference

The discount rate is found by trial and error. Try $i = 15$ percent; this gives a present value (PV) equal to $559/1.15 + 641/1.15^2 + 658/1.15^3 + 808/1.15^4 + 941/1.15^5 = 2333$. For $i = 22$ percent, $PV = 1964$ is obtained. If $i = 21.22$ percent, $PV = 2000$.

3. Make the Economic Choice

The result of the discounted cash-flow evaluation is a 21.22 percent rate of return on the incremental investment. If this rate equals, or exceeds, the hurdle rate established for this investment, Alternative A is the economic choice.

Payback Period

A simple method of assessing a project is its payback period. Use the cash flows of Table 19.15. Simply accumulate the cash flows, year by year, to determine when the accumulated cash flow first becomes positive. Note that the accumulated cash flows become positive between the ends of years 2 and 3. After $2\frac{1}{2}$ yr, any income is profit.

This is a very crude evaluation technique. The payback period for Project A may be shorter than the payback for Project B, whereas the rate of return for Project B may be much larger than the rate of return for Project A. Ignoring the time value of money is the most critical downfall of this method.

Screening Curves

Screening curves (see Sec. 8) are useful for preliminary screening of alternatives with widely different characteristics. Screening curves also may be used for a preliminary evaluation of a new generation concept in comparison with conventional generation.

TABLE 19.16 Developing Screening Curves for Three-Generation Alternatives

Data	Unit A	Unit B	Unit C
Plant cost ($/kW)	900	450	200
Fixed O&M cost ($/kW per year)	10	3	1
Fuel price ($/GJ)	1.5	4	6
($/Mbtu)	1.58	4.22	6.33
Heat rate (kJ/kWh)	10,000	9000	13,000
(Btu/kWh)	9477	8530	12,321
Variable O&M cost ($/MWh)	8	2	3
Levelizing factor, fuel and O&M	2.06	2.06	2.06
Fixed-charge rate (percent)	20	20	20

TABLE 19.17 Levelized Fixed Costs

Plant	Plant cost ($/kW per yr)	Levelized fixed O&M ($/kW per yr)	Total fixed costs ($/kW per yr)
A	180	20.6	200.6
B	90	6.2	96.2
C	40	2.1	42.1

The screening curve is a plot of annual cost of generation expressed in $/kW per year as a function of yearly hours of operation (capacity factor). These curves are straight lines with vertical intercepts at the fixed annual cost and with a slope determined by the variable fuel and operation and maintenance costs; levelized values should be used.

Develop screening curves for the three-generation alternatives in Table 19.16.

Calculation Procedure

1. Find the Levelized Fixed Costs

The fixed costs are the plant cost and the fixed operation and maintenance costs in Table 19.16. The calculated results are given in Table 19.17.

2. Determine the Levelized Variable Costs

From the given values, the calculated values are provided in Table 19.18.

3. Assuming 4000-h/yr Operation, Determine the Total Variable plus Fixed Costs

The values in Step 2 are multiplied by 4000×10^{-3} to obtain the cost in $/kW per year. The results are summarized in Table 19.19.

4. Plot the Screening Curves

The resulting screening curves are plotted in Fig. 19.7.

UNCERTAINTY

The preceding analysis does not take properly into account the uncertainty of forecasted data nor of events that may occur, such as the forced outages of units. One popular

TABLE 19.18 Levelized Variable Costs

Plant	Plant cost ($/kW per yr)	Levelized fixed O&M ($/kW per yr)	Total fixed costs ($/kW per yr)
A	30.9	16.5	47.4
B	74.2	4.1	78.3
C	160.7	6.2	166.9

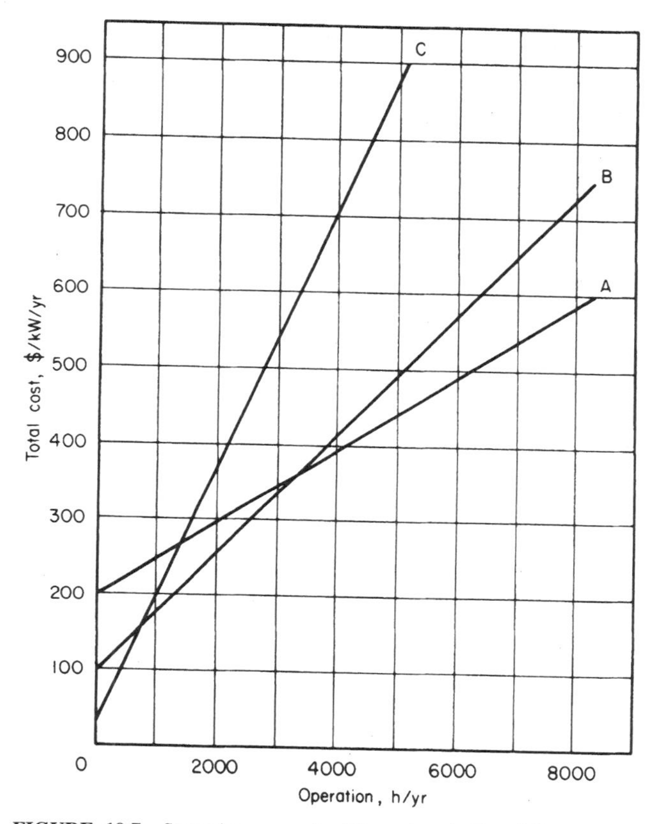

FIGURE 19.7 Screening curves for Alternatives A, B, and C.

method of describing these is to use decision analysis. Decision analysis describes uncertainty by drawing trees to depict the possible outcome of each future event along with a probability (subjective) of that event occurring. The evaluation of such future events includes simulation using Monte Carlo techniques as well as sensitivity analysis. Note that tornado diagrams are most often used to show the importance of uncertain variables in such analysis. These techniques have been applied to many classical problems of uncertainty analysis, including probabilistic production costing (forced outage rates). These techniques are beyond this work.

TABLE 19.19 Total Variable and Fixed Costs

Plant	Total cost at 0 h/yr ($/kW per year)	Total cost at 4000 h/yr ($/kW per year)
A	200.6	390.2
B	96.2	409.4
C	42.1	709.7

APPENDIX: TABLE OF VARIABLES

P	price
Q	quantity
K	capital
Cr	revenue costs
Cp	production cost
$g(K)$	capital allocation to production cost
$h(Cr)$	revenue costs to production allocation
MC_p	marginal cost production
MC	marginal cost
AC_p	average cost production
Π	profit
R	revenue
MR	marginal revenue
E_d	elasticity of demand
$SRMC$	short-run marginal cost
$SRAC$	short-run average cost
$LRMC$	long-run marginal cost
$LRAC$	long-run average cost
i_B	bond interest rate
i_P	preferred stock interest rate
i_C	return on common stock
DR	debt ratio (fraction of bonds)
CR	common-stock ratio (fraction of common stock)
PR	preferred-stock ratio (fraction of preferred stock)
i	weighted average interest rate
f	future values
p	present values
i_i	rate of inflation
j	future period
n	future period for the payment
a	annuity
m	discount periods
i_m	interest for each period
i_{eff}	effective interest rate
i_{nom}	nominal interest rate
N	number of future periods
d_n	depreciation
LF_D	daily load factor
LF_A	annual load factor
R_{WM}	ratio of average daily peak load to monthly peak load
R_{MA}	ratio of average daily peak load to monthly peak load
Os	conversion efficiency
C_{CH}	cost of charging
C_{DC}	cost of discharge energy
S_{ST}	savings from storage operation
FOR	forced-outage rate
SH	service hours
FOH	forced-outage hours
CE	cost of electricity plant cost
CRF	capital recovery factor
T	income tax effect
t	tax rate
R_e	equity weighted return
R	return
d_T	levelized tax depreciation based on sum-of-years digits
d_A	accounting depreciation
FCR	fixed-charge rate
I	insurance
SFF	sinking fund factor
ROR	rate of return
ROI	return on investment

BIBLIOGRAPHY

Baye, M. R. 2000. *Managerial Economics and Business Strategy.* New York: McGraw-Hill Co.

Bergen, A. R., and V. Vittal. 1999. *Power Systems Analysis.* Upper Saddle River, N. J.: Prentice Hall.

Berrie, T. W. 1992. *Electricity Economics and Planning.* London: Peter Peregrinus, Ltd.

Billington, R., and W. Li. 1994. *Reliability Assessment of Electric Power Systems Using Monte Carlo Methods.* New York: Plenum Press.

Binger, B. R., and E. Hoffman. 1988. *Microeconomics with Calculus.* New York: Harper Collins Publishers.

Debs, A. S. 1988. *Modern Power System Control and Operation.* Norwell, M.A.: Kluwer Academic Publishers.

Kahn, E. 1991. *Electric Utility Planning and Regulation.* Washington D. C.: American Council for Energy-Efficient Economy.

Leech, D. J. 1982. *Economics and Financial Studies for Engineers.* Chichester, U.K.: Ellis Horwood Limited.

Newnan, D. G., and B. Johnson. 1995. *Engineering Economic Analysis.* San Jose, C.A.: Engineering Press, Inc.

Sheblé, G. B. 1999. *Computational Auction Mechanisms for Deregulated Electric Power Industry.* Norwell, M.A.: Kluwer Academic Press.

Wang, X., and J. R. McDonald. 1994. *Modern Power System Planning.* Berkshire, U.K.: McGraw Hill.

Wood, A., and B. Wollenberg. 1996. *Power Generation, Operation and Control.* New York: John Wiley.

LIGHTING DESIGN*

Charles L. Amick, P.E., F.I.E.S., L.C.
Lighting Consultant

INTRODUCTION

"Lighting accounts for 20 to 25 percent of all electricity consumed in the United States, and as a nation, we spend billions on it each year."[1] "Electric lighting represents about half of all electricity used in modern commercial buildings."[2]

Lighting design involves:

Establishing proper illuminance levels for the visual tasks to be performed.

Knowledge of the physical characteristics of space to be illuminated.

Selecting and locating luminaries (style appropriate for space, efficacy, color, and lumens per watt of lamps, etc.).

Calculating what portion of the generated light will reach the work plane.

Computing the number of luminaires needed.

Developing a lighting system maintenance plan for the owner to follow.

---

*Original work by John P. Frier, Former Lighting Application Specialist, General Electric Company.

[1]"DOE Expected to Adopt New Energy-Efficient Lighting Standards," *Lighting Design and Application,* Dec. 1999, Vol. 29, No. 12, p. 17.

[2]*Lighting Design Practice*, California Energy Commission, March, 1990, p. 1.

AVERAGE ILLUMINANCE

Many lighting projects call for reasonably uniform illuminance levels for work or other activities located throughout the space. Additional supplementary luminaires can provide higher levels that may be needed for more difficult visual tasks at specific work stations.

Calculation Procedure

The procedure generally employs the *lumen method.* The lumen is the unit of luminous flux, the latter being the integrated product of the energy per unit wavelength and the spectral luminous efficacies. Although luminous flux is the time rate of flow of light, the lumen can be considered the unit of quantity of light, equal to the flux on a square foot of surface, all points of which are one foot from a point source of one candela. Light sources are rated in lumens.

Illuminance is the density of the luminous flux incident on a surface. It is the quotient when the luminous flux is divided by the area of the surface when the flux is uniformly distributed. *Footcandle* is the unit of illuminance when the foot is the unit of length, thus meaning lumens per square foot. *Lux* is the International System (SI) unit of illuminance, where the meter is taken as the unit of length, thus meaning lumens per square meter.

$$\text{Lumen-method calculations: } E = (N)(LL)(\text{CU})(\text{LLF})(\text{BF})(\text{TF})/A$$

where

E = average maintained illuminance, footcandles

N = total number of lamps in all the luminaires contributing to the area being illuminated (lamps per luminaire times number of luminaires)

LL = rated lumen output of the lamp being used

CU = coefficient of utilization; the percentage of lamp lumens reaching the illuminated area, expressed as a decimal

LLF = light loss factor, to estimate the illuminance decline between initial and maintained conditions (multiplying the lamp lumen depreciation, LLD, by the luminaire dirt depreciation, LDD, is often considered sufficiently accurate)

BF = ballast factor

TF = tilt factor

A = lighted area, in square feet

For average initial illuminance, omit LLF in the above formula.

Typical example: A 32,000-ft^2 storage yard is lighted by three floodlights, each having a 400-W high-pressure sodium lamp rated at 51,000 lumens. If the CU was 0.35, the LLF 0.8, and ballast and tilt factors both 1.0, the average maintained illuminance would be 1.3 footcandle.

Power Calculation. Input watts per luminaire: 465 × 3 = 1395 W. Starting/operating current: for 120-V ballasts, 2.8/4.2 A; for 208 V, 1.6/2.45; for 240 V, 1.4/2.15; for 277 V, 1.1/1.85; and for 480 V, 0.75/1.05.

ILLUMINANCE AT A POINT

The lighting system may involve the need to provide specific levels of horizontal, vertical, or other angular illuminance in selected areas, to achieve greater attention or enhance visibility.

Calculation Procedure

The procedure involves a point-by-point method to determine the illuminance at many specific locations. The results provide information for calculating the uniformity throughout all or portions of the lighted area and are helpful in selecting the beam spread and maximum candlepower of the luminaires.

Most point-by-point calculations are based on the inverse square law:

$$\text{Footcandles} = \text{candlepower}/\text{distance}^2$$

when candlepower is expressed in candelas using data from the candlepower distribution curve for the luminaire involved and distances are in feet. As shown in Fig. 20.1, the illuminance on the horizontal surface from beam D at an angle θ from the luminaire axis is a function of D^2, the cosine of θ, and the candlepower at angle θ. Similarly, the illuminance on the vertical surface involves the sine of angle θ.

Luminaire manufacturers provide photometric data showing candela (candlepower: cd) values at various angles. In Fig. 20.1, for example, the light beam identified as D is 90° from the face of the luminaire and would have been pointing straight down if the luminaire had been mounted horizontally. Since the luminaire is tilted up 45°, the number of candelas directed straight down is obtained from the photometric data (published in plotted or tabular form). In Fig. 20.1, the illuminance in footcandles on the horizontal surface directly below the luminaire is equal to the 45° cd divided by the mounting height (in feet) squared, $E = \text{cd}/\text{distance}^2$. If the 45° candela value was 2500, $E = 2500$ divided by the mounting height squared. For a mounting height of 10 ft, E would be 25 footcandles.

If the light strikes the horizontal surface at an angle other than 90°, the area lighted is increased as a function of the cosine of the incident angle—θ in Fig. 20.1. In that example, if the mounting height was 10 ft, and the horizontal H line is also 10 ft, the distance from the luminaire to Point A is 14.14 ft. In this instance, if the zero degree candela value is 4000, the horizontal illuminance at Point A will be

$$E = (\text{cd})(\cos 45°)/(14.14)^2$$

$$E = (4000)(0.707)/(200)$$

$$= 14.14 \text{ footcandles}$$

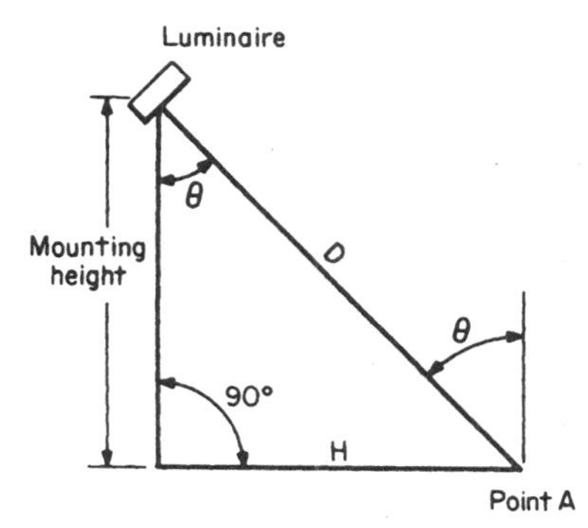

FIGURE 20.1 When illumination strikes a surface at an angle, the area lighted is increased as a function of the cosine of the incident angle θ.

Another, often more convenient formula uses the mounting height, MH, instead of the diagonal distance, D, shown in Fig. 20.1:

$$E = (\text{cd})(\cos^3 \theta)/MH^2$$

For the preceding example,

$$E = (4000)(0.707)^3/100$$

$$E = (4000)(0.3535)/100$$

$$E = (1414)/100 = 14.14 \text{ footcandles}$$

Factors to Consider

Calculations are affected by the light-loss factors and the recommended illumination levels for different activities. Major differences occur because of the types of luminaires and lamps that are normally used, and the manner in which the coefficient of utilization is derived.

Light-Loss Factor. The light-loss factor, also called the maintenance factor, is used to increase the initial illumination level to compensate for the normal deterioration of the lighting system in use. The value can be calculated for the mean illumination level, which usually occurs at the midpoint of the luminaire's cleaning and relamping period. The LLF also can be calculated for the end of the relamping period, which is when the luminaire reaches its minimum output and the illumination level is at its lowest point. LLF values may be recommended by the luminaire manufacturer.

Levels of Illumination. Extensive information on selecting appropriate illumination levels is contained in the *IESNA Lighting Handbook* published by the Illuminating Engineering Society of North America (IESNA). Similar illuminance values, if adopted as American national standards, are also published by the American National Standards Institute (ANSI).

Recommended illuminances should be provided on tasks, whether they are in a horizontal, vertical, or inclined plane. Computations for interior spaces that employ coefficients of utilization give average horizontal illuminance at designated work-plane heights above the floor.

Table 20.1 shows (1) three illuminance categories for typical orientation and simple tasks where visual performance is largely unimportant, (2) three categories for common visual tasks where visual performance is important, and (3) one special category where visual performance is of critical importance. The *IESNA Lighting Handbook*, 9th ed., also reviews situations justifying deviations from Table 20.1, where specific spaces have tasks that are *not* typical with respect to task contrast and/or task size and/or age.

Table 20.2 gives the industrial section of the *IESNA Lighting Design Guide* from the handbook cited above, providing many criteria important to achieve an appropriate quality of the visual environment.

Light-Source Selection. The cost of energy to operate a lighting system is a major factor in selecting the type of lamp and luminaire. The lamp's maintained efficacy (lumens per watt, lm/W) and the luminaire coefficient of utilization and maintained efficiency are the key factors. Low-cost, inefficient systems can be justified only if the annual use is very low.

Table 20.3 provides overall comparison of commonly used lamp types, and Table 20.4 lists a few of the more popular types. Lamp improvements occur at frequent intervals, so recently published catalog material should be used.

Illuminance requirements can also be determined by measurements and analysis. Changes can also be made if it can be determined that improvements result in greater productivity or accuracy. Illuminance levels in adjacent areas should not vary by more than 3:1.

Coefficient of Utilization (CU). The coefficient of utilization is an important factor in the zonal-cavity method of lighting calculation. There are three major factors that influence the CU of an interior lighting system: the efficiency and photometric distribution of

TABLE 20.1 Determination of Illuminance Categories*

Orientation and simple visual tasks. Visual performance is largely unimportant. These tasks are found in public spaces where reading and visual inspection are only occasionally performed. Higher levels are recommended for tasks where visual performance is occasionally important.		
A	Public spaces	30 lx (3 fc)
B	Simple orientation for short visits	50 lx (5 fc)
C	Working spaces where simple visual tasks are performed	100 lx (10 fc)
Common visual tasks. Visual performance is important. These tasks are found in commercial, industrial and residential applications. Recommended illuminance levels differ because of the characteristics of the visual task being illuminated. Higher levels are recommended for visual tasks with critical elements of low contrast or small size.		
D	Performance of visual tasks of high contrast and large size	300 lx (30 fc)
E	Performance of visual tasks of high contrast and small size, or visual tasks of low contrast and large size	500 lx (50 fc)
F	Performance of visual tasks of low contrast and small size	1000 lx (100 fc)
Special visual tasks. Visual performance is of critical importance. These tasks are very specialized, including those with very small or very low contrast critical elements. Recommended illuminance levels should be achieved with supplementary task lighting. Higher recommended levels are often achieved by moving the light source closer to the task.		
G	Performance of visual tasks near threshold	3000 to 10,000 lx (300 to 1000 fc)

*From the *IESNA Lighting Handbook*, 9th ed. The illuminance recommendations are "maintained-in-service" values. Allowances should be made for inaccuracies in calculation, such as erroneous estimates of room-surface reflectances, etc.

the luminaire, the relative shape of the room, and the reflectance of the room surfaces. These factors are combined in a coefficient of utilization table for each luminaire type. Table 20.5 shows sample tables for six commonly used luminaire types.

To account for both uncertainty in photometric measurements and uncertainty in space reflections, measured illuminances should be with ±10 percent of the recommended value. It should be noted, however, that the final illuminance may deviate from these recommended values due to other lighting design criteria.

For calculation purposes, the room is divided into three cavities, as shown in Fig. 20.2. The utilization of the lighting system is a function of the cavity ratio for each section: ceiling-cavity ratio $\text{CCR} = 5h_{\text{CC}}(L + W)/LW$, room-cavity ratio $\text{RCR} = 5h_{\text{RC}}(L + W)/LW$, and floor-cavity ratio $\text{FCR} = 5h_{\text{FC}}(L + W)/LW$, where h_{CC}, h_{RC}, and h_{FC} are as defined in Fig. 20.2.

The ceiling- and floor-cavity ratios are useful in adjusting the actual reflectance of the ceiling and floor surfaces to their effective reflectance based on the size and depth of the cavity. For shallow cavities (2 m or less), the actual surface reflectance can be used with little error. Correction tables are given in lighting handbooks.

TABLE 20.2 Industrial Section of the IESNA Lighting Guide

Very important / Important / Somewhat important

Blank = Not important or not applicable

Industrial locations and tasks[a] / Design issues	Appearance of space and luminaires	Color appearance (and color contrast)	Daylighting integration and control	Direct glare	Flicker (and strobe)	Intrinsic material characteristics	Light distribution on surfaces	Light distribution on task plane (uniformity)	Luminances of room surfaces	Modeling of faces or objects	Reflected glare	Shadows	Source/Task/Eye geometry	Special considerations	Notes on special considerations	Illuminance on task plane[b]	Category or value (lux)	Notes — see end of section	Reference chapter(s)
Basic industrial tasks																			
Raw material processing (cleaning, cutting, crushing, sorting, grading)																			
Coarse																	C		
Medium																	D		
Fine																	E		
Very fine																	F		
Materials handling																			
Wrapping, packing, and labeling																	D		
Picking stock, classifying																	D		
Loading, inside trucks and freight cars																	C		
Component manufacturing																			
Large																	D		
Medium																	E		
Fine																	F		

Machining	
Rough bench or machine work	D
Medium bench or machine work (ordinary automatic machines, rough grinding, medium buffing, and polishing)	E
Fine bench or machine work (fine automatic machines, medium grinding, fine buffing, and polishing)	G
Extra-fine bench or machine work (fine grinding)	G
Assembly	
Simple	D
Difficult	F
Exacting	G
Warehousing and storage	
Inactive	B
Active: bulky items; large labels	C
Active: small items; small labels	D
Inspection	
Simple	D
Difficult	F
Exacting	G
Service spaces	
Stairways, corridors	B
Elevators, freight and passenger	B
Toilets and wash rooms	C
Shipping and receiving	D
Maintenance	E
Motor and equipment observation	D
Control panel and VDT observation	C

continued

TABLE 20.2 Industrial Section of the IESNA Lighting Guide *(Continued)*

Legend: Very important; Important; Somewhat important; Blank = Not important or not applicable

Industrial locations and tasks[(a)] / Design issues	Appearance of space and luminaires	Color appearance (and color contrast)	Daylighting integration and control	Direct glare	Flicker (and strobe)	Intrinsic material characteristics	Light distribution on surfaces	Light distribution on task plane (uniformity)	Luminances of room surfaces	Modeling of faces or objects	Reflected glare	Shadows	Source/Task/Eye geometry	Special considerations	Notes on special considerations	Illuminance on task plane[(b)]	Category or value (lux)	Notes — see end of section	Reference chapter(s)
Welding																			
Orientation																	D		
Precision manual arc-welding																	G		
(Inspection of work after completion of weld)																			
Manual crafting (engraving, carving, painting, stitching, cutting, pressing, knitting, polishing, woodworking)																			
Coarse																	D		
Medium																	E		
Fine																	F		
Exacting																	G		

Notes: (a) For details on specific tasks or spaces refer to Chapter 19, Industrial Lighting. (b) The task may be horizontal, inclined, or vertical.

TABLE 20.3 Comparison of Commonly Used Lamp Types (HID Types Based on 400-W Sizes)

Lamp	Initial lumens per watt (LPW)*	Rated life, h	Lamp lumen depreciation (LLD) mean	CU†	Burning position	Minutes: Warmup	Minutes: Hot restart	Lamp cost	Color temperature
Incandescent	20	1000 2000 (quartz)	0.89	High	Any	0	0	Very low	3000 K
Mercury, clear	52.5	24,000†	0.80*	Medium	Any	5–7	3–6	Low	5700 K
Fluorescent	80	18,000	0.85	Medium-low	Any	0	0	Low	4100 K
Metal halide	85–100	20,000 vert.	0.75–0.80	High	Any	2–4	10–15	Medium	4000 K
High-pressure sodium	125	24,000	0.90	High	Any	3–4	1	High	2100 K

*Based on 16,000 h.

†CU = Coefficient of utilization. Range indoor high, 0.70+; medium, 0.50–0.70; low, less than 0.50. Range outdoor: high, 0.50+; medium, 0.40–0.50; low, 0.40.

TABLE 20.4 Characteristics of Some Popular Lamp Types*

Lamp type	Initial lumens[a]	Mean lumens[b]	Rated life, h[c]	Luminaire line watts[d]
Incandescent				
200 W A21 inside frost 120 V	3920		750	200
500 W PS35 inside frost 120 V	10,850		1000	500
1,000 W PS52 inside frost 120 V	23,100		1000	1000
1,500 W PS52 clear 130 V	34,400		1000	1000
Fluorescent (energy saving types)				
48-in 32 W T8	2950	2800	20,000	61
48-in 40 W T12 ES CW (34 W)	2650	2280	20,000	79
96-in 59 W T8 slimline	5900	5490	15,000	108
96-in 60 W T12 CW slimline	5500	5060	12,000	132
96-in 95 W T12 CW (800ma HO)	8000	6960	12,000	202
96-in 100 W T12 CW (1,500ma)	13,500	10,125	10,000	455
Mercury (phosphor-coated vertical operation)				
400 W BT37	22,600	14,400	24,000+	453
1000 W BT56	58,000	29,000	24,000+	1090
Metal halide				
400 W ED37 clear	41,000	31,200	20,000	460
400 W ED37 phosphor coated	41,000	27,700	20,000	460
400 W ED37 Pulse start clear	42,000[e]	33,600[e]	20,000	452
400 W ED37 Pulse start coated	40,000[e]	30,800[e]	20,000	452
1000 W BT56 clear	115,000	72,300	12,000	1090
1000 W BT56 phosphor coated	110,000	66,000	12,000	1090
High-pressure sodium, clear				
250 W ED18	28,000	27,000	24,000+	295
400 W ED18	51,000	45,000	24,000+	465
600 W T1S	90,000	81,000	12,000+	685
1000 W E25	140,000	126,000	24,000+	1090

*Check manufacturers' technical literature for current data, as values change frequently.

[a] After 100 burning hours for fluorescent and HID lamps.

[b] Not normally published for incandescent lamps, but Lamp Lumen Depreciation (LLD) is often based on light output at 70 percent of rated life (0.89 for 200-, 500-, 1000- and 1500-watt general service lamps). The mean lumens of fluorescent and metal halide lamps occurs at 40 percent of their rated life, and at 50 percent of rated life for mercury and high pressure sodium lamps.

[c] For fluorescent lamps—3 h per start using ballasts which meet industry standards; for HID lamps—10 h per start on ballasts having specified electrical characteristics.

[d] For fluorescent-lamp luminaires, with two energy-saving lamps and one energy-saving magnetic ballast or one electronic ballast.

[e] In open fixtures.

LIGHTING SYSTEM FOR AN INDOOR INDUSTRIAL AREA

A lighting system needs to be designed for a metalworking shop. The area of the shop is 12 m (40 ft) × 60 m (200 ft). (Conversion of meters to feet is approximate.) Total area is therefore 720 m^2 (8000 ft^2). The height of the room cavity, h_{RC}, is 4 m (13 ft). The height of the ceiling and floor cavities is 1 m (3 ft) each. In this facility, medium bench and machine work will be performed. Design an appropriate lighting system.

TABLE 20.5 Coefficients of Utilization (Zonal-Cavity Method) for Six Commonly Used Luminaires*

Typical Luminaire / Typical Intensity Distribution	ρcc→	80			70			50			30			10			0
	ρw→	70	50	30	70	50	30	50	30	10	50	30	10	50	30	10	0
	RCR ↓	EFF = 72.5%			% DN = 97.8%			% UP = 2.2%			Lamp = M400/C/U SC (along, across, 45°) = 1.7, 1.7, 1.7						
Low bay with drop lens, narrow	0	0.86	0.86	0.86	0.84	0.84	0.84	0.80	0.80	0.80	0.76	0.76	0.76	0.73	0.73	0.73	0.71
	1	0.78	0.75	0.71	0.76	0.73	0.70	0.69	0.67	0.65	0.66	0.64	0.63	0.63	0.62	0.60	0.59
	2	0.71	0.65	0.60	0.69	0.63	0.59	0.60	0.56	0.53	0.58	0.54	0.52	0.55	0.53	0.50	0.49
	3	0.64	0.56	0.50	0.62	0.55	0.50	0.53	0.48	0.44	0.51	0.46	0.43	0.48	0.45	0.42	0.40
	4	0.59	0.50	0.43	0.57	0.49	0.42	0.47	0.41	0.37	0.45	0.40	0.36	0.43	0.39	0.36	0.34
	5	0.54	0.44	0.37	0.52	0.43	0.37	0.41	0.36	0.32	0.40	0.35	0.31	0.38	0.34	0.31	0.29
	6	0.49	0.39	0.33	0.48	0.39	0.32	0.37	0.31	0.27	0.36	0.31	0.27	0.34	0.30	0.26	0.25
	7	0.46	0.35	0.29	0.44	0.35	0.28	0.33	0.28	0.24	0.32	0.27	0.23	0.31	0.27	0.23	0.22
	8	0.42	0.32	0.26	0.41	0.31	0.25	0.30	0.25	0.21	0.29	0.24	0.21	0.28	0.24	0.21	0.19
	9	0.39	0.29	0.23	0.38	0.29	0.23	0.28	0.22	0.19	0.27	0.22	0.18	0.26	0.22	0.18	0.17
	10	0.37	0.27	0.21	0.36	0.26	0.21	0.26	0.20	0.17	0.25	0.20	0.17	0.24	0.20	0.16	0.15
		EFF = 83.9%			% DN = 95.2%			% UP = 4.8%			Lamp = M400/C/U SC (along, across, 45°) = 1.6, 1.6, 1.4						
High bay, open metal reflector, medium	0	0.99	0.99	0.99	0.96	0.96	0.96	0.91	0.91	0.91	0.86	0.86	0.86	0.82	0.82	0.82	0.80
	1	0.93	0.90	0.87	0.90	0.88	0.85	0.83	0.81	0.80	0.80	0.78	0.77	0.76	0.75	0.74	0.72
	2	0.86	0.81	0.77	0.84	0.79	0.75	0.76	0.73	0.70	0.73	0.70	0.68	0.70	0.68	0.66	0.64
	3	0.80	0.73	0.68	0.78	0.72	0.67	0.65	0.65	0.61	0.66	0.63	0.60	0.64	0.61	0.58	0.57
	4	0.75	0.67	0.61	0.73	0.65	0.60	0.63	0.58	0.54	0.61	0.57	0.53	0.58	0.55	0.52	0.51
	5	0.70	0.61	0.54	0.68	0.59	0.54	0.57	0.52	0.48	0.55	0.51	0.48	0.54	0.50	0.47	0.45
	6	0.65	0.55	0.49	0.63	0.54	0.48	0.53	0.47	0.43	0.51	0.46	0.43	0.49	0.45	0.42	0.41
	7	0.60	0.51	0.44	0.59	0.50	0.44	0.48	0.43	0.39	0.47	0.42	0.39	0.45	0.41	0.38	0.37
	8	0.56	0.47	0.40	0.55	0.46	0.40	0.44	0.39	0.35	0.43	0.38	0.35	0.42	0.38	0.34	0.33
	9	0.53	0.43	0.37	0.52	0.42	0.36	0.41	0.36	0.32	0.40	0.35	0.32	0.39	0.35	0.31	0.30
	10	0.50	0.40	0.34	0.48	0.39	0.33	0.38	0.33	0.29	0.37	0.32	0.29	0.36	0.32	0.29	0.27

continued

TABLE 20.5 Coefficients of Utilization (Zonal-Cavity Method) for Six Commonly Used Luminaires* *(Continued)*

Typical Luminaire	Typical Intensity Distribution	ρcc→	80			70			50			30			10			0
		ρw→	70	50	30	70	50	30	50	30	10	50	30	10	50	30	10	0
Industrial, white enamel reflector, 20% up		RCR ↓	EFF = 90.5%			% DN = 78.2%			% UP = 21.8%			Lamp = (2) F40T12 SC (along, across, 45°) = 1.3, 1.5, 1.5						
		0	1.03	1.03	1.03	0.98	0.98	0.98	0.90	0.90	0.90	0.82	0.82	0.82	0.74	0.74	0.74	0.71
		1	0.93	0.89	0.85	0.89	0.85	0.81	0.77	0.74	0.72	0.70	0.68	0.66	0.64	0.62	0.61	0.58
		2	0.84	0.77	0.71	0.80	0.74	0.68	0.67	0.63	0.59	0.61	0.58	0.54	0.56	0.53	0.50	0.47
		3	0.77	0.67	0.60	0.73	0.64	0.58	0.59	0.53	0.49	0.54	0.49	0.45	0.49	0.45	0.42	0.40
		4	0.70	0.59	0.51	0.66	0.57	0.50	0.52	0.46	0.41	0.48	0.43	0.39	0.44	0.39	0.36	0.33
		5	0.64	0.53	0.45	0.61	0.51	0.43	0.46	0.40	0.35	0.43	0.37	0.33	0.39	0.35	0.31	0.29
		6	0.59	0.47	0.39	0.56	0.45	0.38	0.42	0.35	0.31	0.38	0.33	0.29	0.35	0.31	0.27	0.25
		7	0.55	0.43	0.35	0.52	0.41	0.34	0.38	0.32	0.27	0.35	0.30	0.26	0.32	0.28	0.24	0.22
		8	0.51	0.39	0.31	0.48	0.37	0.30	0.34	0.28	0.24	0.32	0.27	0.23	0.29	0.25	0.21	0.19
		9	0.47	0.35	0.28	0.45	0.34	0.27	0.32	0.26	0.21	0.29	0.24	0.20	0.27	0.23	0.19	0.17
		10	0.44	0.33	0.26	0.42	0.31	0.25	0.29	0.23	0.19	0.27	0.22	0.18	0.25	0.21	0.17	0.16
Industrial, white enamel reflector, down only			EFF = 86.9%			% DN = 100%			% UP = 0%			Lamp = (2) F40T12 SC (along, across, 45°) = 1.3, 1.5, 1.5						
		0	1.03	1.03	1.03	1.01	1.01	1.01	0.97	0.97	0.97	0.92	0.92	0.92	0.89	0.89	0.89	0.87
		1	0.94	0.90	0.86	0.92	0.88	0.84	0.84	0.81	0.79	0.81	0.79	0.76	0.78	0.76	0.74	0.72
		2	0.85	0.78	0.72	0.83	0.76	0.70	0.73	0.68	0.64	0.70	0.66	0.63	0.67	0.64	0.61	0.59
		3	0.77	0.68	0.60	0.75	0.66	0.59	0.64	0.58	0.53	0.61	0.56	0.52	0.59	0.55	0.51	0.49
		4	0.70	0.60	0.52	0.68	0.58	0.51	0.56	0.50	0.45	0.54	0.48	0.44	0.52	0.47	0.43	0.41
		5	0.65	0.53	0.45	0.63	0.52	0.44	0.50	0.43	0.38	0.48	0.42	0.38	0.47	0.41	0.37	0.35
		6	0.59	0.47	0.39	0.58	0.47	0.39	0.45	0.38	0.33	0.43	0.37	0.33	0.42	0.37	0.32	0.31
		7	0.55	0.43	0.35	0.53	0.42	0.35	0.41	0.34	0.29	0.39	0.33	0.29	0.38	0.33	0.29	0.27
		8	0.51	0.39	0.31	0.50	0.38	0.31	0.37	0.30	0.26	0.36	0.30	0.26	0.35	0.29	0.25	0.24
		9	0.48	0.36	0.28	0.46	0.35	0.28	0.34	0.28	0.23	0.33	0.27	0.23	0.32	0.27	0.23	0.21
		10	0.45	0.33	0.26	0.43	0.32	0.25	0.31	0.25	0.21	0.31	0.25	0.21	0.30	0.24	0.21	0.19

Luminaire	RCR																
2 × 4, 3-Lamp troffer with A12 lens		EFF = 75.6%			% DN = 100%			% UP = 0%			Lamp = (2) F32T8 SC (along, across, 45°) = 1.3, 1.3, 1.4						
	0	0.90	0.90	0.90	0.88	0.88	0.88	0.84	0.84	0.84	0.80	0.80	0.80	0.77	0.77	0.77	0.76
	1	0.83	0.79	0.76	0.81	0.78	0.75	0.75	0.72	0.70	0.72	0.70	0.68	0.69	0.67	0.66	0.65
	2	0.76	0.70	0.65	0.74	0.69	0.64	0.66	0.62	0.59	0.64	0.61	0.58	0.61	0.59	0.57	0.55
	3	0.70	0.62	0.57	0.68	0.61	0.56	0.59	0.54	0.51	0.57	0.53	0.50	0.55	0.52	0.49	0.47
	4	0.64	0.56	0.49	0.63	0.55	0.49	0.53	0.48	0.44	0.51	0.47	0.43	0.49	0.46	0.43	0.41
	5	0.59	0.50	0.44	0.58	0.49	0.43	0.48	0.42	0.38	0.46	0.41	0.38	0.45	0.41	0.37	0.36
	6	0.55	0.45	0.39	0.53	0.45	0.38	0.43	0.38	0.34	0.42	0.37	0.33	0.41	0.37	0.33	0.32
	7	0.51	0.41	0.35	0.50	0.41	0.35	0.39	0.34	0.30	0.38	0.33	0.30	0.37	0.33	0.30	0.28
	8	0.48	0.38	0.31	0.46	0.37	0.31	0.36	0.31	0.27	0.35	0.30	0.27	0.34	0.30	0.27	0.25
	9	0.44	0.35	0.29	0.43	0.34	0.28	0.33	0.28	0.24	0.33	0.28	0.24	0.32	0.27	0.24	0.23
	10	0.42	0.32	0.26	0.41	0.32	0.26	0.31	0.26	0.22	0.30	0.25	0.22	0.29	0.25	0.22	0.21
2 × 4, 3-Lamp parabolic troffer with 3" semi-spec. louvers, 18 cells		EFF = 72.7%			% DN = 100%			% UP = 0%			Lamp = (2) F32TB SC (along, across, 45°) = 1.3, 1.6, 1.6						
	0	0.87	0.87	0.87	0.85	0.85	0.85	0.81	0.81	0.81	0.77	0.77	0.77	0.74	0.74	0.74	0.73
	1	0.81	0.78	0.76	0.79	0.77	0.74	0.74	0.72	0.70	0.71	0.69	0.68	0.68	0.67	0.66	0.65
	2	0.75	0.70	0.66	0.73	0.69	0.65	0.66	0.63	0.61	0.64	0.61	0.59	0.62	0.60	0.58	0.57
	3	0.69	0.63	0.58	0.68	0.62	0.57	0.60	0.56	0.52	0.58	0.54	0.52	0.56	0.53	0.51	0.49
	4	0.64	0.56	0.51	0.62	0.55	0.50	0.54	0.49	0.46	0.52	0.48	0.45	0.51	0.47	0.44	0.43
	5	0.59	0.51	0.45	0.58	0.50	0.44	0.48	0.44	0.40	0.47	0.43	0.40	0.46	0.42	0.39	0.38
	6	0.55	0.46	0.40	0.53	0.45	0.40	0.44	0.39	0.35	0.43	0.38	0.35	0.42	0.38	0.35	0.33
	7	0.51	0.42	0.36	0.50	0.41	0.36	0.40	0.35	0.31	0.39	0.35	0.31	0.38	0.34	0.31	0.30
	8	0.47	0.38	0.32	0.46	0.38	0.32	0.37	0.32	0.28	0.36	0.31	0.28	0.35	0.31	0.28	0.27
	9	0.44	0.35	0.29	0.43	0.35	0.29	0.34	0.29	0.25	0.33	0.29	0.25	0.32	0.28	0.25	0.24
	10	0.41	0.32	0.27	0.40	0.32	0.27	0.31	0.26	0.23	0.31	0.26	0.23	0.30	0.26	0.23	0.22

*Selected from Lighting Calculations Section of IESNA Lighting Handbook, 9th Edition, 2000. CU's assume 20% effective floor-cavity reflectance.
ρcc = percent effective ceiling-cavity reflectance.
ρw = percent wall reflectance.
RCR = room-cavity ratio.
SC = ratio of maximum luminaire spacing to mounting or ceiling height above work plane.

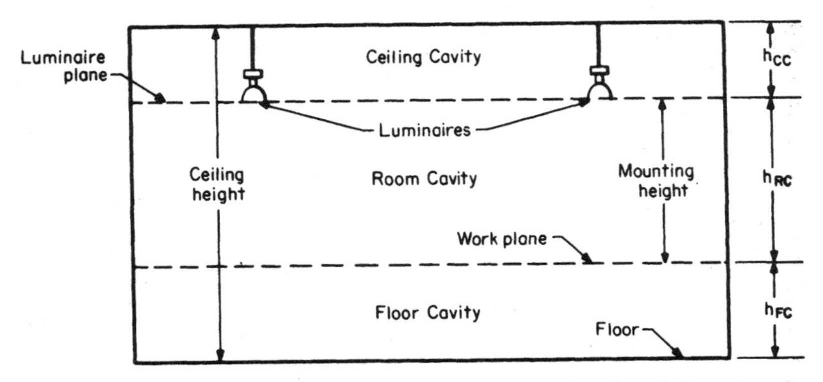

FIGURE 20.2 Utilization of a lighting system is a function of the cavity ratio for each section.

Calculation Procedure

1. Find the Recommended Level of Illumination

The level recommended in Table 20.1 for medium bench and machine work (category E) is 500 lx (50 fc), then adjusted up to 750 lx (75 fc) to allow older workers to see more accurately.

2. Choose a Lamp Type and Lumen Rating

A pulse-start, phosphor-coated metal halide system is a good choice because of its high efficiency, good color, improved lumen maintenance, and long life. For machine shops or any manufacturing space requiring accurate seeing with a minimum of shadows, the maximum luminaire spacing should be close to the mounting height above the work plane. Luminaire spacing beyond $1\frac{1}{2}$ times the mounting height will usually produce poor results in a manufacturing area.

As a rough approximation, half the initial lamp lumens are effective in producing the maintained illumination level. This can be used to calculate the maximum lamp lumens that each luminaire can have. The area per luminaire is equal to the spacing squared.

The maintained illumination lx (maintained) is lx (maintained) = $\frac{1}{2}$(LL/area per luminaire). For a room cavity height of 4 m and a maintained illumination level of 750 lx, the lamp lumens for a luminaire spacing equal to the mounting height would be: $750 = \frac{1}{2}(LL/4^2)$. Solving, find LL = 24,000 lm. For a spacing equal to $1\frac{1}{2}$ times the mounting height, LL = 54,000 lm. A 400-W pulse-start, phosphor-coated metal halide lamp at 40,000 lumens and LLD of 0.77 in an enclosed luminaire would be a good choice, and spacing could be 1.14 times the 4-m mounting height.

3. Select the Luminaire

When shiny metal surfaces are being worked on, it is desirable to have a luminaire with a refractor. The refractor spreads the light over a large area and prevents the lamp from being seen as a bright reflection on the surface of the work. In this problem, a high-intensity discharge (HID) luminaire that combines a reflector and refractor will be used. At a mounting height of 4 m, an enclosed industrial luminaire designed for low mounting heights is the best choice. Its wide beam produces complete overlap of the light from adjacent luminaires.

TABLE 20.6 LDD Values for Some Luminaire Types

Luminaire type	Luminaire dirt depreciation (LDD)		
	Light	Medium	Heavy
Enclosed and filtered	0.97	0.93	0.88
Enclosed	0.94	0.86	0.77
Open and ventilated	0.94	0.84	0.74

4. *Find the Coefficient of Utilization*

Table 20.5 gives the CU value for the chosen luminaire, which has a spacing criterion, SC, of 1.7. Assume a 30 percent ceiling-cavity reflectance, a 30 percent wall reflectance, and a 20 percent floor-cavity reflectance. The room-cavity ratio RCR is equal to $5h_{RC}(L + W)/LW = (5)(4)(60 + 12)/(60)(12) = 1440/720 = 2$. From the table, CU = 0.54 for this 400-W luminaire.

5. *Determine the Light-Loss Factors*

There are several light-loss factors that can cause the illumination level to depreciate in service. Of these, the two most commonly used are the lamp lumen depreciation and the luminaire dirt depreciation. The mean LDD value for a 400-W phosphor-coated pulse-start metal halide lamp is 0.80.

The LLD values vary considerably and can only be predicted accurately by experience with similar-type luminaires under similar service conditions. Suggested values are given in Table 20.6. If conditions within the plant are known to cause greater deterioration, it is more economical to clean the system at frequent intervals than to increase the initial illumination level to compensate for the loss.

For the low-bay 400-W luminaire, the light loss factor is (0.80)(0.86) = 0.69.

6. *Determine the Number of Luminaires* N

Use $N = EA/(\mathrm{LL})(\mathrm{CU})(\mathrm{LLF})$, where E = maintained level of illumination, A = area of space to be lighted, LL = initial-rate lamp lumens, CU = coefficient of utilization, and LLF = light-loss factor. Hence, $N = (750)(720)/(40{,}000)(0.54)(0.69) = 36.2$. Use 36 luminaires. Lamps per luminaire = 1. Ballast and tilt factors = 1.

7. *Determine the Final Luminaire Quantity and Spacing*

The average square spacing S of the luminaires can be determined by $S = \sqrt{A/N} = \sqrt{720/36} = 4.47$ m (15 ft).

The final luminaire quantity is frequently a compromise between the calculated quantity and the shape of the area being lighted. In this case, the length of the area is 5 times the width, so the number of luminaires per row should be approximately 5 times the number of rows. Thirty-six luminaires can be divided only into 3 rows of 12.

There is no fixed rule on luminaire spacing. The final spacing, however, should be as close to the average square spacing as possible. The spacing between luminaires should also be twice the distance of the spacing from the walls.

Related Calculations. Check the luminaire spacing criterion to ensure that it exceeds the actual spacing-to-mounting-height ratio. The luminaire spacing criterion is 1.7 from Table 20.5, so the uniformity should be quite satisfactory.

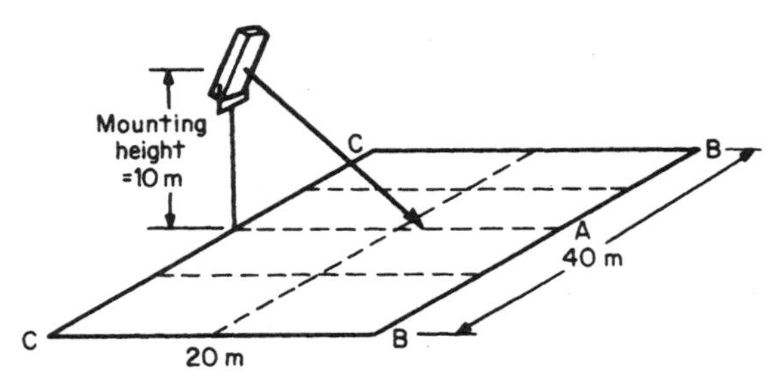

FIGURE 20.3 Layout for the outdoor area-lighting design.

LIGHTING SYSTEM FOR AN OUTDOOR AREA

Determine the illumination level produced by a 400-W high-pressure sodium floodlight on an area 20 m wide × 40 m long. The floodlight is located on a 10-m pole midway along the 40-m side of the area (see Fig. 20.3). Also determine the illumination at points *A*, *B*, and *C*. The area is a construction site where excavation work will be done.

Calculation Procedure

1. Determine the Luminaire Mounting Height (MH)

If possible, the luminaire mounting height should be at least one-half the width of the area being lighted. Lower mounting heights will reduce utilization and uniformity.

2. Calculate the Aiming Angle

Maximum utilization is obtained when the aiming line bisects the angle drawn between the near and far side of the lighted area. This will usually produce poor uniformity because there will be insufficient light at the far edge. The highest illumination level that can be produced away from the floodlight location occurs when the maximum floodlight candela value is directed at an angle of 54.7°, as in Fig. 20.4. This is approximately a 3-4-5 triangle, as shown. For an area which is twice as wide as the mounting height, this would be a logical aiming point. In this case, aim the floodlight 13 m across the 20-m-wide area.

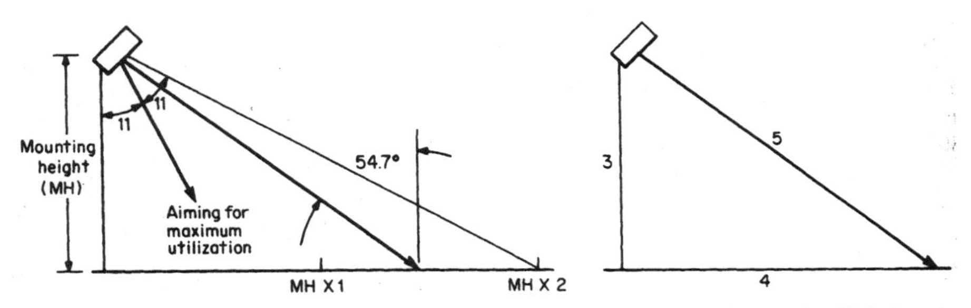

FIGURE 20.4 The highest illumination level that can be produced away from a floodlight location occurs when the maximum floodlight candela value is directed at an angle of 54.7°.

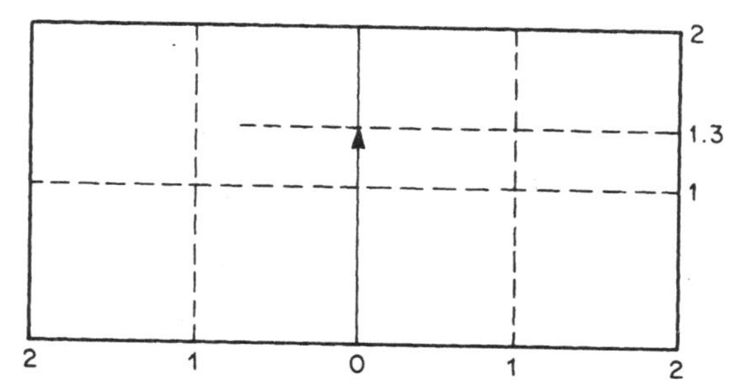

FIGURE 20.5 Lighted area divided into a grid having the same scale as the floodlight mounting height.

3. Determine the Utilization Factor

The most difficult part of area-lighting design is determining the location of the edges of the lighted area in relation to the floodlight aiming line. This is needed to determine the lumens that will fall inside the area. This can be greatly simplified if the area is divided into a grid having the same scale as the floodlight mounting height (Fig. 20.5).

The sample area is 2 mounting heights (2 × 10 m = 20 m) wide and 4 mounting heights (4 × 10 m = 40 m) long. The floodlight is aimed 13/10 = 1.3 mounting heights across the lighted area. The luminaire is mounted in the center, so the area extends 2 mounting heights left and right of the aiming line. In all cases the grid must be drawn in line with and perpendicular to the aiming line, not the edges of the area being lighted. In this example they coincide, but this will not always be the case.

In order to determine the utilization, the dimensions of the lighted area must be translated into the beam dimensions of the floodlight. These are given in horizontal and vertical degrees. Figure 20.6 is used to convert from one to the other.

Plot the lighted area and aiming point on the chart. The area is 2 mounting heights wide and extends 2 mounting heights to each side of the aiming line. The aiming point is 1.3 MH across the area. This corresponds to a vertical angle of 53°.

The 53° aiming line corresponds to the 0.0 line on the photometric data for this floodlight, which is given in Fig. 20.7. The far edge corresponds to 63° or 10° above the aiming line. The lower edge is at the base of the luminaire or 53° below the aiming line. Locate these two lines on the lumen distribution chart in Fig. 20.7.

The right and left edges of the lighted area can be located on the lumen distribution curve by reference to the curved lines on the chart (Fig. 20.6). The far corner makes an angle of 42° with the luminaire. Plot the other points for the 10° increments on the photometric data below the aiming point. For instance, at 53° (0° on the photometric curve) the angle is 50°, and so forth. At the base of the pole, the angle is 53°. A curved line drawn through all the points will show the location of the sides of the area on the photometric curve. In this case, only the right side was plotted. If the two sides are different, each would have to be plotted separately.

The total lumens that fall on the area are the sum of the lumens falling inside the area's boundary. To determine this, add lumen values shown in each block. Where the boundary cuts through the block, estimate the percentage of the value in the block that falls inside the line. This would be proportional to the area. Table 20.7 provides the total lumens falling in the area of the sample problem.

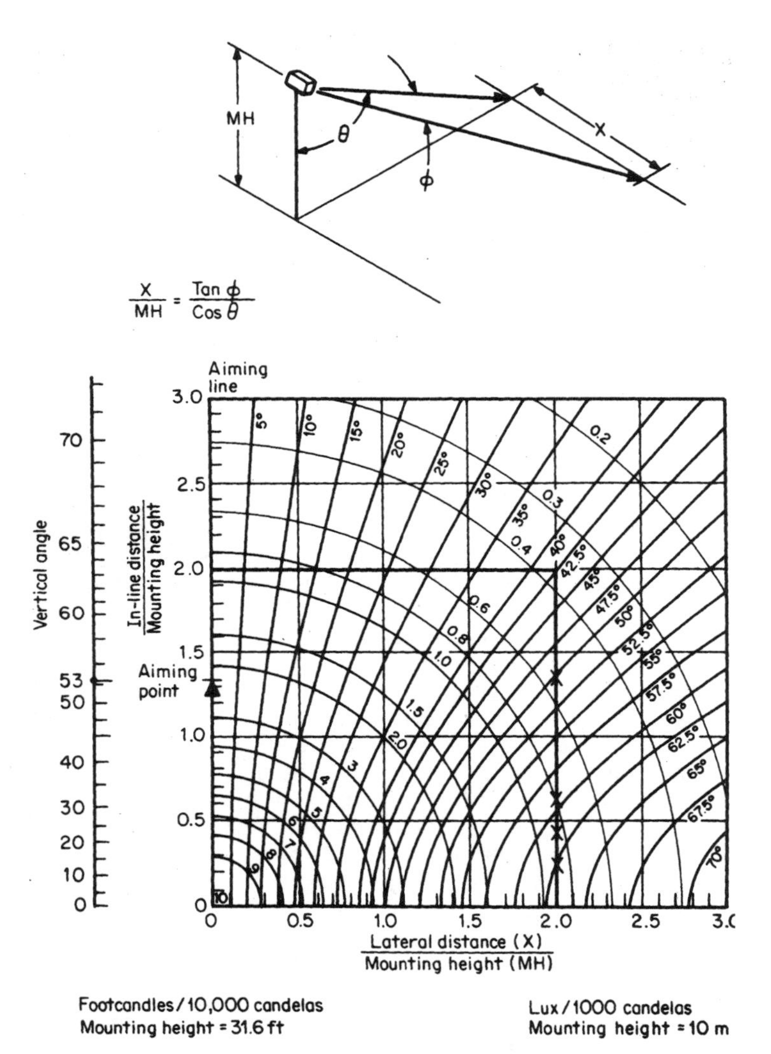

FIGURE 20.6 Chart used to convert dimensions of a lighted area into the beam dimensions of a floodlight. It also can be used to find the level of illuminance at a point.

4. Calculate the Level of Illumination

Average lux = (LL)(CU)(LLD)(LDD)/Area = (51,000)(0.4168)(0.9)(0.95)/(20)(40) = 22.7 lx. Table 20.8 shows that the minimum average recommended lux for this area is 30; therefore, a 600 W model will be needed.

5. Find Point-by-Point Illumination Values

Refer to Fig. 20.6. The circular lines are *isolux* values based on 1000 cd. To determine the illumination at any point, the candela value of the floodlight at the same point must be known. The points can be located the same way as the area boundaries were determined.

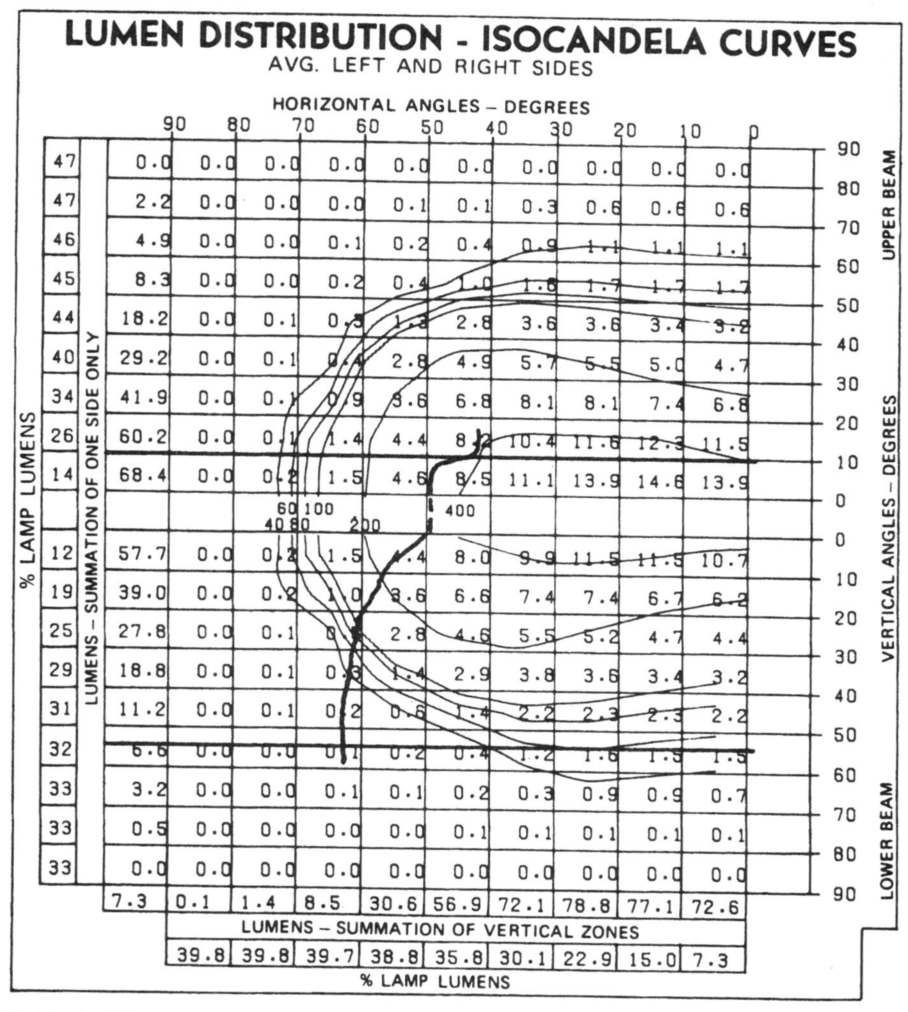

FIGURE 20.7 Photometric data for a floodlight used in the outdoor area-lighting design. Plotted line indicates the right side of the lighted area. (General Electric Co.)

The mounting height used in the chart is 10 m. To calculate the level of illumination at any point, use the following relation:

$$\text{lux} = [\text{lux (from chart)}](\text{cd}/1000)[(\text{LL})/(\text{LF})](\text{LLF})(\text{MHCF}),$$

where

lux (from chart) = illumination in lx/1000 cd (Fig. 20.6)

cd = candela value taken from the photometric data isocandela curves (Fig. 20.7) at the same horizontal and vertical angles as indicated by the chart (must be corrected by dividing by 1000)

LL = lamp lumens

TABLE 20.7 Total Lumens Falling on Outdoor Area-Lighting System

Vertical zone	Horizontal angles 0–10	10–20	20–30	30–40	40–50	50–60	60–70	Total
0–10	13.9	14.6	13.9	11.1	5.1*			58.6
0–10	10.7	11.5	11.5	9.9	8.0	2.2*		53.8
10–20	6.2	6.7	7.4	7.4	6.6	2.9		37.2
20–30	4.4	4.7	5.2	5.5	4.6	2.8	0.1*	27.3
30–40	3.2	3.4	3.6	3.8	2.9	1.4	0.1*	18.4
40–50	2.2	2.3	2.2	1.4	1.4	0.6	0.1*	11.1
50–60	0.5*	0.5*	0.5*	0.4*	0.1*			2.0
							Total right side	208.4
							Left side	208.4
							Total lumens	416.8
							Total lamp lumens 1000	
							Coefficient of utilization 41.68 percent	

*Estimated.

LF = lamp factor, which corrects the lamp lumens used in the photometric data to the rated lamp lumens used in the floodlight (in this case LF = 51)

LLF = light loss factor = lamp lumen depreciation times luminaire dirt depreciation

MHCF = mounting-height correction factor, the ratio of the square of the mounting height used in the chart (Fig. 20.6) to the mounting height used in the problem (in this case, MHCF = $100/MH^2$ = $100/100 = 1$)

Point *A* in Fig. 20.3 is along the aiming line. In Fig. 20.6, it is located at 63°, or 2MH. This point is between the 0.8-lx and 1.0-lx lines, so the value 0.9 can be given. Point *A* is 10° above the aiming point of 53°; in Fig. 20.7, this matches the isocandela curve marked 400. The candela value for this problem is, therefore, 400.

Substituting in the formula for point *A* yields lux = (0.9)(400/1000)(90,000/1000)(0.88)(0.95)(1) = 27.0 lx. Point *B* is located at a horizontal angle of 42° and the vertical angle is 10°. Substituting in the formula, one obtains lux = (0.38)(400/1000)(90,000/1000)(0.88)(0.95)(1) = 11.5 lx.

Point *C* is outside the beam of the floodlight. The only contribution will be spill light from the floodlight. In practice, more than one floodlight can be used at each location to avoid the dark area along the near sideline.

ROADWAY LIGHTING SYSTEM

The information already known is that the street width is 20 m, the mounting height is 12 m, and the overhang of the luminaire is 2 m. The required average maintained level of illumination is 16 lx. It is necessary to determine the staggered spacing required to provide the specified illumination level, as well as the uniformity of illumination with staggered spacing.

TABLE 20.8 Illumination Levels Recommended for Outdoor Lighting

General application	Minimum average recommended lux
Construction	
General	50
Excavation	30
Industrial roadways	7
Industrial yard/material handling	50
Parking areas	
Industrial	2–5
Shopping centers	10
Commercial lots	10
Security	
Entrances (active)	100
Building surrounds	5
Railroad yards	
All switch points	20
Body of yard	10
Shipyards	
General	30
Ways	50
Fabrication areas	300
Storage yards	
Active	100
Inactive	10

Calculation Procedure

1. Find the Level of Illumination

All the information needed to solve for the average illumination level and the minimum level is contained in the photometric data. Each roadway luminaire can be adjusted to produce a number of different light-distribution patterns. In addition, a luminaire can usually use several different lamp wattages. It is necessary, therefore, to obtain from the manufacturer the photometric curves for a family of luminaires. Figure 20.8 provides photometric data for a 250- or 400-W high-pressure sodium luminaire. For this problem, the 250-W source will be used.

The equation for finding average illumination E is: $E = (\text{LL})(\text{CU})(\text{LLF})/WS$ where W is the curb-to-curb street width and the other terms are as defined previously. Values are already given for E and W. For a 250-W HPS lamp, LL = 28,000 lm. Light-loss factors for roadway lighting are always determined for the minimum illumination in service. This occurs at the point where the luminaire is cleaned and relamped. For a 250-W HPS lamp, the LLD is 0.73 at the end of relamping. The suggested LDD for a sealed and filtered luminaire is 0.93 (Table 20.6). After the value for the CU is determined, it is then only necessary to solve the equation for average spacing S.

2. Determine the Coefficient of Utilization

Roadway luminaires are usually mounted over the roadway. They are leveled parallel to the roadway in an orientation perpendicular to the curb, as in Fig. 20.9. The area of the road in front of the luminaire is called the street side (SS) and the area behind is called

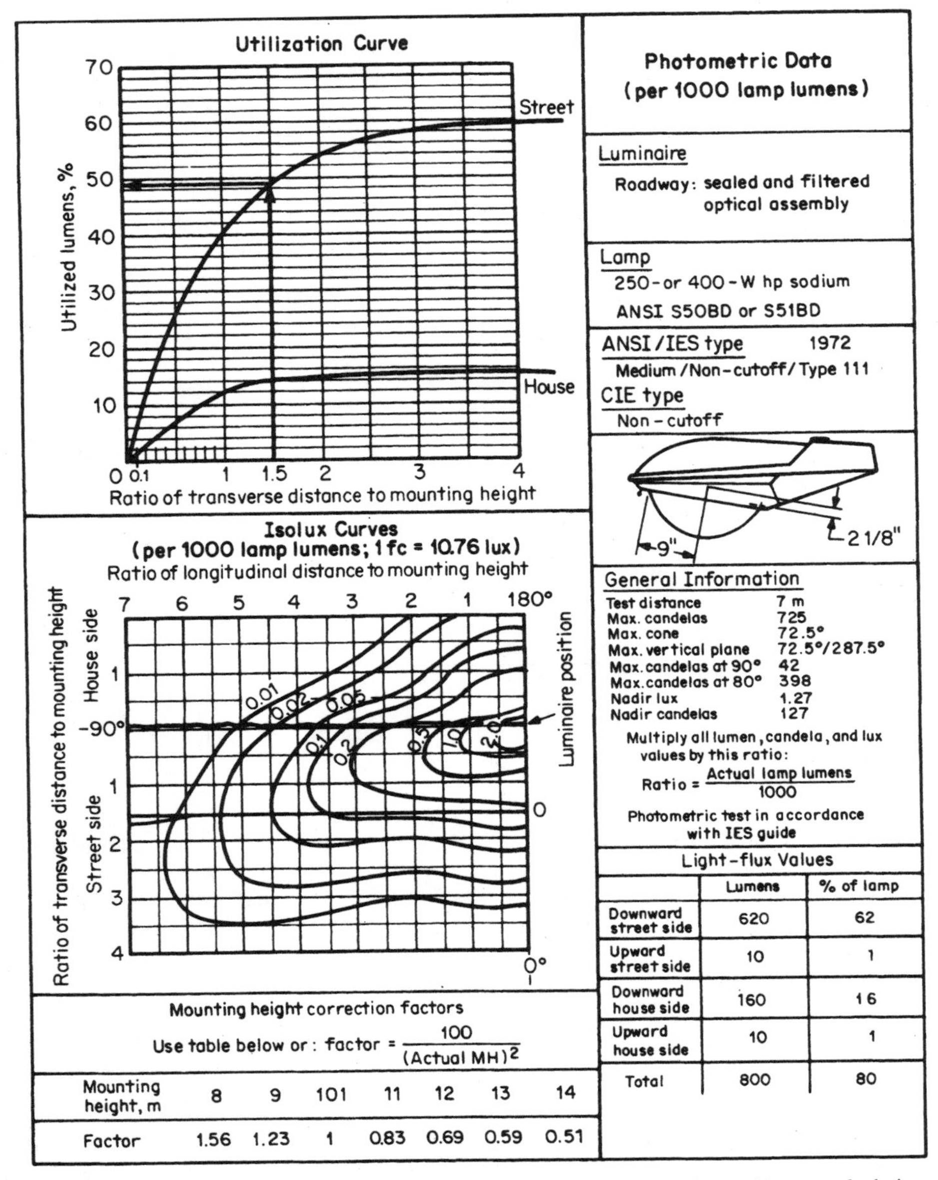

Mounting height, m	8	9	101	11	12	13	14
Factor	1.56	1.23	1	0.83	0.69	0.59	0.51

	Lumens	% of lamp
Downward street side	620	62
Upward street side	10	1
Downward house side	160	16
Upward house side	10	1
Total	800	80

FIGURE 20.8 Photometric data for a high-pressure sodium roadway luminaire used in a sample design. (General Electric Co.)

the curb, or house side (HS). The utilization of the luminaire is a function of the total lamp lumens that fall in each of the triangular sections.

Because the lumens falling on the roadway are the same if the angle subtended by the two curbs is the same, the utilization can be made a function of the ratio of the street side or house side transverse distance divided by the mounting height. The roadway is assumed to be continuous on either side of the luminaire.

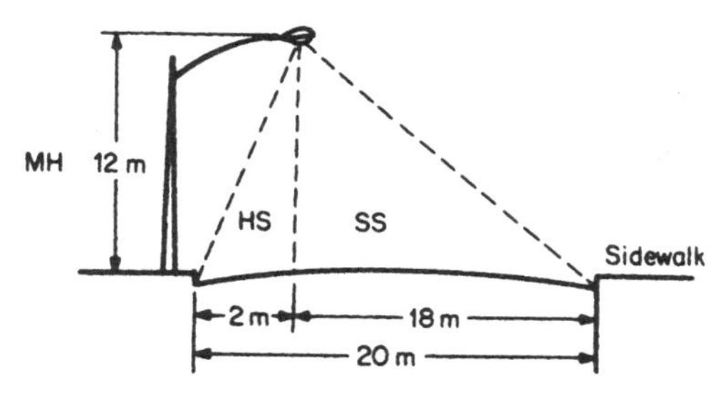

FIGURE 20.9 Luminaire mounted over a roadway.

The coefficient of utilization for each side can be read from the utilization curve in Fig. 20.8:

Ratio 1.35, street side, corresponds to	CU = 49%
Ratio 0.15, house side, corresponds to	CU = 2%
Total	CU = 51%

3. Determine the Staggered Spacing Required

Average staggered spacing can be determined by rewriting the basic equation and solving: $S = (\text{LL})(\text{CU})(\text{LLD})(\text{LDD})/EW = (28{,}000)(0.51)(0.73)(0.93)/(16)(20) = 9694/320 = 30.3$, or 30 m.

4. Determine the Uniformity of Illumination

The uniformity of illumination is usually expressed in terms of the ratio (average lux)/(minimum lux). The average was given as 16 lx at a spacing of 30 m, staggered.

The point of minimum illumination can be found by studying the isolux curves in Fig. 20.8 and taking contributions from all luminaires into account. Generally, the minimum value will be found along a line halfway between two consecutively spaced luminaires. However, this is not always the case, depending upon the geometry of the situation and the distribution pattern of the luminaire.

Related Calculations. All significant illumination contributions from luminaires should be determined for points $P1$ and $P2$, as illustrated in Fig. 20.10, in order to

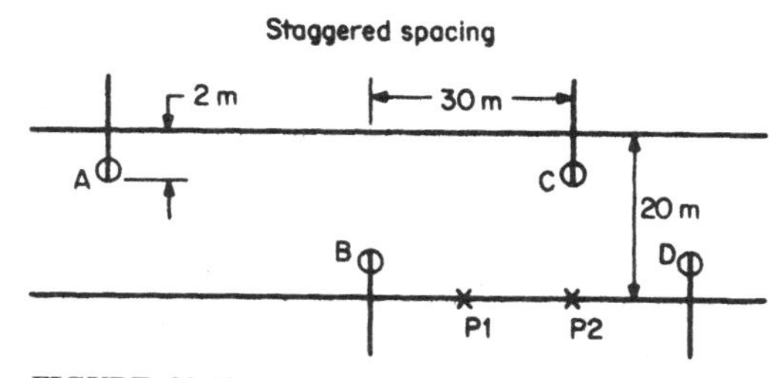

FIGURE 20.10 Determining illumination at points $P1$ and $P2$.

TABLE 20.9 Values of Illumination at Test Points *P*1 and *P*2

Contributing luminaires	Ratios for test points				Illumination at test points	
	Transverse ratio		Longitudinal ratio			
	*P*1	*P*2	*P*1	*P*2	*P*1	*P*2
A	1.5	1.5	3.75	5	0.08	0.03
B	0.167	0.167	1.25	2.5	0.5	0.19
C	1.5	1.5	1.25	0	0.15	0.19
D	0.167	0.167	3.75	2.5	0.05	0.19
				Totals	0.78	0.60

make sure the minimum illumination point is located. First, determine both transverse and longitudinal ratios of distance to mounting height relative to each of the luminaires. Using the ratios in Table 20.9 as coordinates, corresponding illumination values can be read from the isolux plot of Fig. 20.8. These values are tabulated as "illumination at test points" in Table 20.9.

The minimum illumination value of 0.6 lx occurs at point *P*2. This value is the initial value per 1000 lamp lumens. To convert to the actual maintained illumination level, use lux = [lx/(min)](LL/1000)(LLD)(LDD)(MHCF). The mounting-height correction factor MHCF can be read from the chart in Fig. 20.8 as 0.69. Substituting values, find (0.6)(28,000/1000)(0.73)(0.93)(0.69) = 7.87 lx. The average-to-minimum illumination level is therefore 16/7.87 = 2.03:1. Because the recommended maximum ratio is 3:1, this spacing will produce better than minimum uniformity.

BIBLIOGRAPHY

American National Standards Institute. 1993. *American National Standard Practice for Office Lighting.* New York: ANSI.

Bowers, Brian. 1998. *Lengthening the Day: A History of Lighting Technologies.* Oxford; New York: Oxford University Press.

DeVeau, Russell L. 2000. *Fiber Optic Lighting: A Guide for Specifiers.* Lilburn, G.A.: Fairmont Press.

Frier, John P. 1980. *Industrial Lighting Design.* New York: McGraw-Hill.

Hartman, Fred. 1996. *Understanding NE Code Rules on Lighting.* Overland Park, K.S.: Intertec Electrical Group.

Illuminating Engineering Society of North America (IESNA). 2000. *IESNA Lighting Handbook.* New York: IESNA.

Johnson, Glenn A. 1998. *The Art of Illumination: Residential Lighting Design.* New York: McGraw-Hill.

Lechner, Norbert. 1991. *Heating, Cooling, Lighting: Design Methods for Architects.* New York: Wiley & Sons.

Phillips, Derek. 1997. *Lighting Historical Buildings.* New York: McGraw-Hill.

INDEX

DISK WARRANTY

This software is protected by both United States copyright law and international copyright treaty provision. You must treat this software just like a book, except that you may copy it into a computer in order to be used and you may make archival copies of the software for the sole purpose of backing up our software and protecting your investment from loss.

By saying "just like a book," McGraw-Hill means, for example, that this software may be used by any number of people and may be freely moved from one computer location to another, so long as there is no possibility of its being used at one location or on one computer while it also is being used at another. Just as a book cannot be read by two different people in two different places at the same time, neither can the software be used by two different people in two different places at the same time (unless, of course, McGraw-Hill's copyright is being violated).

LIMITED WARRANTY

McGraw-Hill takes great care to provide you with top-quality software, thoroughly checked to prevent virus infections. McGraw-Hill warrants the physical diskette(s) contained herein to be free of defects in materials and workmanship for a period of sixty days from the purchase date. If McGraw-Hill receives written notification within the warranty period of defects in materials or workmanship, and such notification is determined by McGraw-Hill to be correct, McGraw-Hill will replace the defective diskette(s). Send requests to:

McGraw-Hill
Customer Services
P.O. Box 545
Blacklick, OH 43004-0545

The entire and exclusive liability and remedy for breach of this Limited Warranty shall be limited to replacement of defective diskette(s) and shall not include or extend to any claim for or right to cover any other damages, including but not limited to, loss of profit, data, or use of the software, or special, incidental, or consequential damages or other similar claims, even if McGraw-Hill has been specifically advised of the possibility of such damages. In no event will McGraw-Hill's liability for any damages to you or any other person ever exceed the lower of suggested list price or actual price paid for the license to use the software, regardless of any form of the claim.

McGRAW-HILL SPECIFICALLY DISCLAIMS ALL OTHER WARRANTIES, EXPRESS OR IMPLIED, INCLUDING, BUT NOT LIMITED TO, ANY IMPLIED WARRANTY OF MERCHANTABILITY OR FITNESS FOR A PARTICULAR PURPOSE.

Specifically, McGraw-Hill makes no representation or warranty that the software is fit for any particular purpose and any implied warranty of merchantability is limited to the sixty-day duration of the Limited Warranty covering the physical diskette(s) only (and not the software) and is otherwise expressly and specifically disclaimed.

This limited warranty gives you specific legal rights; you may have others which may vary from state to state. Some states do not allow the exclusion of incidental or consequential damages, or the limitation on how long an implied warranty lasts, so some of the above may not apply to you.